BOTANY

Plant Diversity

BOTANY

Plant Diversity

Randy Moore
University of Akron

W. Dennis Clark
Arizona State University

Kingsley R. Stern
California State University–Chico

Darrell Vodopich
Baylor University

WCB Wm. C. Brown Publishers

Dubuque, IA Bogota Boston Buenos Aires Caracas Chicago
Guilford, CT London Madrid Mexico City Sydney Toronto

Book Team

Editor *Margaret J. Kemp*
Production Editor *Kay J. Brimeyer*
Designer *Christopher E. Reese*
Design Assistant *Kathleen F. Theis*
Art Editor *Rachel Imsland*
Photo Editor *Carrie Burger*
Permissions Coordinator *Karen L. Storlie*

Wm. C. Brown Publishers
A Division of Wm. C. Brown Communications, Inc.

Copyedited by *Nick Murray*

Photo Research by *Kathy Husemann*

Cover Photo © Craig Tuttle/The Stock Market

The credits section for this book begins on page Cr-1 and is considered an extension of the copyright page.

A Times Mirror Company

Library of Congress Catalog Card Number: 94–70837

ISBN 0–697–16657–0

Printed in the United States of America by Wm. C. Brown Communications, Inc., 2460 Kerper Boulevard, Dubuque, IA 52001

10 9 8 7 6 5 4 3 2 1

"To Kris—with all my love and
respect—for unfailing love,
patience, and humor."

R. M.

"To Lisa and our children,
Elizabeth and William, whose
love and support are the greatest."

W. D. C.

"To Donna, for her patience.
It just took longer than I thought."

D. V.

A personal library is a lifelong source of enrichment and distinction. Consider this book an investment in your future and add it to your personal library.

Brief Table of Contents

Table of Contents

Foreword

Botany is in the midst of a renaissance. New research methods, powerful instruments, and fresh ideas are transforming how biologists study plants at all levels, from the molecular mechanisms of cells to the ecological dynamics of the biosphere. Developmental biologists are using DNA technology to unravel how genes program a mass of undifferentiated cells to become a flower. Taxonomists are reassessing plant classification by applying new methods for measuring evolutionary relationships. Plant ecologists are analyzing field data by coupling computers to remote environmental sensing instruments placed in tropical forests and other ecosystems. Biotechnologists are engineering plants to improve agricultural productivity. Plant biologists are reinventing every field of botany.

If anything about botany is constant, it is its relevance. The activities of plants are vital to the welfare of nearly all organisms that share Earth, including humans. Renewed interest in environmental issues has brought the importance of plants back into sharp focus at the same time that new approaches are catalyzing botanical research. This is a wonderful time to study plants—a great time to take a botany course!

Botany is a new textbook that captures the excitement of the botanical revolution and reflects how the objectives of botany courses are changing. Randy Moore, Dennis Clark, Kingsley Stern, and Darrell Vodopich are award-winning teachers, distinguished researchers, and gifted writers. More than these talents, it is the authors' shared commitment to emphasize concepts over facts and to engage students in the process of science that sets this book apart. Scan the table of contents and you'll find the requisite topics, including plant structure and function, life cycles, metabolic pathways, and plant diversity. Read any chapter and you'll discover that it is *presentation*, not content, that defines this book. In crafting each chapter, the authors have constructed a strong framework of concepts that places botanical facts and terms in context and discourages rote memorization. Moore, Clark, Stern, and Vodopich build the process of science into every chapter by explaining not only *what* we know about plants, but *how* we know it. They also highlight what we *don't* know, and encourage students to frame questions and pose testable hypotheses of their own.

A good textbook must be clear, current, and correct; a great one also has a point of view. The authors of *Botany* see scientific literacy not as mastery of a vocabulary list, but as development of skills for lifelong enjoyment of nature and for thinking more critically about all subjects. The wonder and significance of plants are not lost along the way. Moore, Clark, Stern, and Vodopich love writing about plants, and their enthusiasm is infectious. Reading one chapter made me realize, more than ever, why it is so much fun to learn about plants. I gave a chapter to a few students to read, and they came back asking for more.

The publication of a new botany textbook is a rare event. This one is worth the wait.

Neil A. Campbell
Visiting Scholar
Department of Botany and Plant Sciences
University of California, Riverside

List of Tables

List of Boxed Readings

To The Student

Why should you study botany? Besides the obvious—a course requirement for your college degree—consider the botanical basis of the following items that appeared recently in newspapers across the country:

The people of France consume more wine and more fat-laden food than the people of most other industrialized countries, yet the French have a lower than expected rate of heart disease caused by high-fat diets. Based on population surveys of these kinds of data, some medical researchers have concluded that wine somehow inhibits heart disease.

The people of France consume more fruits and vegetables with their otherwise high-fat diets than the people of most other industrialized countries. Based on population surveys of these kinds of data, other medical researchers have concluded that fruits and vegetables, not wine, somehow inhibit heart disease.

In a 20-year study of older men in the Netherlands, those who consumed fruits, vegetables, and teas containing the highest amounts of vitamin C, vitamin E, and flavonoids (a group of chemicals produced by plants) had the lowest rate of heart disease. Researchers who did this study concluded that flavonoids somehow inhibit heart disease.

Which of the above are you going to believe: all of the conclusions, some of the conclusions, or none of the conclusions? You are bombarded with these kinds of conclusions in the popular press every day. Whether you accept some and not others may cause you to make decisions that affect your life. In a broad context, therefore, the quality of your life would improve if you could make thoughtful decisions based on how we acquire knowledge in science. Or, more specifically, based on how we acquire knowledge about plants. The basic skill that you need to make these decisions is how to think critically in a scientific context. The main purpose of this textbook is to help you develop that skill.

The abilities to think critically and communicate effectively are the most important skills that a student can develop during his or her formal education. Consequently, we have written this book to help you develop those skills as you learn about plants: what plants are, how they function, how they interact with each other and the environment, where they came from, and how we use them. As is the nature of all textbooks, ours contains an abundance of interesting "facts," but the real emphasis of this book is *how* we know. We have, as much as possible, de-emphasized the details of some of our knowledge and reduced the overwhelming number of new terms that usually appear in a text. In their place, we have substituted more of the process of science.

Our emphasis on the scientific process involves explaining botany as botany is done. Specifically, we describe the competing hypotheses that botanists have devised to answer questions about botanical phenomena, the experiments done by botanists to test these hypotheses, interpretations of data, and the many unanswered questions and unresolved conflicts that remain. This approach differs significantly from that of merely presenting definitions and the conclusions of experiments (i.e., the "facts" of botany).

We also pose many questions, some of which we don't answer and others of which we answer by showing you how botanists have addressed those questions. We do this to mirror the nature of botany: botanists generate knowledge by posing and testing hypotheses. Nevertheless, there are many things that we still don't know about plants. We hope that these questions, both answered and unanswered, will trigger your curiosity and help you sharpen your critical thinking skills.

In addition to learning about how we know, throughout this book you'll also learn about *what* we know. We present this information in the context of various principles (e.g., the

relation of structure and function) and integrate it into an informational framework. In *Botany*, that framework is created by the following themes:

The importance of plants. Plants affect virtually everything that we do. If you understand plants, you can better appreciate their influence on your life.

The process of science. We stress the process of science throughout the textbook. For example, you'll not just read about the products of evolution (i.e., the diversity of plants). Rather, you'll see the dynamics of evolutionary thought, as well as how botanists are applying various techniques to test competing hypotheses that explain what evolution is and how it operates.

Interactive learning. We designed this book to be a private tutor that will not only help you learn, but will also make you a more effective learner. Consequently, we have included a variety of learning aids. For example, chapter overviews introduce the topics of each chapter, artwork is integrated with and complements the text, and in-text concepts summarize themes. We hope that these features will encourage you to learn by making you an active participant in your learning.

We've used these features to present information with authority, breadth, and quality. Although we've treated some topics in depth, we've not written a botanical encyclopedia. That is, this text is not a compendium, but rather a stimulus. Because the first step to knowing is the curiosity of a question, we hope you'll generate your own questions to study. Indeed, we'll help you develop those questions in every chapter.

Unit Openers, Chapter Outlines, and **Chapter Openers** will stimulate your curiosity and prepare you for what you'll learn. These features will also motivate, orient, and tell you what you'll learn, and why.

Unit Seven
Evolution . . .
Throughout recorded history, people have wondered how the great variety of earth's organisms came to be. Among the earliest thinkers was Aristotle, who thought that each species arose independently from inorganic matter and did not change. His scala naturae (scale of nature) stretched from nonliving matter, through lower forms of life, to humans at the top. Although each living species held a permanent place on the scale, species were viewed as imperfect. Perfection reflected the work of a perfect God, and was found only in the transcendent world of ideas.
Today our ideas about the diversity of life differ markedly from those of Aristotle and his contemporaries. Specifically, we view populations of organisms as entities that change as they adapt to changing environments. The foundation for understanding how biological change occurs in populations and how such change forms new species rests on the theories of organic evolution. The earliest of these theories are most closely identified with Charles Darwin, the greatest naturalist-philosopher of the nineteenth century. His ideas about evolution have impacted a wider array of human endeavors than any other scientific advancement of the past 150 years.
How Darwin arrived at his ideas and how they have affected science, philosophy, religion, and human attitudes is one of the most fascinating stories of human achievement. This subject occupies a significant portion of hundreds of books spanning the past century. For this textbook, the most important aspect of the story is how Darwin's work provided the first scientific explanation for the diversity of life.
Chapter 31
Population Dynamics and Community Ecology
Chapter Outline
Introduction
Populations, Communities, and Ecosystems
Populations
Communities
Ecosystems
Nutrient Cycles
Succession
Primary Succession Initiated on Rocks or Lava
Primary Succession Initiated in Water
Secondary Succession
Box 31.1
A Rogue's Gallery of Pests
Fire Ecology
Humans in the Ecosystem
The Greenhouse Effect
Box 31.2
Curbing Methane Emissions to Curtail the Greenhouse Effect
Carbon Dioxide
Methane
Acid Rain
Ozone
Contamination of Water
Chapter Summary
Questions for Further Thought and Study
Suggested Readings
Chapter Overview
This chapter deals with basic ecological concepts, and includes an introduction to populations, communities, and ecosystems. You'll read about the interactions of producers, consumers, and decomposers, as well as the factors involved in succession. The chapter concludes with discussions of the results of human disruption of ecosystems; the topics covered include the greenhouse effect, acid rain, ozone depletion, and water pollution.

Throughout the text you'll also encounter four features that will help you extend your learning beyond the classroom and textbook:

The Lore of Plants

will tell you about the fascinating—and even bizarre—history of plants and their uses by humans.

What are Botanists Doing?

will challenge you to learn the newest ideas about various botanical subjects.

Boxed Readings will expand the scope of botany to include people, controversies, curiosities, and new developments.

Chapter Summaries will help you integrate what you've learned.

Questions for Further Thought and Study will help you apply what you've learned. Because these questions emphasize critical thinking rather than the mere recall of isolated "facts," you'll find no true-false or multiple-choice questions at the end of chapters. Nor will you have to fill in the blanks of someone else's thinking. Instead, you'll be asked to put *your* ideas together.

Suggested Readings provide sources of additional information about topics that you've read about.

BOXED READING 19.1
TRACKING THE SUN

One of the most unusual and adaptive movements in plants is solar tracking, also known as heliotropism (from the Greek word helios, meaning "sun"). Solar tracking is how the organs of plants track the sun across the sky, much as a radio telescope tracks stars or satellites. Like several other plant movements, solar tracking is caused by turgor changes in motor cells in pulvini located at the base of leaves and leaflets.

The tracking movements of leaves depend on environmental and physiological conditions. For example, the so-called compass plant (Silphium lacinatum) orients its leaves parallel to the sun's rays, thereby decreasing leaf temperature and minimizing desiccation. Other plants often orient their leaves perpendicular to the sun's rays, thereby increasing the amount of light intercepted by the leaf for photosynthesis. Finally, some plants orient their leaves more obliquely to the sun when it is hot than when it is cooler. This variation in solar tracking helps keep the leaf temperature near the optimal temperature for photosynthesis.

Solar tracking occurs in many plants (e.g. cotton, alfalfa, beans, and soybeans) and is not restricted to leaves. What happens on cloudy, overcast days? On these days, leaves are oriented horizontally in their "resting" position, if the sun appears from behind the clouds late in the day, leaves rapidly reorient themselves: they can move up to 60 degrees per hour, which is four times faster than the movement of the sun across the sky.

Solar tracking is controlled not only by the sun's position: leaves begin orienting themselves toward the direction of sunrise several hours before sunrise. How plants "remember" the position of the last sunrise is a mystery, but IAA may be involved: IAA from leaves can alter turgor of motor cells. Whatever the mechanism, plants are fast learners: only four sunrises are needed to entrain solar tracking.

BOX FIGURE 19.1 Sunflowers (Helianthus annuus) oriented in the same direction, toward the sun. Helianthus means "sunflower."

BOX FIGURE 19.2 Time-lapse photograph of a buttercup (Ranunculus ficaria) tracking a source of light.

Chapter Summary

Cells are the simplest units of a plant that can live independently. Each plant cell consists of a cell wall that surrounds a plasma membrane, which encloses the contents of the cell. The contents of a plant cell usually include a nucleus and the cytoplasm. The cytoplasm includes all organelles except the nucleus, plus all internal membranes and the cytosol.

The study of cells is aided by instruments such as the light microscope and the electron microscope. The quality of microscopic observation depends both on magnification and on resolving power. Resolving power is limited by the wavelength of the light or the electron beam that is used for observation. Our study of plant cells has also been aided by examining the functions of organelles that have been isolated by cell fractionation and by selectively staining different parts of cells prior to microscopy.

Membranes divide cells into many interconnected compartments. These compartments are connected by membranes that either move between or are attached to the nucleus and other organelles. All membranes consist of a double layer of phospholipids that has enzymes attached to or embedded in it. The composition of membranes varies from one organelle to another.

The distribution and movement of membranes and other cell compartments are affected by the cytoskeleton, a network of filaments that includes microtubules, actin filaments, and intermediate filaments. All three types of cytoskeletal filaments are made of different kinds of proteins.

Young cells and actively growing cells have flexible primary cell walls. As cells mature, however, they often form rigid secondary cell walls just inside the primary walls. Cell-wall synthesis is controlled by arrays of microtubules just inside the plasma membrane.

Plant cells are connected to one another by plasmodesmata. These connections contain cytoplasm and a plasma membrane that is continuous between cells.

The nucleus, chloroplasts, and mitochondria are organelles that are surrounded by double membranes. Microbodies have single membranes, and ribosomes have no membranes. Dictyosomes and the endoplasmic reticulum are composed mostly of membranes that enclose relatively little internal fluid. Many of the chemical reactions in a cell are catalyzed by enzymes that are in or on membranes, including the plasma membrane.

Sperm cells in plants and animals, as well as other kinds of cells in algae and water molds, have flagella that enable them to swim. The structure and function of the flagella are identical in these organisms.

The cell theory has been the dominant theory of how living things are organized. It explains how cells are the basic units of organization. An alternative to this idea is the organismal theory, which holds that the basic unit of organization is the whole organism. Some evidence supports both theories, but observational and experimental evidence supports the organismal theory better in plants.

Questions for Further Thought and Study

1. Discuss the importance of having both proteins and lipids as components of membranes.
2. How does the function of rough ER differ from that of smooth ER?
3. Where are microtubules in cells? What are their roles?
4. How does microscopy provide evidence for the functions of cellular organelles?
5. What evidence supports the organismal theory?
6. The cell theory is an excellent example of inductive reasoning. How do you use inductive reasoning in your life?

Suggested Readings

ARTICLES

Albersheim, P. 1975. The walls of growing plant cells. Scientific American 232 (April):81–95.

de Duve, C. 1983. Microbodies in the living cell. Scientific American 248 (May):74–84.

Kaplan, D. R., and W. Hagemann. 1991. The relationship of cell and organism in vascular plants. BioScience 41:693–703.

Lane, M. A., et al. 1990. Forensic botany. BioScience 40:34–39.

Niklas, K. J. 1989. The cellular mechanisms of plants. American Scientist 77:344–349.

Rothman, J. E. 1985. The compartmental organization of the Golgi apparatus. Scientific American 253 (September):74–89.

Storey, R. D. 1990. Textbook errors and misconceptions in biology: Cell structure. American Biology Teacher 52:213–218.

Symons, M., et al. 1989. The shifting scaffolds of the cell. New Scientist (18 February):44–47.

Vogel, S. 1987. Mythology in introductory biology. BioScience 37:611–614.

BOOKS

Becker, W. M., and D. W. Deamer. 1991. The World of the Cell. 2d ed. Redwood City, CA: Benjamin/Cummings.

de Duve, C. 1984. A Guided Tour of the Living Cell. Scientific American Library. New York: W. H. Freeman.

Gunning, B. E. S., and M. W. Steer. 1986. Plant Cell Biology: An Ultrastructural Approach. Copyright M. W. Steer.

CHAPTER THREE Structure and Function of Plant Cells

71

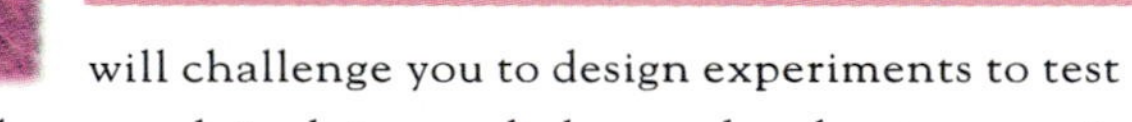

Doing Botany Yourself

will challenge you to design experiments to test hypotheses and, in doing so, help you develop an experimental approach to botany.

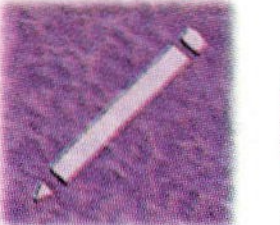

Writing to Learn Botany

will encourage you to use writing as a tool to learn botany. This feature will help you develop your communications skills, which are crucial to success in virtually any profession.

We have tried to make this textbook entertaining and accessible; we believe these features are critical to making students better learners, and instructors better teachers. To accomplish this, we've rejected the awkward, stoic writing style typical of "scientific writing." In its place, you'll find an engaging text that stresses communication and understanding.

We hope that *Botany* helps you learn about and appreciate plants. If you have questions, suggestions, or comments about what you read, let us know.

Randy Moore
Akron, Ohio

Kingsley Stern
Chico, California

Dennis Clark
Tempe, Arizona

Darrell Vodopich
Waco, Texas

April, 1994

From the Editor

It's been three decades since you've had a real choice in textbooks for your introductory botany course. Now you have the opportunity to peruse the most exciting and innovative new text in the field. Written by the dynamic author team of Randy Moore, Dennis Clark, Kingsley Stern, and Darrell Vodopich, *Botany* is the text that will instill an understanding of ***how*** science really works. The authors offer many examples of research: How questions are generated; How experiments are designed; How they succeed or fail; How slow the process can be; Why science is not about writing "laws" but about forming "hypotheses" and offering the evidence to support the hypotheses.

Botany is a fresh, new approach that offers a comprehensive discussion of general botany in a lively writing style. Your students will learn about the "Green Revolution" and why plant biotechnology is so important to our future. They will learn about the most current research in molecular biology and genetics. They will be fascinated by the anecdotes presented in "Lore of Plants," a feature presented throughout the chapters. They will be introduced to the vast diversity of plant life within a solid evolutionary framework. *Botany* stresses *what* we know, *how* we know it, and *why* we should care about plants. This text is more than a source of information. It will help the students understand the nature of science, how to think critically, and how to communicate effectively.

The text is logically organized to introduce the student to basic biological principles (Chapters 2–11), followed by Plant Form and Function (Chapters 12–17), Plant Growth and Development (Chapters 18–19), Nutrition and Transport (Chapters 20–21), Evolution, Diversity, and Ecology (Chapters 22–32), and ending with an Epilogue titled "Plants and Society."

To better serve your needs and to respond to the wide variety of approaches to the course, *Botany* will be available in two separate volumes as well as the complete, casebound text (ISBN 03775). Volume 1, Chapters 1–21 (ISBN 16656), comprises the units covered in a basic botany course—cellular biology, energy, genetics, plant form and function, development, and nutrition (approximately 532 pps). Volume 2 (ISBN 16657), Chapters 22–32, discusses evolution, plant diversity and ecology (approximately 336 pps). Both volumes will be accompanied by a full table of contents, glossary, and index. Volumes 1 and 2 are also available as a packaged set (ISBN 16957).

List of Ancillaries

Instructors Manual/Test Item File. Prepared by Rebecca McBride DiLiddo, Suffolk University, the manual features Chapter Outlines, Teaching Goals, Vocabulary Words, Mastery Concepts, Teaching Tips, Puzzles, and Projects. The Test Item File contains approximately 40–50 objective questions/answers per chapter for the instructor's use in preparing exams.

Micro Test III. User-friendly computerized software available to the instructor for testing and grading. It can be ordered in either IBM DOS or Windows format and MacIntosh.

Laboratory Manual. Prepared by Randy Moore and Darrel Vodopich, it offers 32 exercises for the general botany course. Each exercise begins with a list of objectives and a brief review of key concepts. There are numerous illustrations and diagrams to aid in lab explorations. The exercises conclude with a set of questions that can be used for homework assignments (ISBN 03777-0).

Laboratory Resource Guide. Is available to instructors and contains a list of equipment needs for each exercise and helpful hints for completing the labs safely and correctly.

Student Study Guide. Prepared by Rebecca McBride DiLiddo, this guide will aid students in testing their comprehension of important concepts and principles. The study guide contains Learning Objectives, Concept Maps, Multiple-Choice and Fill-in-the-Blank Questions (ISBN 03776–2).

Transparencies. A set of 100 key figures from the text are available free to instructors.

Transparency Masters. An additional set of 150 figures from the text are available free to instructors.

Student Study Art Notebook. Available to students to facilitate note-taking during lecture. The notebook comprises the figures from the text that are used as acetates or transparency masters and offers space for taking class notes (ISBN 24309–5).

Slides. A set of 200 additional images not found in the book are available free to instructors.

Writing to Learn Botany. Prepared by Randy Moore, this supplement will provide students with useful guidelines needed to formulate research papers and laboratory reports (ISBN 17455).

How to Study Science. Written by Fred Drewes, Suffolk County Community College, this workbook is an excellent guide for students enrolled in a science course. It offers tips on how to take notes, now to utilize laboratory time, and how to overcome science anxiety (ISBN 14474).

In order to better serve your needs, *Botany* is available with the following packaging options. The text and all its ancillaries have been printed on recycled paper in an effort to minimize the environmental impact of our products. If you have any questions about *Botany* or its supplements and packaging options, feel free to call us at 1–800–228–0459.

VOLUME 1 PLANT FORM AND FUNCTION ISBN 0–697–16656–2	Chapters 1–21 Appendix (Chemistry, Metric System) Glossary Index	Topical coverage includes an introduction to botany, cellular biology, energy, genetics (with the most current information available), plant form & function, reproduction, growth and development, and plant nutrition.
VOLUME 2 PLANT DIVERSITY ISBN 0–697–16657–0	Chapters 22–32, Epilogue Appendix (Diversity) Glossary Index	Topical coverage includes evolution, plant diversity, ecology, and an epilogue on "plants and society."
COMPLETE VERSION BOTANY (casebound) ISBN 0–697–03775–4	Chapters 1–32 Epilogue Appendix Glossary Index	The complete text provides comprehensive coverage of traditional botany topics as noted above.
VOLUMES 1 & 2 PACKAGED SET ISBN 0–697–16957–X	Chapters 1–21 (Vol. 1) Chapters 22– 32 Epilogue (Vol. 2)	Shrinkwrapped set containing Volumes 1 and 2 as noted above.

Acknowledgments

I cannot adequately thank my many teachers, friends, colleagues, reviewers, and students; their suggestions, teachings, and encouragement have improved this book immensely. I also thank Marge Kemp, Carol Mills, Kathy Husemann, Kay Brimeyer, and Kevin Kane for their continued help and guidance all writers should be so lucky to have editors and colleagues like these people.

I greatly appreciate the help of Robert Bergad in preparing this manuscript. His comments improved the accuracy and readability of the text.

I thank Jill Hammann and Flo Fiehn for their ongoing (and usually undeserved) patience, Marc Low and David Jamison for their encouragement, and Doyle and Tillie for their helpful advice. I also thank Stella, who has never said anything bad about me.

R. M.

List of Reviewers

Many instructors have contributed to the making of this text. We gratefully acknowledge the invaluable assistance of the many reviewers, whose constructive criticism was integral to the development of this text.

Bonnie Amos — *Angelo State University*
Robert Antibus — *Clarkson University*
Michael Barbour — *University of California-Davis*
Linda Barham — *Meridian Community College*
Jerry Baskin — *University of Kentucky*
William Beasley, Jr. — *Paducah Community College*
Rolf Benseler — *California State University-Hayward*
David Bilderback — *University of Montana*
Allan Bornstein — *Southeast Missouri State University*
Jack Bostrack — *University of Wisconsin-River Falls*
David Buckalew — *Xavier University of Louisiana*
Linda Chalker-Scott — *SUNY College at Buffalo*
Lee Anne Chaney — *Whitworth College*
Ronald Cheetham — *Worchester Polytechnic Institute*
William Chrouser — *Warner Southern College*
Tommy Cole — *Phillips College*
Bill Cook — *Midwestern State University*
Jonathan Cumming — *University of Vermont*
Ronald Darey — *SUNY-Cortland*
Archie Dickey — *Yavapai College*
Rebecca McBride DiLiddo — *Suffolk University*
David Domozych — *Skidmore College*
Max Dunford — *New Mexico State University*
Roland Dute — *Auburn University*
G. F. Estabrook — *University of Michigan-Herbarium*
Lance Evans — *Manhattan College*
Gerald Farr — *Southwest Texas State University*
Mary Fields — *Ursinus College*
Rosemary Ford — *Washington College*
Ronald Frank — *University of Missouri-Rolla*
Janice Glime — *Michigan Technological University*
Sue Harley — *Weber State University*
Marcia Harrison — *Marshall University*
Jon Huston — *Murray State College*
Julius Ikenga — *Mississippi Valley State University*

Susanne James — *Berry College*
George Johnson — *Arkansas Tech. University*
Ken Kilborn — *Shasta College*
Mark Knauss — *Shorter College*
Craig Landgren — *Middlebury College*
Peter Lindeman — *Madisonville Community College*
University of Kentucky Community College System
William Lipke — *Northern Arizona University*
Karen Lustig — *Harper College*
M. A. Madore — *University of California-Riverside*
George Manaker — *Temple University*
Lawrence Matten — *Southern Illinois University*
James Matthews — *University of North Carolina at Charlotte*
Jack McDonald — *Monroe County Community College*
Conley McMullen — *West Liberty State College*
L. Maynard Moe — *California State University-Bakersfield*
Lytton Musselman — *Old Dominion University*
Craig Nessler — *Texas A&M University*
Rebecca Nystrom — *Jamestown Community College*
Wayne Owen — *Boise State University & U.S. Forest Service*
Lois Peck — *Philadelphia College of Pharmacy & Science*
R. Gordon Perry — *Fairleigh Dickinson University*
Jerry Pickering — *Indiana University of Pennsylvania*
William Pietraface — *SUNY-Oneonta*
Paul Redfearn — *Southwest Missouri State University*
Christa Schwintzer — *University of Maine*
Fred Searcy, Jr. — *Broward Community College*
Stephen Shimmel — *Chipola Junior College*
J. Kenneth Shull, Jr. — *Appalachian State University*
Linda Spessard-Schueth — *University of Nebraska-Kearney*
Stephen Stocking — *San Joaquin Delta Community College*
Richard Storey — *Colorado College*
William Suder — *Brevard College*
Lawrence Swails — *Francis Marion University*
Herbert Tepper — *SUNY College of Environmental Science & Forestry*
Stephen Timme — *Pittsburgh State University-Kansas*
Dwain Vance — *University of North Texas*
Terry Vassey — *Horry-Georgetown College*
Judith Verbeke — *University of Arizona*
Dennis Walker — *Humboldt State University*
Leon Walker — *University of Findlay*
James Wallace — *Western Carolina University*
Frank D. Watson — *St. Andrews Presbyterian College*
Cherie Wetzel — *City College of San Francisco*
Erleen Whitney — *Clark College*
Charles Wimpee — *University of Wisconsin-Milwaukee*
George H. Wittler — *Ripon College*
Dan Wujek — *Central Michigan University*
Steve Adams — *Lake City Community College*
Charla Rae Armitage — *Lees-McRae College*
Marilyn Barker — *University of Alaska Anchorage*
Norlyn Bodkin — *James Madison University*
Cecile Boehmer — *Alice Lloyd College*
Frank Bowers — *University of Wisconsin-Stevens Point*
Joe Bruner — *Columbus College of Art & Design*
Richard Churchill — *Southern Maine Technical College*

Opal Dakin *Hinds Community College*
Jill Deikman *Pennsylvania State University*
Roger del Moral *University of Washington*
C. Ron Dillon *Clemson University*
Ronald Doney *SUNY at Cortland*
Robert Dorrance *Herkimer County Community College*
Jim Ebersole *Colorado College*
Daniel Flisser *Cazenovia College*
Royce Granberry *Texarkana College*
Danny Ingold *Muskingum College*
Jody Jellison *University of Maine*
J. Morris Johnson *Western Oregon State College*
William Kinnison *Central Arizona College*
M. Joseph Klingensmith *Rochester Institute of Technology*
Penelope Koines *University of Maryland*
Duane Kolterman *University of Puerto Rico-Mayaguez Campus*
Vaughnda Larson *West Virginia Northern Community College*
Robert Lebowitz *Mankato State University*
Jerri Dawn Martin *Sue Bennet College*
William Mathena *Kaskaskia College*
Richard Alan Mayes *Greensboro College*
W. D. McBryde *Central Texas College*
Sheila McCormick *Plant Gene Expression Center*
Robert McGuire *University of Montevallo*
Jerry Melaragno *Rhode Island College*
Robert Mellor *University of Arizona*
Jeanette Oliver *Flathead Valley Community College*
P. C. Pendse *Cal Poly University*
Ed Perry *James Faulkner State Community College*
James Rasmussen *Southern Arkansas University*
Maralyn Renner *College of the Redwoods*
Herbert Robbins *Dallas Baptist University*
Joseph Rohrer *University of Wisconsin-Eau Claire*
Robert Ross *Linn-Benton Community College*
Robert Rupp *Ohio State University/Agricultural Technical Institute*
Barbara Schumacher *San Jacinto College*
John Sehloff *Bethany Lutheran College*
Steven Selva *University of Maine-Fort Kent*
Richard Sims *Jones County Junior College*
Ken Smith *Montcalm Community College*
Richard Stalter *St. John's University*
Richard Storey *Colorado College*
Jon Stucky *North Carolina State University*
Nicholas Sturm *Youngstown State University*
Jerry Tammen *Rochester Community College*
Rahmona Thompson *East Central University*
Michael Timko *University of Virginia*
Thomas Vierheller *Prestonsburg Community College*
Joyce Weary *Oakland Community College*
Edgar Webber *Keuka College*
David Williams *Ann Arundel Community College*
Donald Williams *Sterling College*
Kathryn Wilson *Indiana University-Purdue University at Indianapolis*
Kenneth Wilson *Miami University*
Robert Winget *Brigham Young University-Hawaii*
James Winsor *Penn State University*
Todd Christian Yetter *Cumberland College*

About the Authors

Dr. Randy Moore is currently Dean of Arts and Sciences and professor of biology at The University of Akron. In addition to writing numerous articles and textbooks, he is Editor-in-Chief of the *American Biology Teacher*. Recently, Dr. Moore was awarded the 1993 Teacher Exemplar Award by the Society for College Science Teachers. Other awards and honors include many research grants (e.g., NSF, Nasa) a Fulbright Scholarship, a listing in the *Who's Who in Science and Engineering*, 1986 Most Outstanding Professor from Baylor University, and 1993 Most Outstanding Professor at Wright State University. He received his Ph.D. in plant development from UCLA.

Dr. W. Dennis Clark is currently a professor at Arizona State University in Tempe, where he has taught for the past 17 years. A prolific researcher, Dr. Clark has authored numerous scientific papers, while at the same time being active in many professional societies. He received his Ph.D. in botany from the University of Texas at Austin.

Dr. Kingsley R. Stern has been teaching botany courses for over 40 years. He received his Ph.D. in botany from the University of Minnesota with minors in zoology and horticulture. He has received numerous awards, including the 1989 Outstanding Professor Award from California State University, Chico. His test, *Introductory Plant Biology*, now in its sixth edition, is the best-selling nonmajors botany textbook in the world.

Dr. Darrell Vodopich is a co-author of *Volume 2*, with specific contributions in the discussion of evolution. He is currently teaching general biology at Baylor University in Waco, Texas. He has also co-authored the best-selling laboratory manual that accompanies *Biology* by Raven/Johnson.

Unit Seven

Evolution . . .

Throughout recorded history, people have wondered how the great variety of earth's organisms came to be. Among the earliest thinkers was Aristotle, who thought that each species arose independently from inorganic matter and did not change. His *scala naturae* (scale of nature) stretched from nonliving matter, through lower forms of life, to humans at the top. Although each living species held a permanent place on the scale, species were viewed as imperfect. Perfection reflected the work of a perfect God, and was found only in the transcendent world of ideas.

Today our ideas about the diversity of life differ markedly from those of Aristotle and his contemporaries. Specifically, we view populations of organisms as entities that change as they adapt to changing environments. The foundation for understanding how biological change occurs in populations and how such change forms new species rests on the theories of organic evolution. The earliest of these theories are most closely identified with Charles Darwin, the greatest naturalist-philosopher of the nineteenth century. His ideas about evolution have impacted a wider array of human endeavors than any other scientific advancement of the past 150 years.

How Darwin arrived at his ideas and how they have affected science, philosophy, religion, and human attitudes is one of the most fascinating stories of human achievement. This subject occupies a significant portion of hundreds of books spanning the past century. For this textbook, the most important aspect of the story is how Darwin's work provided the first scientific explanation for the diversity of life.

Evolution may be the grandest puzzle in all of natural science; our clues about this puzzle are sometimes ancient and sometimes modern. This fossilized *Ginkgo* leaf dates from the Paleocene epoch over 65 million years ago. A comparison with its modern counterpart raises tantalizing questions about change. How much? How little? What happened during those 65 million years?

Evolution

Chapter Outline

Chapter Overview

Evolution is the most comprehensive theme in biology, because it explains how life began and how it diversified into the organisms of today. Evolution implies change, but it specifically refers to the modification and descent of successful forms of life. The foundation for modern evolutionary thought was described by Charles Darwin more than a century ago; Darwin also described natural selection as the main force that makes evolution work. Although Darwin provided evidence for evolution by the mechanism of natural selection, he and his contemporaries were unaware of how parental traits could be changed and passed to offspring. Beginning with the work of Gregor Mendel, however, genetics provided some clues as to how genetic variation arose and how it was inherited. After the turn of the century, evolutionary biologists began to learn how genes behave at the population level. More recently, explanations of how genes evolve have come from information about their molecular biology.

Introduction

Organic evolution implies change in living things, and its definition in texts and dictionaries is usually something like this: The development of complex forms of life from simple ancestors. Unfortunately, this definition is oversimplified to the point of being only partially correct. There is no simple definition of evolution; a full description of it should include the main components, or postulates, of a theory of evolution. Such a theory is popularly attributed to Charles Darwin, although the word *evolution* did not appear in his famous book, *Origin of Species*.[1] It is more accurate to think of Darwin's work as a set of three theories. Two of his theories are still accepted: the theory of *descent with modification*, which explains the pattern of biological diversity, and the theory of *natural selection*, which explains the primary mechanism by which evolution works. The third theory proposed by Darwin, the theory of *pangenesis*, was an incorrect explanation of inheritance. This theory was soon discarded and later replaced by Gregor Mendel's theory of *inheritance* and the *chromosome theory of heredity* (see Chapter 8).

A set of theories to define evolution seems cumbersome, so evolutionary biologists have spent a lot of time trying to refine, reinterpret, and update Darwin's work into a more modern and more streamlined form. As a result, we now have the genetic or synthetic theory of evolution.[2] The postulates of this theory include many of Darwin's views on evolution, as well as modern ideas about inheritance and a theoretical foundation for genetic change within populations. Ideas about population genetics have been especially important in understanding how populations evolve.

As you read this chapter, keep table 22.1 handy as a reference for the major postulates of currently accepted theories of evolution. It is especially important to note that Darwin proposed separate explanations for the *pattern* of evolution (descent with modification) versus the main *process* of evolution (natural selection). Also note that the synthetic theory incorporates ideas from Darwin, Mendel, modern genetics, and population genetics.

Like all important ideas, evolution attracts controversy; it has attracted more controversy than most scientific ideas because it has affected not only science but also philosophy, religion, and human attitudes. Unfortunately, evolution is often seen as competing with religious ideas as a way of explaining the origin and diversity of life. This competition is inappropriate, since evolutionary theory is scientific, and religions are not. Evolution is evaluated according to the scientific method; the existence of a divine creator is a product of faith and cannot be evaluated scientifically.[3]

Evolution Before Darwin

Greek Philosophy

Several people, including some classical Greek philosophers, recognized the gradual evolution of life as early as 2,000 years before Darwin. However, the influential teachings of Plato, Aristotle, and others denied the concept of an evolving world and profoundly influenced Western culture. Plato's philosophy of idealism stated that living organisms were merely variations of ideal forms. Each species was created by God, with modern individuals tracing their ancestry to the individual created by God. Aristotle, the natural historian, also wrote of the divine creation of eternal forms. He recognized

1. The full title of Darwin's book was *On the Origin of Species by Means of Natural Selection, or the Preservation of Favoured Races in the Struggle for Life.*

2. *Synthetic* refers to the synthesis of information and ideas from different fields of biology into a cohesive framework of evolutionary thought.

3. This distinction has not stopped some religious groups from inventing a discipline called *scientific creationism*, which has nothing to do with scientific thinking.

Table 22.1

Major Postulates of Theories That Encompass Evolution

Theory of Descent with Modification	Theory of Natural Selection	Genetic (Synthetic) Theory of Evolution
1. All life evolved from one simple kind of organism or from a few simple kinds.	1. A population of organisms has the tendency and the potential to increase at a geometric rate.	1. Evolution is the change of gene (allele) frequencies in a gene pool over many generations.
2. Each species, fossil or living, arose from another species that preceded it in time.	2. The number of individuals in a population remains fairly constant.	2. Each species is an isolated pool of genes possessing regional gene complexes that are connected by gene flow.
3. Evolutionary changes were gradual and of long duration.	3. The conditions supporting life are limited.	3. An individual contains only a portion of the genes in the gene pool, and the portions are different for each individual.
4. Over long periods of time, new genera, new families, new orders, new classes, and new phyla (divisions) arose by a continuation of the kinds of evolution that produced new species.	4. The environments of most organisms have been in constant change throughout geologic time.	4. The kinds of genes and gene combinations in an individual that reproduces sexually came from the transmissible halves of the parents, from recombination, and from mutation.
5. Each species originated in a single geographic location.	5. Only a fraction of the offspring in a population will live to produce offspring.	5. An individual with a phenotype that favors the production of more offspring will contribute a larger proportion of genes and gene combinations to the gene pool.
6. The greater the similarity between two groups of organisms, the closer is their relationship, and the closer in geologic time is their common ancestral group.	6. Individuals in a population are not all the same; they have heritable variations.	6. Isolation that restricts gene flow between a subpopulation and its parent population is essential if the subpopulation is to evolve into a new species.
7. Extinction of old forms (species, etc.) is a consequence of the production of new forms or of environmental change.	7. The struggle for existence determines which traits are favorable or unfavorable by determining the success of the individuals who possess the traits.	7. Changes of gene (allele) frequencies come about by natural selection, migration, gene flow, mutation, and random genetic changes. Natural selection is the most important cause of changes in gene (allele) frequency.
8. Once a species or other group has become extinct, it never reappears.	8. Individuals having favorable traits will, on the average, produce more offspring, and those with unfavorable traits will produce fewer offspring.	8. Evolution of a species may result in a chronological sequence of species without an increase in the number of species, in a group of new species, or in combinations of these two possibilities.
9. Evolution continues today in generally the same manner as during preceding geologic eras.	9. Natural selection causes the accumulation of favorable traits and the loss of unfavorable traits such that a new species may arise.	9. Speciation is completed when variations have accumulated in a species subpopulation to the extent that genetic exchanges with the parent population, or with "sister" populations, do not occur, even though the two populations meet.
10. The geologic record is very incomplete.		10. Mutations are the ultimate source of new genes in a gene pool.

as well that organisms ranged from relatively simple to complex, and he organized them on a scale of increasing complexity called *scala naturae* (scale of nature). That scale implied a linear increase in complexity and was completely uniform: there were no missing rungs and no movement up or down the ladder. However, the ladder implied nothing about the origin of various groups. This organization of static, ideal forms endured for 2,000 years. According to Aristotle, organisms did not evolve.

The Judeo-Christian culture also promoted prejudice against evolution. Most people believed in the special creation of each species, as described in the Bible's book of Genesis. At least through the 1600s, people believed that plants and animals were created in their current form during the six days of creation. This doctrine of fixed species was never convincingly challenged before Darwin. Even during Darwin's time, biology in the Western world was dominated by natural theology, the attempt to discover God's plan by studying God's works. To natural theologians, variation and adaptation in organisms "proved" that each species was designed by God for a particular purpose. Besides, these theologians believed that the earth was only a few thousand years old. Clearly, this wasn't long enough for significant evolutionary change.

Catastrophism

The study of fossils was begun by Georges Cuvier, a French anatomist working in the Paris Basin of France. He realized that the history of living organisms was recorded in layers of rock containing a succession of fossil species trapped and preserved in chronological order (fig. 22.1). Cuvier noted that each layer contained fossils from a different set of ancient species. Furthermore, deeper layers revealed plants and animals that were increasingly different from today's organisms. This evidence implied change, but Cuvier still opposed the concept of evolving species. As an anatomist, he noted similarities among fossils as well as differences, and believed that similar structures reflected a grand design by a Creator. He also believed that boundaries between geological strata of fossils represented catastrophic events such as floods or drought. These events destroyed some of the resident species which, therefore, would be absent in the next stratum and would not be living today. This view of natural history is called **catastrophism.** To Cuvier, fossils were organisms that had died in a series of catastrophes, after which the extinct plants and animals were replaced by the immigration of distant species to the devastated region. Through the eighteenth century, most naturalists invoked catastrophism rather than evolution to explain fossils. However, evidence that species changed over time was steadily accumulating.

FIGURE 22.1

The layers of rock in this canyon hold snapshots of plants of the past, and retain a chronological record of ancient organisms. The deeper the layer, the older the fossils it contains. These layers of preserved impressions of ancient plants hold the most tangible clues to evolutionary history.

Gradualism

In 1795, James Hutton argued that the earth was much older than a few thousand years. Hutton explained that profound geological change occurred slowly but continuously by the process of **gradualism.** Sedimentary rock that encased fossils formed by the gradual accumulation of sediments in lakes, rivers, and oceans (fig. 22.2). Hutton's explanation was widely debated by geologists, including Scottish lawyer-turned-geologist Charles Lyell. Lyell believed that Hutton's evidence for gradualism indicated that the earth was millions rather than thousands of years old. Furthermore, Lyell believed that even slow and subtle processes could cause substantial change over a long time.

Catastrophists and gradualists such as Cuvier, Hutton, and Lyell were good geologists convinced of an ancient earth, but were at odds over how to explain the appearance and disappearance of species in the fossil record. They were all creationists. Lyell, in particular, invoked an unspecified Creator as the source of new species that were added gradually to the earth's flora and fauna. Later, he only grudgingly accepted Darwin's ideas about evolution.

FIGURE 22.2

This fossil of a Triassic seed plant, *Dicroidium* sp., is about 200 million years old. The gradual accumulation of sediments that surrounded and fossilized such plants indicated to Charles Lyell in the early 1800s that the earth was millions rather than thousands of years old.

CONCEPT

Prior to Darwin's development of evolutionary theory, species were believed to be divine and unchanging creations that appeared on earth a few thousand years ago. Studies by Cuvier, Hutton, and Lyell of fossils and geology indicated that the earth was old and slowly changing, and that life existed millions of years ago.

Lamarckism

In 1809, the year Charles Darwin was born, Jean Baptiste Lamarck (1744–1829) proposed that modern species had descended from other species (fig. 22.3). Lamarck proposed some of the ideas found later in Darwin's work, such as the origin of species from preexisting species and the ability of organisms to adapt to changing environments. However, he is most remembered for his theory of the inheritance of acquired characteristics. According to this theory, traits acquired by an individual during its life were passed to its offspring, a theory caricatured by many detractors in drawings showing giraffes obtaining their long necks from previous giraffes who stretched to eat the leaves of high tree branches. According to Lamarck, stretching increased the length of their necks, and this acquired characteristic was passed to the next generation. Their necks were stretched further in each successive generation. Using similar reasoning, Lamarck also proposed that the use and disuse of a feature governed the fate of that feature in successive generations.[4] Organs of the body that were used extensively to cope with the environment become larger and stronger, while organs that were not used deteriorated. Instead of the static scheme of Linnaeus, Lamarck provided a dynamic one.

Lamarck was the first naturalist to present a unified theory that attempted to explain the changes in organisms from one generation to the next. Lamarck was a dedicated and observant biologist, and his theories stimulated inquiry into the history and diversity of organisms. However, he had no insight into the laws of heredity, and the mechanisms he proposed for change were wrong. Individuals do not inherit acquired characteristics. Inheriting acquired characteristics would mean that a feature acquired during an organism's lifetime would somehow have to be transmitted to the genes of that organism's gametes. There is no evidence that this happens.

Lamarck recognized change in organisms, but incorrectly explained it as the inheritance of acquired characteristics. Acquired characteristics are not heritable.

4. Lamarck either ignored or was unaware that hundreds of generations of Jewish males had been circumcised without giving rise to a single line of already-circumcised males.

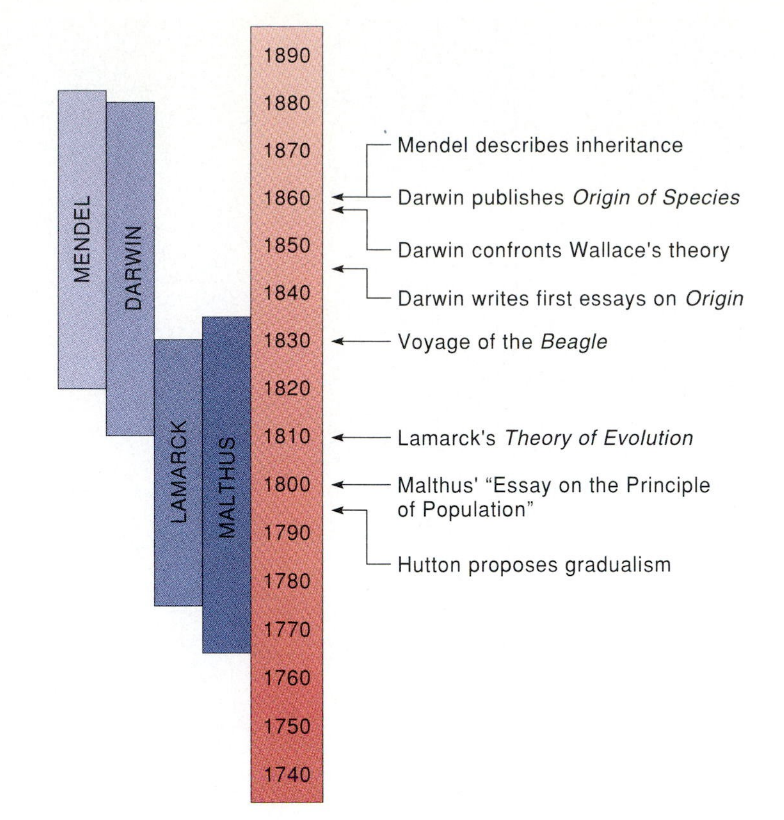

FIGURE 22.3

A historical time-line of the major people and events of the development of evolutionary theory.

DARWIN

As a boy in Shrewsbury, England, Charles Darwin was fascinated with nature. His well-to-do father pressured Charles to pursue the family tradition of a medical career at the University of Edinburgh, but Charles was soon bored and discouraged by medicine and the ghastly nature of surgery, so he quit school. He then enrolled at Christ College of Cambridge University to study theology, which appealed to him because most naturalists at that time were also clergymen. As a clergyman, Charles's pursuit of natural history was accepted by his peers, and he graduated from Cambridge in 1831.

One of Darwin's botany professors at Cambridge was John Stevens Henslow, who became one of his lifelong friends. Henslow recommended Darwin as a keen observer, naturalist, and companion to Captain Robert FitzRoy for his voyage aboard the H.M.S. *Beagle* to survey lands around the world. Darwin's father initially refused to let Charles go on the trip. Fortunately, Charles's father was swayed by the supportive appeal of his uncle Josiah Wedgewood, son of the famous porcelain maker and father of Darwin's future wife, Emma. Darwin's father finally allowed Charles to go on the voyage.

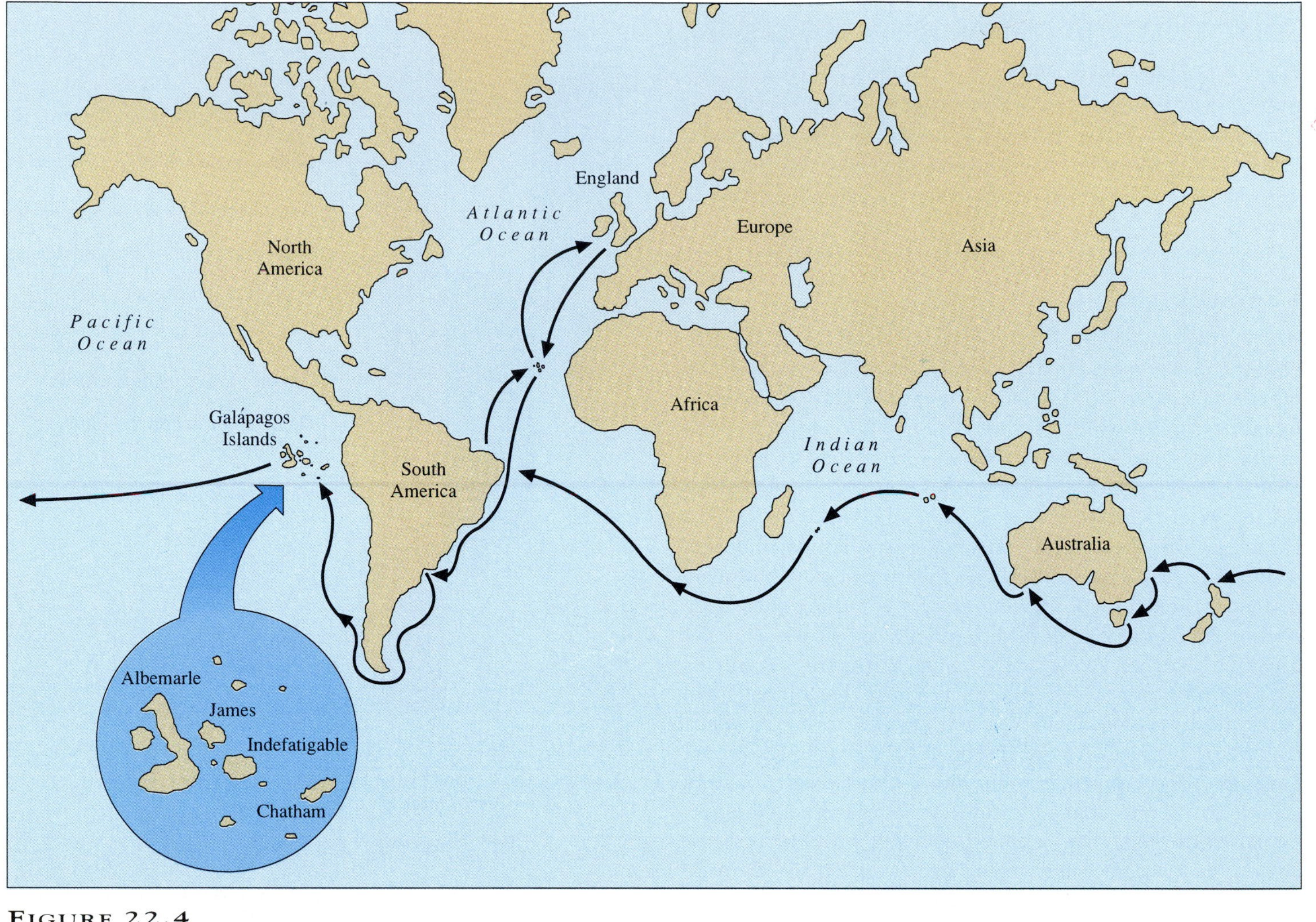

FIGURE 22.4

Voyage of the *Beagle*. Between 1831 and 1836, the voyage covered 60,000 km.

The Voyage of the *Beagle*

The *Beagle* sailed from England in 1831 to chart some of the remote islands and coastline of South America (fig. 22.4). Darwin was twenty-two years old, and spent much of his time on shore collecting plants and animals and making geological and biological observations. The flora and fauna of South America were remarkably more diverse and exotic than those of Europe, and they gave Darwin examples of the varied adaptations of organisms to life in mountains, jungles, grasslands, and desolate islands such as the Galápagos Islands and Tierra del Fuego of South America. The geographic distribution of South American organisms both fascinated and puzzled Darwin, and he pondered why they were so diverse. During the voyage, Darwin also read Lyell's *Principles of Geology* and realized that the earth was very old and constantly changing. This idea, along with his collections and observations, swayed Darwin from his previously staunch, literal interpretation of the Bible. Slowly, he embraced the idea that the organisms that fascinated him may have changed or evolved along with a slowly changing environment.

Darwin made some of his most profound observations on the Galápagos Islands. These islands are near the equator and about 900 kilometers west of South America. Most of the plants and animals found there are endemic to the islands (i.e., they live nowhere else), but they strongly resemble species from the closest mainland. Among the more famous examples of unusual organisms on these islands were finches (fig. 22.5). Darwin collected fourteen types of finches that were similar but appeared to be different species. Furthermore, these finches were curiously similar to the mainland species, yet slightly different. If species were separate and unchanging creations, then why, Darwin wondered, did so many of the Galápagos finches resemble nearby South American species rather than European or African organisms? If finches on the islands had been specially created, why didn't they all look the same? Darwin later concluded that the island finches had descended from a population of ancestral finches from the mainland. The population of mainland finches changed and diversified after becoming isolated on the islands, giving rise to the fourteen species endemic to the islands.

FIGURE 22.5

Darwin discovered that finches of the Galápagos Islands arose from mainland species; the island species diverged in part based on adaptations to different sources of food.

Since the time of Darwin, many other groups of organisms have been discovered to be descended from a single ancestral type. One of the best-known examples of such diversification in plants is the genus *Phlox*, which evolved from an ancestral bee-pollinated species to species that undergo pollination by beetles, flies, bats, hummingbirds, moths, or butterflies (fig. 22.7b).

The Formulation of Darwin's Theories

After returning to England in 1836, Darwin did not immediately begin writing his book on the origin of species. Instead, he methodically organized his notebooks and data and did an extensive analysis of a tremendous number of specimens and data. During his voyage, Darwin had shipped back to England many crates of carefully packed specimens.

In 1838, two years after returning to England, Darwin encountered an inspiration for his theory of natural selection: *An Essay on the Principle of Population*, by Thomas Malthus. Malthus was an economist and clergyman who wrote that populations had an inherent tendency to increase in geometric proportions. Malthus also claimed that resources to support this growth may increase slowly or not at all. He therefore reasoned that because continued growth of a species would outstrip needed resources (especially the food supply), growth would be limited (fig. 22.6). Specifically, Malthus warned of an explosive growth of the human population. Malthus pessimistically warned that humans were reproducing so fast that war, famine, disease, and other human sufferings must eventually limit growth. He believed that humans were doomed to suffering, and he foresaw the need for birth control and social change, such as delayed marriage.

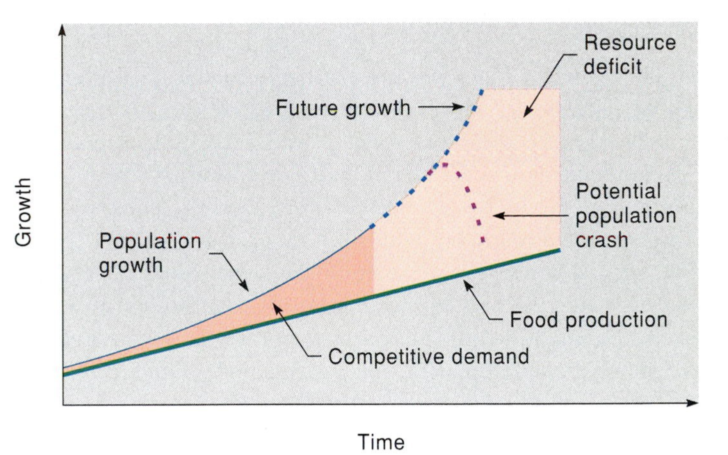

FIGURE 22.6

The predictions of Malthus guided Darwin's formulation of natural selection theory. Malthus believed that an expanding resource deficit between population growth and food production forced competition and would later result in a population crash.

Writing to Learn Botany

Thomas Malthus was pessimistic about the future of the human race. What factors have diverted humanity from the fate predicted by Malthus? Or, have we actually been diverted?

Darwin disagreed with Malthus's social views, but adopted his ideas that populations tended to increase geometrically and outstrip their resources. Logically, some newborn organisms will die in competitive environments. Darwin observed that wild organisms have variable traits, and he reasoned that in a resource-limited environment, the hardiest or best-suited individuals had a competitive and reproductive advantage over weaker individuals. As a result, traits occurring in well-adapted individuals would increase in successive generations, and the traits of poorly adapted individuals would decrease. Darwin called this process **natural selection,** and he saw a similar mechanism in the modern world among animal breeders who practiced **artificial selection.** These breeders changed the population by selecting organisms with desirable traits and destroying those with undesirable traits.

In Darwin's view, the history of changes of organisms could be represented as a tree with multiple branches (fig. 22.7). The trunk and the bases of the branches represented common ancestors that gave rise to all the subsequent species represented by the more distal branches, some of which were common ancestors of other diverging branches and species. The diversity we see today is represented by all the tips of the branches, which may in the future become extinct or produce new branches through speciation. Most branches of the evolutionary tree (possibly as many as 99%) have been dead ends; that is, they did not produce new species, but became extinct. Darwin often referred to "descent with modification," and the pattern of descent included many branches.

In Darwin's view, a common ancestor could have given rise to a new species by accumulating adaptations to a new environment. New and different adaptations were especially probable when an ancestral species became divided into populations isolated by geographical barriers. Each of the separated populations might have reproduced slowly and changed in response to local environmental conditions. As an isolated population adapted to its new environment, its members would deviate more and more from the appearance and adaptations of the ancestral population. After many generations, an isolated population would be different enough to be a new species. Darwin believed that this had happened to the Galápagos finches. The populations on the Galápagos Islands gradually adapted to the island environment and its food supply, and became different from their mainland ancestors. In addition, the populations on different islands diverged from one another.

Within five years Darwin had established the major postulates of his theory of natural selection. But he hesitated to publish his idea that species arose by adaptations of organisms to their environment. This idea strongly contradicted public opinion, even though the concept of a changing earth and changes in its inhabitants was gaining credibility among naturalists. Darwin overcame his hesitation in 1858 after receiving a special letter from Alfred Russel Wallace containing a manuscript about a theory of natural selection that was remarkably similar to his own. Darwin was shocked and discouraged. He wrote to Lyell, "I never saw a more striking coincidence; . . . all my originality, whatever it may amount to, will be smashed." However, Darwin and Wallace were cooperative scientists interested in the truth. Their theory was presented jointly at a scientific conference (fig. 22.8), after which Darwin quickly finished and published his *Origin of Species* on Thanksgiving Day in 1859. The first edition sold out on the first day it appeared. Darwin explained and documented his ideas so much more extensively than Wallace that Darwin is known as the leading author of the theory of natural selection. The theory was revolutionary, not only because it was unique, but also because Darwin presented a strong argument based on overwhelming evidence. Interestingly, the first edition of *Origin of Species* did not contain the word *evolution*.

Darwin's *Origin of Species* shook the foundations of Western culture because his explanation for the origin of and change in species did not require a supernatural power. He proposed that species are not divine, immutable creations but have gradually evolved from ancestral species over many thousands of years. To Darwin, evolution was the gradual and selective accumulation of adaptations that were beneficial to a species in its environment. Even more important, he proposed a mechanism for evolution (i.e., natural selection), which he deduced in an organized fashion from evidence.

According to Darwin, evolution includes gradual and selective accumulation of the beneficial adaptations of a population to its environment and a reduction of detrimental characteristics. Natural selection is a mechanism that accounts for heritable changes in populations.

What are Botanists Doing?

Piecing together the phylogenetic tree that maps the evolutionary path of organisms can sometimes be tedious and far removed from everyday events. Interview a taxonomist and ask specifically what relevance his or her work has to current topics such as biodiversity or pharmacology.

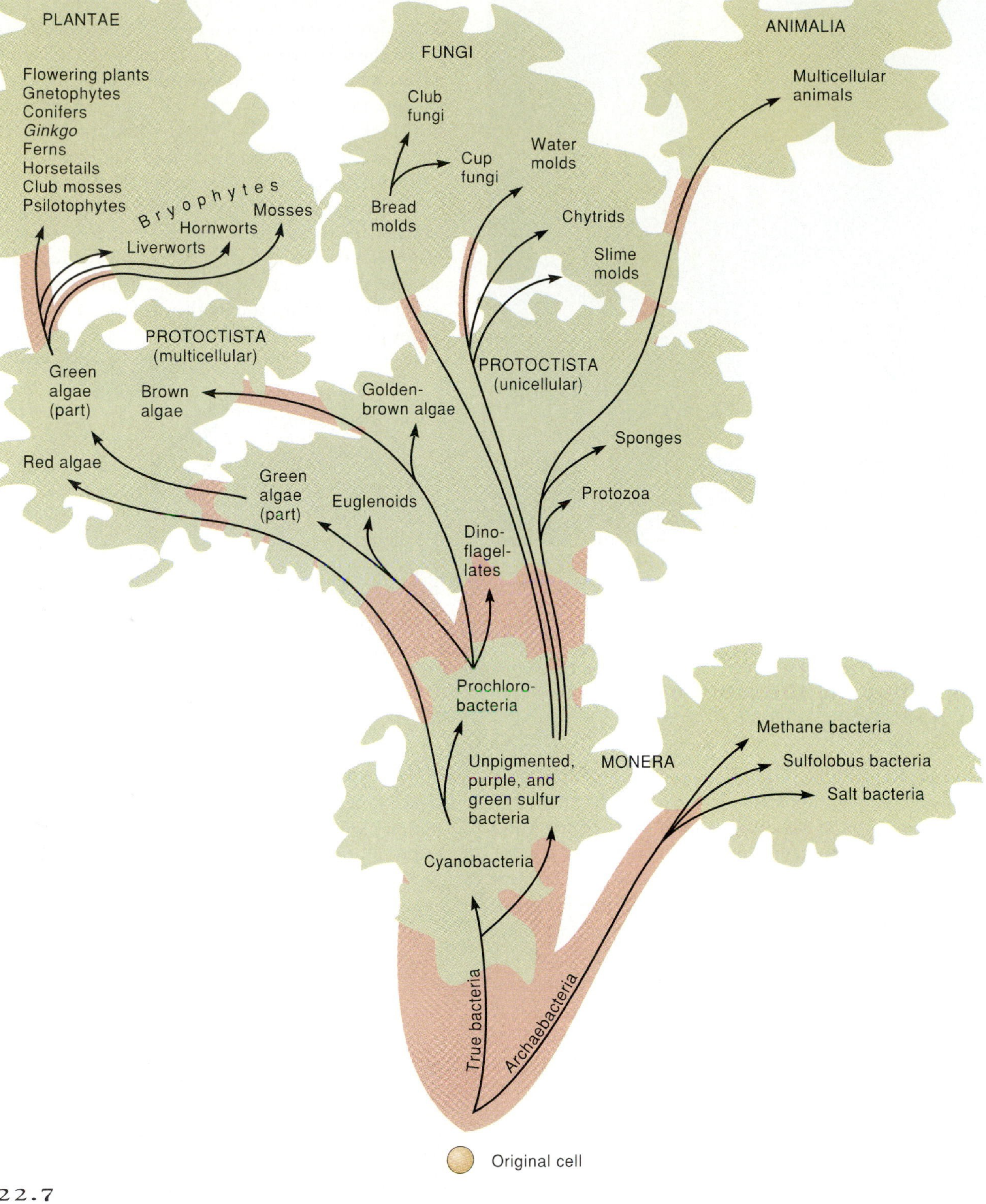

FIGURE 22.7

The evolutionary history of organisms can be portrayed as a tree growing through time. Ancient forms are low on the tree. At the tips of the branches are modern forms. The base of each branch represents the common ancestry of the newly evolved kinds of organisms.

FIGURE 22.8

Alfred Wallace's paper on natural selection (coauthored with Darwin) was presented to the Linnaean Society on July 1, 1858.

EVIDENCE FOR DESCENT WITH MODIFICATION

Darwin meant to publish a full explanation of evolution based on a thorough analysis of the evidence he had gathered, but he never did. The *Origin of Species* was an abstract of his work, which he rushed into publication when Wallace's work appeared. Although not complete, Darwin's work provided several kinds of evidence for the theory of descent with modification. Work since Darwin's has added much new information, some of which is summarized below. Note that evidence for evolution by descent with modification is observational and indirect; thus, the theory of evolution is not based on direct experimentation. Nevertheless, taken together, different kinds of indirect evidence provide the basis for strong inference of the process of evolution.

The Age of the Earth

If evolution occurs gradually, then the earth must be old enough to allow it to happen. Many physicists in Darwin's day believed that the earth was only a few thousand years old—a period too brief for all living things to have evolved from a single common ancestor.

We now estimate the age of the earth by comparing ratios of radioactive to nonradioactive elements in rocks. For example, biologists assume that the ratio of uranium-238 (U^{238}) to lead when the solar system was formed was about double the current ratio. We know that the **half-life** of U^{238}—that is, the amount of time in which half of the U^{238} in a sample decays to something other than U^{238}—is about 4.5 billion years. This indicates that the earth is between 4.5 and 5.0 billion years old, since the ratio of U^{238} to lead has decreased to about half of what it was when the earth was formed. Biologists believe that this is long enough for the evolution of all of the organisms of today to have occurred from a single ancestor. Incidentally, meteorites are also estimated to be about 4.5 billion years old, which supports the idea that the earth was formed simultaneously with other parts of the solar system.

Midnight:
Earth is formed; the geological clock starts ticking. Each second of the clock passes 50,000 years.

Fossil Evidence

The oldest organisms are probably either absent or their preservation in rocks is unrecognizable. The oldest recognizable fossils are of bacteria, some of which were photosynthetic, that lived at least 3.5 billion years ago (fig. 22.9). This age establishes a minimum time span for organic evolution.

5:20 A.M.:
First life? Bacteria appear in the fossil record.

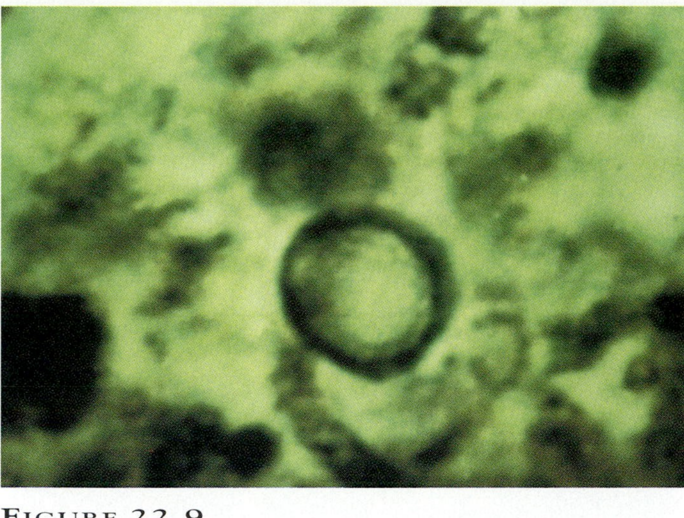

Figure 22.9

Fossilized bacterial cells such as these have been found in ancient rocks of Australia and South Africa. These rocks are about 3.5 billion years old and predate the earliest known fossils of eukaryotes by at least 2 billion years. Chemical evidence within the bacterial fossils suggests that photosynthesis was occurring 3.5 billion years ago.

Figure 22.10

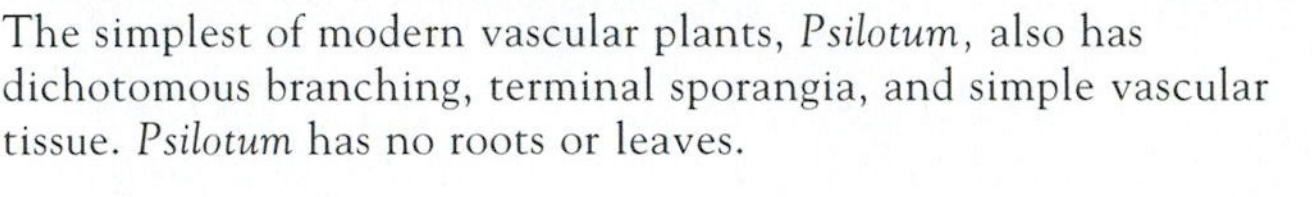

The simplest of modern vascular plants, *Psilotum*, also has dichotomous branching, terminal sporangia, and simple vascular tissue. *Psilotum* has no roots or leaves.

Darwin predicted that the fossil record would contain links among related organisms, from ancient to progressively more recent species. Much of the information about fossils that has subsequently been gathered agrees with Darwin's prediction. Some of the best fossils are from bones, especially teeth, so zoologists can construct nearly complete series for some groups, such as horses, humans, other mammals, and reptiles. The most complete record of plants is of those with vascular tissue; however, there is no clear lineage of plants, one giving rise to the other, as there seems to be with other organisms. Nevertheless, 400-million-year-old fossils of the first vascular plants, if not directly ancestral to present-day groups, look like they share a common ancestry with certain groups of living vascular plants (fig. 22.10). More details about the relationships of fossilized and living vascular plants are presented in Chapters 28–30.

9:52 P.M.:
The first vascular plants appear.

Homology

Features that are similar because of common descent show **homology.** Many organisms have features that appear to have come from the same ancestor. In flowering plants, for example, the ovary tissue surrounding the seeds of magnolia resembles that of lily, poppy, oak, and orange, because all of these plants inherited it from a common ancestor. Likewise, the red pigments of cactus flowers, beet roots, *Portulaca* stems and flowers, and spinach petioles are homologous. These pigments are of the same chemical type, called *betalains* (see Chapter 2), which is unlike other kinds of red pigments. Most of the major features of plants that we've discussed, such as their vascular structure and their hormonal responses to stimuli, probably descended from a common ancestor of the plant kingdom. Furthermore, plants have features that are homologous with features of other kinds of organisms. Cells with nuclei, chromosomes with histones, and genes with introns were inherited from the common ancestor of plants, animals, and fungi. Similarly, photosystem I in plants and in certain photosynthetic bacteria probably came from a common ancestor.

Convergence

Whether similar features are homologous is often unclear because not all similarities are due to homology. The issue of homology is clouded by features that arise in distantly related plants that live in similar environments. One of the most common examples, which often fools all but the most astute observers, is the stem succulence of cacti and euphorbias

A.

B.

FIGURE 22.11

Stem succulents: (a) cactus (b) euphorbia. The similar adaptations to arid environments by these two plants are the products of convergence as unrelated plants adapt to similar conditions, rather than due to common ancestry.

(fig. 22.11). Instead of homology, this feature shows **convergence,** which is one explanation for a similarity that does not come from a common ancestor. In this case, descendants of a precactus ancestor became adapted to arid environments in the Americas by developing succulent stems. Likewise, descendants of a preeuphorbia ancestor evolved succulent stems in response to arid environments in Africa. Entire plant communities in different geographical areas also appear similar because they are adapted to similar environments. Shrublands around the Mediterranean Sea, the coast of southern California, and the coast of central Chile are similar, even though the plants that comprise them are not closely related to one another.

Similarities among distantly related organisms are not likely to result from coincidence. Instead, nonhomologous features of plants and plant communities indicate that different organisms have evolved in similar ways to similar environments in different areas of the world.

Biogeography

Biogeography is the study of the geographical distribution of organisms. The distribution of today's organisms provides strong evidence for evolution because biogeography reflects the history of living species. That is, it reveals their movement and change. From this information we can determine the ecological factors that control the extent of a species's distribution and the events that led to this distribution. In other words, the biogeography of a species partially records its evolution.

Darwin and Wallace both questioned the apparent discrepancy between divine creationism and the biogeography of species that they observed. Why wouldn't similar environments in distant parts of the world have the same divinely created species? What explained the oddly disjunct distributions of organisms such as *Sanicula crassicaulis* (a member of the carrot family), found only in California and southern Chile; or creosote bushes (*Larrea*), which live in the deserts of western North America and southern South America; or skunk cabbage (*Symplocarpus foetidus*) and the tulip tree (*Liriodendron tulipifera*), which occur in eastern Asia and eastern North America; or cacti, which are all American except for one species in Madagascar? How have these patterns of distribution come about? The answers lie in the study of biogeography.

Island biogeography has been particularly informative about the movement and evolution of organisms. *Island* refers not only to a habitat surrounded by water, but also to any area

A.

B.

C.

Figure 22.12

Adaptive radiation is striking and rapid for plants on the Hawaiian Islands. The three species shown here, (a) *Argyroxiphium sandwicense*, (b) *Wilkesia gymnoxiphium*, and (c) *Dubautia ciliolata*, and 25 other closely related species probably descended from a single species originating on the Pacific Coast of North America. Adaptive radiation of Hawaiian species has produced diverse morphologies even though the species differ in only a few critical genes.

surrounded by an aberrant habitat such as a mountain surrounded by desert or a clump of trees surrounded by prairie. Islands often have species found nowhere else in the world, and their closest relatives may inhabit a different environment on the nearby mainland. This is best explained by migration, adaptation to a new environment, and subsequent speciation. Populations from an ancestral group living in a particular place (e.g., the mainland) often spread or radiate to other habitats (e.g., islands). These new habitats include new environmental conditions that promote the evolution of new adaptations. Migration to new environments followed by adaptation and speciation is called **adaptive radiation** (fig. 22.12).

The finches that Darwin found on the Galápagos Islands are an example of adaptive radiation. Botanists have also found many examples of adaptive radiation in plants, most notably on the islands of Hawaii. A well-studied example involves the sticktights (*Bidens* species), so named because their fruits stick to socks or other clothing when you walk among the plants. This explains how they may have attached and been transported by seabirds from the mainland to the Hawaiian Islands. The sticktights in Hawaii apparently radiated from a single colonizer into the more than forty species that now occur only on the islands. This radiation was accompanied by a loss of colonizing ability: the first weedy sticktights produced progressively larger and less weedy descendants. The most recently appearing species may be *Bidens ctenophylla*, which is a long-lived, woody tree whose fruits are larger and fewer than those of its weedy ancestor and do not stick to anything (fig. 22.13). It no longer has the weedy lifestyle that characterized its earliest island ancestor.

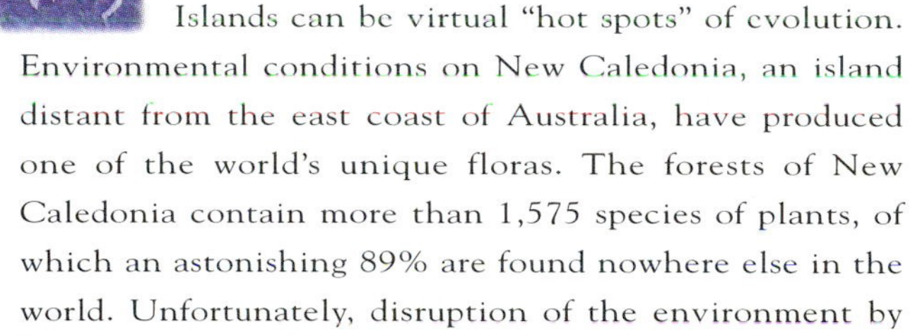

The Lore of Plants

Islands can be virtual "hot spots" of evolution. Environmental conditions on New Caledonia, an island distant from the east coast of Australia, have produced one of the world's unique floras. The forests of New Caledonia contain more than 1,575 species of plants, of which an astonishing 89% are found nowhere else in the world. Unfortunately, disruption of the environment by logging, mining, brush fires, and so on has reduced the amount of undisturbed forest to less than 1,500 square kilometers that cover less than 9% of the island.

Molecular Evidence

Since the late 1960s, evolutionary biologists have been comparing proteins and genes of various species, much as they have done with morphology and anatomy for more than a century. The most widely used comparisons have been based on cytochrome c oxidase, an enzyme of oxidative electron transport (see Chapter 6). The universal occurrence of this enzyme indicates that it evolved early in the origin of life and was inherited by all of the descendants from a single ancestor; thus, it is evidence of a single origin of life. In addition, its amino acid sequence has changed during evolution and now contains information about patterns of descent among organisms. Fewer differences in amino acids between organisms indicate a more recent common ancestor and a closer evolutionary relationship. The oldest ancestor gave rise to fungi, plants, and animals, and

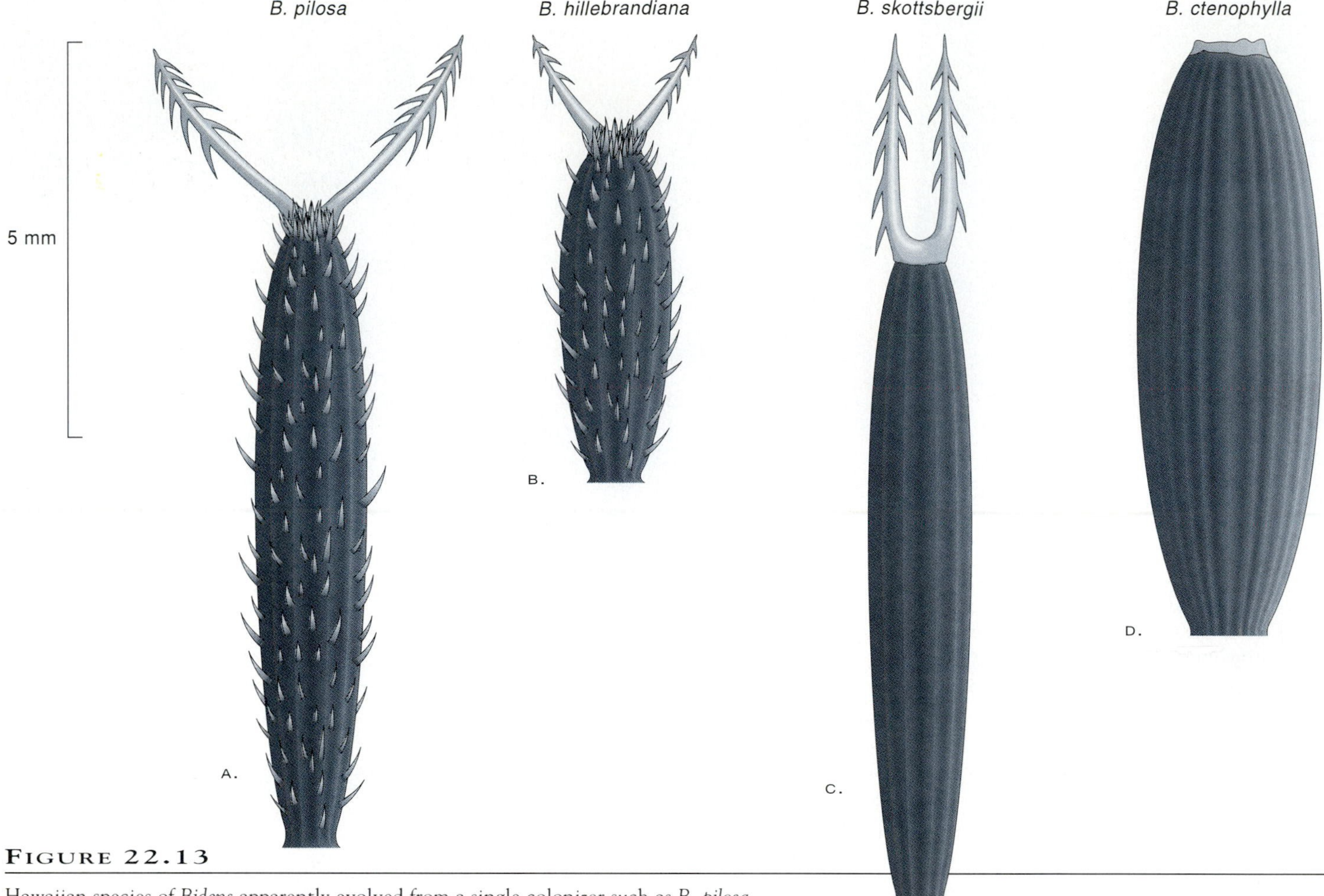

Figure 22.13

Hawaiian species of *Bidens* apparently evolved from a single colonizer such as *B. pilosa*. (a) *B. pilosa* may have been easily dispersed from the American mainland because its fruits have upward-pointing bristles and spreading, barbed awns that enabled the fruits to stick to migratory birds; (b) easily dispersed fruits of weedy Hawaiian species, such as *B. hillerbrandiana*, represent coastal colonizing species on the islands; (c) the loss of upward-pointing bristles and the partial closing of awns have reduced the dispersability of fruits in *B. skottsbergii* and other low elevation, inland species; (d) the fruits of *B. ctenophylla* show that the invasion of mountain habitats was accompanied by a complete loss of features for animal dispersal.

the most recent ancestors are the one shared by donkeys and horses and another shared by humans and monkeys. The only plant in this comparison, wheat, seems to be more closely related to animals than to fungi.

More recently, evolutionary biologists have also begun to compare sequences of DNA and RNA, and many other kinds of molecular information, to deduce patterns of evolutionary descent. In general, the results of these comparisons are consistent with the patterns of other homologous features. Specific examples of molecular evolutionary comparisons are discussed for each group of organisms presented in Chapters 25–30.

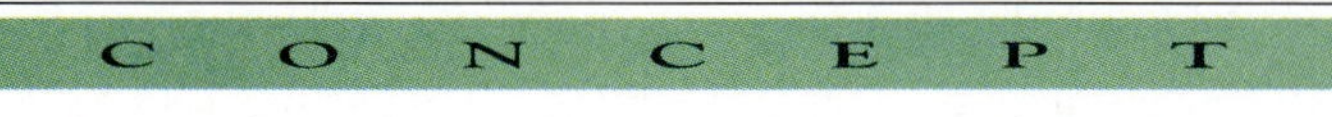

Modern evidence for evolutionary descent includes the age of the earth, the fossil record, homology, the convergence of adaptations of different species in similar environments, the similarity of species in different parts of the world, and similarities of nucleotide and amino acid sequences among species descended from a common ancestor.

Evidence for Natural Selection

Although the theory of descent with modification was revolutionary, even during Darwin's time it was supported by convincing evidence. Darwin's greatest contribution to evolutionary biology, however, was not the concept of change, but his explanation of a *mechanism* for change. To Darwin, observable evidence verified the results of evolution, but the specific *process* was apparently so subtle that it had eluded previous naturalists. That process was natural selection. Other processes change populations, but natural selection is the most important.

Even if the advantages of some characteristics are only slightly greater than those of others, the favorable characteristics will accumulate *after many generations* of being "selected," or disproportionately reproduced. This accumulation of characteristics is synonymous with long-term genetic change, which is the heart of evolution. Of course, the same mechanism decreases the occurrence of unfavorable characteristics. Remember that natural selection and evolution are not identical: Natural selection is a mechanism that results in evolution. The important result of natural selection is that it adapts a population to its environment.

C O N C E P T

Natural selection entails the differential reproduction of different genotypes. Genetic variation provides traits that determine an organism's success in the environment. Natural selection is the mechanism that fosters the reproduction of advantageous traits in the next generation.

Comparison with Artificial Selection

Darwin observed that farmers often selected certain animals or plants for reproduction based on their desirable traits. Humans have practiced the selection and controlled mating of plants and animals for centuries, a practice we now call *artificial selection*. Animal breeders know that to produce fatter hogs, they must allow only the fattest hogs to mate. Plant breeders know that to grow the largest tomatoes, they must allow cross-pollination only between plants that produce large tomatoes. These ideas are not new: most of our fruits and vegetables have been artificially selected for size, yield, pest-resistance, and so on since agriculture began more than 10,000 years ago (see box 22.1, "Selection for Monsters").

The practice of artificial selection led Darwin to assume that a similar process occurs in nature, which he called *natural selection*. The main difference is that humans are the agents of selection in the former, and the forces of nature are the agents of selection in the latter. Many of the postulates of natural selection, therefore, can be deduced from the process of artificial selection. Furthermore, unlike the evidence for descent with modification, the evidence for several of the postulates of natural selection is direct or experimental. Some of the postulates of natural selection can be compared with their counterparts in artificial selection as follows (postulate numbers come from table 22.1).

POSTULATE #1

A population of organisms has the tendency and the potential to increase at a geometric rate.
How many seeds does a tomato have? How many does a tomato plant or a whole field of tomatoes produce? If the field of tomatoes is analogous to a population in nature, then many more seeds are produced than ever grow into new plants. If all seeds produced new plants, the earth would be knee-deep in tomato plants by the end of one growing season. The overabundance of tomato seeds is evidence that the population has the potential to increase. Imagine the potential for certain orchid populations to grow when each fruit produces more than 20,000 seeds!

FIGURE 22.14

Fruits of the saguaro produce dozens of small black seeds. Most of these seeds are viable and easily cultivated, but few survive as seedlings in their natural environment.

POSTULATE #5

Only a fraction of the offspring in a population will live to produce offspring.
In agriculture, most plants and animals are harvested before they reproduce. Humans choose which individuals to harvest; that is, we are the agents of selection that limit reproduction. In nature, most plants die as seeds or seedlings. Conditions for germination or growth may not be good enough, pests may destroy the seeds or seedlings, or a freeze or drought may wipe them out. Saguaro cacti have been extensively studied in this regard. Fruits of the saguaro produce dozens of tiny black seeds (fig. 22.14), most of which will germinate and grow under cultivation. A population of saguaros probably produces millions of seeds, yet in most populations very few seedlings appear during any one year. Moreover, only a few new seedlings live more than a few years, and a saguaro must be 10–20 years old before it can form flowers. Some populations have no seedlings for several years in a row.

POSTULATE #6

Individuals in a population are not all the same; some have heritable variations.
Even in a field of corn, which may seem uniform, certain features vary from one plant to another. This was shown in a long-term experiment in artificial selection that was begun in 1896 at the Illinois Agricultural Experiment Station. At the beginning of

Boxed Reading 22.1

Selection for Monsters

Monstrous plants are plants that differ markedly from the norm. We have bred and selected for giant seeds, fruits, flowers, and vegetative organs. For example, we have selected and perpetuated cultivars of banana, grape, sugarcane, and other plants for increased bulk, sugar content, and visual appeal. Most of these varieties can no longer produce viable seeds, but our production of tasty tissue has been remarkable.

Nowhere has our rampage through the genome of a plant been more apparent than in the genus *Brassica* of the mustard (Brassicaceae; also called Cruciferae) family. This genus contains perennial, biennial, and annual herbs, with 100 wild species found in north temperate parts of the world. You can easily recognize the family by its characteristic flower having four petals which form a cross (hence the family name), six stamens of which two are short, and its special fruit, the silique. Although the cultivated brassicas are probably derived from several wild species, a reasonable ancestor is colewort (*B. oleracea*), a scrubby perennial native to Europe and Asia. From colewort, we have selected and developed a wide array of common vegetables. One line produced kale, cauliflower, and broccoli. A second line gave us kohlrabi, with a short stem enlarged into an above-ground tuberous vegetable, and rutabaga, whose tuberous storage stem develops below ground. The turnip's (*B. rapa*) storage organ is also a tuber, and its true root extends down from the swollen stem. Although all the leafy vegetable members of the genus may not be derived from colewort, they are all closely related. Several species of mustard grow wild, and we exploit their edible leaves (*B. juncea*) and particularly their seeds (*B. nigra* and *B. alba*), which are used as mustard. Pakchoi (*B. chinensis*) is an important leafy vegetable in the Orient, as are the closely related petsai (*B. pekinensis*) and false pakchoi (*B. parachinensis*). Rape or colza (*B. napus*) is grown primarily for its seeds, which yield an edible oil.

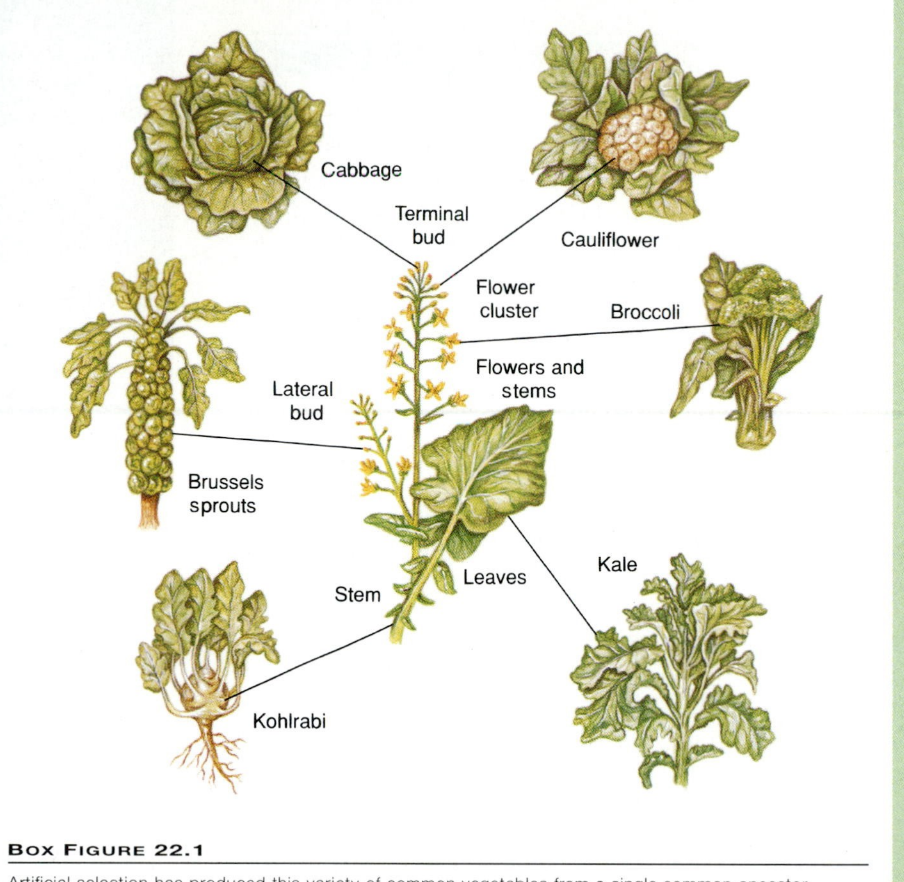

Box Figure 22.1

Artificial selection has produced this variety of common vegetables from a single common ancestor. All of these vegetables are of the same genus, *Brassica* spp. Various morphological features have been enhanced.

Adapted from Richard M. Klein, *The Green World: An Introduction to Plants and People*, 2d ed. Copyright © 1986 Harper & Row, Publishers, Inc. New York, NY.

the experiment, kernels of two types were selected for reproduction, one of slightly higher protein content than the other. Every year, the plants producing the highest amount of protein were hybridized with one another, and plants producing the lowest amount of protein were hybridized with one another. At the beginning of the experiment, the average protein content per kernel was 10.9%. After fifty generations, the low-protein line was down to 4.9%, and the high-protein line was up to 19.4%. In this case, heritable variation occurred in protein content, and humans selected only certain individuals for breeding.

The evidence for heritable variation in corn is that the protein content changed over time because of strong selection pressure by humans. What evidence do we have that heritable variation occurs in nature? In other words, is natural variation due to genetic differences (heritable), or is it due to different responses of individuals to their environment (nonheritable)? We know that the sizes of leaves and fruits, the number of seeds, and other features of cultivated plants can be influenced by how much fertilizer and water we give them. If water or nutrient availability is uneven in a certain habitat, then certain plants could be bigger merely because they get more water or more nitrogen from the soil.

Transplant experiments reveal evidence for heritable variation. Genetically controlled features do not change after individuals are transplanted from one environment to another. A classic example of this kind of experiment was done on yarrow (*Achillea lanulosa*), a highly variable species that grows along the Pacific Coast and in nearby mountains. When transplanted to a garden at Stanford University near the coast and to a garden at timberline in the Sierra Nevada, plants from different habitats maintained their basic features, sometimes even to their demise. Plants from the coast, for example, normally bloom in the fall. In the timberline garden they still flowered in the fall, but fall frost caught them before they could set seed. This experiment shows that the adaptive features of yarrow in different habitats are genetically controlled.

Not all features are heritable, however. In another transplant experiment, clones of the same individual of cinquefoil (*Potentilla glandulosa*) were grown in different environments. Since all plants were identical genetically, their subsequent variability was attributed to individual responses to different environments, not to heritable variation.

POSTULATE #8

Individuals having favorable traits will, on the average, produce more offspring, and those with unfavorable traits will produce fewer offspring. High-protein corn, fat hogs, and large tomatoes are favorable traits for artificial selection. Organisms with these traits produce more offspring because people select them for breeding. In nature, if a plant is rugged, disease-resistant, a good competitor for resources, and an efficient photosynthesizer, then we also expect it to reproduce more successfully than plants that are not so rugged or disease-resistant. For example, a plant having a larger leaf-surface area is adapted to a low-light environment and will have greater reproductive success than small-leaved plants in the same environment. Random events such as storms and floods may cause the unexpected, but under normal circumstances the organisms that are most attuned to their environment will be most successful.

A well-documented example of natural selection involves the work of Janis Antonovics with plants growing in soil containing high concentrations of copper, zinc, or lead (i.e., the waste products from mines). Most plants can't grow in soils contaminated with these metals; however, some species of grasses and weeds do grow around these mines. Surprisingly, they are the same species that grow in nearby pastures. Apparently, some members of the species can tolerate toxic metals. In one experiment, Antonovics planted seeds from the pasture population of grasses in soil that was rich in copper. Only 1 in 7,000 survived. Thus, copper was intensely selecting against the intolerant phenotype.

In a related experiment, Antonovics gathered seeds from metal-tolerant adults on contaminated soil and planted them in the same contaminated soil. Not all the offspring survived. Moreover, the offspring varied more in their tolerance to metals than did their parents. This limited generation of seedlings revealed that the parents still housed genetic variation for metal tolerance. Alleles for intolerance were still present, even though metal in the soil continually selected against intolerant individuals. Natural selection was operating in this population, and genetic variation was still present.

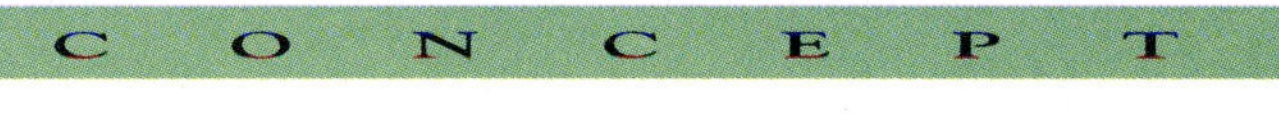

Direct evidence exists for several of the postulates of natural selection: the overproduction of seeds indicates that populations can increase at a geometric rate; a small fraction of seeds grow into reproductively mature plants; different individuals in a population have variable, genetically controlled features; plants that are better-adapted to a habitat have more reproductive success in that habitat than plants that are not so well adapted to it.

Subtle Features of Natural Selection

Populations

Evolution is a genetic change in populations over many generations. Thus, the concept of a **population** is critical to understanding natural selection and evolution because a population is the smallest unit that can evolve.

A population is a group of interbreeding individuals of the same species sharing the same territory at the same time. Members of the same species may be widely distributed and divided into many populations, but a population is specific to a location, and members of a population must be close enough to interbreed. Environmental conditions are also location-specific, and natural selection of traits adapted to an environment occurs in local populations.

The concept of a population was alien to many of the original readers of Darwin's work. Indeed, the idea of a population that included unique and varied individuals was in stark contrast to the constancy associated with divinely created species. The population concept implied that variation occurred within a species and that each individual was unique; there was no average or ideal form of a species. In natural populations an average individual or the average for a characteristic (e.g., the average American family has 2.2 children) is merely a statistical abstraction.

Survival of the Fittest or Reproduction of the Fittest?

The phrase *survival of the fittest* is not entirely appropriate to natural selection. It implies that the singular direction of natural selection is to increase survival, but this is inaccurate. More precisely, natural selection promotes traits that increase the passage of genes to the next generation. That is, it promotes successful *reproduction*. Of course, survival is a prerequisite to reproduction, but reproductive success can be enhanced by traits not associated with long survival times. For example, annual flowering plants are diverse and abundant, even though individuals do not live as long as perennial plants. Their survivorships are different, but they both have successful reproductive strategies. The term *survival* may be appropriate if one thinks in terms of the survival, over generations, of species or genetic traits rather than of individuals.

Successful reproductive strategies are highly varied and strongly selected. Indeed, many strategies have evolved that replenish the population and pass advantageous traits according to natural selection. For example, mushrooms and cottonwood trees produce thousands or millions of spores and seeds, respectively, whereas an avocado tree produces only a few. The seeds of an orchid may be the size of a pinhead, while those of a coconut tree may be as large as a grapefruit and loaded with reserves for the developing embryo. Early reproduction, abundant reserves for the embryo, easily dispersed seeds, noxious-tasting seeds, large numbers of seeds, and conspicuous flowers all promote reproduction in appropriate environments. Such strategies are selected if they ensure production of the next generation. Reproductive traits are more critical than survival traits to natural selection.

Natural Selection in the Context of the Environment

Successful adaptations and natural selection are inseparable from the environment. For example, an adaptation of succulent leaves for water storage is neither positive nor negative unless put in the context of an environment. Succulent leaves may promote reproductive success in a desert, but not in a rain forest. That is, the same trait can be positively selected in one environment and negatively selected in another environment. Thus, environmental conditions are the forces that determine the nature and outcome of natural selection. These environmental conditions are called **selection pressures.** For example, fungal disease is a selective pressure. A dispersed seed may have a tough, protective seed coat that prevents fungal invasion. However, the seed coat may also prevent the penetration of water needed for germination. Thus, thicker seed coats may be advantageous in moist, fungus-rich environments, while thinner seed coats are selected in drier, less fungus-rich environments. If the embryo is disease-resistant, however, then a thin seed coat may be best in moist as well as dry habitats.

Organisms must deal with various and often contradictory environmental factors. For example, warmth may increase metabolism for rapid growth, but it may also dry the soil rapidly. Different species solve problems with a variety of strategies promoted by natural selection in the context of the environment.

Selection as a Noncreative Force

Natural selection does not create new alleles, traits, or adaptations. Instead, it channels genetic variation that is already present in a population by differentially reproducing traits with negative or positive adaptive value. If the alleles and genetic combinations for traits such as color, competitiveness, or starch storage are absent in the population, then these adaptations cannot develop, regardless of the intensity of natural selection.

A common misconception is that natural selection brings organisms toward an ideal, maximum fitness. Natural selection does not lead to perfection; rather, it promotes genetic combinations that "work" in the local environment. If environments are different, then naturally selected organisms will be different. Variation is an expected consequence of natural selection among populations.

Natural selection operates on populations and their interactions with the local environment. Rather than create new adaptations, natural selection promotes the most adaptive, available variants through increased reproduction and retards nonadaptive variants through decreased reproduction.

Kinds of Natural Selection

Selection can alter the occurrence of a trait in a population in various ways. In a simple case, positive selection for tolerance to toxic metals can increase the number of tolerant versus intolerant plants in a population over many generations. But genetic traits are usually not in clear categories such as tolerant and intolerant. Instead, populations contain a continuum of values for a trait. That is, some individuals may be slightly more or slightly less tolerant than others. Thus, natural selection can move the distribution of values for a phenotype along a continuous scale (fig. 22.15). We can graphically represent this scale as the x-axis—a continuum of values for the alleles and the trait they affect. The y-axis is the frequency (i.e., number) of organisms expressing each value. This simple representation is more ideal than realistic, but it illustrates different ways that natural selection alters the genetic composition of a population. The peak of the curve represents the most common

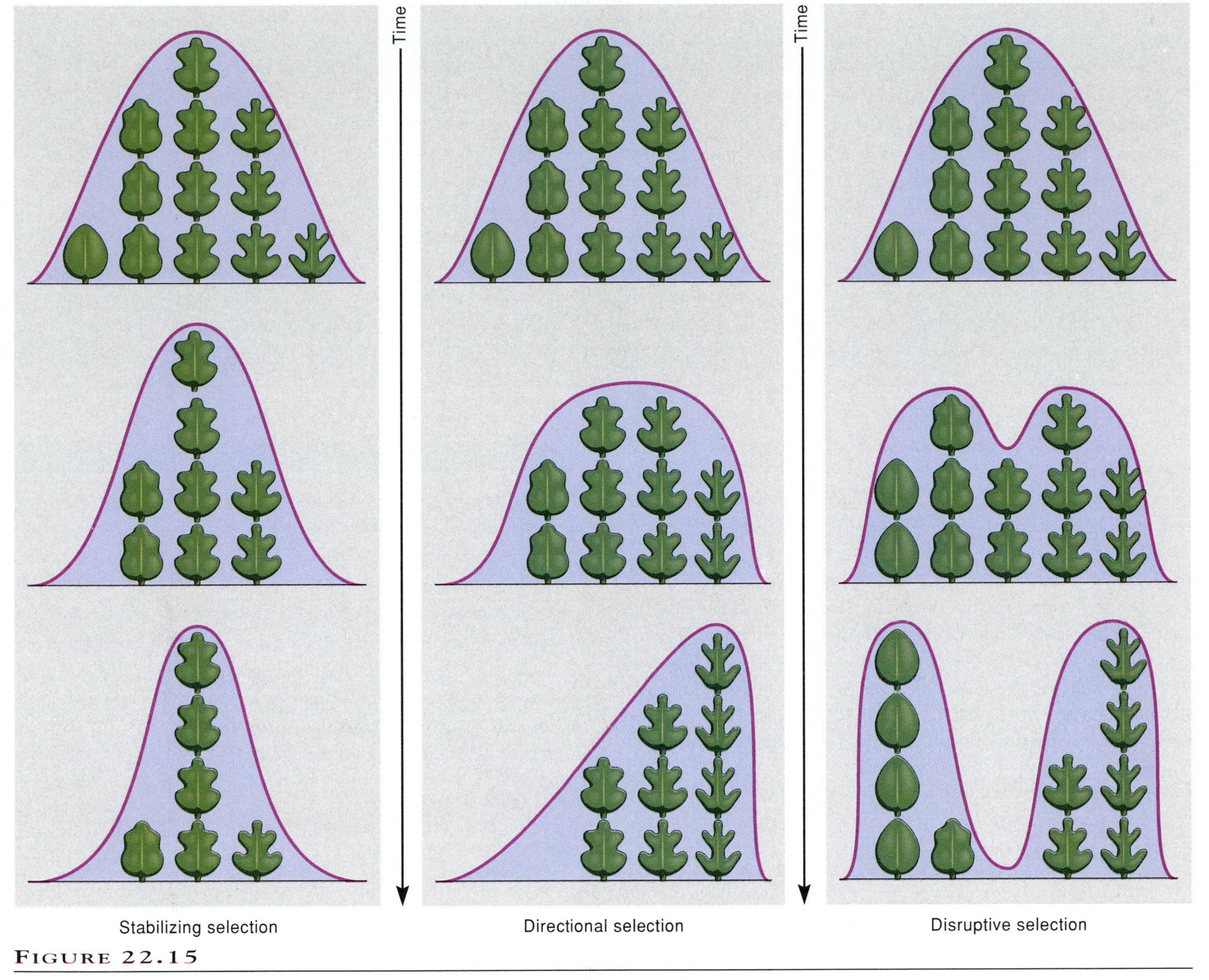

Figure 22.15

Stabilizing, directional, and disruptive selection for leaf margination. The height along the curve represents the number of individuals with that phenotype. The shaded area represents negatively selected phenotypes. Over time, stabilizing selection reduces variation and preserves a narrow phenotype best adapted to the environment. Directional selection promotes phenotypes at one end of the variation, and the distribution of phenotypes moves toward that end. Disruptive selection promotes extremes of a characteristic. Strong disruptive selection may produce two distinct populations.

value for the trait, and the breadth of the curve represents the degree of variation in the population. Different kinds of selection alter the shape of the curve.

Stabilizing Selection

Stabilizing selection promotes the norm and reduces both extremes. This kind of selection operates in most populations, and individuals with traits near the peak are positively selected. Organisms near the extremes of variation are selected against and rarely pass their extreme traits to the next generation. As a result, the frequency of alleles producing the optimal phenotype continues to increase, while alleles interacting with the environment to produce extreme or variable phenotypes decrease. Stabilizing selection in most populations eliminates extreme individuals, including mutant forms. The resulting stable or predictable distribution of traits is most common in stable, unchanging environments.

Directional Selection

In changing environments, **directional selection** is a primary mechanism of change. In directional selection, individuals at one extreme of population variation are negatively selected, whereas those at the other extreme are positively selected. Over many generations, the average condition moves toward the positively selected extreme.

Boxed Reading 22.2

Darwin's Big Mistake

In seeking a genetic mechanism for traits passing from one generation to another, Darwin addressed the previously unquestioned assumption that the parent's body somehow influenced the form of the offspring. In this regard, Darwin subscribed to Lamarck's theory of the inheritance of acquired characteristics. But Darwin took the idea of somatic influence on heritable traits a step further. He knew that there must be some physical connection between parent and offspring. He explained this connection in his theory of pangenesis, which was the greatest mistake of his scientific career. According to this theory, every part of a mature individual produces tiny packets of heritable information, which he called **gemmules.** To be transmitted to offspring, gemmules from all over the body were somehow transported to the reproductive organs and packed into the gametes before fertilization. The gemmules of the two parents would blend together in the offspring, where they would sprout wherever appropriate to determine the features of each part of the new organism. Darwin's ideas about pangenesis probably arose from his belief that the use and disuse of organs influenced heredity, which was a widely held view at the time. In that context, pangenesis was a perfectly logical theory, although it was inaccurate.

Directional selection can cause change over a relatively short period of intense selection. For example, plants studied in a grazed pasture in Maryland were shorter than the same species in ungrazed pastures. Were they short due to grazing or due to genetic differences? To separate the effects of grazing from genetics, some of the short plants were planted away from the selection pressure of grazing herbivores. Surprisingly, some of the transplants remained short even without grazing. Apparently, directional selection pressure by grazers had genetically altered the grazed population in the direction of short plants less susceptible to herbivores. Directional selection is the main type of selection used in plant breeding.

Diversifying Selection

Diversifying selection (also called *disruptive* or *catastrophic* selection) is common in environments that are suddenly or drastically changing. These environments become unfavorable to the previously successful and adaptive peak of a trait. As a result, peak values for a phenotype are selected against because extreme values in an altered environment are just as likely to enhance successful reproduction. Sometimes organisms with extreme values escape catastrophic negative selection because they live at the periphery of the population. They survive to reproduce, while normal, centrally located organisms are negatively selected. As you might suspect, drastic environmental changes are usually short-lived, but they significantly increase genetic diversity, and temporarily decrease population size.

Genetic diversity of sugar maples in Ohio is maintained by opposing or diversifying selective pressures. The southern strain of sugar maples can survive occasional droughts and cool winters, while the northern strain is better adapted to the coldest winters. In Ohio a cool, dry winter may promote the southern strain and its alleles, but the following year a severe winter may select for the northern strain. Diversity is maintained as both types intermingle in the Ohio population. Sugar maples with intermediate tolerances are not favored in either cold or warm climates.

C O N C E P T

Stabilizing selection in constant environments reduces variation and promotes the normal, most prevalent phenotype. Directional selection is common in changing environments, and fosters an extreme phenotype while reducing the occurrence of the opposite phenotypic condition for a trait. This shifts the population's distribution of phenotypes toward an extreme. Diversifying selection occurs in drastically changing environments and retards the normal, intermediate phenotype while promoting the extremes, thereby promoting overall variation.

Doing Botany Yourself

Artificial selection can improve strains of agricultural species. How would you design an experiment to determine which kind of selection (stabilizing, directional, diversifying) would be the most valuable?

The Genetic Basis of Evolution

Among the most important developments in evolutionary biology since Darwin is the application of genetics to the theory of natural selection. A major problem with the initial understanding and acceptance of evolution by natural selection was a lack of knowledge of the genetic causes of variation (see box 22.2, "Darwin's Big Mistake"). Mendel's work was unknown. Most biologists believed that parental traits "blended" to produce intermediate traits in offspring.

Evolutionary theory came of age when biologists stopped thinking in terms of individual organisms and began thinking in terms of populations and frequencies of genes and alleles. The integration of Mendelian genetics and evolution by natural selection is called **population genetics.** For example, dandelions

FIGURE 22.16

Large populations such as these white prickly poppies (*Argemone albiflora*) share a common gene pool. Population geneticists study this gene pool rather than the genetics of an individual.

growing in a valley or water hyacinths growing in a lake constitute populations. The unifying property of a population is its *gene pool.* The gene pool is the total genes in a population at any one time; it consists of all genes at all loci in all individuals of the population. Members of the next generation get their genes from this pool, and individual organisms temporarily contain a small portion of the gene pool. To population geneticists and to evolutionary theory, the gene pool, its variations, and the mechanisms causing change are more important than the individual. Indeed, because populations often include thousands or millions of organisms, the fate of one individual usually has little influence on the fate of the whole population (fig. 22.16).

Evolutionary theory presumes that some alleles of natural populations increase in frequency from generation to generation, while others decrease. Favorable combinations of alleles in a genotype are more likely to occur in greater proportion in the next generation. In contrast, unfavorable alleles are less likely to be passed to the next generation, and their frequency will be reduced. The change in the frequency of alleles within a gene pool over many generations is the geneticist's definition of evolution.

Fitness is a measure of an organism's evolutionary success, but does not strictly refer to its health, survival, or adaptation to the environment. Rather, an individual's fitness is measured by its number of surviving offspring relative to the number surviving from other individuals in the population. Fitness represents the extent to which an individual's alleles will occur in succeeding generations. Highly fit individuals produce many surviving offspring.

The Hardy-Weinberg Equilibrium

Population genetics is the mathematical study of the events occurring in gene pools that modify gene frequencies. Among the most significant models of gene frequencies is the **Hardy-Weinberg** equilibrium, named for G. H. Hardy and W. Weinberg, who independently developed and proposed this concept in

1908.[5] This model describes a gene pool whose various allelic frequencies are at equilibrium under certain conditions. Hardy-Weinberg equilibrium is not concerned with the genotypes or phenotypes of individuals, but with the composition of the entire gene pool. It predicts genetic equilibrium (nonevolution). Such equilibrium is rarely attained in natural populations, but it is a good starting point for explaining how gene pools change.

To illustrate Hardy-Weinberg equilibrium, let's consider a simple, hypothetical population of diploid wildflowers carrying only two alleles, A and *a*, for flower color. The allele A is a dominant allele for blue flowers, and its frequency in the gene pool is designated as p. The allele *a* for white flowers is recessive, and its frequency is q. Since A and *a* are the only alleles occurring at this gene locus, the sum of their frequencies (i.e., their proportion of the total) must equal 1.00 (i.e., 100%). For example, if 60% of the alleles are A, then 40% of the alleles must be *a*. That is,

$$p + q = 1.00$$
$$q = 1.00 - p = 1.00 - 0.60 = 0.40$$

In this example, 60% of all gametes (sperm cells and egg cells) produced by the population carry A, while 40% of all gametes carry *a*. For a randomly mating population, we can portray the genotypes of the offspring and their relative frequencies from the fusion of eggs and sperm as follows:

Genotypes of the next generation

	A (Eggs)	*a* (Eggs)
A (Sperm)	AA	A*a*
a (Sperm)	A*a*	*aa*

Phenotypic frequencies of the next generation

	A = 0.60 (Eggs)	*a* = 0.40 (Eggs)
A = 0.60 (Sperm)	0.36	0.24
a = 0.40 (Sperm)	0.24	0.16

This table is a Punnett square (see Chapter 8), which helps biologists visualize all mating possibilities. The genotypes resulting from random mating within this population are 36% AA, 48% A*a* (i.e., 24% + 24%), and 16% *aa*. These same results can be calculated using the following Hardy-Weinberg notation:

$$p + q = 1.00$$
$$\text{Therefore} \quad (p + q)^2 = (1.00)^2$$
$$\text{Therefore} \quad p^2 + 2pq + q^2 = 1.00$$

This equation is an expansion of the binomial $(p + q)^2$, which is a model of the frequencies of genotypes. Rather than using the Punnett square, we can directly calculate genotypic frequencies for the next generation from the allelic frequencies p and q:

$$\text{Frequency of AA} = p^2 = (0.6)(0.6) = 0.36$$
(blue flowers)

$$\text{Frequency of A}a = 2pq = (2)(0.6)(0.4) = 0.48$$
(blue flowers)

$$\text{Frequency of } aa = q^2 = (0.4)(0.4) = 0.16$$
(white flowers)

The frequencies of the two possible phenotypes (A __ and *aa*) can also be calculated for the next generation.

$$\text{Blue flowers} = \text{AA} + \text{A}a = p^2 + 2pq$$
$$= 0.36 + 0.48 = 0.84$$
$$\text{White flowers} = aa = q^2 = 0.16$$

This raises an important question: How have the frequencies of the two alleles in this gene pool changed after production of the next generation? To answer this, we must solve for p and q from the new genotypic frequencies and compare our results with the original values of p and q. If 36% of the offspring are AA (i.e., p^2), and half of the alleles of the A*a* offspring (i.e., $2pq$) (48%) are A, then the total frequency of A (i.e., p) is $0.36 + [(1/2)(0.48)] = 0.60$. In a similar calculation the new frequency of *a* ($=q$) $= 0.40$. Thus, *the allelic frequencies have not changed*. In fact, even after a second, third, or fourth generation of random mating, the allelic frequencies do not change. Thus, *the Hardy-Weinberg equilibrium predicts that the proportional gene frequencies of dominant and recessive alleles are retained from generation to generation in a randomly mating population*. The genotypic frequencies will also be in equilibrium after at least one generation. There will be just as many white flowers after multiple generations as in the previous generation, even though white flowers are a recessive trait. However, there are constraints and assumptions associated with maintaining this equilibrium. Specifically, Hardy-Weinberg equilibrium makes the following assumptions:

1. The population is large. Changes in the gene frequencies of large populations are less likely to occur by chance. In a small population, the random loss of one or more individual genotypes (such as by failure to breed) can eliminate one or more alleles from the population.

5. Others had proposed the model earlier, including W. E. Castle and S. S. Tschetverikov, but the tradition of naming it for Hardy and Weinberg is too deeply ingrained to overcome. It is often inappropriately referred to as the Hardy-Weinberg law.

2. Individuals do not migrate in or out of the population. The net movement of individuals between populations (which occurs via seed dispersal in plants) must not be extensive enough to change gene frequencies. In natural populations the degree of migration varies considerably.
3. Mutations do not occur. In natural populations the mutation rate varies considerably. Mutations are often spontaneous.
4. Reproduction is random. That is, every member of a population must have an equal opportunity to reproduce with any other member. In the terms of Hardy-Weinberg population genetics, genotypes mate in proportion to their frequency.
5. There is no natural selection; that is, all alleles and combinations of alleles have equal fitness. The number of offspring must be independent of genotype.

C O N C E P T

Population genetics deals with entire gene pools rather than individuals. Hardy-Weinberg equilibrium predicts that allelic frequencies will not change over multiple generations in a large, isolated, randomly mating population that is not subject to selection or mutation.

The Significance of Hardy-Weinberg Equilibrium

Hardy-Weinberg equations are valuable models, even though no natural populations are in Hardy-Weinberg equilibrium for all of their gene loci. (Populations may maintain Hardy-Weinberg equilibrium for specific genes that are not adaptive.) Nevertheless, by understanding equilibrium, we can study how natural populations deviate from this condition. Although the Hardy-Weinberg model is of an unchanging population, it allows us to measure change and examine the forces causing that change.

We can investigate basic questions about simple population genetics and Hardy-Weinberg equilibrium. For example, why don't dominant alleles eventually drive recessive alleles from the population? Is a lethal recessive allele quickly eliminated from a population? Are less frequently occurring alleles lost from the gene pool after many generations?

The answer to all these questions is no. Hardy-Weinberg equations predict that the relative frequencies of alleles do not change; even rare alleles will not disappear just because they are rare. Natural selection alters their frequencies regardless of whether they confer an advantage or a disadvantage. Nevertheless, selection can hardly ever change the frequency of a recessive allele because such alleles are so infrequently expressed in the phenotype. For example, if the frequency of a recessive allele is 0.01 (1%), then the probability of homozygous expression is one in ten thousand. Well-known examples of such rare alleles include those that cause simple metabolic defects that are easily observable: e.g., deficiency of the amino acid cysteine in the mouse-ear cress (*Arabidopsis thaliana*), deficiency of the vitamin thiamine in tomato (*Lycopersicon esculentum*), reduced ability to use magnesium in celery (*Apium graveolens*), and deficiency of the hormone gibberellin in maize (*Zea mays*). Such recessive alleles do not occur in the homozygous state, and natural selection only operates on phenotypes. In summary, Hardy-Weinberg equilibrium shows that genetic diversity tends to be maintained, thereby making evolution possible.

Hardy-Weinberg equilibrium also demonstrates that the recombination of genes during sexual reproduction does not alter the total gene pool of a population. Rather, sexual reproduction provides progeny with different genetic combinations of the alleles *already present*. Segregation by meiosis and recombination by fertilization neither add alleles to nor remove alleles from the population. Changes in allelic frequencies occur through agents such as migration, mutation, and selection.

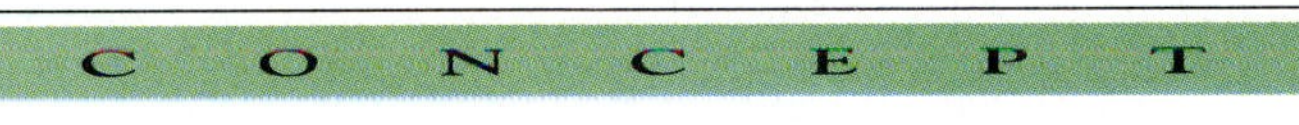

Hardy-Weinberg equilibrium offers population geneticists a model of stability for comparison with more natural, changing populations. Understanding deviations from equilibrium promotes investigation of the causes of change.

Sources of Genetic Variation

Selection is probably the most important cause of deviation from Hardy-Weinberg equilibrium because it promotes genetic change in the context of the environment and requires genetic variation. Natural selection can change the frequency of phenotypes, but it must have genetic variation as a raw material to effect change. Where does genetic variation originate? Where do new alleles come from? In this section we describe sources of variation upon which natural selection can operate. Understanding these processes is essential to evolutionary theory, because natural selection does not affect all genes in the same way. Changes that occur at the level of genes and chromosomes are mutations, which were discussed in Chapter 8. Genetic variation within populations comes from two main sources: gene flow and genetic drift.

Gene Flow

Gene flow is the transfer of genetic material from one population to another (fig. 22.17). This typically occurs through migration of individuals or movement of seeds or pollen to neighboring populations or, in some cases, to distant populations. For example, the grass *Bothriochloa intermedia* seems to have incorporated genes from many other grasses, including *B. ischaemum* in Pakistan, *B. insculpta* in eastern Africa, and *Capillipedium parviflorum* in northern Australia.

Gene flow minimizes geographic variation in gene pools; that is, it decreases genetic differences between populations. Genes flow frequently between neighboring populations, and significantly minimize the differences between these populations. Gene flow between distant or isolated populations is rare, which allows their gene pools to diverge over time. Reduction in gene flow partially explains why islands isolated by water are geographically more varied and more likely to produce new species than are vast expanses of grassland, and why lakes and streams contain more geographic variation among populations than oceans. Separated populations of a species are seldom genetically identical, and the differences coincide with the distance between populations. As you'll see in the next chapter, geographic barriers that prevent gene flow are important in the formation of new species.

Gene flow in plant populations is difficult to measure, but it can be experimentally estimated by planting recessive homozygotes at various distances from a strain marked with a dominant allele and then examining the distribution of heterozygous progeny. Using this technique, A. J. Bateman measured pollen dispersal in wind-pollinated (e.g., corn) and insect-pollinated (e.g., radish) crops. The proportion of corn plants receiving the dominant allele by gene flow decreased exponentially with distance and was reduced to 1% at only 13–16 meters from the pollen source. Similarly, most pollen of insect-pollinated plants is carried only a short distance; however, the small proportion that is carried farther may contribute importantly to gene flow.

Genetic Drift

Genetic drift refers to changes due to chance in the gene pool of a small population. In small populations, chance events such as mutation, mating, or pollination may significantly affect the gene pool and change gene frequencies independently of natural selection (fig. 22.18). If, for example, one individual in a small population carries the only copy of an allele, then the passage of that allele to the next generation may depend largely on the vagaries of insect pollination or random, lethal storms rather than natural selection. Favorable alleles in a small population can be eliminated by chance alone. Similarly, catastrophic damage to or death of well-adapted individuals may increase the frequency of the alleles of less fit but surviving individuals. Current research indicates that genetic drift may be a more significant force for changing gene frequencies than previously assumed. This would be especially true for the frequencies of genes that are not subjected to heavy selection pressure.

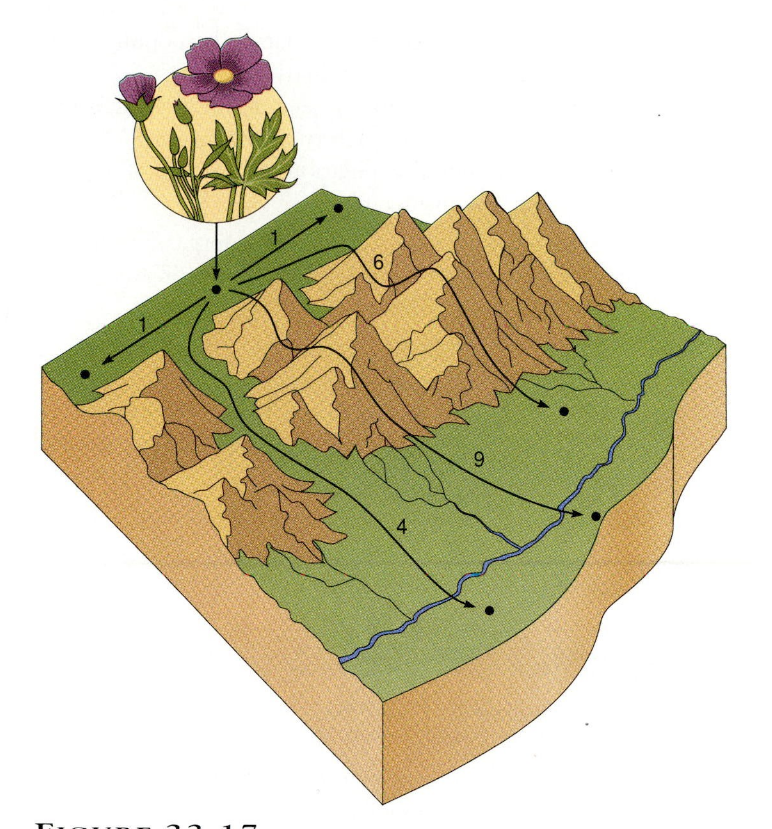

FIGURE 22.17

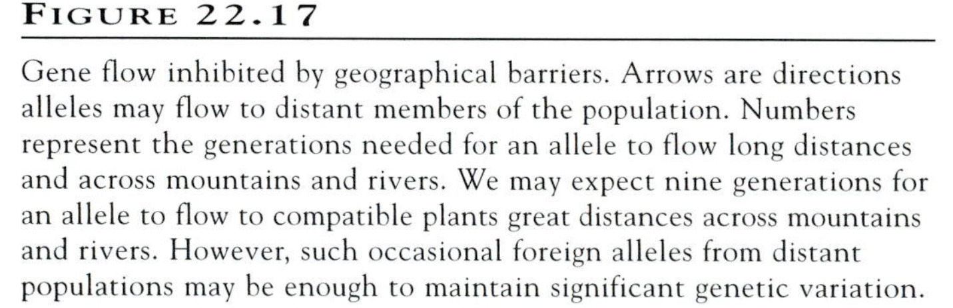

Gene flow inhibited by geographical barriers. Arrows are directions alleles may flow to distant members of the population. Numbers represent the generations needed for an allele to flow long distances and across mountains and rivers. We may expect nine generations for an allele to flow to compatible plants great distances across mountains and rivers. However, such occasional foreign alleles from distant populations may be enough to maintain significant genetic variation.

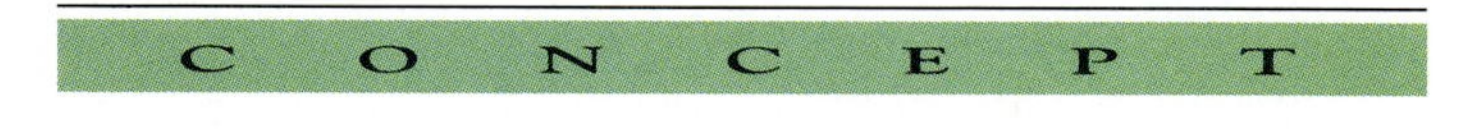

C O N C E P T

Genetic variation is introduced to populations mostly through gene flow and genetic drift.

Maintaining Genetic Variation

If natural selection continually promotes genes that confer high degrees of fitness and adaptation to the environment, we might expect variation to slowly decrease. During stabilizing and directional selection, a population's gene pool may become increasingly homogeneous or consistent. This doesn't usually happen, however, and the reasons are not well understood. Apparently, there are forces that promote mixing and maintain genetic variation.

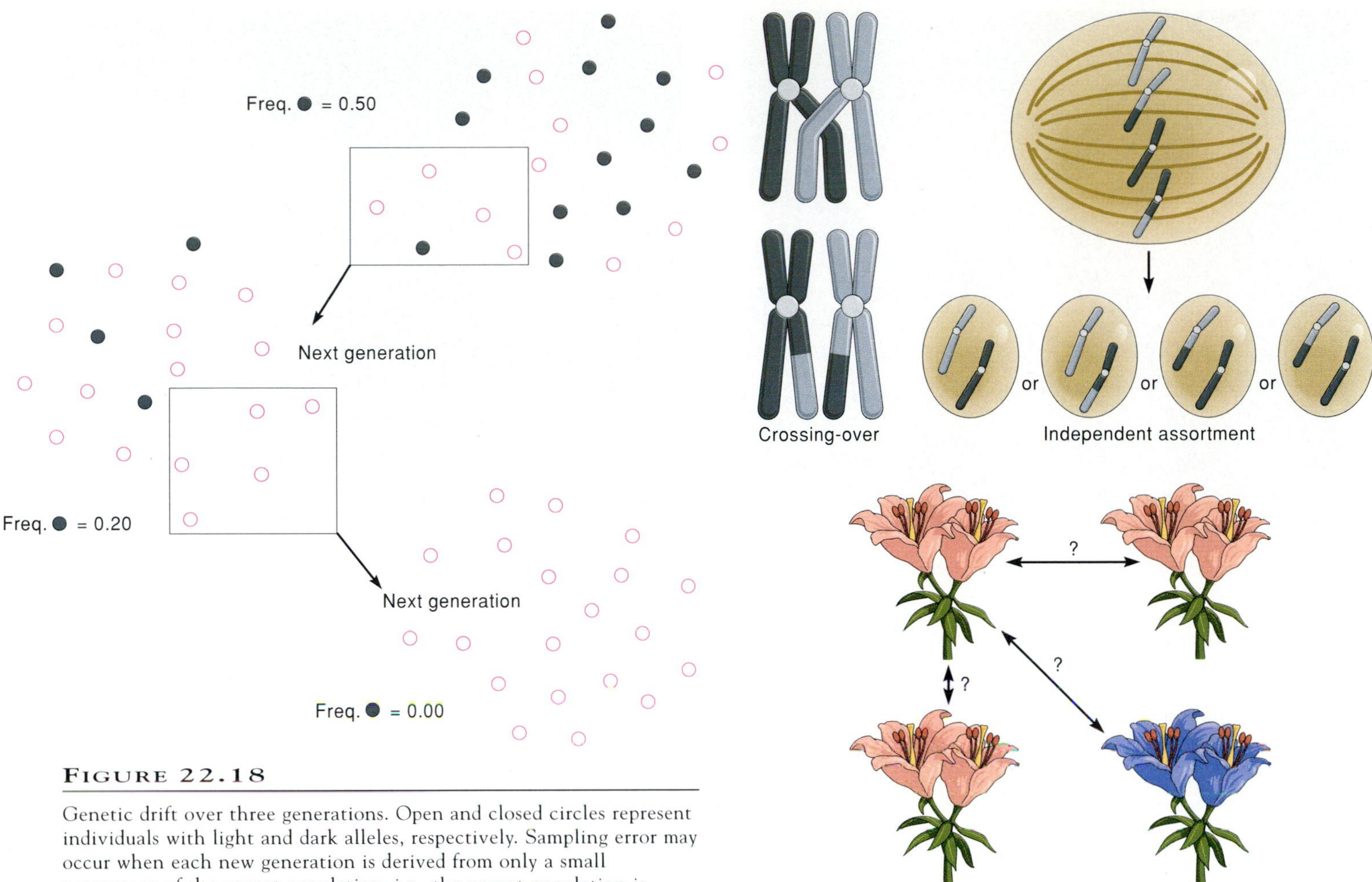

FIGURE 22.18

Genetic drift over three generations. Open and closed circles represent individuals with light and dark alleles, respectively. Sampling error may occur when each new generation is derived from only a small percentage of the parent population, i.e., the parent population is poorly "sampled." Sampling error between these generations has caused the frequency of dark alleles to drift from 0.50 to 0.00.

FIGURE 22.19

Three mechanisms of genetic recombination during sexual reproduction. Crossing over produces new combinations of alleles on a chromosome. Independent assortment of chromosomes during meiosis produces gametes with differing combinations of chromosomes. Matting pattern is subject to a variety of pressures including selection, chance and spatial distribution. Different patterns produce different genetic combinations.

Sexual reproduction is probably the most important factor that promotes genetic variation in plants and animals (fig. 22.19). Specifically, sexual reproduction (1) assorts chromosomes during meiosis, (2) includes crossing-over and genetic recombination after chromosome synapsis, and (3) combines genetic material from two different parents. All three of these mechanisms produce new combinations of genes. In contrast, asexual organisms acquire genetic variation only through mutation. Their offspring are clones of the parents.

Outbreeding, meaning reproduction with individuals other than the self, promotes new genetic combinations. Although many plants are self-compatible, several mechanisms have evolved to ensure sexual reproduction with other individuals (fig. 22.20). For outbreeding to occur, pollen must land on the stigma of a flower of another individual. Holly trees and date palms, for example, have male flowers on one tree and female flowers on another. In other plants, such as avocado, gamete production is separated by time: when a flower is producing pollen, the stigma of the same flower is not receptive. Therefore, pollen is only viable from other plants. Stigmas are commonly receptive to pollen only after the flower, or all flowers on the plant, have shed their pollen. Other plants have anthers and nectaries arranged in variable patterns on the flowers of different individuals. Flowers of the primrose (*Primula vulgaris*) and the bogbean (*Menyanthes trifoliata*) include two arrangements of styles and anthers (fig. 22.21). The "pin" flowers have a long style and low-set anthers, whereas the "thrum" flowers

FIGURE 22.20

The bizarre flower of the bird-of-paradise (*Strelitzia reginae*) is decorated with colors and shapes to attract bird pollinators. Successful attraction of flying pollinators promotes outbreeding.

FIGURE 22.21

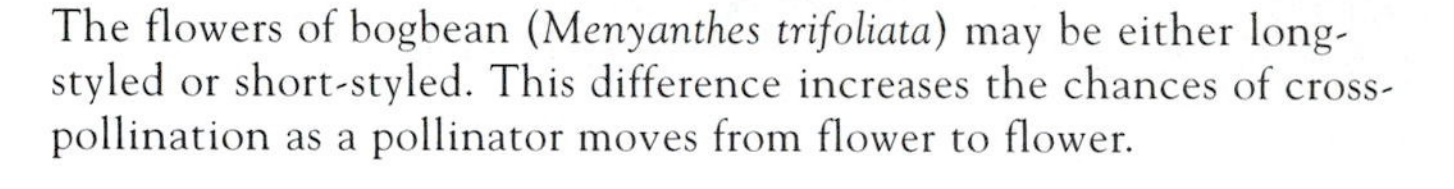

The flowers of bogbean (*Menyanthes trifoliata*) may be either long-styled or short-styled. This difference increases the chances of cross-pollination as a pollinator moves from flower to flower.

have a short style and elevated anthers. The contrasting positions of style and anthers ensure that the insects collecting pollen on their abdomens while drinking nectar from a plant with pin flowers will be more likely to pollinate a plant with thrum flowers.

Alleles for self-sterility in plants such as evening primrose (*Oenothera*) also promote outbreeding. These plants cannot fertilize themselves, because both flower and pollen have the same allele for self-sterility. They can only fertilize another individual if it has a different allele at the locus for self-sterility. This system works well if the population has many different alleles, so that many different genotypes can fertilize many other genotypes. A plant having a relatively infrequent allele for self-sterility will probably pollinate another plant (i.e., a plant with a different allele). A population of red clover may have over 200 such alleles.

Diploidy also preserves variation. In haploid individuals all alleles are immediately expressed in the phenotype, but in diploid organisms, genetic variation in the form of "hidden" recessive alleles can be carried and reproduced but not expressed or subjected to intense selection. For example, if recessive allele *a* has a frequency of 0.01 in the gene pool, it will be expressed as *aa* in the phenotype in only one of every 10,000 individuals ($q^2 = 0.0001$). Furthermore, it would take 100 generations of complete negative selection to diminish the frequency by half. Because removal or promotion of this allele by selection is very slow, it will remain in the population for a long time.

Also associated with diploidy and the maintenance of genetic variation is **heterozygote superiority.** As stated earlier, recessive alleles can be hidden in the heterozygous genotype. For reasons beyond the scope of this chapter, the heterozygous condition is often more fit than either the homozygous dominant or the homozygous recessive condition. This phenomenon is called *heterozygote superiority,* and may increase the frequency of a recessive allele carried in the heterozygote, even if the recessive allele in the homozygous condition is harmful.

Heterozygote superiority is also a benefit of outbreeding and is responsible for **heterosis** (sometimes called *hybrid vigor;* fig. 22.22). Hybrid corn, for example, bred from two different strains, is extremely vigorous and hardy because it is more heterozygous at many of its loci than either of the parents. The controlled process of corn improvement initially involves five to seven years of inbreeding and selection of desired features. This inbreeding produces purebred and strongly homozygous strains. After repeated inbreeding, two purebred strains are crossed to produce a hybrid. Frequently, another hybrid is produced from two other purebred, homozygous strains, and the two hybrids are then crossed. The result is a double-hybrid strain of seeds with strong hybrid vigor.

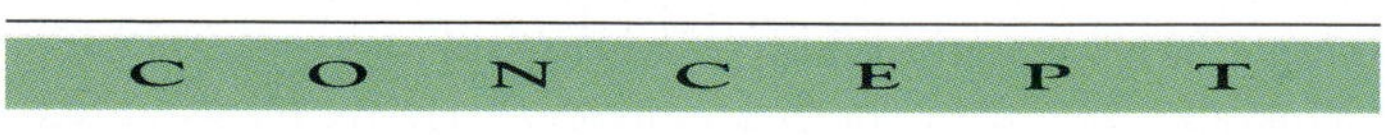

C O N C E P T

Genetic variation is maintained by sexual reproduction, outbreeding, diploidy, heterozygote superiority, and some forms of selection.

FIGURE 22.22

(a) Hybrids such as those growing behind the inbred strains show heterosis and are often taller and more robust. (b) Corn ears produced by the hybrids are much larger.

MACROEVOLUTION VERSUS MICROEVOLUTION

The foregoing discussion of evolution is a brief synopsis of some of the best-supported ideas on the subject. It is not the final word, however, since evolutionary biology itself is evolving. More recent ideas about evolution involve concepts such as the following: *molecular drive*, which is an alternative to natural selection as the main force behind evolution; the *transposon hypothesis* (see Chapter 10), which explains why organisms reproduce sexually and has little to do with natural selection of phenotypes; and the *neutral allele hypothesis*, which explains why some genes might have no positive or negative selective value. The most heated debate about evolution, however, examines *punctuated equilibria*, which unites ideas about catastrophism and gradualism, and *macroevolution* versus *microevolution*.

The fossil record shows that there have been relatively short periods of explosive diversification in large groups of organisms, followed by long periods of gradual change. Some groups change slowly for millions of years, then almost disappear, only to be replaced over tens of thousands of years by other groups. Cone-bearing plants and dinosaurs, for example, existed for tens of millions of years and dominated the earth, but these plants dwindled rapidly, and the dinosaurs disappeared completely about 65 million years ago. A large diversity of flowering plants and mammals quickly took their place. This pattern of evolutionary change is one of slow change punctuated with short periods of rapid change—that is, gradualism followed by catastrophism. Such a pattern is referred to as *punctuated equilibria*.

Macroevolution is the origin of taxonomic categories above the species level, and microevolution is the origin of species **(speciation).** These two concepts are sometimes contrasted because of the suggestion that the processes of macroevolution are somehow fundamentally different from those of microevolution. However, direct evidence is lacking for such ideas about macroevolution. Conversely, microevolution, which occurs by natural selection and other mechanisms described in this chapter, is a robust experimental discipline that draws on an abundance of evidence. We therefore offer no further discussion of macroevolution; the next chapter explains the processes of microevolution.

Chapter Summary

Evolution refers to all the changes of life on earth from its earliest beginnings to the diversity of today. Today's organisms arose through modification of ancient forms of life. Evolution is fueled by accumulations of genetic changes through many generations. Evolution involves a change in the frequencies of alleles in a population's gene pool from one generation to the next.

Evolution was an intimidating idea; it violated preconceived notions that species were divine and unchanging creations. Georges Cuvier examined fossils as evidence of change, but he proposed that catastrophes accounted for changes in ancient assemblies of species. James Hutton and Charles Lyell argued that the earth was millions rather than thousands of years old and that organisms had changed through time. Lamarck tried to explain changes in species by inheritance of acquired characteristics; however, such characteristics are not heritable from generation to generation.

Charles Darwin voyaged to South America for five years and observed remarkably diverse adaptations and distributions of species that eventually caused him to question divine creation. He adopted Lyell's contention that the earth was old and gradually changing. He also subscribed to Malthus's observation that species increase geometrically and face limited resources. From these observations and ideas, Darwin conceived of a mechanism called *natural selection* to explain the development of diverse species through time. At the same time, Alfred Russel Wallace also developed a theory of natural selection, and both naturalists presented their work in 1858. The next year Darwin published his extensive documentation and explanation for natural selection in *Origin of Species*.

Support for evolutionary theory includes both indirect and direct evidence. Indirect evidence for descent with modification comes from the age of the earth, the fossil record, homology, convergence, biogeography, and molecular biology.

The essence of natural selection is the differential reproduction of genetic variants. Organisms produce more offspring than needed to replace themselves and more than environmental resources can support. Some organisms die, and those with genes that confer the best adaptation to their environment will reproduce the most. The frequency of reproduced genotypes increases with each generation. Unfavorable genotypes will be reduced. Natural selection functions at the population level, and a population, not an individual, is the smallest unit that can evolve. Populations evolve in the context of their environment, which confers success or failure to phenotypes. Selection does not create new adaptations but promotes the most adaptive genetic variation available through increased reproduction. Direct evidence for natural selection comes from experimentation and from comparison with artificial selection.

Stabilizing selection in constant environments promotes the normal, most prevalent phenotype and reduces variation. Directional selection is common in changing environments, and fosters an extreme phenotype while reducing the occurrence of the opposite phenotypic condition for a trait. This shifts the population's distribution of phenotypes for a characteristic toward an extreme. Diversifying selection occurs in drastically changing environments; it reduces the normal, most frequent phenotype and promotes the extremes, thereby maintaining overall variation.

Evolution and natural selection are genetic processes. Population genetics deals with entire gene pools rather than individuals. The Hardy-Weinberg equilibrium predicts that allelic frequencies will not change over multiple generations in a large, isolated, randomly mating population that is not subject to selection or mutation. Hardy-Weinberg equations allow the prediction of allelic, genotypic, and phenotypic frequencies. The Hardy-Weinberg equilibrium is important because it provides population geneticists with a model of stability for comparison to more natural, changing populations. A measure of deviation from equilibrium is needed to investigate causes of change.

Genetic variation as the raw material for natural selection is introduced into populations by mutation, gene flow, and genetic drift. Mutations are sudden, heritable changes in genes or chromosomes. Gene flow is the transfer of genetic material from one population to another by migration of individuals or reproductive cells. Genetic drift is random, but it may produce significant changes in allelic frequencies in small populations that are subject to chance events such as natural catastrophes, mutation, and nonrandom mating.

Genetic variation is maintained by sexual reproduction, which assorts chromosomes during meiosis and recombines genetic information during crossing-over and the combining of genes from two different individuals. Outbreeding promotes new genetic combinations by ensuring sexual reproduction with other individuals rather than within the same individual. Diploidy preserves variation by effectively hiding recessive alleles in the heterozygous condition and isolating these alleles from selection, which operates only on phenotypes. Heterozygote superiority also maintains diversity by promoting the heterozygous condition over either dominant or recessive homozygotes.

More recent ideas about evolution include alternatives to natural selection (e.g., molecular drive and the transposon hypothesis) as the main driving force of evolution. Other ideas involve neutral alleles and macroevolution versus microevolution. Microevolution explains the origin of species, while macroevolution explains the origin of higher taxonomic categories.

Questions for Further Thought and Study

1. In earlier chapters of this book you've seen examples of evolution—for example, how plants form symbioses with other organisms (e.g., fungi, ants, and bacteria). You've also seen more subtle examples of biochemical evolution, such as how plants make compounds that deter herbivores, and even how other organisms such as grasshoppers have evolved ways of using plants' defenses for their own survival. What would be the selection pressures for these examples of biochemical evolution?
2. Lamarck was an insightful evolutionist. What was the major problem with his theory?
3. How are gametes, eggs, and embryos subject to environmental selection pressures?
4. Assume that a randomly mating population of blue-flowered and white-flowered plants produces offspring consisting of thirty-one blue-flowered plants and seventy-six white-flowered plants. Explain these results in terms of Hardy-Weinberg equilibrium and its assumptions.
5. Describe how stabilizing selection changes the gene pool.
6. What would be the consequences if a population reproduced without the influence of natural selection?

7. Assume heterozygote superiority operates for a gene locus of two alleles. How does selection for the heterozygote affect the relative frequencies of the two alleles?
8. What would be the most likely situation in which a single mutation could significantly affect a population?
9. How would you explain positive selection for survival beyond the oldest reproductive age?
10. How is evolution different from natural selection?
11. Why can't individuals evolve?
12. What was the importance of artificial selection to Darwin's formulation of his theory of natural selection?
13. If natural populations are rarely, if ever, in Hardy-Weinberg equilibrium, why is this model important to population geneticists?
14. How would you determine if most mutations were harmful?
15. Herbert Spencer was an eighteenth-century English sociologist who believed in applying Darwin's concepts to human society. He popularized the phrase *survival of the fittest* and suggested that the unemployed, the sick, and other "burdens on society" be allowed to die rather than be objects of public assistance and charity. What do you think of this idea and of the validity of his interpretation of the concept of natural selection?
16. Hypotheses to account for evolutionary events such as the pace of evolution and the cause of mass extinctions are often conflicting. Those who do not believe that evolution has taken place sometimes perceive these conflicts among scientists as signs of doubt that evolution occurs. Scientists, however, maintain that conflicting hypotheses do not argue against evolution at all but demonstrate the process of scientific thinking. What is your opinion? Why are hypotheses particularly important in understanding evolutionary processes compared to other fields of biology?
17. Mimicry is a common biological phenomenon. For example, the Notodontidae moth (*Antaea jaraguana*) from Brazil has a wing pattern that resembles a dead leaf and its venation. How would an insect benefit by resembling a plant?
18. James Watson, who along with Francis Crick discovered the structure of DNA (see Chapter 8), said the following: "Today, the theory of evolution is an accepted fact for everyone but a fundamentalist minority, whose objections are based not on reasoning but on doctrinaire adherance [sic] to religious principles." Do you agree with Watson? Why or why not?

Suggested Readings

Articles

Ayala, F. J. 1978. The mechanisms of evolution. *Scientific American* 239(3):56–69.

Eiseley, L. C. 1959. Alfred Russel Wallace. *Scientific American* 200(2):70–84.

Hartman, H. 1990. The evolution of natural selection: Darwin vs. Wallace. *Perspectives in Biology and Medicine* 34:78–88.

Niklas, K. J. 1994. One giant step for life: simple law-abiding plants led the invasion of hostile lands. *Natural History* 103:22–25.

Rensberger, B. 1982. Evolution since Darwin. *Science* 82:40–45.

Schluter, D., T. D. Price, and P. R. Grant. 1985. Ecological character displacement in Darwin's finches. *Science* 227:1056–1059.

Stebbins, G. L., and F. J. Ayala. 1985. The evolution of Darwinism. *Scientific American* 253(1):72–82.

Wilson, A. C. 1985. The molecular basis of evolution. *Scientific American* 253:164–173.

Books

Ayala, Francisco J., and James W. Valentine. 1979. *Evolving: The Theory and Process of Organic Evolution*. Menlo Park, CA: Benjamin/Cummings.

Brent, P. 1983. *Charles Darwin. A Man of Enlarged Curiosity*. New York: W. W. Norton.

Brown, J. H., and A. Gibson. 1983. *Biogeography*. St. Louis: C. V. Mosby.

Darwin, C. 1967. *On the Origin of Species*. Facsimile 1st ed. of 1859. New York: Atheneum.

Endler, J. A. 1986. *Natural Selection in the Wild*. Princeton, NJ: Princeton University Press.

Hancock, J. F. 1992. *Plant Evolution and the Origin of Crop Species*. Englewood Cliffs, NJ: Prentice Hall.

Stebbins, G. L. 1982. *Darwin to DNA, Molecules to Humanity*. San Francisco: W. H. Freeman.

A sample of the diversity of flowering plants.

CHAPTER 23

Speciation

Chapter Outline

Chapter Overview

Speciation is distinguished from broad evolutionary change by the formation of distinctive, interbreeding groups of organisms called species. The origin of new species is the central phenomenon of evolution. A species is broadly defined as a group of organisms capable of breeding successfully with each other but not breeding with members of other groups. Formation of a new species is a genetic event involving isolation of the gene pool of a population. Reproductive isolation, the most critical characteristic of a species, occurs via geographic isolation, temporal isolation, poor development, or sterility of hybrid offspring from mated individuals, or any other means of preventing fertilization by gametes from different species. Despite reproductive barriers, sympatric plants hybridize frequently, and hybrids may occasionally produce new species.

Models of allopatric speciation rely on the likelihood of gradual genetic divergence caused by mutation, genetic drift, and natural selection occurring in populations separated in space. Sympatric speciation frequently involves cellular nonseparation of chromosomes resulting in polyploidy (i.e., increased number of chromosome sets). Even though such polyploids and some hybrids may be sterile, they may represent a new, reproductively isolated, asexually reproducing species. In some cases, polyploidy may restore fertility of hybrids.

Introduction

Evolutionary theory focuses on speciation—that is, the origin of new species—to explain the diversity of life. New species are the foundation of biological diversity. Today, 2–3 million species of living organisms have been named, and 2–10 times that many remain to be discovered. This diversity may appear vast, but extrapolations from the fossil record indicate that today's species probably represent less than 1% of all those that have ever existed; that is, more than 99% of all species have become extinct. Obviously, speciation has occurred countless times in all sorts of environments; it is as much a property of life as growth, reproduction, and death.

Speciation is the evolution of new species from already existing species. This explanation, first proposed by Charles Darwin, is now uniformly accepted in biology. The widespread agreement that speciation has occurred and continues to occur is, however, spiced with plenty of arguments about how speciation works. The theory of natural selection (Chapter 22) provides some of the answers, at least in a general way. Specifically, though, biologists cannot seem to agree on what a species is in the first place. Furthermore, many possible mechanisms of speciation have been described, some of which are based on scientific evidence, and some of which are rational explanations with little or no evidence.

Biologists are alternately excited or frustrated by the *species problem*—that is, by the difficulty of defining a species. No single definition fits all species. This does not keep biologists from using the term *species* or from using species names, as we have done throughout this text. At this point, however, we can explain that our frequent use of species names in earlier chapters hides one of the biggest problems in biology: many different kinds of species have been described, but we don't know enough about most species to understand how they fit different definitions of species. Nevertheless, knowing what kinds of species exist is important for understanding how they might arise. The first objective of this chapter, therefore, is to define different kinds of species.

The second objective of this chapter is to explain how species are formed. If you suspect that our understanding of the mechanisms of speciation is muddied by the species problem, you are right. Ideas about how species arise have produced even more heated arguments than ideas about how to define a species. Many of these arguments are about the relative importance of the different mechanisms in nature. Furthermore, several different mechanisms for the origin of species are strongly supported by evidence; some of them have been used by artificial selection to produce new species. The creation of new species by human activities is indirect evidence that the same mechanisms occur in nature. However, most of our evidence for natural mechanisms does not show how significant each mechanism might be.

What Is a Species?

The species is the fundamental category of biological classification; species are the building blocks of the classifications that are presented in subsequent chapters on diversity. Beginning with Darwin, species recognition has commanded a great deal of attention from biologists because of its importance in classifying and understanding speciation. Perceptions of species (*species concepts*) remain a hot topic today: among botanists, spirited arguments on this topic are still heard every year at the annual meeting of the Botanical Society of America. At least a dozen species concepts have been proposed during the past two centuries. The most widely used are morphological species, biological species, genetic species, paleontological species, and evolutionary species concepts. We focus on these five species concepts in the following discussion.

The Morphological Species Concept

The **morphological species concept** holds that species are the smallest groups of organisms that can be consistently distinguished by their morphology. This is the most practical and

A.

B.

C.

FIGURE 23.1

(a) *Lilium regale* (b) *Lilium auratum* (c) *Lilium martagon*. These three lilies are easily recognized by color, shape, and general morphology. The morphological species concept works well to distinguish these organisms.

widely used species concept among taxonomists who deal with the identification and classification of organisms. Taxonomists use this concept mostly by default, since they know most of the species on earth only from their morphology. This concept is often called the *classical species concept*, since it generally has described species since people first began to classify organisms.

New species are named on the basis of morphological species concept; morphological descriptions are all that is required for distinguishing new species from previously known ones. Virtually all of the names of plant species used in this text are of morphological species. Likewise, species that are listed in books such as the *Manual of the Vascular Plants of Texas*, the *Manual of the Vascular Plants of the Northeastern United States and Adjacent Canada*, and *An Illustrated Flora of the Pacific States* are morphological species.

Most of the morphological species that you are familiar with are intuitively recognizable as distinct from one another. For example, you can distinguish species of lily, pine, or oak. With a little study, you could also learn to distinguish some of the 100 species of *Lilium* (lilies), the 93 species of *Pinus* (pines), and the 600 species of *Quercus* (oaks) from one another (fig. 23.1). In so doing, you would be learning these plants as morphological species.

At this point it might seem that the morphological species concept allows you to recognize all the species that you want to know. So why go any further? The answer is that speciation has not always formed well-defined species; the lines between species are often blurred. In such cases, different taxonomists often disagree on which plants are the same or different species. For example, the morphological species concept of the cactus genus *Opuntia* includes anywhere from 400 to 1,000 species, depending on whose judgment is followed (fig. 23.2).

A.

B.

FIGURE 23.2

These two forms of prickly pear were at one time classified as the same species, *Opuntia chlorotica*. Currently, they are classified as two species: (a) *O. chlorotica*; (b) *O. curvospina*.

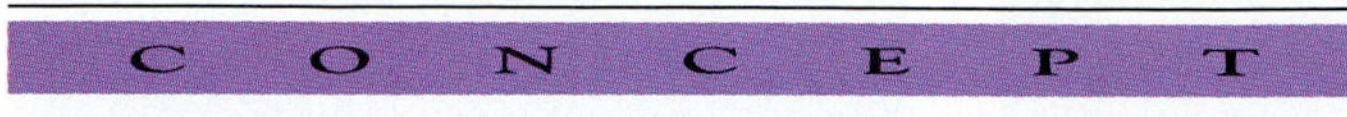

Ideas about speciation are based on what kinds of species exist and on the different ways that species arise. Biologists argue about how to define species and about which mechanisms of speciation might best explain biological diversity. The morphological species concept defines species by their observable morphological features. It is the most widely used species concept, because most species are known only by their morphology.

The Biological Species Concept

In contrast to a morphology-based concept of species, the **biological species concept** uses reproductive biology to define species. According to this concept, a species is a group of interfertile populations that are reproductively isolated from other such groups. The biological species concept was first proposed by ornithologists (bird specialists), because birds commonly occur in interfertile, reproductively isolated populations that are hard to distinguish morphologically. Most animal species probably fit this concept. Many plants do not fit the biological species concept, however, so most botanical discussions of this concept focus on how poorly it applies to plants. For example, geographically isolated morphological species may hybridize readily when they are grown together (fig. 23.3).

Exceptions to the biological species concept are common among morphological species that form extensive hybrid populations where the two parent species overlap (fig. 23.4). Furthermore, plants often reproduce asexually or by self-pollination, which means that interfertility, or at least interbreeding, may be absent. Dandelions (*Taraxacum officinale*) and blackberries (*Rubus* species), for example, have played particular havoc with the biological species concept because they produce seeds without fertilization. The biological species concept is useless in defining certain crop species, such as the commercial banana or the navel orange, because many of these species are exclusively asexual. The biological species concept is also useless with fossils, because information about the reproductive biology of extinct organisms is usually missing.

The biological species is evolutionarily important because it requires reproductive isolation. It emphasizes a fundamental assumption about speciation: A population is a distinct species only if it does not mix genes with other populations. Speciation can begin only when reproductive isolation occurs, and it may occur in several ways, ranging from geographical isolation to genetic or behavioral isolation (mostly in animals). Although plants often hybridize in nature, their hybrids are frequently weak and less fit. The reduced fitness of some hybrids provides some reproductive isolation, although it is incomplete.

A.

B.

C.

FIGURE 23.3

These two species of *Liriodendron* are geographically isolated from one another, but they hybridize when they are grown together. (a) *L. tulipifera* is a common tree of the eastern U.S. (b) *L. chinensis* occurs in small populations in eastern China and Vietnam. (c) The hybrid occurs in botanical gardens where trees of both species are cultivated near one another.

FIGURE 23.4

The interior live oak (*Quercus wislizenii*, left) hybridizes in nature with the California black oak (*Q. kelloggii*, right). Their hybrid is called the oracle oak (Q. X *morehus*, center; the "X" refers to its hybrid origin).

A biological species is a group of interfertile populations that are reproductively isolated from other populations. Plants often do not fit this species concept because they may be interfertile, self-pollinating, or predominantly asexual.

The Genetic Species Concept

The **genetic species concept** defines a species by its genetic uniqueness, that is, by how different it is genetically from its nearest relative. In this regard, the genetic species concept is like the morphological species concept, except that the recognition of genetic species is based on genetic data instead of morphological data.

The uniqueness of a species is typically expressed as its **genetic distance** from other species. The main drawback to a genetic distance measure is that the distance is based on only a small part of the genome. For example, allele frequencies can be obtained by electrophoresis for only about thirty different enzymes; several thousand other enzymes cannot be used to measure genetic distance because we do not yet know how to detect them. Recently, genetic distance has been calculated by comparing of DNA fragments (RFLPs; see Chapter 11) and gene sequences. Like allele frequencies, however, these measures are limited to a small fraction of the genome.

Because the genetic species concept provides a quantifiable definition of species, it is important for estimating the difference between a species and its nearest relative. This information is becoming increasingly useful for evaluating endangered species, that is, for determining whether an endangered species is either very different or not so different from its closest relative (see box 23.1, "Genetic Distance: A Tool for Conservation").

The genetic species concept is based on a quantifiable comparison of genetic data. The definition, or uniqueness, of a species is determined by its genetic distance from its nearest relatives.

The Paleontological Species Concept

Paleobotanists cannot deal directly with species concepts that involve genetic distance or reproductive isolation. Species in the fossil record are represented from few localities, and they are often fragmentary. Fossils best fit the morphological species concept but with a time dimension, because they appear and disappear in the fossil record. Such species are sometimes called **paleospecies;** sometimes paleospecies seem to precede existing species, which indicates a gradual change from one species to another (fig. 23.5) (see "Gradualism" in Chapter 22).

The Evolutionary Species Concept

None of the species concepts mentioned so far refer directly to evolution. For this reason, evolutionary biologists have proposed that an ancestral-descendant sequence of populations be called an **evolutionary species.** For example, a paleospecies that precedes a living species represents an ancestor-descendant sequence. They are, therefore, a single evolutionary species.

The paleontological species concept is based on distinguishing morphological species over geological time. A sequence of paleontological species along one evolutionary line is an evolutionary species.

Writing to Learn Botany

What are the comparative advantages and disadvantages of each of the various species concepts?

Boxed Reading 23.1

Genetic Distance: A Tool for Conservation

Hundreds of plant and animal species are endangered in the United States: their habitat is disappearing, and they are in danger of extinction. Many of the species are endangered because of human activities such as land development, logging, and pollution. The cost of protecting endangered species involves restricting such activities, but such restrictions often slow economic development. Extremists on one side of the argument say that humans are the most important species, so let the bulldozers loose. Extremists on the other side say that species should be protected at all costs. Although it seems arrogant for one species, humans, to judge the value of another species, this is necessary when too many people demand too few resources. Which species should we value the most and spend the most to protect? One suggestion is to use genetic distance to determine how distinctive an endangered species is from its nearest neighbor. The more distinctive it is, the more protection it deserves.

More recently, conservation biologists have discovered that protecting endangered species is an inadequate, stopgap strategy. Species are endangered because of habitat destruction. Fortunately, measurement of genetic distance can indirectly compare habitats of different populations just as it compares endangered species. The uniqueness of a habitat is measured as the sum of genetic distances between each occupant species and its nearest relative. If a habitat is filled with populations that are highly distant from their nearest relatives, then the habitat is probably unique. This idea is still relatively new, however, and it has been used mostly for comparing animal species. Most plant conservationists have yet to adopt the idea in any significant way.

A.

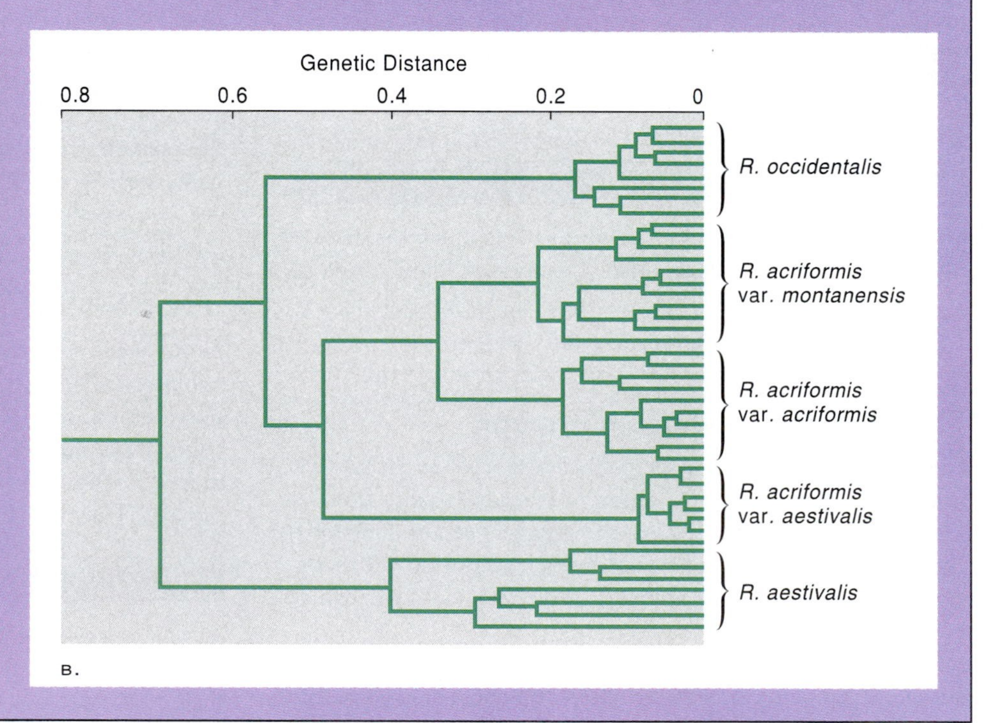

B.

Box Figure 23.1

(a) This autumn buttercup (*Ranunculus aestivalis*) is endangered species that is known only from the Sevier River Valley in Utah; (b) comparisons of DNA fingerprints among populations of the autumn buttercup and other kinds of buttercups show how genetically distinctive this endangered species is.

Mechanisms of Reproductive Isolation

Reproductive isolation is the foundation of the species concept. Reproductive isolation protects the integrity of a species by preventing the gene pool from being contaminated by genes from other species. Mechanisms that prevent the production of fertile offspring (hybrids) are **reproductive barriers** that promote reproductive isolation. The most obvious barrier is geographic separation. If two populations are separated by a great distance, they will not interbreed. However, most barriers to hybridization are not absolute—other biological barriers operate to varying degrees in different species, even if the species ranges overlap.

Reproductive barriers that isolate the gene pools of species may be prezygotic or postzygotic. Prezygotic mechanisms function before the formation of a zygote (i.e., they prevent successful fertilization). For example, a prezygotic barrier might block pollination or chemically impede fertilization of the egg. Postzygotic mechanisms operate after zygote formation, by obstructing development of the hybrid zygote into a fertile adult.

A.

B.

Figure 23.5

The paleospecies *Lebachia piniformis* (a) shares many of the features of a modern species *Araucaria hunsteinii* (b). This implies a lineage of gradual change. Paleospecies can be classified only by morphological features rather than reproductive or genetic characteristics.

Development of reproductive barriers is critical for speciation, and pre- and postzygotic barriers are promoted by natural selection. For example, the hybrid offspring resulting from parents that somehow circumvent reproductive barriers have genes from both parents. These genes are typically incompatible. They provide the growing embryo with mixed signals, confuse development, and may kill the embryo. Even if the embryo survives, the mature organism is often weak and sterile. Parents that circumvent barriers often produce unfit offspring. In contrast, parents that do not overcome reproductive barriers produce more viable, healthy offspring and pass their genes to the next generation. In this way, natural selection favors organisms that neither mate nor reproduce outside their species. The prezygotic and postzygotic barriers are summarized in table 23.1.

C O N C E P T

Reproductive isolation is critical to speciation and is caused by the formation of a variety of reproductive barriers. Prezygotic barriers prevent successful fertilization. Postzygotic barriers prevent successful development of a zygote.

Prezygotic Barriers

There are several types of prezygotic barriers.

Geographic Isolation

The gametes of individuals will not meet if the organisms are separated by significant distances.

Table 23.1

Mechanisms of Reproductive Isolation

1. Prezygotic mechanisms prevent mating or successful fertilization.
   a. Geographical isolation: Populations live in distant geographical areas.
   b. Microhabitat isolation: Populations live in different habitats and do not meet.
   c. Temporal isolation: Individuals are receptive to mating or flowering at different times.
   d. Mechanical isolation: Flowers are structurally different, and pollination is prevented.
   e. Gametic isolation: Female and male gametes are incompatible.
2. Postzygotic mechanisms prevent production of fertile adults.
   a. Hybrid inviability: Hybrid zygotes do not develop to sexual maturity.
   b. Hybrid sterility: Hybrids do not produce viable gametes.
   c. Hybrid breakdown: The progeny of fertile hybrids are weak or infertile.

Microhabitat Isolation

Species living in different microhabitats within the same area may not encounter each other. Although they are not geographically isolated, they are significantly separated in space. For example, scarlet oaks (*Quercus coccinea*) of the eastern United States grow in the same area as black oaks (*Q. velutina*) over a broad range. Although both species are wind-pollinated and will hybridize if brought together artificially, natural hybrids are rare. Scarlet oaks grow in moist, slowly draining, acidic habitats, whereas black oaks grow in dry, well-drained habitats. The separation of these two habitats within the same area is sufficient to discourage hybrid formation. Hybrids often are successful only in disturbed areas.

Figure 23.6

This night blooming *Cereus* (*Cereus* sp.) is temporally isolated because its flowers open only at night for a limited time.

Temporal Isolation

Two species will not interbreed if they reproduce during different times of the day, season, or year (fig. 23.6). For example, three orchid species of *Dendrobium* live in the same tropical forest but produce flowers at different times. Flowering in all three species is triggered by the same event—a thunderstorm; however, one species opens its flowers eight days after the storm, a second species flowers nine days after the storm, and the third flowers ten days after the storm. Pollination occurs only on a single day because the flowers open in the morning and close the following evening. Consequently, all three species are reproductively isolated in time.

Mechanical Isolation

Closely related species may begin the reproductive process, but mating or pollination fails because the plants are anatomically incompatible. For example, the flowers of some plants have mechanical barriers specifically adapted to certain pollinating insects or other animals (fig. 23.7). As a result, pollen is transferred only between plants of the same species with the same floral anatomy.

Gametic Isolation

Flowers can distinguish between pollen of the same species and pollen of different species. Even if pollination should occur, germination of the pollen grain may be retarded by components in the stigma and style. Surfaces of the egg and sperm may have compatibility factors that promote or prevent fertilization.

Figure 23.7

The flower morphology of this hammer orchid makes its pollen accessible to only a few kinds of pollinators such as this wasp. The flower is mechanically isolated from the pollen of other flowers frequented by different pollinators.

Postzygotic Barriers

Several postzygotic barriers also ensure reproductive isolation.

Hybrid Inviability

If a hybrid zygote forms, the embryo's development may be prevented because the combined genetic instructions are unworkable (e.g., the wrong gene-products are produced) or because the control mechanisms are inoperable.

Hybrid Sterility

If two species produce healthy hybrid offspring, the offspring are often sterile (fig. 23.8). In most sterile hybrids, meiosis produces abnormal gametes because the sets of chromosomes contributed by the parents are different in number or structure.

Hybrid Breakdown

Sometimes the first generation of hybrids is healthy and fertile, but when they mate with each other or with members of parent populations, their offspring are frail or sterile. For example, some species of cotton produce fertile hybrids, but the progeny of the next generation die as embryos or as weak, vulnerable plants.

Figure 23.8

Many hybrids such as this orchid, *Dendrobium crepidatum*, are sterile because they cannot produce normal gametes. This is a postzygotic barrier to successful reproduction.

Population Divergence and Speciation

To form a new species, the gene pool of a population must diverge from that of the parental (i.e., ancestral) population. Individuals in a diverging population become different—so different that they usually look different from the ancestral population, exploit different habitats, and respond to the environment differently. The mechanisms controlling this divergence are not predictable or easily observed, because some forces homogenize, mix, and minimize variation, while other forces enhance and promote divergence (table 23.2). The first step in speciation, however, is partial reproductive isolation, which allows genetic divergence. Once the gene pool of a population is isolated, it inevitably evolves further, and its gene frequencies change because of selection, genetic drift, and mutations in subsequent generations. For example, because mutations can occur randomly, isolated populations of the same species can randomly accumulate different mutations, even if the environments are identical. Enough mutations may accumulate that the two populations diverge and no longer interbreed. Similarly, genetic drift and environmental variation can enhance the divergence of recently isolated populations. Sexual reproduction does not usually promote divergence directly, because it mixes genetic information within a population and can slow isolation; however, it does supply new combinations of genes. For example, a new genetic combination arising in a splintered population may allow some individuals to exploit a new or different patch of environment. Once established in a new environment, a population's morphology, competitiveness, or reproductive life history may change and enhance isolation from the parent population.

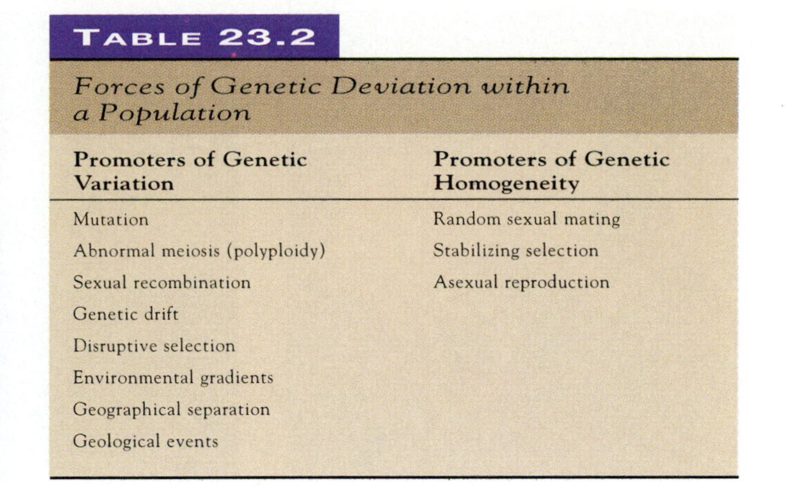

Table 23.2

Forces of Genetic Deviation within a Population

Promoters of Genetic Variation	Promoters of Genetic Homogeneity
Mutation	Random sexual mating
Abnormal meiosis (polyploidy)	Stabilizing selection
Sexual recombination	Asexual reproduction
Genetic drift	
Disruptive selection	
Environmental gradients	
Geographical separation	
Geological events	

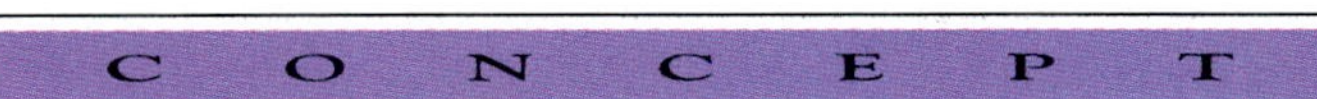

Speciation involves reproductive isolation of a gene pool, and results from genetic divergence of populations. Reproductively isolated populations continue to evolve through mutation, genetic drift, and selection. Therefore, physical or temporal separation of populations allows genetic divergence to increase.

A crucial event in the origin of a species occurs when the gene pool of a population is separated from other populations of the parent species. With its gene pool isolated, the splinter population can evolve on its own as changes in allele frequencies caused by mutations, genetic drift, and selection occur undiluted by gene flow from other populations.

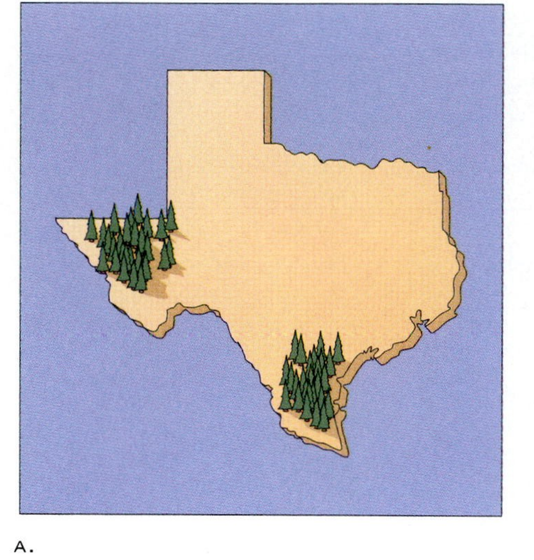

A.

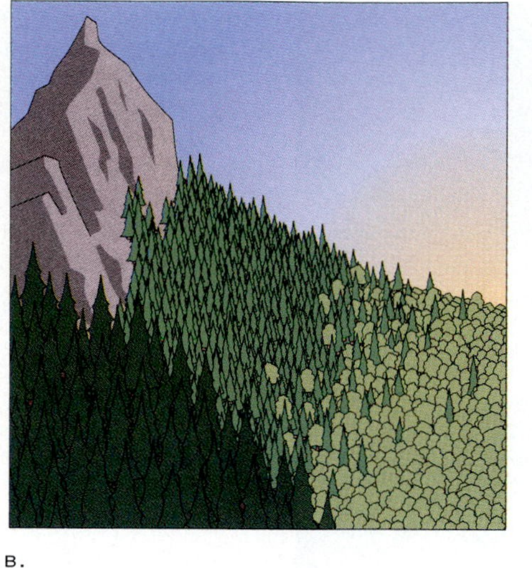

B.

C.

FIGURE 23.9

Modes of speciation: (a) Allopatric speciation occurs when populations are geographically isolated. (b) During parapatric speciation, populations share a common border and hybridize somewhat. But subtle differences in the adjacent environments are distinctive and lead to speciation. (c) During sympatric speciation, reproductive isolation occurs in the midst of intermingled populations. Isolation is usually achieved by abrupt genetic changes of an individual within the population.

The Lore of Plants

The peak of global plant diversity is the flora of Colombia, Ecuador, and Peru. Over 40,000 species occur on just 2% of the world's land surface. In the rain forest near Iquitos, Peru, Alwyn Gentry found about 300 tree species in each of two 1-hectare (2.5-acre) plots. Peter Ashton discovered over 1,000 species in a combined census of ten selected 1-hectare plots in Borneo. In contrast, only about 700 native species may be found in all of the United States and Canada.

E. O. Wilson, *The Diversity of Life*
W. W. Norton and Co. New York. 1992. 424 pp.

There are three modes of speciation, each based on the geographical relationship of a new species to its ancestral species: allopatric speciation, sympatric speciation, and parapatric speciation (fig. 23.9).

Allopatric Speciation

One obstacle to interbreeding among organisms is geographic isolation. Populations that are separated geographically are **allopatric** (meaning "different homeland"); similarly, **allopatric speciation** is the genetic divergence of two populations that are geographically separated. Because separated populations do not share the same range, they are likely to accumulate genetic differences, and their phenotypes will diverge. This divergence initiates speciation and is promoted by different selective pressures in different environments. The rate of genetic divergence also relates to unequal selection pressures in different locations. Isolated populations diverge as they accumulate different genetic characteristics and may become different species.

Geological processes can segment a population. For example, mountain ranges, canyons, lakes, and rivers can separate organisms. The formation of islands and deserts creates geographic barriers (fig. 23.10). The effectiveness of a barrier depends on the ability of an organism or its spores, pollen, or seeds to cross the barrier.

Small geographically isolated populations often become extinct, but those that survive diverge from their ancestors faster than large populations. Recall that genetic drift and mutations are more significant in small populations; large populations have great genetic momentum and change more slowly. Their genetic drift is insignificant, and natural selection may require many generations to replace alleles and change the phenotype of a large population. For example, large populations of North American and European sycamore trees have been allopatric for 30 million years, but they can still interbreed; that is, they have not evolved into separate species. In contrast, the gene pool of a small population can change in relatively few generations. A few reproductively successful individuals with favorable gene combinations can greatly affect the evolution of a small population and may lead to speciation in as few as hundreds or thousands of generations. Small populations become a new species faster because they are genetically more transient and erratic.

Allopatric populations are subject to speciation, but it is not easy to detect. In nature, reproductive isolation is not tested because separated organisms do not have the opportunity to sexually reproduce. An experiment using artificial conditions to test compatibility is not a definitive measure of speciation. Nevertheless, if successful hybrids are not produced, biologists usually designate the allopatric populations as true species. The

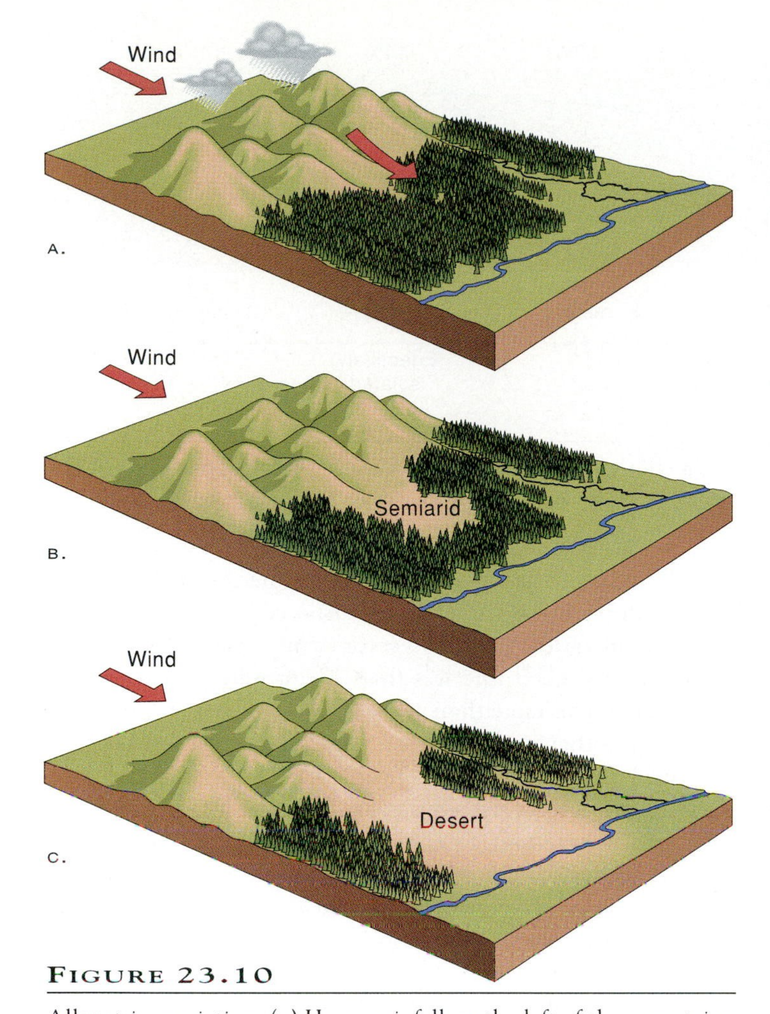

Figure 23.10

Allopatric speciation. (a) Heavy rainfall on the left of the mountains leaves a dry wind on the leeward side. (b) A desert can form and expand to divide a population into reproductively isolated subpopulations. (c) Accumulations of adaptations and incompatibilities by the two allopatric populations can form two isolated species.

best measure of genetic divergence in allopatric populations occurs when geographical barriers are naturally removed and these populations come together. Interbreeding will occur readily if divergence was only slight.

Another possible outcome when populations are brought together is limited hybridization. Surprisingly, the production of a few hybrids may reinforce genetic divergence because hybrid fertility is usually low. In this case, selection is against parents that are likely to hybridize and produce weak, unfit offspring. Selection favors parents that are less likely to interbreed. In other words, parents with characteristics that prevent hybridization (usually prezygotic barriers) would probably leave more viable offspring than would those that occasionally produce hybrids. This would reinforce prezygotic barriers and enhance divergence. The populations would diverge as characteristics that promoted hybridization were slowly eliminated. Increased reproductive isolation by selection against hybrids is difficult to demonstrate in natural populations.

Figure 23.11

A genetic deviation such as this variation in petal color of wild bluebells (*Mertensia virginica*) may arise within a sympatric population. This deviation could prevent cross-breeding with surrounding individuals and genetically isolate the deviant individual.

Parapatric Speciation

Populations in adjacent areas sometimes have separate ranges that meet along a common border, which often follows a discontinuity in some important environmental feature (fig. 23.9b). These populations are called **parapatric** (meaning "parallel native land") populations. For example, parapatric populations of a plant species that are adapted to different soils may abut along a zone where the soil changes. There, the gene pools of the two populations on either side of the border would differ somewhat because of the plants' adaptations to different soils. Because most plants breed only with their neighbors, gene flow between populations would seldom go far beyond the area where the populations of plants meet. Nevertheless, there would be a limited flow of genes between the populations via interbreeding at the contact zone. **Parapatric speciation** could occur if, despite the contact, strong reproductive isolating mechanisms develop as the populations slowly split into ecological units.

Sympatric Speciation

Populations that overlap geographically are **sympatric** (meaning "same homeland"). **Sympatric speciation** is the production of new species within a single population or overlapping populations (fig. 23.9c). Speciation of sympatric populations may seem unlikely, but there are mechanisms that reproductively isolate groups in the same environment. For example, a slight change in color, shape, or chemical attractants in a flower may make it unattractive to a pollinator (fig. 23.11). Such a mutation prevents the individuals from being pollinated by the same pollinator that visits the surrounding plants, and may effectively isolate them reproductively from the parent population.

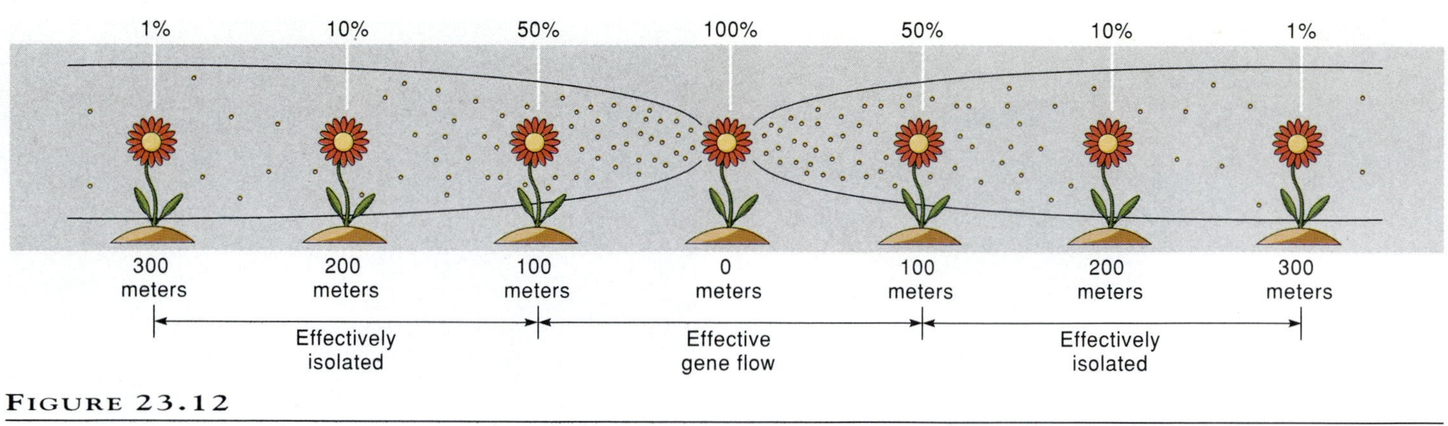

FIGURE 23.12

Pollen dispersal is limited; less than 1% will travel as far as 300 m. Effective gene flow occurs between individuals no farther apart than 100 m.

The isolated plants may survive by self-fertilization (reproduction with limited members of their own kind) or hybridization with a nearby population of a different species. Once the small subpopulation is isolated it may diverge genetically and form a new species.

Allopatric populations are separated geographically (e.g., by mountains or rivers) and will speciate as genetic variations accumulate and reproductive barriers form. When successful hybridization can no longer occur, the allopatric populations represent different species. However, the potential for hybridization between two separated populations is difficult to test in natural conditions. Sympatric speciation occurs within a single population or among overlapping populations. Mechanisms for sympatric speciation include mutations, hybridization, and polyploidy. Also, genetic divergence may occur along an environmental gradient across the range of a population. Parapatric speciation can occur at a boundary between two populations where some gene flow occurs, but at a rate too slow to overcome divergence of the gene pools of the two neighboring populations.

Plants are especially prone to sympatric speciation. Their adaptive radiation has been extensive and is attributed to a tendency toward gradual formation of clines and ecotypes and to more instantaneous processes of hybridization and polyploidy.

The Formation of Clines and Ecotypes

Patchy distributions of plant species and populations within species are produced by complex environments and sharp boundaries such as lakes, streams, and soil types. Plant populations are immobile and are sensitive to local selective pressures in these patches. Limited movement of their alleles emphasizes the importance of selection and the heterogeneity of the environment.

Partial genetic isolation of one population from another may be effective at relatively short distances. Even though pollen dispersal varies greatly by vector (wind, insect, etc.) and by species, as a general rule less than 1% of released pollen will reach individuals more than 300 m away (fig. 23.12). Considering all the other constraints on gene flow, isolation is often effective beyond 100 m. This means that pressures of the local environment may overshadow gene flow from dispersed pollen. Separated populations or extremes of a dispersed population will diverge due to differences in local selection pressures. However, occasional gene flow by long-distance fertilization may revive genetic compatibility and interbreeding, and thus integrate the populations.

One response to local conditions is modification of morphology. **Morphological plasticity** means that plants display different growth forms in response to different environmental conditions even without genetic change (fig. 23.13). As a result, organisms with the same genes often look different according to the environment. For example, a plant limited by available water may respond to increased moisture by growing taller and producing larger leaves. This growth form may be maintained as long as moisture is available. The genomes of most plants allow enough morphological variability for the environment to promote considerable variation.

Plants are morphologically plastic, meaning that their genome allows for great morphological variation according to the local environmental conditions.

Some populations or species range over a large geographic area with a gradient of climatic features that cause organisms at extremes to change morphologically and diverge genetically. A north-to-south gradient includes a range of temperatures and possibly a gradient of plants—taller ones at the southern end grade into shorter plants toward the north. This continuum of

Figure 23.13

Morphological plasticity may be evident in different regions of the same plant.

a characteristic is a **cline** of morphology. Members of a true cline are reproductively compatible and gene flow can occur throughout the cline (fig. 23.14).

Different morphologies along a cline may be due to genetics as well as morphological plasticity. That is, short, northern plants transplanted to the south may not grow as tall as the southern plants. Cline formation is a simple form of genetic divergence, and the degree of divergence is proportional to the length of the cline. Few species form smooth clines.

A cline consists of plants that diverge morphologically along an environmental gradient, but gene flow occurs throughout the population. Plants at the extremes of a cline are morphologically plastic and usually look different.

Environmental patchiness can further isolate segments of a cline over a range of microhabitats. These discontinuous subpopulations will diverge further because they are subject to local selection pressures and develop combinations of alleles that are best suited to the local environment. The differences among subpopulations will have a genetic basis. Fragmented subpopulations that have genetically diverged in association with their environments are called **ecotypes.** Significant genetic differences combined with inherent morphological plasticity may cause ecotypes to look markedly different, even though they are all reproductively compatible and produce viable offspring. Ecotypes may also differ physiologically (e.g., frost tolerances, time of flowering, photosynthetic rates). Such ecotypes may not look different, but they react differently to their environments. Well-defined ecotypes occur where habitats have well-defined boundaries.

A well-known description of an ecotypic variation was presented by Jens Clausen, David Keck, and William Hiesey, who studied the genetic adaptation of isolated populations along a variety of Californian climatic zones. These zones included a sea-level station near the Pacific coast, a mid-altitude station

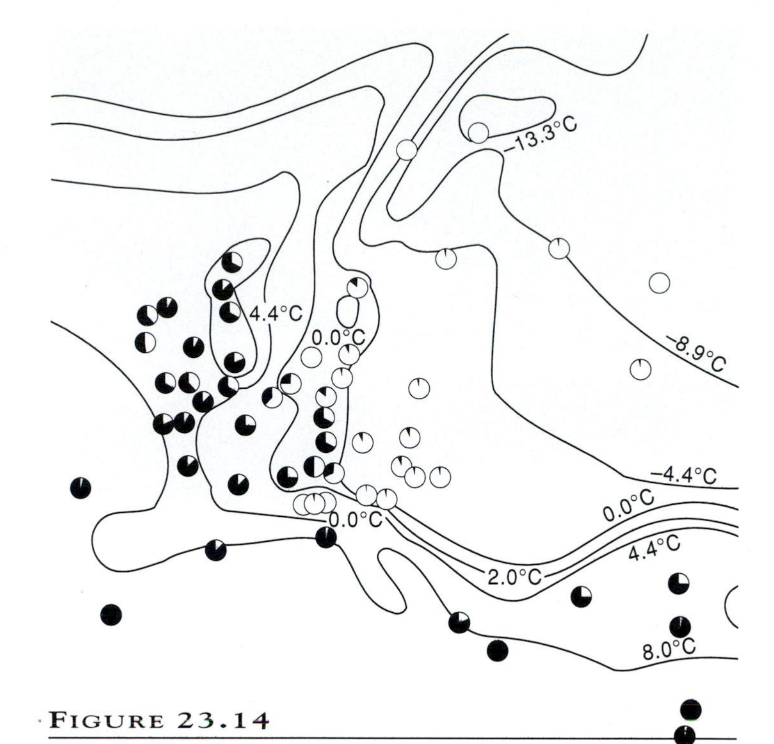

Figure 23.14

A European species of clover produces cyanide to discourage herbivores. As shown in this diagram, a cline of the cyanogenic phenotype and genotype occurs along a temperature gradient. This distribution of cyanogenic plants is apparently determined by a balance between the advantage they derive from being unpalatable to herbivores and the disadvantage they suffer when frost ruptures their cellular membranes, releasing cyanide in the plant's tissues.

in the Sierra Nevada Mountains, and a timberline station at 3,000 m elevation. One of their experimental plants, *Potentilla glandulosa,* a member of the rose family, grew at all three locations and showed genetic adaptations to each habitat. Because this plant reproduces asexually by runners (as well as sexually), clones could be used to control for genetic variation and emphasize habitat-induced variation. Individual plants were gathered from a variety of habitats along the gradient and grown side by side at a field station in each climatic zone. At least three distinct ecotypes became apparent. Each had distinct morphological traits that were adaptive to its original environment; that is, these subpopulations had diverged genetically. Besides different morphological traits, each of the ecotypes flowered at a different time of the year in response to the length of the growing season. The coastal ecotype flowered in mid-April, the foothills ecotype flowered in May–June, and the alpine ecotype flowered in late August.

Populations of aspen (*Populus tremuloides*) high in the mountains of Utah have become a genetically based ecotype (fig. 23.15), producing leaves early in summer and reproducing only vegetatively. Most aspens can produce viable seeds, but they also reproduce asexually by producing suckers from roots. In the Utah

FIGURE 23.15

This stand of aspen is probably a clone of asexually reproducing organisms.

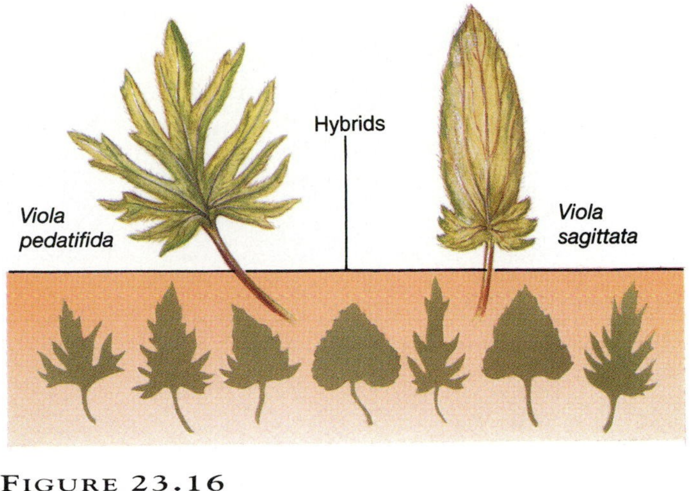

FIGURE 23.16

Hybridization can produce individuals with phenotypes that do not occur in either parent and represent a variety of intermediate forms.

mountains aspen seeds rarely germinate, because the climate has rainfall early in the summer; here aspens reproduce only vegetatively. Under these conditions an ecotype producing leaves earlier in the growing season and reproducing asexually has circumvented the dry season and successfully colonized higher elevations of the Utah mountains. Considering that the climate of summer drought began about 8,000 years ago in Utah, it is possible that some clones of aspen are 8,000 years old.

In a patchy or graded environment, fragmented subpopulations with significant genetic divergence, called *ecotypes*, may form. Gene flow between ecotypes is limited but greater than that along a cline.

Hybridization

Hybridization is the production of offspring from parents of different species or genetically distant strains within a species. Hybridization doesn't occur often, but it can produce individuals with new genetic combinations (fig. 23.16). If these combinations are successful, they represent evolutionary change and possibly a new species.

In 1760, Josef Kölreuter successfully conducted the first controlled hybridization of two plant species that resulted in fertile offspring. He crossed two species of tobacco and produced offspring that were considerably different from the parents. More crossings among the hybrids produced highly variable offspring. Some resembled one grandparent, some resembled the other grandparent, and some looked different from either of the original species. He concluded that hybridization mixed the genome from the original parents and provided considerable genetic variation.

Hybridization occurs in animals but is far more common in plants. Many species of oaks (*Quercus*), cottonwoods (*Populus*), aspens (*Populus*), and other trees and shrubs readily hybridize, but their offspring are often sterile (fig. 23.17). Plants such as Kentucky bluegrass (*Poa pratensis*) have frequently hybridized with a variety of species within the genus and produced hundreds of genetic strains that are well adapted to their local environments.

Plant hybrids are often sterile because their chromosomes cannot synapse during meiosis (fig. 23.17). Since the two haploid sets of chromosomes come from different parents, they have no homologues, and meiosis without synapsis will not produce viable gametes. (Later in this chapter we discuss how polyploidy may restore fertility to sterile hybrids.) Many sterile hybrids are **apomictic,** meaning that they reproduce vegetatively rather than sexually. However, sterility or infrequent outcrossing does not mean lack of success. For example, vegetative reproduction may be particularly appropriate in environments such as the Arctic that have a poor climate for pollination.

Hybridization may be a major mechanism for the production of new species suited to different habitats. However, sterile individuals do not interbreed and therefore do not fit the biological definition of a species. Some hybrid varieties of dandelions produce seeds, but the embryo is produced asexually. All the individuals in an immediate area may be clones. Nevertheless, hybrids may be considered true species because they are reproductively isolated from their parents.

Hybrid offspring may be superior and outcompete their parents in a variety of ways. For example, frequent disturbance

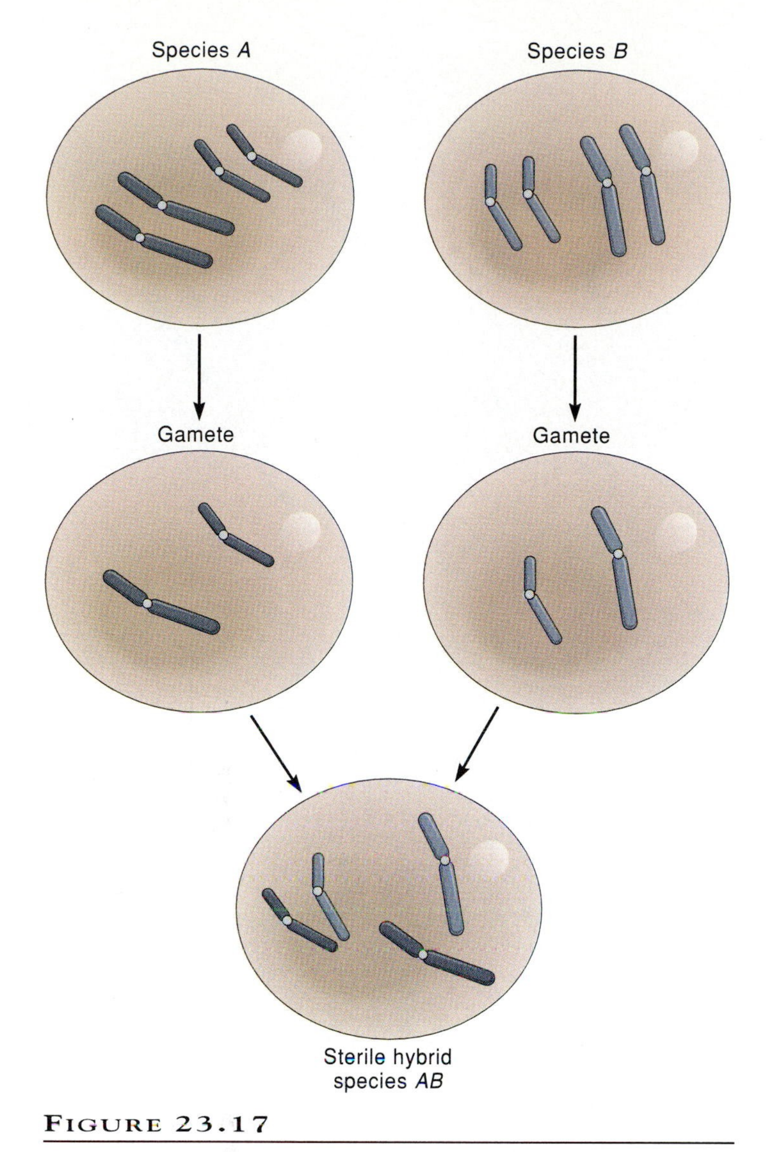

FIGURE 23.17

Hybrids are often sterile because the chromosomes of the hybrid are unpaired and cannot synapse during meiosis.

or environmental change such as cooler temperature or less rainfall may favor new hybrid genotypes and select against parental genotypes. Some hybrids may be genetically varied enough to inhabit environments that are unsuitable for the parents. Even in stable environments, hybrids may be superior due to heterozygote advantage (see Chapter 22). Multiple alleles may be favorable; for example, an organism may be better off with alleles coding for two different enzymes instead of one. Hybrids often have **allozymes,** which are enzymes controlled by alternate alleles (fig. 23.18). Even if the enzymes catalyze the same reaction, they may have different tolerances, optimal reaction temperatures, or substrate affinities. They can be useful at different stages of an organism's life, or they may broaden the range of potential environments.

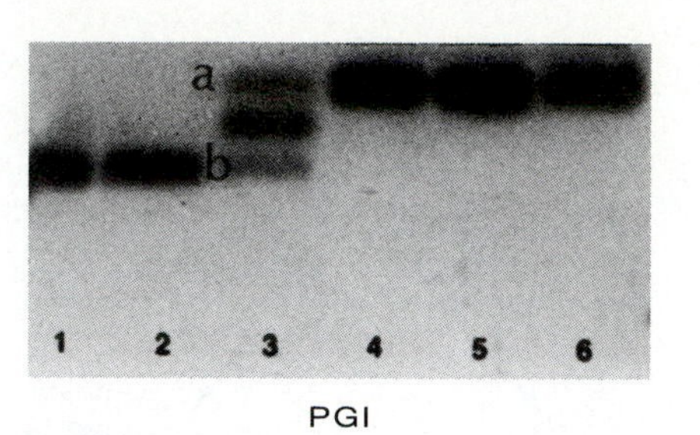

FIGURE 23.18

Each allozyme of an individual migrates on an electrophoretic gel at a different rate. The different combinations of stained bars represent different combinations of allozymes in each individual.

Hybrids are especially superior to inbreeding parents that are usually homozygous at many loci. Specifically, these inbreeders may have many homozygous recessive loci. In contrast, the more heterozygous hybrids will have a dominant and often advantageous allele at many of the previously recessive loci. This contributes to the vigor of many first-generation hybrids.

Hybridization may also move genes between species or populations. For example, a hybrid will occasionally backcross with one of its parents. In this way alleles from one species enter the gene pool of another species (fig. 23.19). This backcrossing is called **introgression.** Introgression is the flow of genes that bypass reproductive barriers and move between populations via the mating of fertile hybrids with parent populations. Hybrids of *Poa* frequently cross with other grasses and introduce new alleles to these grasses. Some genes of corn (*Zea mays*) can be traced to teosinte (*Zea mexicana*), a related grass; this gene flow probably occurred through a hybrid. Introgression provides crop geneticists with additional genetic variation for artificial selection and crop improvement.

Hybridization between two species is more common in plants than animals and typically produces sterile offspring. Hybrids may be successful in disturbed or new environments and may initiate a new species. Hybrids that can reproduce with other species can cause introgression, a type of gene flow between two species.

Polyploidy

The most widespread cellular process affecting plant evolution is polyploidy. Organisms with more than two sets of chromosomes are **polyploids** (**poly** = many, **ploid** = sets) and result from aberrant

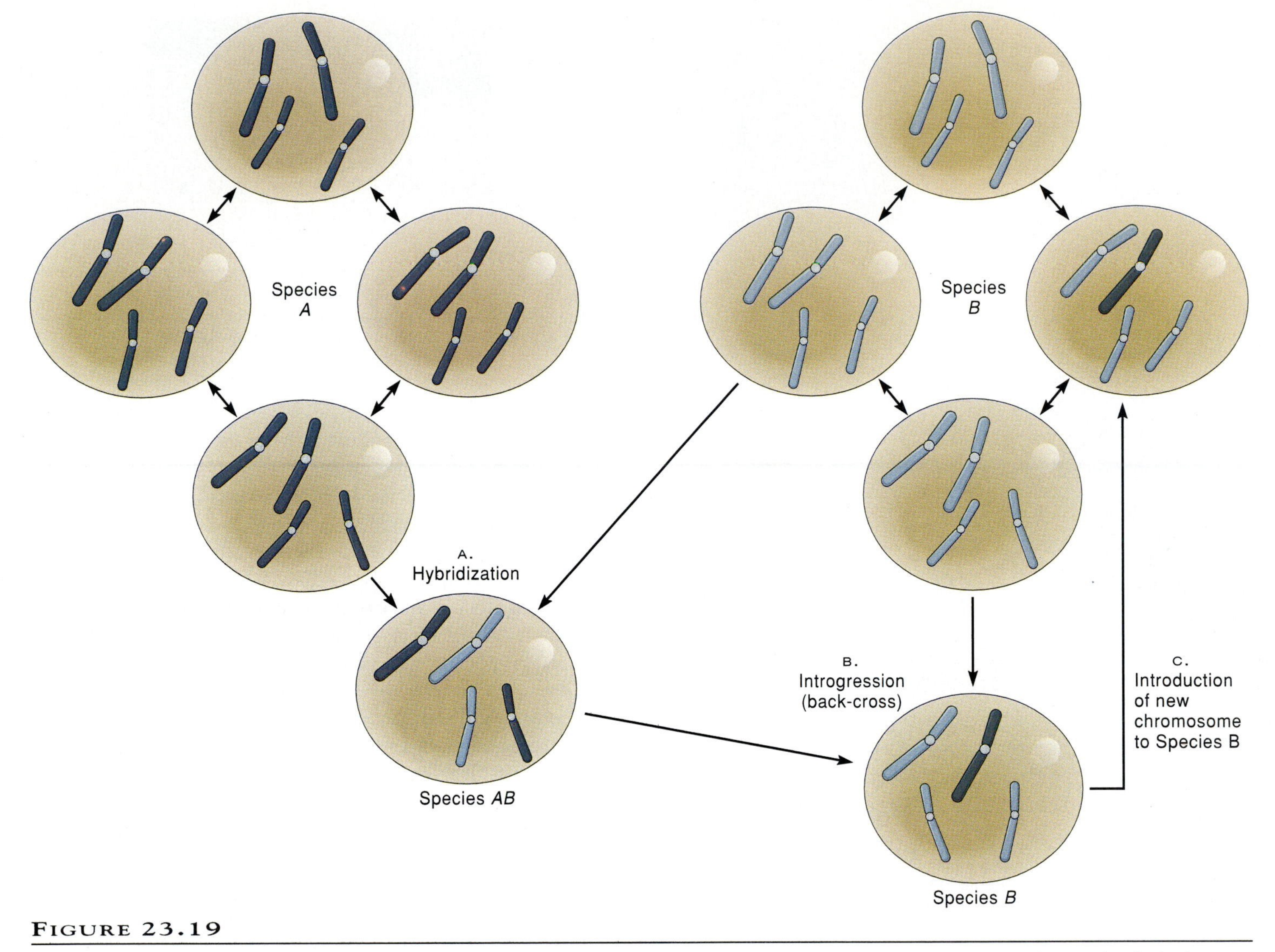

FIGURE 23.19

(a) In this schematic example, Species A hybridized with Species B to form Species AB. (b) This hybrid backcrossed with an individual of Species B. (c) This offspring subsequently carried a chromosome from Species A and introduced it back to the Species B population as it continued to reproduce.

chromosomal separation during cell division (fig. 23.20). Domesticated grains such as durum wheat (tetraploid), oats (hexaploid), and rye (hexaploid) are polyploids, as are cotton, tobacco, and potato. Polyploids also include garden flowers such as chrysanthemums, pansies, and daylilies. Polyploidy is rare among animals, fungi, and most gymnosperms, but occurs regularly among angiosperms and ferns. The rarity of polyploidy among animals may be related to their having distinct sex chromosomes. Polyploid sets of sex chromosomes in animals may fatally disrupt hormonal and sexual development.

Estimates of polyploidy in angiosperms have ranged from 30% to 80%. In 1994, Jane Masterson of the University of Chicago tested these estimates by studying a feature affected by polyploidy: cellular size. Because cellular size correlates with DNA content and thus with chromosome number, Masterson could estimate ploidy levels by studying cellular size in fossils of extinct plants. Masterson's clever research allowed her to infer that the haploid chromosome number in many angiosperms is 7 to 9, and that approximately 70% of angiosperms have polyploidy in their history.

The most common polyploids are tetraploids, with four sets of chromosomes. Different species in the same genus often have different numbers of chromosome sets. For example, various species of wheat (*Triticum*) are diploid, tetraploid, and hexaploid (see box 10.1 "Polyploidy in Plants" on p. 219; also see Epilogue).

Sexually reproducing polyploids usually have an even number of sets of chromosomes that allow successful separation during meiosis. Successive generations remain fertile and maintain a constant number of chromosomes. Triploids and pentaploids are usually sterile due to unsuccessful meiosis. Evolutionarily they are rarely successful, but they may be sought after as novelties. For example, some orchids are sterile triploids or pentaploids that flower more profusely without the demands of sexual production of seeds and fruit (fig. 23.21).

Polyploids with many copies of the same chromosome set are **autopolyploids.** These plants usually form within a single species when chromosomes are duplicated. However, if spindle fibers are disrupted, the cell may not divide cleanly (fig. 23.20a). The resulting cell has twice the original number of chromosomes. This nondisjunction of chromosomes can be induced chemically by colchicine, a drug that is derived from the autumn crocus (*Colchicum autumnale*). Colchicine disrupts spindle formation and prevents chromosome separation during mitosis.

Autopolyploidy occurs naturally in many genera, including *Sedum* and *Galax;* some domestic plants are induced to become autopolyploids to promote desired characteristics. For

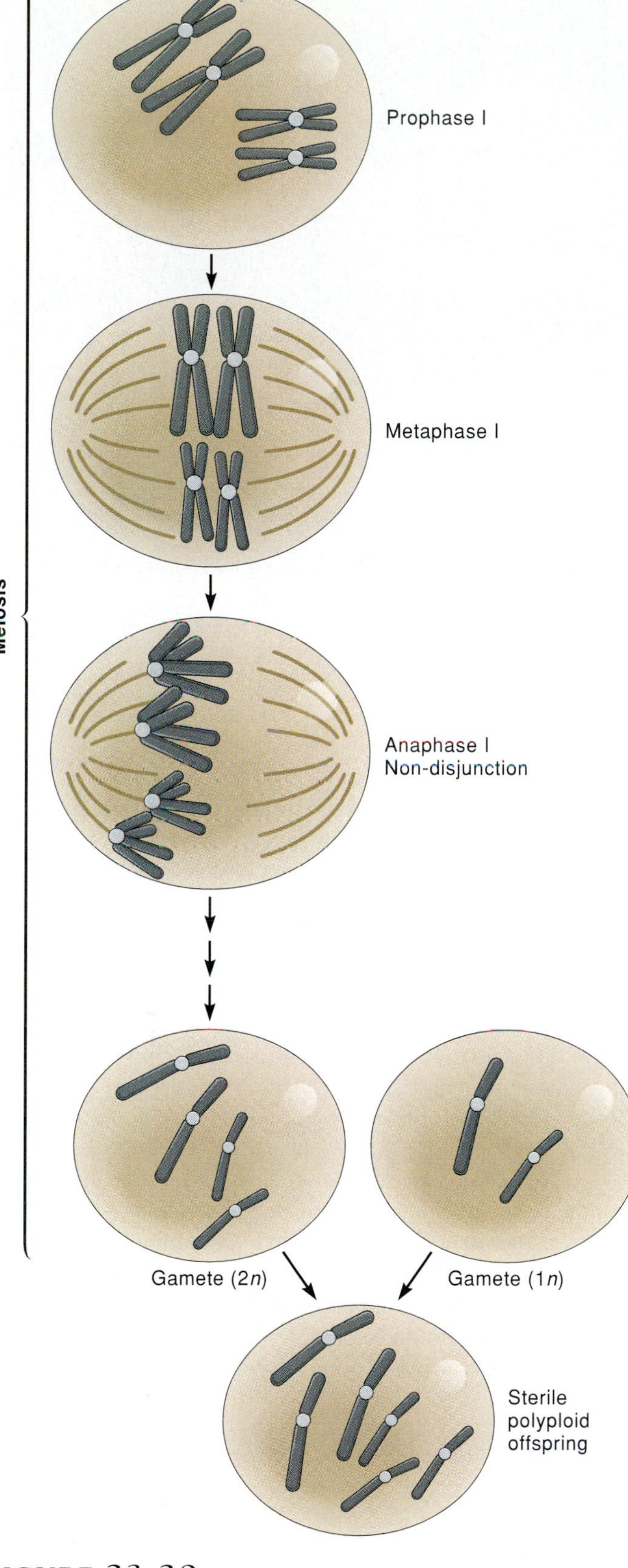

FIGURE 23.20

(a) Polyploids have nuclei with more than two sets of chromosomes. This condition may arise when chromosomes fail to separate properly during meiosis, and a diploid gamete is produced. Fusion of this diploid gamete with a haploid gamete produces a polyploid offspring. If chromosome numbers are multiplied in such a manner within a species it is autopolyploidy. Multiple sets of chromosomes from fusion of gametes from different species is allopolyploidy. (b) Three species of *Ranunculus* live in the Northern Island of New Zealand. Two of these species have 24 pairs of chromosomes ($2n$=48). Allopolyploidy probably produced the third species with twice the chromosome number ($2n$=96) as the two hybridizing parent species ($2n$=48).

FIGURE 23.21

Hybridization of orchids to produce new color patterns often results in sterile polyploid offspring. Fortunately, orchids such as this *Dendrobium* sp. can be cloned asexually, and the loss of fertility in a triploid or pentaploid hybrid is not of great concern to horticulturists.

FIGURE 23.22

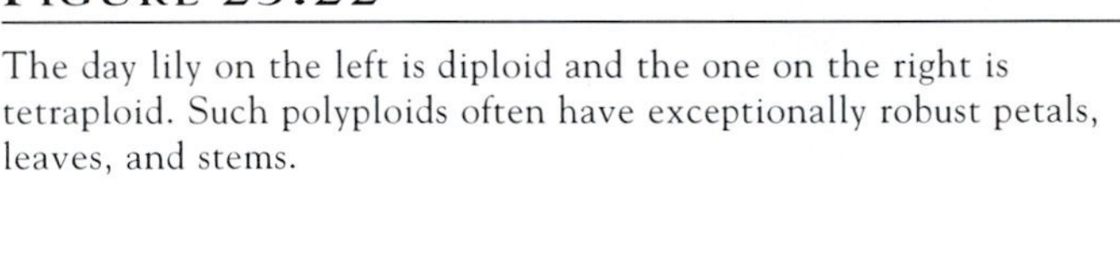

The day lily on the left is diploid and the one on the right is tetraploid. Such polyploids often have exceptionally robust petals, leaves, and stems.

example, polyploids often have larger cells, thicker leaves, increased water retention, slower growth, delayed flowering, and flowering over a longer season. Cold tolerance is also enhanced, and the number of polyploid species increases with increasing latitude. For these reasons, natural autopolyploids are most common in harsh environments where selection pressures are intense. Enhanced tolerance to harsh environmental factors is often the competitive edge that allows a species to establish a permanent population.

Polyploids with two or more distinct sets of chromosomes are **allopolyploids.** Allopolyploidy usually arises by duplication and nondisjunction of one or more of the chromosome sets in hybrids, and can restore sexual viability to an otherwise sterile hybrid (recall that hybrids without homologues cannot complete meiosis). Allopolyploidy solves this problem by providing duplicate sets of chromosomes that synapse during meiosis. The diploid hybrid between the primroses *Primula verticillata* and *P. floribunda* is sterile, but tetraploid hybrids called *P. kewensis* have been produced artificially and are fertile. Most plant hybrids are allopolyploids, including cultured hybrids that have been artificially selected for their large cells, plump plant parts, high water content, and drought resistance (fig. 23.22).

Polyploidy, especially hybridization followed by allopolyploidy, can be an immediate mechanism of sympatric speciation in higher plants (recall that a hybrid between two species is often sterile because chromosomes are not paired). Polyploidy restores homologues and fertility (fig. 23.23). The resulting allopolyploid cannot readily reproduce with either parent and immediately becomes a new species. Bread wheat, *Triticum aestivum*, is an allopolyploid species with 42 chromosomes that probably arose about 8,000 years ago by hybridization between a wheat species with 28 chromosomes and a grass with 14 chromosomes. The sterile hybrid had 21 chromosomes, but doubling of the chromosomes following hybridization produced a fertile allopolyploid wheat (see box 10.1, "Polyploidy in Plants" on p. 219; also see Epilogue). Rutabaga (*Brassica napus*, 38 chromosomes, $n = 19$) arose as an allopolyploid of a hybrid between cabbage (*Brassica oleracea*, 18 chromosomes, $n = 9$) and turnip (*Brassica rapa*, 20 chromosomes, $n = 10$).

Similar sympatric speciation by backcrossing of autopolyploids is apparently rare. Recall that autopolyploidy produces multiple sets of chromosomes that are often tetraploid. If a tetraploid plant backcrosses with a diploid individual, the hybrid will probably be sterile due to unbalanced sets of chromosomes. If the hybrid can successfully reproduce asexually, it may become a new species as it diverges from the parental population.

C O N C E P T

Polyploids have more than the typical diploid number of chromosomes. Polyploidy is a common evolutionary mechanism that can produce immediate speciation and restore fertility to hybrids.

Boxed Reading 23.2

Polyploids in the Kitchen

In 1924, a Russian geneticist had a tasty idea: crossing a cabbage and a radish to produce a new kind of vegetable called a *cabbish,* having the crispy leaves of cabbage and the red, sharp-tasting root of radish. The geneticist knew that he could probably produce a hybrid, because cabbages and radishes each have eighteen pairs of chromosomes. Unfortunately, the hybrid that he produced was sterile because the chromosomes of radishes differed so much from those of cabbage that they did not pair during meiosis. Consequently, the hybrid could not produce viable gametes. However, the hybrid spontaneously produced a tetraploid having a total of seventy-two chromosomes: two sets from radish and two from cabbage. In this tetraploid, each chromosome of cabbage and radish had a homologous partner, which allowed pairing to occur during meiosis. The gametes produced by the tetraploid each had a complete haploid set of chromosomes from radish and cabbage. Therefore, the hybrid was an allopolyploid, a polyploid containing the chromosomes of different species. Although the allopolyploid was fertile, it never reached our dinner tables because it was a "radbage," not a "cabbish": it had distasteful, radishlike leaves and uninteresting, cabbagelike roots.

Although the cabbish experiment was a culinary failure, other allopolyploids have been successful. For example, the development of Western civilization depended largely on wheat, an allopolyploid that is a combination of three species that each contributed two sets of chromosomes. Thus, while you may never sink you teeth into a cabbish, your probably eat an allopolyploid every day.

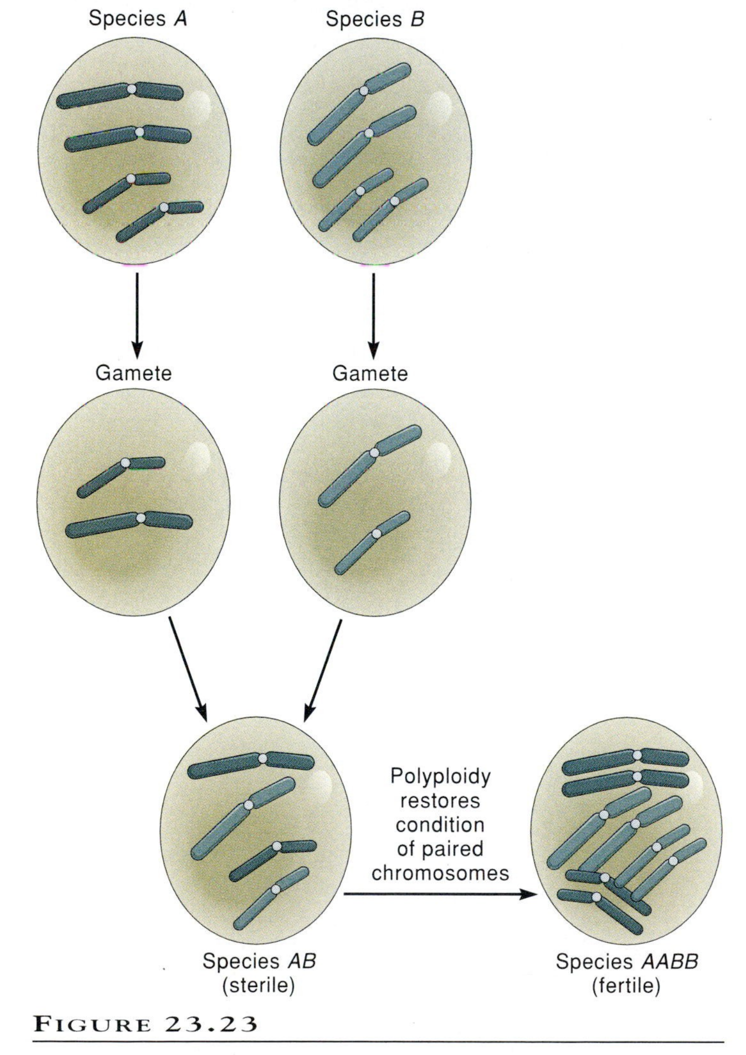

Figure 23.23

Restoration of fertility by polyploidy. Hybrids are often sterile because their chromosomes are unpaired. Polyploidy can restore the condition of paired chromosomes necessary for successful meiosis and sexual fertility.

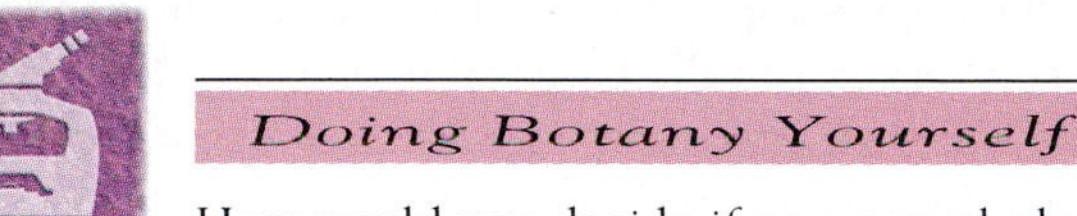

Doing Botany Yourself

How would you decide if an unusual plant that you've discovered is a new species?

Chapter Summary

Speciation—the origin of new species—is the central concept of evolutionary theory and is the mechanism that explains the vast diversity of organisms. Speciation is difficult to describe because it includes a variety of mechanisms; moreover a species is difficult to define clearly. The most general definition of a species is a group of individuals that can successfully breed with one another but not with members of other groups. This emphasizes reproductive isolation. Distinguishing between species and between the mechanisms that produce new species is further confounded because all members of a species are not identical.

During speciation, segments of the gene pool of a plant population diverge—sometimes rapidly through processes such as polyploidy and hybridization, and sometimes slowly through responses to natural selection pressures. As segments of the population become different, they lose some features and accumulate others that prevent one segment from interbreeding with another. This genetic accumulation and divergence is promoted by environmental variation, different mutations, and different rates of gene flow and genetic drift. Ultimately, members of two different populations acquire different traits and are reproductively isolated. They no longer produce fertile offspring (hybrids), and the two populations represent two species.

Biologists need a working definition of a species as a framework for questions and research. Morphologically, a species is a group of organisms that look the same and appear significantly different from other organisms. This definition is easily applied and communicated. Biologically, a species is one or more populations whose members can reproduce with each other in nature to produce fertile progeny, but cannot successfully reproduce with members of other populations and species. This reproductive isolation is frequently violated by hybridization. Genetically, a species is a group of organisms that exchange genes.

Reproductive isolation is maintained by prezygotic barriers that prevent fertilization of gametes from individuals of two different species, or by postzygotic barriers that result in sterile or unfit hybrid offspring. Prezygotic barriers include (1) geographical isolation of populations in distant areas, (2) microhabitat isolation of populations in the same general area, (3) temporal isolation of organisms receptive to mating at different times, (4) mechanical isolation of organisms that are structurally incompatible for reproduction, and (5) gametic isolation from incompatible gametes. Postzygotic barriers include (1) inviability of hybrid zygotes that cannot develop to sexual maturity, (2) sterile hybrids, and (3) the breakdown of successive hybrid generations that become weak or infertile.

For a subpopulation to form a new species, the gene pool must diverge from that of the parental population. Divergence during allopatric speciation of geographically separated populations is usually gradual and results from genetic differences accumulated through mutation, genetic drift, and selection. When divergence is great enough to eliminate interbreeding, then the two populations are different species. However, it is difficult to test the reproductive isolation of naturally separated populations, because the bringing together of potential parents by researchers introduces artificial conditions.

Populations that overlap geographically are sympatric. Speciation in these populations may begin with the formation of a cline, which is the pattern of variation among individuals of the same species distributed along an environmental gradient. Demands of the local environment along a cline may overshadow gene flow and lead to variation. Plants are morphologically plastic and display different growth forms in response to different environmental conditions, even without significant genetic change. If the environment is distinctly patchy, then the cline may also develop ecotypes, which are subpopulations subject to local selection and which diverge both genetically and morphologically. Significant divergence can lead to speciation.

Hybridization is the production of offspring from parents of different species, which produces new genetic combinations. New combinations may lead to speciation if the hybrids are fit and reproductively isolated from the parent populations. Plant hybrids are often sterile because sets of chromosomes are typically unbalanced and cannot pair during meiosis. However, many hybrids reproduce vegetatively and are especially successful in stressful or disturbed environments. Successful hybrids may be ecologically superior to their parent populations and become a new species. Hybrids that backcross with parent populations may introduce genes from one species to another by introgression.

Polyploids have more than two sets of chromosomes; they result from the nondisjunction of chromosomes during cell division. Polyploidy, which is common in higher plants, is an instantaneous mechanism of speciation because polyploids are reproductively incompatible with the parent population. Polyploids with multiple copies of the same sets of chromosomes are autopolyploids. Polyploids with two or more distinct sets of chromosomes are allopolyploids. Polyploidy may restore fertility to a hybrid by providing paired chromosomes.

Questions for Further Thought and Study

1. Shown at bottom of next page are the distributions of *Encelia* in the wash and slope habitats of the Sonoran Desert. How would you interpret these data?
2. Describe how reproductive contact between two populations may reinforce their genetic divergence.
3. When would the production of an offspring not necessarily indicate that the two parents were of the same species?
4. What role could a mountain range play in speciation?
5. What are the basic differences between sympatric and allopatric speciation?
6. Why are many hybrids sterile?
7. In what ways does the environment affect speciation?

8. The orchid shown below, *Angraecum sesquipedale*, has a long spur with nectar at its base. Darwin, aware that this unusual anatomy must have evolved with a particular pollinator able to reach it, predicted the existence of a moth with a long proboscis. The moth was discovered years later. What unusual flowers are you familiar with that have unusual structures and specific pollinators?

9. Why are hybrids often more successful in disturbed environments?

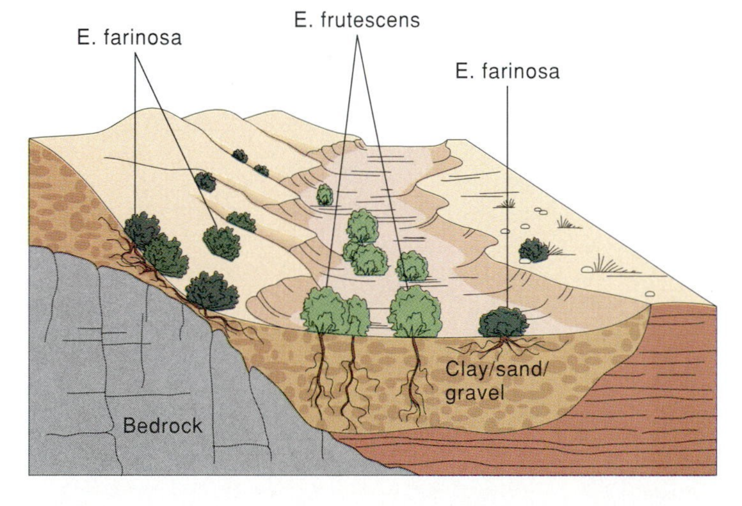

Suggested Readings

Articles

Elisens, W. J. 1989. Genetic variation and evolution of the Galápagos shrub snapdragon. *National Geographic Research* 5:98–110.

Masters, J. C. 1989. Why we need a new genetic species concept. *Systematic Zoology* 38:270–279.

Masterson, J. 1994. Stomatal size in fossil plants: Evidence for polyploidy in majority of angiosperms. *Science* 264:421–423.

May, R. M. 1992. How many species inhabit the earth? *Scientific American* 267:42–49.

Mayr, E. 1989. Speciational evolution or punctuated equilibria. *Journal of Social and Biological Structures* 12:137–159.

___. 1992. A local flora and the biological species concept. *American Journal of Botany* 79:222–238.

Rose, M. R., and W. F. Doolittle. 1983. Molecular biological mechanisms of speciation. *Science* 220:157–162.

Slatkin, M. 1987. Gene flow and the geographic structure of natural populations. *Science* 236:787–792.

Templeton, A. R. 1981. Mechanisms of speciation—A population genetics approach. *Annual Review of Ecology and Systematics* 12:23–41.

Books

Grant, V. 1981. *Plant Speciation*. 2d ed. New York: Columbia University Press.

Hancock, J. F. 1992. *Plant Evolution and the Origin of Crop Species*. Englewood Cliffs, NJ: Prentice Hall.

Margulis, L., and R. Fester, eds. 1991. *Symbiosis as a Source of Evolutionary Innovation: Speciation and Morphogenesis*. Cambridge, MA: MIT Press.

Otte, D., and J. A. Endler, eds. 1989. *Speciation and Its Consequences*. Sunderland, MA: Sinauer Associates.

White, M. J. D. 1978. *Modes of Speciation*. San Francisco: W. H. Freeman.

Unit Eight

Diversity . . .

Plants dominate our planet and our lives, thanks largely to their diversity and remarkable adaptations. In the previous unit we discussed how evolution produces this diversity. In this unit, we discuss the nature and extent of the diversity of plants. Specifically, we discuss the major groups of plants: bryophytes, ferns and their "allies," gymnosperms, and angiosperms. We also discuss algae, because of their presumed relationship to plants, and certain other organisms—bacteria and fungi—that are traditionally included in botany textbooks. Although bacteria and fungi are no longer classified as plants, botanists once thought they were.

Our treatment of diversity includes discussions of the distinguishing features of each group of organisms. You will also see how botanists use some of the techniques described in this chapter to construct evolutionary trees to understand how the diversity of different groups of organisms might have arisen. Moreover, you will learn about some of the unresolved problems concerning each group. We include these discussions to emphasize the changing nature of our understanding of plant diversity.

CAROLI LINNÆI

S:Æ R:GIÆ M:TIS SVECIÆ ARCHIATRI; MEDIC. & BOTAN.
PROFESS. UPSAL; EQUITIS AUR. DE STELLA POLARI;
nec non ACAD. IMPER. MONSPEL. BEROL. TOLOS.
UPSAL. STOCKH. SOC. & PARIS. CORESP.

SPECIES PLANTARUM,

EXHIBENTES

PLANTAS RITE COGNITAS,

AD

GENERA RELATAS,

CUM

DIFFERENTIIS SPECIFICIS,
NOMINIBUS TRIVIALIBUS,
SYNONYMIS SELECTIS,
LOCIS NATALIBUS,

SECUNDUM

SYSTEMA SEXUALE

DIGESTAS.

TOMUS I.

Cum Privilegio S. R. M:tis Sueciæ & S. R. M:tis Polonicæ ac Electoris Saxon.

HOLMIÆ,
IMPENSIS LAURENTII SALVII.
1753.

—Title page of Linnaeus' *Species Plantarum*.

CHAPTER

24

Systems of Classification

Chapter Outline

Chapter Overview

Classification is often viewed by nonspecialists as the discipline of biology that places organisms in their proper categories, where they remain permanently. Nothing could be further from the truth. Taxonomists, whose specialty is the discovery and classification of organisms, throughout history have had to revise and modernize classification systems as new species were discovered and as new knowledge about plants and other organisms was obtained in different disciplines of biology. The first classifications of plants, for example, made more than 2,000 years ago, dealt with hundreds of species and were based on features that could be seen with the naked eye. The number of known plant species is now in the hundreds of thousands, and taxonomists must integrate knowledge from genetics, ecology, anatomy, chemistry, physiology, and other areas into their classifications.

For many centuries, the first major dividing line among all living things was between plants and animals. Now we also recognize bacteria, fungi, and other kinds of organisms as equals to plants and animals at the highest levels of classification. The history of classification, therefore, is one of continual change. The subject of classification is more dynamic now than at any time in the past.

INTRODUCTION

Perhaps the first requirement for classifying plants—putting them into groups—is to name them. Once a plant has a name anyone using that name can expect other people to associate it with a specific plant. Although there are few records of plant names from more than about 4,000 years ago, you can nevertheless imagine that plant names were important to early humans for talking about plants that were used for medicines and foods. Plant names are still important for these reasons, but they are also important for tracking the diversity of the plant kingdom, for monitoring the effects of environmental change, for reporting botanical research, for cultivating crops and ornamental plants, and for all other aspects of the uses of plants.

Taxonomy is the science of discovering, describing, naming, and classifying organisms. If naming plants is a natural human endeavor, then everyone is to some degree or another a plant taxonomist. Many people also share a tendency to organize plants into groups. Edible plants, medicinal plants, and roses, for example, are recognized as groups. Rationales for grouping plants have varied over time, with some systems of classification having been influenced by religious or ethnic considerations. In northern Mexico, for example, the Tarahumara tribe classifies some plants on the basis of ceremonial uses involving hallucinogens. Tarahumarans group botanically unrelated plants such as *hikuli*—known to us as peyote cactus (*Lophophora williamsii*)—and *dowaka* (*Tillandsia benthamiana*—a member of the bromeliad family) because these plants are used together in ceremonies in the belief that their combined pyschoactivity is greater than that of either plant by itself.

Although informal classifications are widespread, they are generally not recorded, nor are they used outside of the tribes or small groups of people who invented them. The focus of this chapter is on classifications of plants that have been recorded and can be used to follow the historical development of taxonomic thought up to the present.

The history of classification is one of continual change. Change, which still occurs today, has come from the discovery of new species, the acquisition of new knowledge about the features of plants and other kinds of organisms, and the development of new philosophies and methods of classification. After reading this chapter, you will probably realize that classification has been, and continues to be, one of the most dynamic aspects of biology. You will also see that, partly because of its dynamic nature, the subject of classification tolerates a large diversity of ideas and opinions on how organisms should be classified.

THE EARLY HISTORY OF CLASSIFICATION

Any biological records of discoveries before the fourth century B.C. have either been lost or are too fragmentary to be useful. During the fourth century B.C., however, the famous Greek philosopher Aristotle wrote several scientific essays, primarily about animals. Unfortunately, all of Aristotle's writings about plants have been lost, and it was his brilliant student, Theophrastus of Eresus, who is credited with giving us the first known classification system for living organisms. In his *Historia Plantarum*,[1] Theophrastus described the parts, uses, and habitats of plants, and sorted them primarily by the different forms of their leaves and whether they were trees, shrubs, or herbs. In

1. *Historia Plantarum* is Latin for "History of Plants." Theophrastus and other early scientists of Western civilization wrote in Latin because it was the language of scholars. Major essays and books were published in Latin as recently as the late nineteenth century. Today, descriptions of new species of organisms are still published in Latin.

so doing he classified nearly five hundred plants. For several centuries after Theophrastus, many new classifications were written by Greek and Roman scholars. Probably the most important of these was *De Materia Medica,* meaning "about materials medicinal," written in the first century A.D. by the Greek physician Dioscorides. The purpose of that book, which included all six hundred medicinally useful plants known at the time, was to improve medical service in the Roman Empire. Nevertheless, it became the principal book of plant classification in Western civilization for nearly 1,500 years. Several European botanists and physicians used the work of Dioscorides during the fifteenth and sixteenth centuries as a basis for their *herbals,* which were illustrated books on the presumed medicinal uses of plants. Although some of the herbals had excellent drawings, they contained much folklore and many stories that became legends. The herbals also led to the development of the *Doctrine of Signatures,* which held that if a plant part resembled a part of the human body it would be useful in treating ailments of that part. For example, walnut meats, which resemble tiny brains, were used to treat brain disorders, and *Hepatica* leaves, whose lobes resemble those of the liver, were used for liver ailments.

Beginning in about the fifteenth century, classification became more complicated as European explorers returned to Europe with many new plants from other continents. As the plants accumulated, botanists began to turn their attention from medicinal plants alone to classifying and cataloging all types of plants. This effort was aided by the practice of categorizing each plant by a few unique features. By 1623 the Swiss botanist Gaspard Bauhin had used this method to classify about 6,000 kinds of plants. Many of the plants he classified had two-part Latin names. All of this occurred more than one hundred years before the great Swedish botanist Linnaeus introduced the binomial system of nomenclature, which we still use today.

Many botanists of the late seventeenth and early eighteenth centuries focused nearly all of their work on classification. To these scientists, to name a plant was to know a plant. Among these botanists was John Ray, whose three-volume *Historia Plantarum* was a detailed classification of more than 18,000 kinds of plants. In that book, Ray divided the flowering plants into dicotyledons and monocotyledons. Ray's classification was significant because it grouped plants on the basis of multiple similarities rather than on just a few key features. This was a notable advance in thinking, because it started to show the natural relationships of plants—what we would now call their evolutionary relationships—even though Ray still believed species to be unchanging. Ray's natural classification, coming more than a century before Charles Darwin's ideas about evolution, was prophetic; many of the plant families defined by Ray are still recognized today.

C O N C E P T

The earliest classifications of plants were primarily utilitarian classifications; that is, they included mostly medicinal or other useful plants. Beginning in about the fifteenth century, classifications became more complicated as thousands of new plants were discovered during worldwide exploration. These plants were classified on the basis of a few key features until the seventeenth century, when John Ray developed the first classification based on multiple features. Ray's classification showed natural relationships among plants.

Carolus Linnaeus

Carolus Linnaeus merits special attention in any discussion of the history of classification because he is credited with giving two-part scientific names to organisms, which taxonomists still do today (Linnaeus even converted his original Swedish name, Carl von Linné, to the two-part Latin version we use today). His system is called the **binomial system of nomenclature** because all of the scientific names of organisms have two parts; that is, they are *binomials.*

Linnaeus's classification was published in 1753 in a two-volume set called *Species Plantarum* ("Species of Plants"), which included about 7,300 kinds of plants. Linnaeus tried to do more than simply publish long lists of plants with Latin binomials. He also organized plants into twenty-four classes. These classes were based primarily on the features of stamens, including the number of stamens per flower, whether or not they were fused together, and whether or not they occurred on the same flower as the carpels. For example, plants in the class Triandria had three stamens per flower, and those of the class Monadelphia had fused stamen filaments. In addition, within each class of plants with flowers, Linnaeus recognized subunits based on the number of carpels. Organisms such as fungi and algae, which lack flowers, were placed in a class of their own.

Since Linnaeus's classification was based on a few reproductive features, it often did not reflect natural relationships, and is therefore referred to as an artificial classification. Nevertheless, the system was more convenient and more comprehensive than any other system available at that time.

When Linnaeus began his work, it was customary to use descriptive Latin phrase names for both plants and animals. The first word of the phrase constituted the **genus** (plural: **genera**) to which the organism belonged. For example, all known poplars were given phrase names beginning with the word *Populus.* Similarly, the phrases for willows began with *Salix,* those for roses with *Rosa,* and those for mints with *Mentha.*

FIGURE 24.1

Peppermint plants (*Mentha piperita*).

The complete phrase name for peppermint (fig. 24.1) was *Mentha floribus capitatus, foliis lanceolatis serratis subpetiolatis*, or "Mentha with flowers in a head, leaves lance-shaped, saw-toothed, with very short petioles." Although such more-than-a-mouthful names were specific, they were far too cumbersome to be useful.

In addition to including a referenced list of all the Latin phrase names previously given to plants, Linnaeus also changed some of the phrases to emphasize similarities among groups of plants, and he limited the phrases to a maximum of twelve words. Furthermore, he used such similarities for placing groups of plants in the same genus; each member of genus was called a **species.**[2]

In the margin next to the phrase, Linnaeus listed a word that, when combined with the genus name, formed a convenient abbreviation for a species. For example, Linnaeus adopted *Vitis*, the first word in the Latin phrases for grapes, as the genus name for grapes. The word *vinifera* was placed in the margin next to the phrase describing the common wine grape, and the word *vulpina* next to the phrase for the winter grape. In doing so, Linnaeus designated the abbreviated names for two species of grapes as *Vitis vinifera* and *Vitis vulpina*. Similarly, peppermint was designated *Mentha piperita*.

Although Linnaeus originally considered the phrase names to be the scientific names of plant species, he and those who followed him eventually replaced all the phrase names with abbreviated ones; that is, with binomials. Today all scientific names of plants are binomials. In addition, the complete scientific name also includes the initials or name of the person or persons who first described the species.[3] Accordingly, the scientific names of plants that were first described and named by Linnaeus still bear an *L.* after the binomial. Thus *Mentha piperita* L., *Plantago major* L. (common plantain), *Populus alba* L. (silver poplar), and *Hedera helix* L. (English ivy) are current scientific names that came from Linnaeus. Plant species that were discovered after Linnaeus, such as *Arctostaphylos canescens* Eastw. (hoary manzanita; fig. 24.2), bear reference to more recent taxonomists.[4] In this case, *Eastw.* is a standard abbreviation for Alice Eastwood, who was curator of botany at the California Academy of Sciences for the first half of the twentieth century. For more information about plants' names, see box 24.1, "How to Name a New Plant Species" on the following page.

FIGURE 24.2

Hoary manzanita (*Arctostaphylos canescens*) of California.

2. Note that *species* is like the word *sheep* in that it is spelled and pronounced the same way in either singular or plural usage. There is no such thing as a plant or animal specie.

3. In contrast to plant names, the full scientific names for animals do not include references to authors of species names.

4. In scientific publications the author of a plant species is supposed to be included with the scientific name the first time the binomial is mentioned. Thereafter it is not necessary to include the author. However, the editors of some journals prefer omission of the name's authority altogether. This usually does not cause confusion unless the same scientific name has been used by two different authors for two different plants, or unless two authors have given different names to the same plant.

Boxed Reading 24.1

How to Name a New Plant Species

New species are discovered every day, usually in little-explored regions of the world such as tropical rain forests. New plant species, however, may be found almost anywhere that wild plants grow. When a botanist finds a plant that does not match the description of any known plant species, a new species may have been discovered. Because a new species usually fits into an existing genus, the botanist needs to show how the new plant differs from other species in the genus, and give it a name. Botanists are guided in such a task by the *International Code of Botanical Nomenclature,* which is the rulebook for naming new plants. The new name must be one that has not been used for another plant and must be accompanied by a description of the species in Latin and also a description in the language of the species' author. The name and description must be published in an acceptable journal to establish the validity of the new name. Editors of such journals send the description to specialists to verify that the species is a new one and that the name is unique.

New species must also be represented by a single specimen, called the type specimen, that was used to describe the species. Specimens (including type specimens) are generally made by pressing the freshly collected plants as flat as possible between sheets of newspaper and cardboard, drying them, and then mounting them on large sheets of paper for safekeeping in a herbarium. A herbarium is like a library of preserved plant specimens. Large herbaria, such as that of the New York Botanical Garden or the Gray Herbarium of Harvard University, maintain millions of specimens, thousands of which are type specimens.

Herbarium collections are vital for scientific research. They are especially useful for comparing new species with previously known ones, for keeping records of the discovery of new species, for keeping track of species migration and extinction, and for knowing which species are becoming rare or endangered due to human activities.

Box Figure 24.1

A type specimen and a description of a new species.

Many plants have two authorities listed after the Latin binomial. The full scientific name of the giant sequoia, for example, is *Sequoiadendron giganteum* (Lindl.) Buchh. In this instance John Lindley, a nineteenth-century English botanist first described the giant sequoia and gave it the binomial *Wellingtonia gigantea*. However, the name *Wellingtonia* had already been used for a genus of dicots, and so Illinois botanist John Buchholz later transferred the giant sequoia to the genus *Sequoiadendron*. Since both botanists helped form the scientific name by which the giant sequoia is currently known, both are cited after the Latin binomial. The author of the original description is given in parentheses.

C O N C E P T

Linnaeus is credited with the first use of the binomial system of nomenclature. His classification of plants was based primarily on features of stamens, which he used to organize plants into twenty-four classes. Because Linnaeus's system was based on just a few reproductive features, it was considered to be an artificial system of classification.

The Lore of Plants

Honoring a "Disregarded" Botanist

When he was 25 years old, Swedish botanist and explorer Carolus Linnaeus (1707–1778) was paid by the Swedish Royal Society of Science to make the first scientific survey of the plants, animals, and geology of Lapland, a vast region of northern Europe almost entirely within the Arctic Circle. North of Gävle on the central east coast of Sweden, Linnaeus came upon the twinflower, a woodland plant common in the cold, swampy, coniferous forests of that area. The twinflower (see photo at right) is easily recognized as a creeping, dainty, shade-loving plant that, during May and June, produces pairs of pink to purplish, nodding flowers.

Although twinflower had already been named by another botanist, it was later renamed after Linnaeus. Today twinflower remains known as *Linnaea borealis* and is the only species of the genus (in North America two varieties of the species are recognized). The Latin binomial of twinflower denotes its taxonomic history: *Linnaea* for Linnaeus, and *borealis* for its boreal (i.e., northern) location. Although Linnaeus is now regarded as a giant of biology's past, with his work having been the foundation for the science of modern taxonomy, he may not have had great self-esteem. As he later wrote, "*Linnaea* ... a plant of Lapland, lowly, insignificant, disregarded, flowering for but a brief space—from Linnaeus who resembles it."

Early Post-Linnaean Classifications

By the end of the eighteenth century, botanists had begun to oppose the artificial system of Linnaeus because his system often placed unrelated plants together. For example, cherries and cacti were placed in the same class because they both have many stamens per flower. Instead, the development of a natural system became a major goal of classification. As in Ray's earlier classification, however, the idea of a natural system had no evolutionary foundation. Rather, natural systems of classification were meant to reflect the divine creation of related groups.

The most significant of the natural systems of classification after Linnaeus included *Genera Plantarum*, "Genera of Plants," by French botanist Antoine Laurent de Jussieu; *Prodromus Systematis Naturalis Regni Vegetabilis*, "Forerunner of the Natural System of the Vegetable Kingdom," which was started by Swiss botanist Augustin Pyrame de Candolle in 1824 and finished by others fifty years later; and another *Genera Plantarum*, this one published from 1862 to 1883 by two Englishmen, George

Twinflower (*Linnaea borealis*).

Boxed Reading 24.2

Benefits of a Phylogenetic Classification of Life

Systematics is the scientific study of organisms and their phylogenetic relationships. Scientists who specialize in systematics are called *systematists.* Recently, systematists worldwide have begun to organize a campaign to classify all organisms on earth in a comprehensive phylogenetic system before the year 2020. This monumental effort has at least six goals that should have major benefits to science and human society:

1. To guide the search for new crop species, novel genes, more biological control agents, and new products from organisms.
2. To help resource managers, conservation biologists, and policy makers set environmental priorities.
3. To show how all fields of biology can be linked through comparative, interdisciplinary studies encompassing all groups of organisms.
4. To provide a basis for managing biological knowledge and for communicating it across science.
5. To establish a framework for estimating rates of extinction and of global environmental change.
6. To provide the evolutionary context for understanding the history of life.*

The goals of this multinational project are especially ambitious because there may be several million species yet to be discovered, hundreds of thousands of which may be plants.

Systematists estimate that the research needed to obtain a phylogenetic classification of life within the next twenty-five years will cost about 3 billion dollars per year. Currently about 500 million dollars are spent annually on global systematics. At the present rate of research funding and species extinction, systematists estimate that more than half of the species or organisms now living will be extinct before the project can be completed. Moreover, they estimate that most extinctions will be of undiscovered species.

* Adapted from "Systematics Agenda 2,000: Charting the Biosphere," a pamphlet written by internationally renowned systematists to publicize their concerns about the rapid worldwide extinction of species due to human activities.

Bentham and Sir Joseph Dalton Hooker. De Jussieu's *Genera Plantarum* is credited with creating wide acceptance of the idea of natural systems. De Candolle's *Prodromus* was essentially a large-scale expansion of Linnaeus's *Species Plantarum*. Nevertheless, the *Prodromus* included many new families and original descriptions of plants, several of which remain the only worldwide treatment of certain groups. Bentham and Hooker's classification was based on the natural systems of de Jussieu and de Candolle, but Bentham and Hooker used their own descriptions of plants rather than relying on previously published ones. Partly for this reason, Bentham and Hooker's treatment remains an important resource for plant taxonomists today.

Modern Classifications

Until the latter half of the nineteenth century, virtually all classification systems were based on the belief that living organisms had not changed since their creation and would not change in the future. As the ideas of Charles Darwin and Alfred Wallace spread and became accepted, however, these systems fell into disfavor, and classifications based on evolutionary relationships began to be developed.

Classifications that try to reflect evolution are said to be **phylogenetic.** Taxonomists consider phylogenetic systems to be superior for several reasons (see box 24.2). However, taxonomists do not agree on which phylogenetic system is the best. Accordingly, various phylogenetic classifications have been proposed, each differing in its basic assumptions about the features that best reflect the phylogeny of plants.

The first major phylogenetic systems of plant classification were proposed around the turn of the twentieth century. The most complete is *Die natürlichen Pflanzenfamilien*, "The Natural Plant Families," first published from 1887 to 1915 by the German botanists Adolf Engler and Karl Prantl. Engler and Prantl recognized about 100,000 species of plants, which were organized in their presumed evolutionary sequence.[5] To begin such a sequence, Engler and Prantl first assumed that flowers lacking petals, such as those of walnuts, willows, and oaks, were primitive, and that monocots were more primitive than dicots (fig. 24.3). Engler and Prantl's system is still widely used; the twelfth edition of their classification was published in 1964. However, current ideas about the nature of primitive flowers differ considerably from those of Engler and Prantl (see Chapter 30).

Classifications of the twentieth century also marked the beginning of another trend in major treatments of plants: the taxonomic focus in major works changed from being comprehensive for all plants to being specialized for different groups of plants. The flowering plants have received the most attention because they are the largest and most diverse of all plant groups. For this reason, the most significant new ideas about plant classification in the twentieth century have come from studies of flowering plants.

5. Note that by the beginning of the twentieth century, major scientific papers were no longer written exclusively in Latin. Engler and Prantl's work, for example, was published in German.

Figure 24.3

Male flowers of oak (*Quercus*). These flowers have no petals.

Figure 24.4

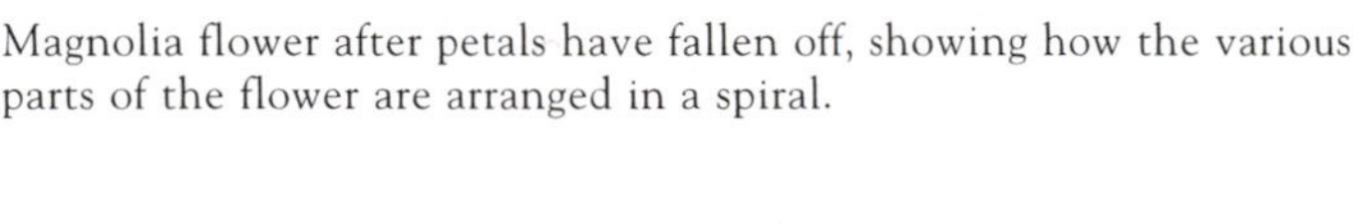

Magnolia flower after petals have fallen off, showing how the various parts of the flower are arranged in a spiral.

The first American to contribute significantly to the development of systems of classification was Charles Bessey, a botanist at the University of Nebraska. Bessey rejected some of Engler and Prantl's basic assumptions about matters such as the nature of primitive flowers, and replaced them with his own. Bessey assumed, for example, that flowers such as magnolias, which have many separate spirally arranged parts, were the primitive ancestors of all other flowers (fig. 24.4).

Although Bessey's system was published in 1915, many of his ideas are still followed today. Nevertheless, many new phylogenetic systems of classification have been published since Bessey's system was first proposed. All of these systems stress what their authors believe are the primitive features of plants versus those that have undergone evolutionary change; that is, they emphasize primitive features versus derived features. These more recent classifications also draw lines of descent between related groups of plants (fig. 24.5).

At this point you may be wondering why there are so many phylogenetic classifications of plants. Why, in other words, if there is general agreement about the desirability of phylogenetic systems, is there no such general agreement about which system is best? There are at least two good answers to this question. One is that classification involves a subjective selection of characters (features) and a subjective evaluation of their importance for designating species, genera, families, or any other taxonomic grouping. A good example of this subjectivity is the classification of flowering plants whose fruits are *legumes* (see Chapter 17). Legume-producing plants can be divided into three groups on the basis of different flower types. One type is exemplified by the common pea (*Pisum*) flower, which is bilaterally symmetrical and has five petals, the middle petal being larger and exterior in the bud to the other four (fig. 24.6). A second type, represented by honey locust (*Gleditsia*), also has a bilaterally symmetrical flower with five petals, but the middle petal is

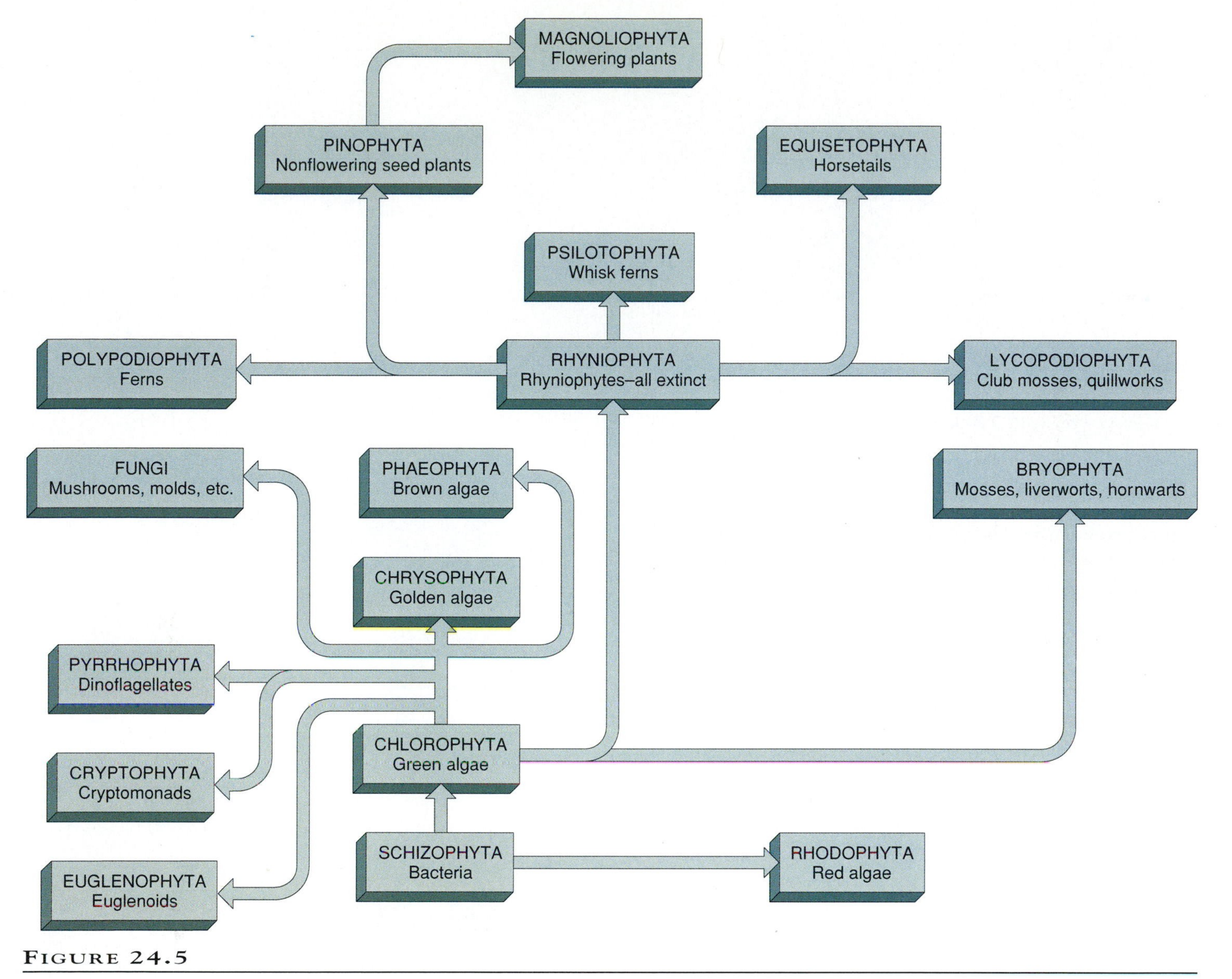

FIGURE 24.5

Recent example of a traditional classification and phylogeny of plants (by the late Arthur Cronquist, formerly of the New York Botanical Garden).

not larger and is interior in the bud to the other four. Flowers of the third type, represented by *Acacia* and *Mimosa,* are radially symmetrical. Taxonomists who consider the fruit (i.e., legume) the unifying feature recognize a single family, the Fabaceae, with three subfamilies based on flower type. Other taxonomists, who consider differences in flower types more significant, divide the group into three families: the Fabaceae (*Pisum*-type), the Caesalpiniaceae (*Gleditsia*-type), and the Mimosaceae (*Mimosa*-type). Taxonomists who habitually take the more conservative view, such as considering all leguminous plants in one family, are often called *lumpers.* Those who divide families or other groups more liberally are often called *splitters* (fig. 24.7). Both views are widely accepted, although heated debates over lumping or splitting specific groups have been common.

A second explanation for the number of phylogenetic classifications of plants comes from differing views of taxonomists on how to derive evolutionary relationships from the features, or characters, of plants. Some make intuitive judgments, whose underlying assumptions about the evolution of characters often are not obvious and cannot easily be scientifically evaluated. Many taxonomists, therefore, consider traditional classifications to be more artistic than scientific, despite the fact that most of the recent classifications of flowering plants as a whole have been traditional. Currently, four of these traditional classifications are widely accepted: those of Americans Robert Thorne and Arthur Cronquist; that of Russian botanist Armen Takhtajan; and that of Swedish botanist Rolf Dahlgren. These four systems differ from earlier ones primarily in their use of sophisticated types of data, which were unavailable to

FIGURE 24.6

Diagrams of the three flower types of Fabaceae: (a) *Pisum*-type, (b) *Gleditsia*-type, and (c) *Acacia*-type.

FIGURE 24.7

This diagram shows how Dr. Lumper and Prof. Splitter would differ in their classifications of the legume-bearing plants shown in figure 24.6.

earlier taxonomists. The newer types of data, for example, include information from genetics, ecology, anatomy, chemistry, and physiology of plants. Despite the sophistication of such data, however, intuitive assumptions about the significance of characters have been made in the same way assumptions were previously made.

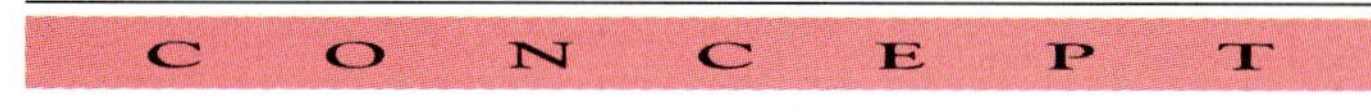

Post-Linnaean classification systems were said to be natural on the basis of nonevolutionary views of relationships among plants. Conversely, modern systems have attempted to be phylogenetic. Several different phylogenetic classifications of flowering plants are currently used widely by different taxonomists.

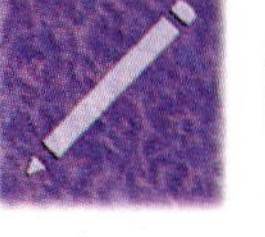

Biologist Stephen Jay Gould said the following about taxonomy:

> *Taxonomy is often regarded as the dullest of subjects, fit only for mindless ordering and sometimes denigrated within science as mere "stamp collecting." ... If systems of classification were mere hat racks for handling the facts of the world, this disdain might be justified. But classifications both reflect and direct our thinking. The way we order represents the way we think. Historical changes in classification are fossilized indicators of conceptual revolutions.*

What was Gould saying? Do you agree or disagree? Why?

CLADISTICS

Systematists are now developing new approaches to phylogenetic classifications. These approaches are more explicit in their assumptions and are scientifically testable. Perhaps the most widely adopted of the new approaches of the past two decades is one called **cladistics.** Cladistics is generally defined as a set of concepts and methods for determining **cladograms,** which are the branching patterns of evolution (fig. 24.8).

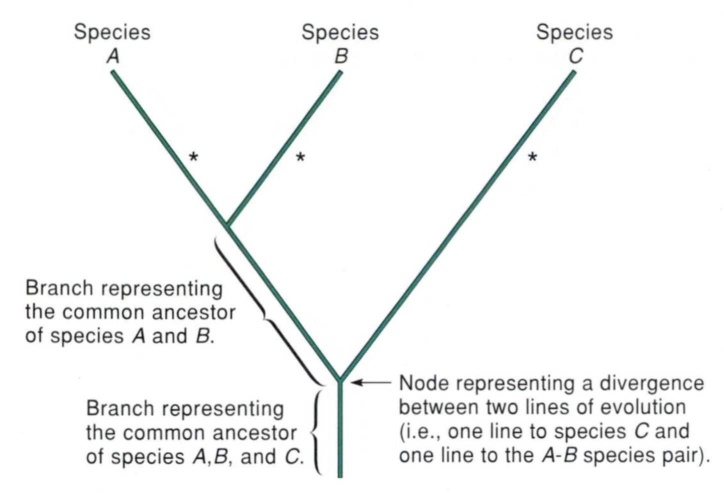

FIGURE 24.8

This cladogram, shows the phylogenetic relationships of three species, all of which arose from a single ancestor. As shown here, Species A and Species B are more closely related to each other than either is to Species C.

A.

B.

FIGURE 24.9

Two possible states of the character, *chemical type of red pigment*. (a) The flowers of roses (*Rosa* sp,) contain an anthocyanin-type red pigment. (b) The red flowers of this hedgehog cactus (*Echinocereus triglochidiatus*) contain a betalain-type red pigment.

For cladistics, a *character state* is the form or value of a character. For example, the character could be flower color, and the state could be red or blue. More specifically, the pigment chemical-type for red pigments is a character, whose states are either a *betalain-type* chemical or an *anthocyanin-type* chemical (fig. 24.9). The states, or values, of each character must be explicit. Furthermore, once characters and their states are declared, much emphasis in cladistics is placed on assumptions about which states are primitive (ancestral) and which are derived (evolved). If the anthocyanin type of pigment chemical is assumed to be primitive, then the betalain type is derived, and vice versa. By assuming anthocyanin type to be primitive we can see that plants with betalain-type pigments have a recent common ancestor in which betalain pigments first appeared (fig. 24.10a). In this case, the occurrence of betalain pigments indicates an evolutionary change that was passed from an ancestral betalain producer to its descendant betalain producers.

What if, instead, we assume the betalain type to be primitive? The same cladogram would then show two evolutionary origins of the anthocyanin type, one in the flowering plants and one in the nonflowering seed plants. Alternatively, another cladogram based on the evolution of pigment chemical-types could be made (fig. 24.10c). The importance of this alternative cladogram is that it represents an alternative hypothesis about phylogenetic relationships: it minimizes the number of times the anthocyanin type arose, and shows instead that flowering plants had two evolutionary origins. As an alternative hypothesis, it can be tested by evaluating additional characters. In cladistics, testing hypotheses entails finding additional characters whose states may corroborate some cladograms but not others. Those that are not corroborated by additional data can be rejected, and those that are corroborated can be maintained for further testing.

Assume for the moment that figure 24.10a represents the correct direction of the evolution of pigment chemical-types in seed plants. If so, it shows that of the two character states for pigment type, only the derived state indicates relatedness. All of the plants with betalains are said to have the *shared-derived* character state, which shows that plants producing betalains are more closely related to each other than they are to plants producing anthocyanins. Conversely, the *shared-primitive* character state does not show such a close relationship between plants producing anthocyanins; these plants had already diverged into flowering

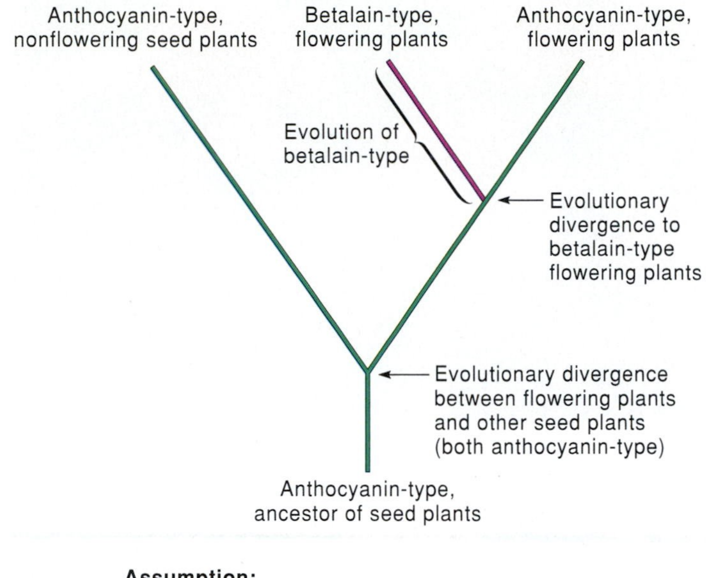

Assumption:
Anthocyanin-type (primitive) → Betalain-type (derived)

A.

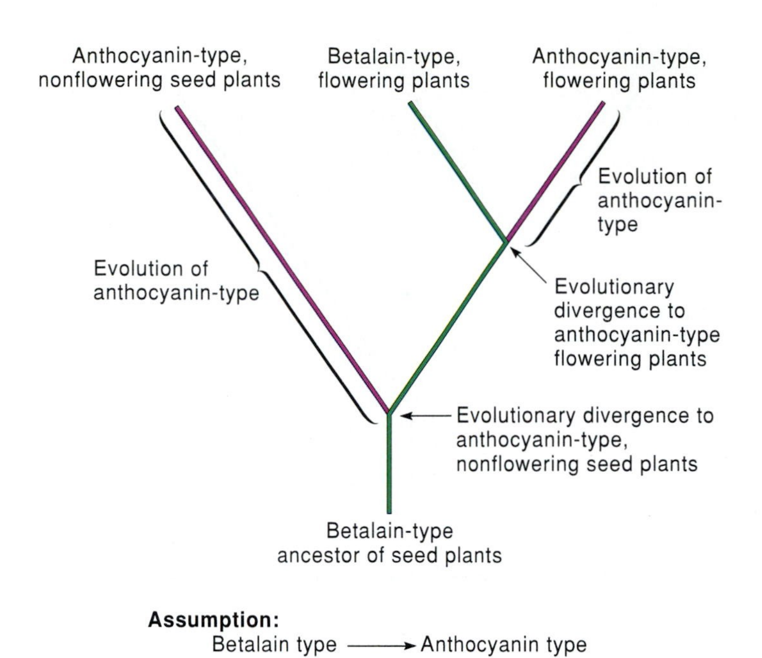

Assumption:
Betalain type (primitive) → Anthocyanin type (derived)

B.

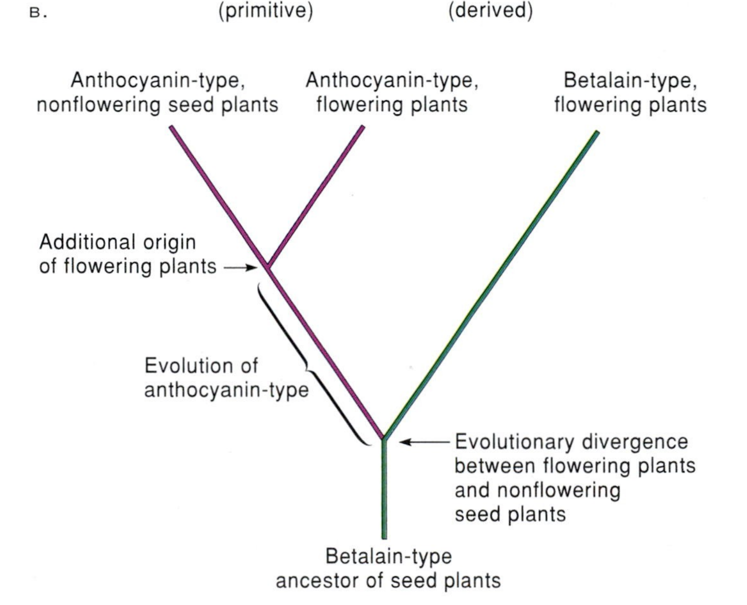

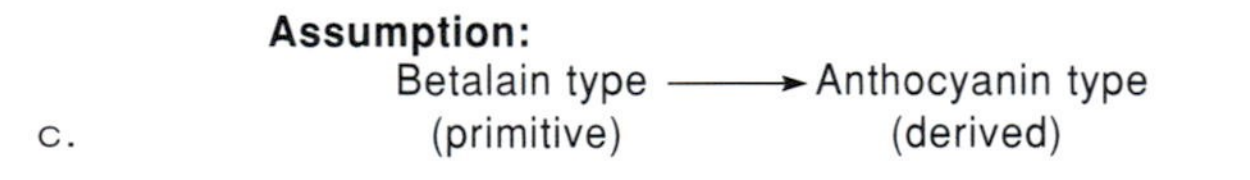

C.

Figure 24.10

Alternative cladograms for plants with anthocyanins or betalains. (a) Cladogram built by assuming that anthocyanins were primitive and betalains were derived. (b) Cladogram showing how character evolution would look if betalains were primitive and anthocyanins were derived twice. (c) Alternative cladogram showing that betalains were primitive and anthocyanins were derived once.

plants and nonflowering seed plants before betalains evolved. This example illustrates a significant assumption of cladistics and one that distinguishes cladistics from other phylogenetic methods: Only shared-derived character states indicate phylogenetic relationships; shared-primitive character states do not. This contrasts with the assumption that overall similarity can be used to show evolutionary relatedness—an assumption that remains common in most large-scale phylogenetic classifications.

Cladistics is a set of concepts and methods that are represented in evolutionary branching diagrams called *cladograms*, which express hypotheses about evolutionary relationships among organisms. The evolutionary relationships shown in cladograms are based on shared-derived character states.

Doing Botany Yourself

Devise a system for classifying plants around the building in which your botany class meets. What features did you use to classify the plants? Which, if any, of these features were more important than others? Why? How could your system be improved so that it would apply to plants around other buildings on campus?

How Do We Choose the Best Cladogram?

The alternative cladograms based on pigment chemical-type show that betalain biosynthesis arose once in evolution (fig. 24.10a), that anthocyanin biosynthesis arose twice in evolution (fig. 24.10b), or that anthocyanin biosynthesis arose once, but flowering arose twice (fig. 24.10c). Anthocyanin biosynthesis involves a complex metabolic pathway, and it is intuitively hard to believe it could have two independent evolutionary origins, or that flowers and all their associated features arose twice. The simpler scenario, that each of these arose only once, is somehow more satisfactory. Such intuition is consistent with a principle of logic called Occam's razor, which is attributed to a fourteenth-century English philosopher named William of Occam. This principle states that a person should not make more assumptions than the minimum needed to explain anything. The principle is often called the principle of **parsimony,** which is what systematists call it as it is applied to cladistics.

In cladistic terms, the principle of parsimony can be restated to say that the best cladogram requires us to make the fewest assumptions about evolutionary changes. Parsimony is especially powerful when there can be many possible cladograms for a group of organisms. The number of possible cladograms would be easy to evaluate for three kinds of organisms (3 possibilities), for four kinds (26 possibilities), or maybe even for five kinds (125 possibilities). However, the number of imaginable cladograms increases more than exponentially as the number of different organisms increases. Indeed, there are more than 7 *trillion* possible cladograms that can be made for fifteen kinds of organisms.

Cladistics and Classification

The goal of cladistics or any other approach to making phylogenies is to evaluate and improve classifications. Opinions vary, however, as to how the branching points on a cladogram should be used to designate taxonomic levels, and usually will not convince splitters to lump or lumpers to split. In spite of its newness and the competing ideas about how it should be applied to classification, however, cladistics has been used to infer the phylogenetic relationships of many different plant groups. Several of these phylogenies will be helpful in learning about different groups as they are discussed in the following chapters in this unit.

FIGURE 24.11

The red-capped *Amanita* (*A. muscaria*) mushroom. Although the red pigment in this fungus resembles that in the plants shown in figure 24.9, this feature alone does not mean that fungi are closely related to plants.

A THEORETICAL FOUNDATION OF BIOLOGICAL CLASSIFICATION

Several basic assumptions, or *postulates*, about classification can be derived from the foregoing sections of this chapter. Some of these postulates are more obvious than others, but the main ones so far are as follows:

1. Organisms exhibit different degrees of similarities and differences among individuals and groups (fig. 24.11).
2. Those organisms that have many similarities are a species.
3. Species that share some of their features comprise a genus.
4. On the basis of their shared features, similar genera can be organized into a family; likewise, families and larger groups can be organized into successively higher levels of a taxonomic hierarchy.
5. The greater the similarity among organisms and among groups, the closer is their evolutionary relationship.

   OR

   The greater the number of shared-derived character states among organisms, the closer is their relationship.

These postulates function together as the theoretical foundation of biological classification. As such, they can be called the *theory of biological classification*. Like all modern theories of biology, the theory of biological classification is constantly being tested and modified. Systematists, for example, are especially interested in what a species is (postulate 2). There is no general agreement about how many similarities are enough to call a group of organisms a species. Postulate 2 is often modified to include a reproductive component: Species are an interbreeding or potentially interbreeding population. This addition to the postulate is more applicable to animal species, however, because plant species often hybridize with one another but are still considered separate species (see Chapter 23).

TABLE 24.1

The Taxonomic Hierarchy for Four Species of Plants

Kingdom:	Plantae	Plantae	Plantae	Plantae
Division:	Anthophyta	Anthophyta	Anthophyta	Coniferophyta
Class:	Liliopsida	Liliopsida	Magnoliopsida	Coniferopsida
Order:	Zingiberales	Commelinales	Fagales	Coniferales
Family:	Musaceae	Poaceae	Fagaceae	Pinaceae
Genus:	*Musa*	*Hordeum*	*Quercus*	*Pinus*
Species:	*Musa acuminata*	*Hordeum vulgare*	*Quercus alba*	*Pinus ponderosa*

Note: The common names of these species are, left to right, banana, barley, white oak, and ponderosa pine. The Anthophyta are the flowering plants; the Coniferophyta are the conifers. Liliopsida is the current name for the monocots; Magnoliopsida is the name for the dicots.

Postulate 4 is most often applied to form a hierarchy that has species at the lowest level and kingdom at the highest level. According to such a hierarchy, species are organized into genera, genera into families, families into orders, orders into **classes,** classes into **divisions,**[6] and divisions into kingdoms.

Each category is a **taxon** (plural, **taxa**), which is a general term for any level of classification, such as species, genus, or family. A good way to envision the postulate of a taxonomic hierarchy is to see how it looks for a few examples, as shown in table 24.1 and figure 24.12.

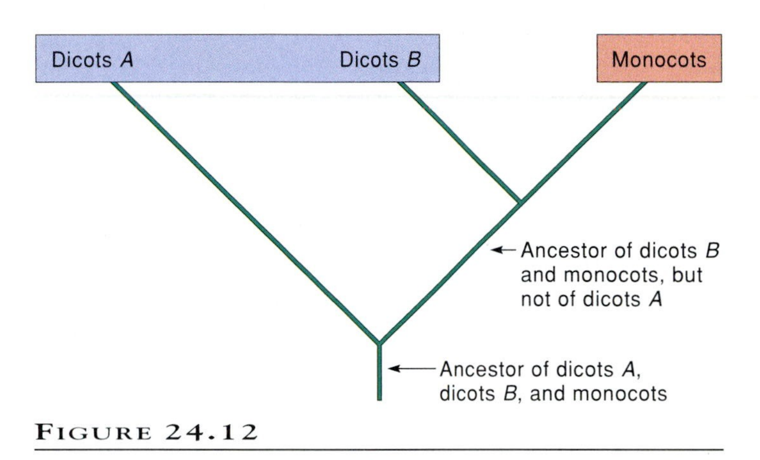

FIGURE 24.12

Cladogram of flowering plants. What do you conclude from this cladogram?

CLASSIFICATION OF KINGDOMS

Botany does not exist in a vacuum; plant biology is best understood as it relates to the biology of other kinds of organisms. With regard to diversity, this means that the classification of plants as a group is best understood as it relates to the classification of animals, bacteria, and other groups of organisms. In finding out what plants are and where they came from—that is, in defining the plant kingdom—it is helpful, therefore, to consider relationships of members of the plant kingdom with members of other kingdoms.

A good starting place for evaluating relationships of plants to organisms of other kingdoms might be to determine what kingdoms there are and how the organisms are similar to or different from plants. Before the middle of the twentieth century almost everyone viewed living organisms as either plants or animals. Those who didn't, and who classified living organisms in three or four kingdoms, were ahead of their time, and were generally not followed by their colleagues. In the 1950s, however, biologists began to consider features of the biochemistry and ultrastructure of cells as fundamental to classification, and finally started accepting the idea that living things comprised more than two kingdoms. Some even went so far as to divide all living things into thirteen kingdoms.

The Five-Kingdom System of Classification

Currently, classification at the kingdom level generally includes plants, animals, fungi, and bacteria in four separate kingdoms, with a hodge-podge of other kinds of organisms thrown together into a fifth kingdom. The first four kingdoms

6. Division is equivalent to the term **phylum,** which is used in classifying animals. Although the use of the word *phylum* for all living organisms is supported by many taxonomists, the *International Code of Botanical Nomenclature*, which governs the use of plant names, presently permits only the use of the word *division* for that taxonomic level of plants.

Table 24.2

Major Characteristics of the Five Kingdoms

	Monera	Protoctista	Fungi	Plantae	Animalia
Cell type	Prokaryotic	Eukaryotic	Eukaryotic	Eukaryotic	Eukaryotic
Genetic material	DNA not associated with protein in chromosomes	DNA associated with protein in chromosomes in most groups	DNA associated with protein in chromosomes	DNA associated with protein in chromosomes	DNA associated with protein in chromosomes
Gene structure	Introns absent in most groups, present in some	Introns present	Introns present	Introns present	Introns present
Nuclear envelope	Absent	Double or single	Double	Double	Double
Membrane-bound organelles	Absent	Present	Present	Present	Present
Chloroplasts	Absent	Present or absent	Absent	Present	Absent
Cell wall	Noncellulosic	Absent or present; cellulosic or various types	Chitinous or cellulosic	Cellulosic	Absent
Means of genetic recombination	Conjugation, transduction, or plasmid transfer	Fertilization and meiosis in most groups	Fertilization and meiosis	Fertilization and meiosis	Fertilization and meiosis
Mode of nutrition	Autotrophic or heterotrophic by absorption	Autotrophic or heterotrophic by absorption or by phagocytosis	Heterotrophic by absorption	Autotrophic	Heterotrophic by ingestion
Flagella/cilia	Solid, rotating	9+2 microfibrils	Absent	Absent or 9+2 microfibrils in gametes of some groups	9+2 microfibrils
Multicellularity/cell specialization	Absent	Absent in most, present in algae and water molds	Present	Present	Present
Nervous system	Absent	Absent or simple	Absent	Absent	Present, often complex
Respiration	Aerobic or anaerobic	Aerobic	Aerobic or anaerobic	Aerobic	Aerobic
Life cycle	Haploid; fission	Mostly haploid, some forms diploid	Mostly haploid; some alternation of generations	Alternation of generations	Diploid except for gametes
Unicellular spore formation	Present in some groups	Present in some groups	Present in all groups	Present in all groups	Absent

are referred to as the **Kingdom Plantae,** the **Kingdom Animalia,** the **Kingdom Fungi,** and the **Kingdom Monera,** respectively. Organisms in the hodge-podge group are in the **Kingdom Protoctista.** Robert H. Whittaker is credited with the suggestion, in 1969, that all living things could be divided into five kingdoms. His classification, based on numerous features, is summarized in table 24.2. A more recent and more comprehensive five-kingdom classification, proposed by Lynn Margulis and Karlene Schwartz, is shown in figure 24.13; it provides a phylogeny that was based on traditional methods of systematics.

Other Hypotheses for the Classification of Kingdoms

Phylogenetic hypotheses like that of Margulis and Schwartz are continually being scrutinized, especially by systematists who use molecular data for inferring evolutionary relationships. Molecular data, such as nucleotide sequences of genes or amino acid sequences of proteins, can be compared in the same way that morphological data are compared, either by overall similarity or by cladistic methods. Nucleotide or amino acid sequences are obtained for the same gene or the same protein from each of the organisms in the group. The characters in nucleotide sequences are the positions of the nucleotides along the sequence; the character states are whatever the nucleotide is at each position (i.e., adenine, guanine, cytosine, or thymine). For proteins, amino acid position is the character, and the character state can be any one of twenty amino acids.

A phylogeny based on molecular data is called a *molecular phylogeny.* Several molecular phylogenies have been proposed for the kingdoms of organisms each differing in the gene

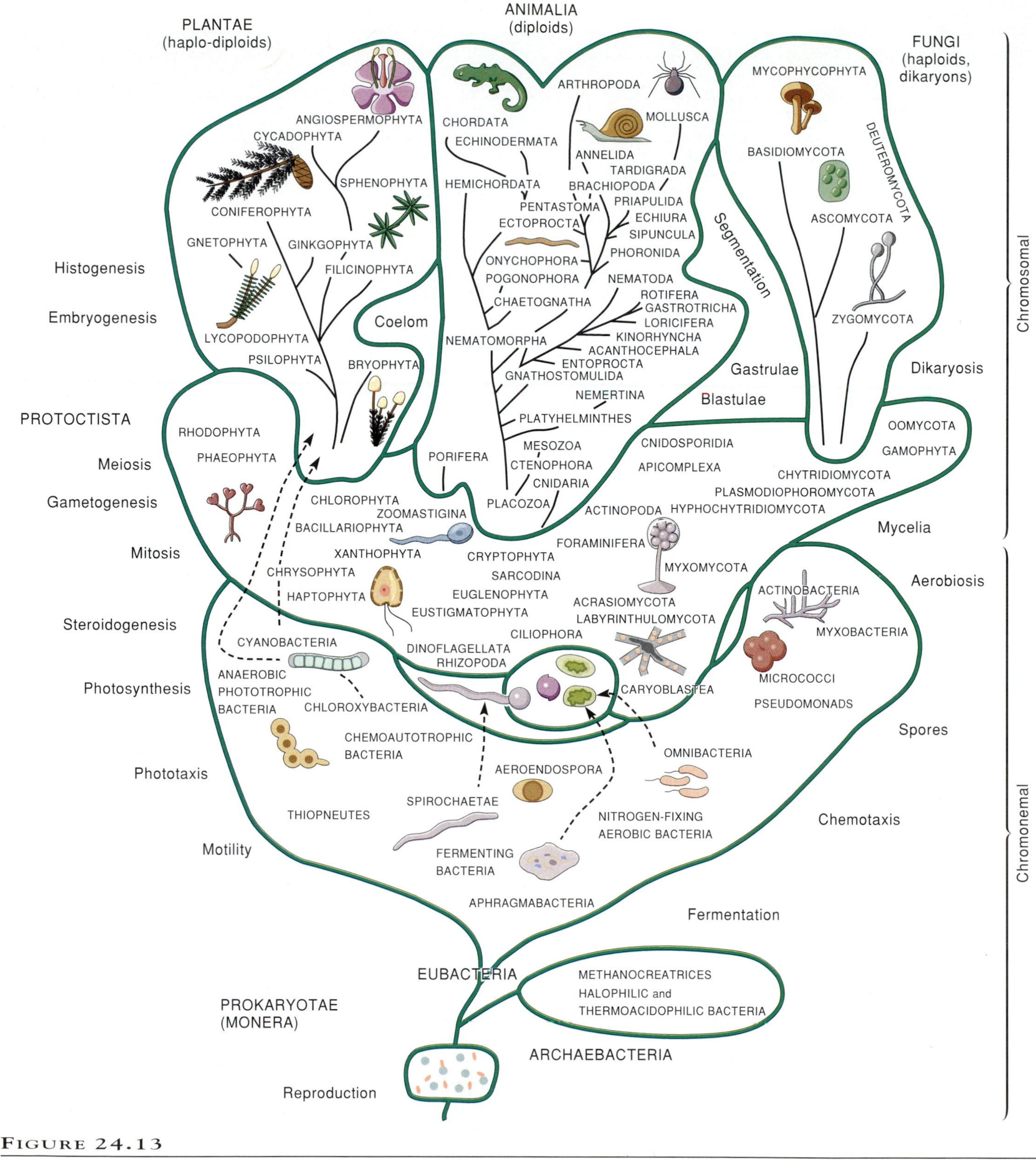

FIGURE 24.13

Classification and phylogeny of five kingdoms by Margulis and Schwartz.

From Five Kingdoms 2/e by Margulis and Schwartz. Copyright © 1988 by W. H. Freeman and Company. Reprinted with permission.

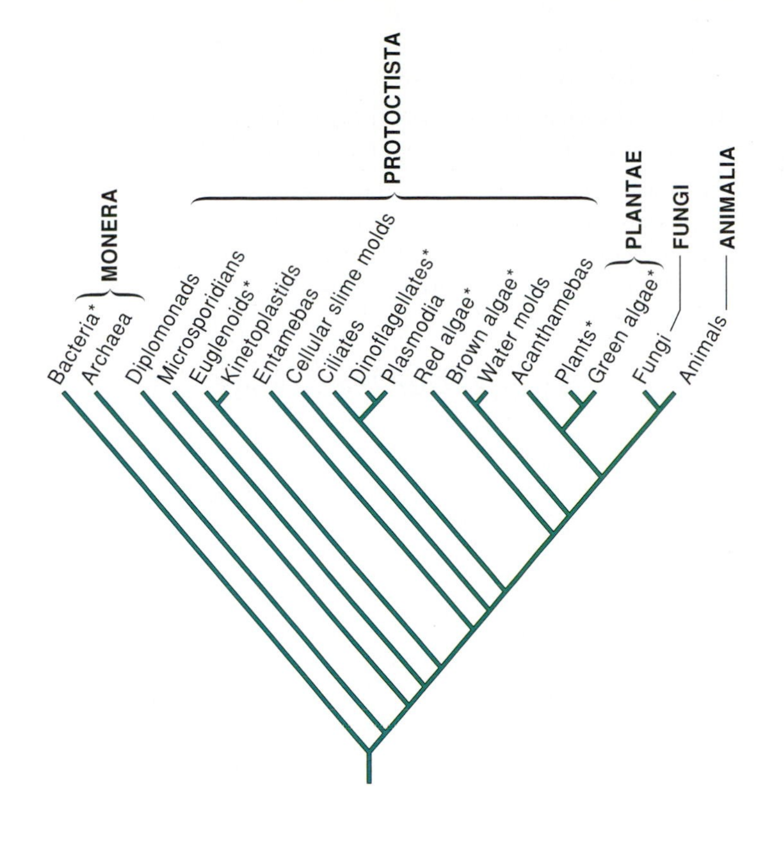

FIGURE 24.14

RNA-based molecular phylogeny of life. Kingdoms are indicated according to the five-kingdom system. Organisms with chlorophyll *a* were classified as plants until recently; some botanists still classify them in the plant kingdom, with the exception of chlorophyll *a*-containing bacteria.

or protein of choice in the method of data analysis (fig. 24.14). Although the details vary, the different phylogenies generally show that the Kingdom Monera consists of two main groups and that the protoctists are scattered, sometimes between branches leading to other kingdoms.

What are Botanists Doing?

Several systems of plant classification now depend on techniques of molecular biology (e.g., gene sequencing). Go to the library and read an article about how molecular biology is changing our ideas about plant classification. Are molecular data more important than traditional data such as leaf shape and flower morphology? Why or why not?

What Kinds of Organisms Should a Botany Text Include?

Traditionally, botany texts have presented overviews of all organisms containing chlorophyll *a:* green plants, all groups of algae, dinoflagellates, euglenoids, and certain kinds of bacteria. Furthermore, for comparison with bacteria containing chlorophyll *a*, all other groups of bacteria have often been included as well. Fungi also have received significant coverage in botany texts, mostly because fungi were previously classified as plants. Finally, for good measure, botany texts have usually provided brief surveys of the diversity of viruses. If we stand back and take a look at this odd assortment of diversity, we see that it encompasses all but the animals, with at least four kingdoms plus a dash of the nonliving (i.e., viruses) represented.

The traditional inclusion of nonplants in botany texts also gets indirect approval from the Botanical Society of America. This organization maintains special sections for mycology, microbiology, and phycology (algae), and uses space in its journal, the *American Journal of Botany*, for research papers on these subjects. Nevertheless, a botany text is supposed to be about plants, which implies that a unit of diversity such as this should exclude nonplants. Such treatment would delight some instructors and students but dismay others. We have therefore chosen a compromise. The chapters on diversity that follow exclude the viruses, dinoflagellates, euglenoids, and flagellated molds entirely, and include briefer than usual overviews of bacteria, fungi, and protoctists. This enables us to provide a more expansive coverage of the plant kingdom.

Chapter Summary

The first record of a system of classification is that of Theophrastus, who is credited with classifying nearly 500 plants in the fourth century B.C. In the first century A.D., Dioscorides wrote a book that listed all known medicinally useful plants; this was the principal book of plant classification for nearly 1,500 years. Worldwide exploration and the discovery of new plants inspired many revisions of plant classification, beginning in about the fifteenth century. Subsequent major works included those of Gaspard Bauhin (1623), John Ray (ca. 1700), and Carolus Linnaeus (1753). Linnaeus's system is especially important because it divided all plants into twenty-four classes, and it designated each species by a Latin binomial. The binomial system of nomenclature invented by Linnaeus is still used today for all species of organisms.

Classifications after Linnaeus continued to use the binomial system but adopted Ray's view that taxonomic categories should be natural; that is, that they should be based on natural relationships among plant groups. Natural classifications before Charles Darwin and Alfred Wallace, however, were meant to reflect the divine creation of related groups.

Once the ideas of Darwin and Wallace became widely accepted, however, classifications were revised to show the evolutionary relatedness among plant groups. These classifications are called *phylogenetic classifications*. Although these systems are still important, they have been modified and revised repeatedly throughout the twentieth century. These modern classifications are based largely on the intuition of their authors, whose assumptions about character evolution are usually not obvious and cannot be easily evaluated scientifically.

The abundance of different phylogenetic classifications of plants has prompted taxonomists to develop more explicit methods of building phylogenies. The most widely adopted of these newer methods comes from cladistics, whose goal is to determine evolutionary branching patterns among hierarchies of organisms. Cladistics is based on determining the evolutionary direction of character-state changes and showing phylogenetic relatedness based on shared-derived character states.

After the 1950s, as classification systems were being further developed for plants, it became apparent that many kinds of organisms traditionally included in the plant kingdom were not plants. Defining the plant kingdom, therefore, has involved determining the phylogenetic relatedness of plants to organisms of other kingdoms. At present, biologists classify organisms into five kingdoms by traditional phylogenetic methods. In contrast, the methods of cladistics, especially as they are applied to molecular data (e.g., DNA, RNA, proteins), provide phylogenies that show the five-kingdom system to be inadequate for making phylogenetic classification of all living things.

Questions for Further Thought and Study

1. Few, if any, intelligent persons would mistake a giraffe for an elephant, or a tomato for a potato, nor would most know or care about the scientific names of the organisms. Why, then, should we be concerned about all organisms having scientific names?
2. Contrast and explain the similarities and differences among artificial, natural, and phylogenetic systems of classification.
3. How might the arrangement of plants in Linnaeus's *Species Plantarum* have shown or not have shown relationships among plants at different taxonomic levels?
4. Most early taxonomists used the form and structure of major plant parts in their systems of classification. What additional features of plants can modern taxonomists use? How might they be used for classification?
5. How might you explain the abundance of classifications of plants that exist today?
6. What explanations might there be for the current absence of a cladistic classification of all plants?
7. Isaac Asimov wrote that "The card-player begins by arranging his hand for maximum sense. Scientists do the same with the facts they gather." How does Asimov's analogy relate to plant classification? Explain your thinking.

Suggested Readings

Articles

Doyle, J. J. 1993. DNA, phylogeny, and the flowering of plant systematics. *Bioscience* 43:380–389.

Funk, V. A., and T. F. Stuessy. 1978. Cladistics for the practicing taxonomist. *Systematic Botany* 3:159–178.

Hillis, D. M., J. P. Huelsenbeck, and C. W. Cunningham. 1994. Application and accuracy of molecular phylogenies. *Science* 264:671–677.

Knoll, A. H. 1992. The early evolution of eukaryotes: A geological perspective. *Science* 256:622–627.

Leedale, G. F. 1974. How many are the kingdoms of organisms? *Taxon* 23:261–270.

Wainwright, P. O., G. Hinkle, M. L. Sogin, and S. K. Stickel. 1993. Monophyletic origins of the Metazoa: An evolutionary link with fungi. *Science* 260:340–342.

Whittaker, R. H. 1969. New concepts of kingdoms of organisms. *Science* 163:150–160.

Books

Crawford, D. J. 1990. *Plant Molecular Systematics: Macromolecular Approaches*. New York: John Wiley.

Hillis, D. M., and C. Moritz, eds. 1990. *Molecular Systematics*. Sunderland, MA: Sinauer.

Jones, S. B., Jr., and A. E. Luchsinger. 1986. *Plant Systematics*, 2d ed. New York: McGraw-Hill.

Margulis, L., and K. V. Schwartz. 1988. *Five Kingdoms: An Illustrated Guide to the Phyla of Life on Earth*, 2d ed. New York: W. H. Freeman.

Scagel, R. F., R. J. Bandoni, J. R. Maze, G. E. Rouse, W. B. Scholfield, and J. R. Stein. 1984. *Plants: An Evolutionary Survey*. Belmont, CA: Wadsworth.

Stuessy, T. F. 1990. *Plant Taxonomy: The Systematic Evaluation of Comparative Data*. New York: Columbia University Press.

Wiley, E. O. 1981. *Phylogenetics: The Theory and Practice of Phylogenetic Systematics*. New York: John Wiley.

Turkey tail bracket fungi.

CHAPTER

25

Bacteria and Fungi

Chapter Outline

Chapter Overview

Many kinds of organisms have had an important impact on the classification and biology of plants. Fungi and certain kinds of bacteria have even been classified as members of the plant kingdom. Most scientists agree now, however, that neither of these groups of organisms should be called plants. Nevertheless, any presentation about the diversity of plants cannot be complete without some mention of these organisms and their influence on plant biology. Accordingly, this chapter presents a brief overview of bacteria and fungi.

Introduction

Ideas about the biology of plants are often derived from the biology of other organisms, especially the chlorophyll-containing bacteria, unicellular fungi, and algae. Knowledge about plants gained from other organisms includes understanding how the mitotic spindle works in a diatom, how genes work in bacteria and in yeasts, and how carbon is fixed during photosynthesis in unicellular algae. Botanists extrapolate this information to plants and, whenever possible, seek evidence that plants are similar to or different from other organisms. Accordingly, botanists think that the push and pull of spindle fibers is the same in plants as it is in isolated spindles of diatoms, that well-known mechanisms of gene action in fungi are the same as in plants, and that the Calvin cycle for carbon fixation in green algae also occurs in plants.

Besides their importance in basic plant science, other kinds of organisms are also important to plants. The ecology, physiology, chemistry, reproductive biology, and almost all other aspects of plant biology are heavily influenced by symbiotic and disease-causing bacteria, fungi, and other kinds of organisms.

Bacteria

Everyone knows something about bacteria because some bacteria make people sick. This limited awareness of bacteria gives them a bad reputation, because most bacteria do not cause diseases. For example, Robert J. Price, a seafood technology specialist at the University of California at Davis, told the press in 1990 that about once a month he receives a report of seafood that glows in the dark. The glow usually comes from *Photobacterium phosphoreum,* a bacterium that produces light instead of heat as it respires. Such bacteria occur on the skin and in the intestines of many fish and shellfish, and they glow when salt is added to cooked seafood. These luminescent bacteria are harmless; no one has ever reported getting sick from eating seafood on which the bacteria were growing.

Harmless bacteria also live in our intestines, and we depend upon them for normal digestion. There may be two or three dozen species of bacteria in the human digestive tract. They are usually so abundant that the dry weight of normal feces can consist of up to 80% bacterial cells.

Features of Bacteria

The main features of bacteria were mentioned briefly in Chapter 24 (see table 24.2). The distinguishing feature of these organisms is that they are prokaryotic, which means that they have no membrane-bound organelles—that is, no nucleus, no mitochondria, and no chloroplasts. Metabolic reactions that occur in the organelles of eukaryotes occur instead in folds of the plasma membrane in bacteria. Also, because bacterial DNA is not associated with histone proteins in chromosomes, the "chromosomes" of bacteria are sometimes referred to as **genophores** to distinguish them from true chromosomes. Multicellular forms of bacteria are rare except among the cyanobacteria. Cell walls, which occur in most bacteria, are usually made of **peptidoglycans,** in which carbohydrate polymers are interconnected with short chains of amino acids. Different bacteria may also have various layers outside their walls and may have flagella or shorter, thinner filaments called **pili.** Genetic material can move from donor to recipient through pili, but reproduction is asexual by means of fission or budding. Many bacteria are strictly anaerobic and are killed by oxygen. Others can live anaerobically or aerobically, and still others require oxygen for respiration.

Although most bacteria are 1–5 μm in diameter (much smaller than plant cells), bacteria known as mycoplasmas are often only 0.2 μm across (fig. 25.1a).[1] At the other extreme, some cyanobacteria (fig. 25.1b) are up to 3 mm long, and the recently discovered *Epulopiscium fishelsonii*—the largest known nonphotosynthetic bacterium—is as large as a small hyphen (0.3 mm long) (see fig. 25.1c).

Bacteria are mostly simple unicells, but there are some exceptions to such simple organization. For example, some species have cells that cohere in chains, filamentous forms occur among the cyanobacteria, and others have flagella and live as motile individuals or colonies (fig. 25.1d). Furthermore, a group called the **myxobacteria** often form upright, multicellular reproductive bodies. These are thought to be the most complex forms among bacteria.

Several groups of bacteria contain chlorophyll or other photosynthetic pigments and are capable of either plantlike photosynthesis or photosynthesis that is unique to bacteria. Many aquatic bacteria have a membranous structure, called a

1. Mycoplasmas are the smallest free-living microorganisms. Unlike other bacteria, mycoplasmas lack a cell wall and are too small to be seen with a light microscope. They cause a variety of conditions ranging from kidney stones to premature labor. Mycoplasmas also cause several plant diseases that are associated with symptoms of yellowing and stunting (e.g., citrus stubborn disease).

A.

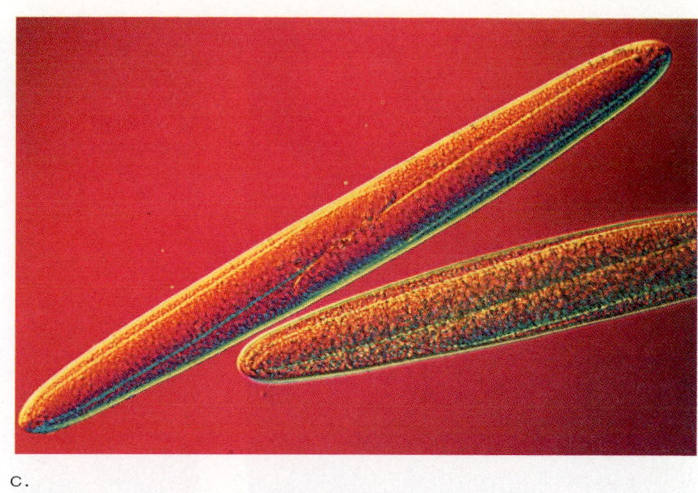

C.

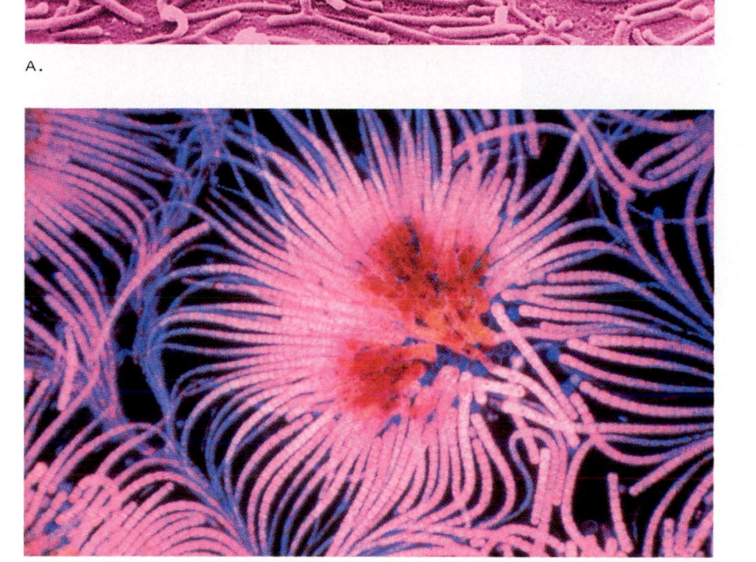

B.

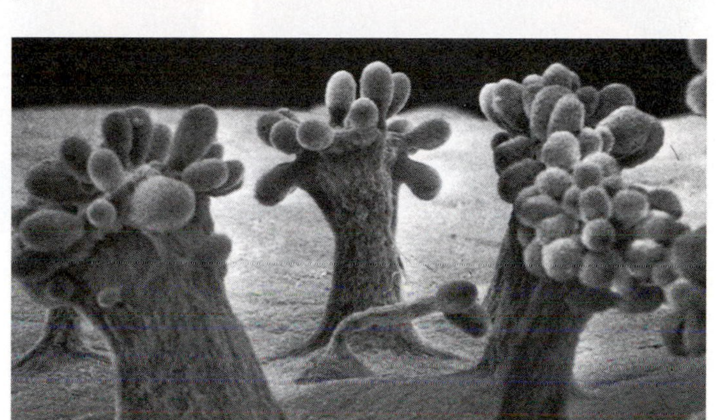

D.

FIGURE 25.1

Mycoplasmas and bacteria. (a) Pneumonia mycoplasma bacteria, ×62,000. (b) Strands of the cyanobacterium *Gloeotrichia echinulata*, ×63. (c) *Epulopiscium fishelsonii*, a very large bacterium discovered in the guts of surgeonfish. Each cell is about 0.3 mm long. (d) *Chondromyces crocatus*, a myxobacterium.

gas vacuole, in each of their cells. As the name suggests, these vacuoles are filled with gas and enable the bacteria to float.

The distinguishing features of different bacteria are their shapes, metabolism, and chemical composition. Bacterial cells typically have one of three shapes (fig. 25.2a): spherical (also called *cocci*), rod-shaped, or corkscrew-shaped (also called *spirilla*). Bacteria can be further distinguished by a diagnostic test called the **Gram stain,** named after Danish physician H. C. Gram. This stain distinguishes two bacterial groups: Gram-positive bacteria and Gram-negative bacteria (fig. 25.2b). **Gram-positive** bacteria take up a purple stain in their thick, peptidoglycan-containing cell walls. **Gram-negative** bacteria cannot retain the stain, either because they have thin cell walls surrounded by lipids, or they have no cell walls at all.

Bacteria are far more diverse metabolically than all of the eukaryotes. This diversity is reflected in the range of energy sources that bacteria can use. Different bacteria can use sulfide, iron, methane, or carbon dioxide in energy metabolism. Moreover, many bacteria can fix nitrogen, which means they can convert atmospheric nitrogen (N_2) into metabolically useful forms (e.g., nitrates) for other organisms. Cyanobacteria have both photosystems I and II, although their only chlorophyll is chlorophyll *a*. Purple sulfur bacteria and green sulfur bacteria have a different type of chlorophyll, called **bacteriochlorophyll,** and a photosynthetic pathway that is analogous but not identical to photosystem I. Bacteria can also ferment a variety of sugars and organic acids, thereby producing important commercial products such as ethanol, methanol, acetone, lactic acid, acetic acid, and propionic acid. Bacteria have also been used to clean up oil spills because they can digest many of the chemicals in petroleum oil. Still other bacteria can detoxify polychlorinated biphenyls (PCBs), a class of human-made pollutants that have become a worldwide environmental nightmare.

Some bacteria form special, durable endospores (fig. 25.3). These spores, which are relatively thick-walled, enable bacteria to survive adverse conditions. Endospores, including those produced by the notorious botulism bacteria (*Clostridium botulinum*), are exceptionally resistant to heat: some can survive

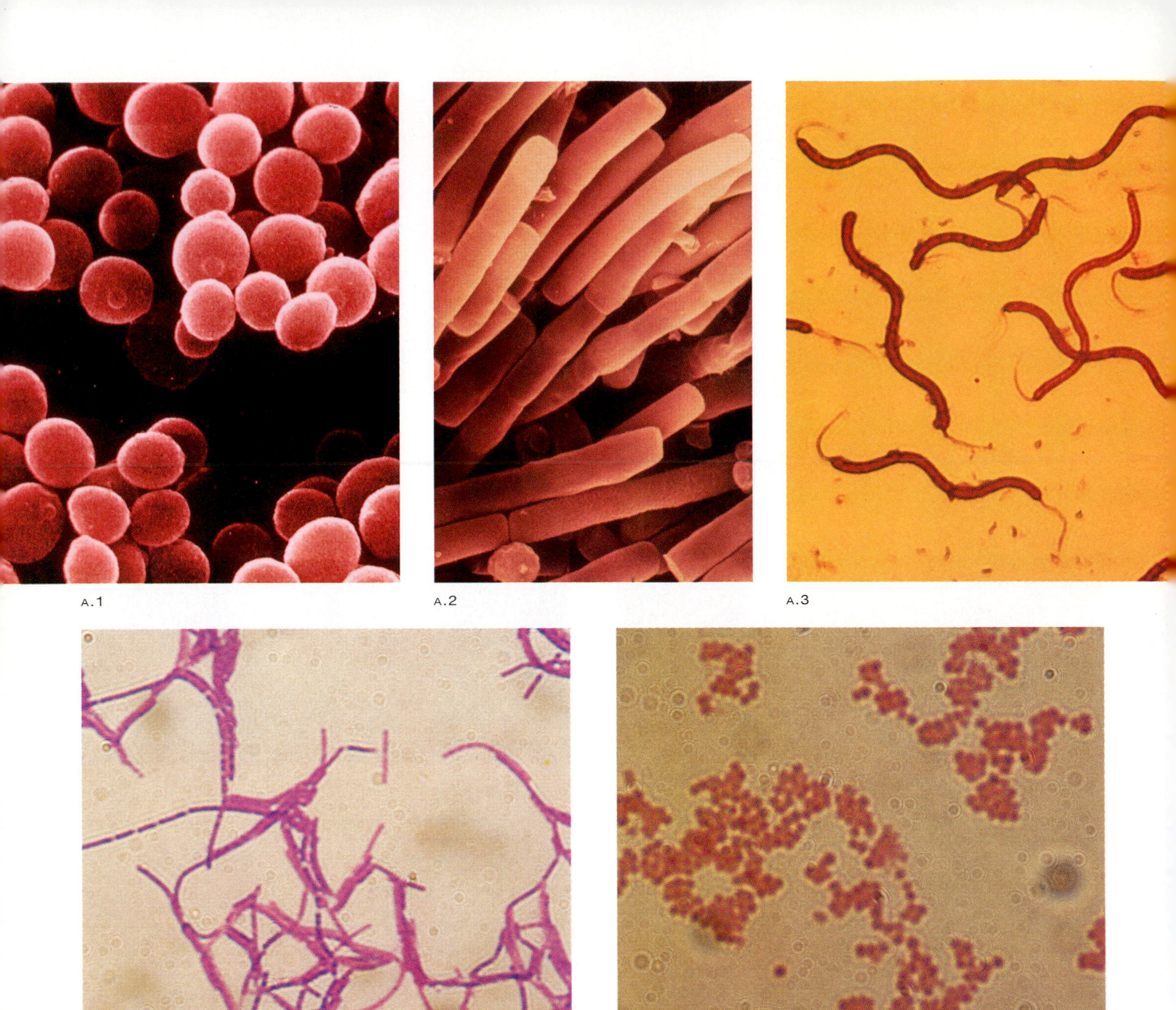

Figure 25.2

The shapes and Gram staining of bacteria. (a) The three basic shapes of bacteria: [1] spherical (*Micrococcus*), ×40,000; [2] rod (*Bacillus*), ×60,000; [3] corkscrew (*Spirilla*), ×450. (b) [1] Gram-positive bacteria, ×400; [2] Gram-negative bacteria, ×1,000.

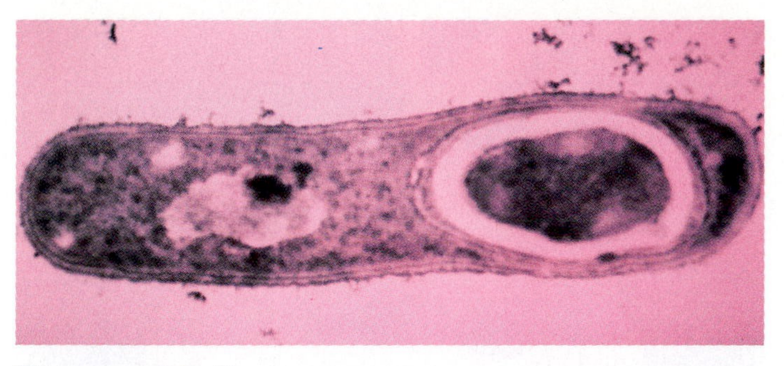

FIGURE 25.3

Electron micrograph of *Bacillus subtilis* bacterium showing endospores, ×49,000.

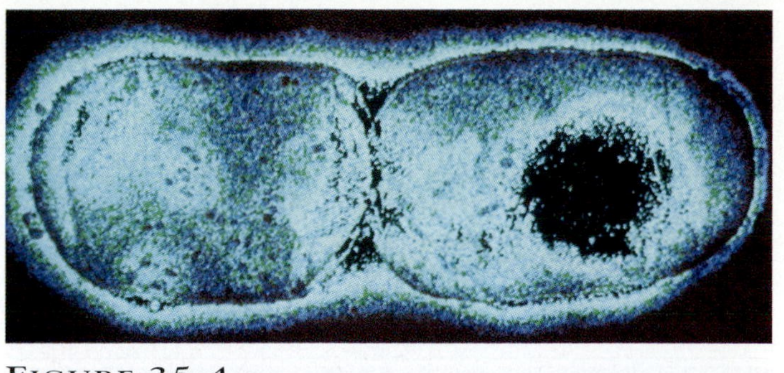

FIGURE 25.4

Electron micrograph showing fission of *Escherichia coli*, a common bacteriuum.

boiling water for over an hour, while others can survive desiccation, sunlight, a vacuum, and various chemicals. Many may remain dormant for hundreds of years.

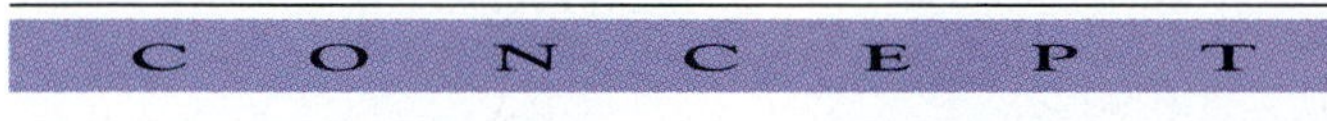

Bacteria are metabolically the most diverse of all groups of organisms. They use a variety of chemicals as sources of metabolic energy, and they have different kinds of photosynthesis.

The Generalized Life Cycle of Bacteria

Bacteria reproduce mainly by **fission,** which occurs when a cell reaches a certain size and then pinches in half to form two cells (fig. 25.4). During fission, the genophore attaches to the plasma membrane, where it replicates. Cell division occurs between the two genophores, thereby ensuring that each new cell gets an identical molecule of DNA.

Because bacteria have no nuclei, they cannot undergo meiosis and fertilization. However, genetic material can move from one cell to another through a **conjugation pilus** (fig. 25.5). Genetic material that moves in this fashion usually involves plasmid DNA but not the genophore.

Two other types of genetic transfer also occur in bacteria. One is **transformation,** which occurs when free DNA is taken up by cells, after which the genes of the absorbed DNA are expressed in their new host. The second type is **transduction,** in which viruses enter a bacterial cell, and any DNA from the previous bacterial host of the virus can replace the DNA of the new host cell. Both of these processes have immense potential for genetic engineering because they can be manipulated in the laboratory (see Chapter 11).

In spite of the absence of sexuality, and a consequent lack of genetic recombination from sexual reproduction, bacteria can evolve quickly. There are about 5,000 genes present in the DNA of a typical bacterium. Mutations occur randomly approximately once in every 200,000 genes. This means that about

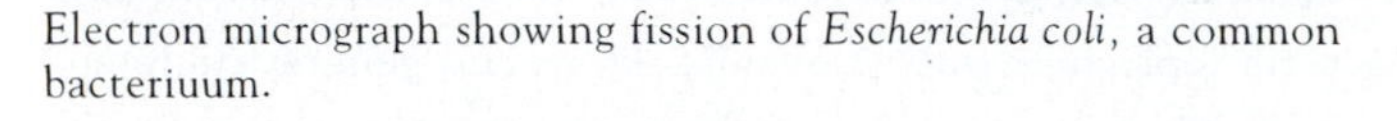

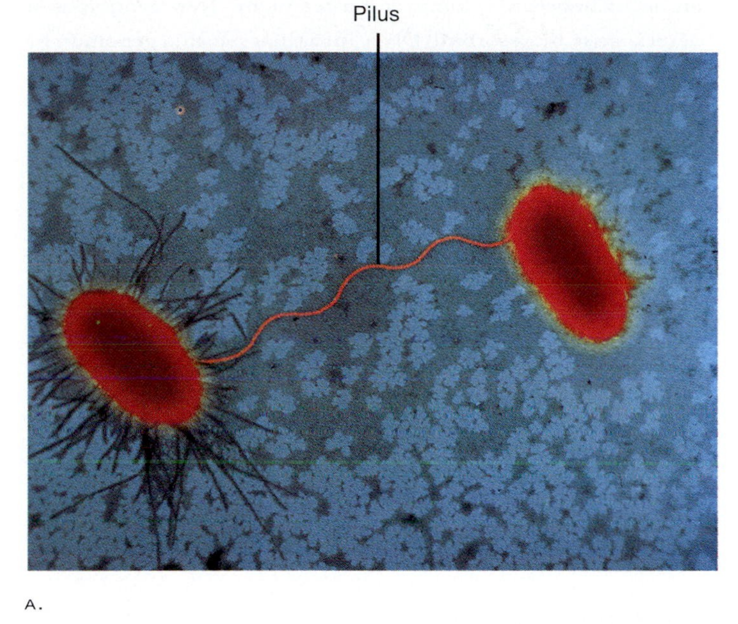

A.

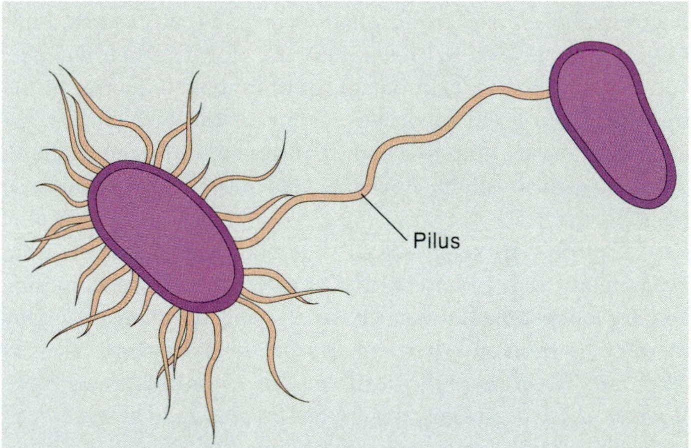

B.

FIGURE 25.5

Electron micrograph (a) and accompanying diagram (b) of bacterial cells (*E. coli*) with a conjugation pilus.

one out of every 40 bacteria in a given population may spontaneously develop a mutant characteristic. At this rate, as many as 50 million mutant bacteria could be present among the more than 2 billion bacteria inhabiting a single gram of garden soil. This high rate of mutation in bacteria is important to the medical industry, because random mutations often enable bacteria to become resistant to antibiotics and disinfectants. Such drug-resistant bacteria often spread when susceptible strains are eliminated by the overuse of antibacterial agents.

C O N C E P T

Bacteria reproduce only asexually. However, genetic exchange can occur between a donor and a recipient. New genotypes of bacteria arise by absorbing DNA into their plasmid genomes, by genetic transfer from viruses, and by high rates of mutation.

Where Bacteria Grow

Bacteria are everywhere. They grow in such extreme environmental conditions that they must be considered the most hardy of all organisms. They live in the intestines of all kinds of animals, in soil, in clouds, in water, on airborne dust particles, and in smog. Many can tolerate hot acids, others can survive temperatures below freezing for years, and still others thrive in boiling hot springs. Their tolerance of high temperature requires methods of sterilization that include a combination of high heat (120° C) and high pressure. However, there are deep-sea bacteria that live near volcanic vents, where the temperature approaches 360° C and the pressure can reach more than 26 megapascals (ca. 260 atmospheres).

Bacterial Ecology

Bacteria play an important role in every habitat on earth. Photosynthetic bacteria help maintain the global carbon balance, and nitrogen-fixing bacteria account for about a quarter of the total nitrogen fixed in oceans. Some of the most important bacteria in agriculture live and fix nitrogen in the root nodules of crops such as alfalfa, soybean, and pea (fig. 25.6; also see pp. 479–481).

Saprobic bacteria—those that obtain nourishment from dead organic matter—are partially responsible for decomposing and recycling organic material in the soil. Decomposition by saprobic bacteria and their cohorts, saprobic fungi, produces as much as 90% of the biologically made CO_2 in the atmosphere. Without these organisms, the earth's surface would be knee-deep in dead plants and animals in a matter of weeks. So much carbon would be tied up in organic matter that less atmospheric CO_2 would be available, which would limit photosynthesis by plants.

The Diversity and Relationships of Bacteria

Taxonomic disputes abound regarding Kingdom Monera. Different biologists estimate the number of bacterial species to be

FIGURE 25.6

Root nodules containing nitrogen-fixing bacteria.

anywhere from 2,500 to 10,000. The most widely used manual for identifying bacteria, *Bergey's Manual of Systematic Bacteriology*, divides them into four divisions and seven classes (table 25.1), but this arrangement is artificial because it classifies groups based on a few similarities that may not be derived from a unique common ancestor. For example, Class Anoxyphotobacteria includes all bacteria that photosynthesize without producing oxygen. The likelihood that nonoxygenic photosynthesis fails to indicate relatedness is supported by a phylogeny of bacteria based on sequence analyses of ribosomal RNA (fig. 25.7). That analysis suggests that nonoxygenic photosynthetic bacteria have widely scattered phylogenetic relationships among nonphotosynthetic bacteria. This means that groups such as the purple bacteria are more closely related to nonphotosynthetic bacteria such as *Escherichia coli* (colon bacterium) and *Rickettsia typhi* (typhus pathogen) than they are to other photosynthetic bacteria.

TABLE 25.1

Proposed Higher Taxa for Bacteria, According to Bergey's Manual of Systematic Bacteriology

Kingdom *Prokaryotae*

- Division I. *Gracilicutes* (Gram-negative, typical bacteria)
  - Class I. *Scotobacteria* (nonphotosynthetic bacteria)
  - Class II. *Anoxyphotobacteria* (nonoxygenic photosynthetic bacteria)
  - Class III. *Oxyphotobacteria* (oxygenic photosynthetic bacteria)
- Division II. *Firmicutes* (Gram-positive, typical bacteria)
  - Class I. *Firmibacteria* (simple, Gram-positive bacteria)
  - Class II. *Thallobacteria* (branching, filamentous bacteria)
- Division III. *Tenericutes* (bacteria lacking a cell wall)
  - Class I. *Mollicutes*
- Division IV. *Mendosicutes* (bacteria with defective cell walls or lacking peptidoglycan)
  - Class I. *Archaebacteria*

According to Carl Woese at the University of Illinois, who is one of the leading proponents of using molecular phylogenetics for classifying prokaryotes, traditional classifications of bacteria like that in *Bergey's Manual* rely heavily on characters that are almost useless for determining phylogenetic relatedness. Nevertheless, traditional views and molecular phylogenies do agree on the classification of prokaryotes into two main groups, the **Eubacteria** (*eu* = true) and the **Archaebacteria** (*archae* = ancient). (Molecular systematists prefer to call them **Bacteria** and **Archaea,** respectively.) These two groups are often classified as subkingdoms of the Monera, although there is growing sentiment to separate the Eubacteria and Archaebacteria into their own kingdoms. Such a split is supported by molecular phylogenies, which show the prokaryotes to be a paraphyletic group, with the Archaebacteria being more closely related to eukaryotes (see Chapter 24).

Distinguishing features of the Eubacteria and Archaebacteria are still being discovered, but at present the Archaebacteria differ from the Eubacteria in the following ways:

1. The major lipids of Archaebacteria are ether-linked instead of ester-linked, as in the triglycerides of Eubacteria and eukaryotes.
2. Archaebacteria have cell walls of glycoproteins and polysaccharides but lack peptidoglycans.
3. Archaebacteria have a single type of DNA-dependent RNA polymerase, which is more similar to the RNA polymerases of eukaryotes than to those of Eubacteria.
4. Archaebacterial ribosomes have a distinctive shape (fig. 25.8)
5. Archaebacteria can metabolize methane, grow at high temperatures, or grow in high concentrations of salt.
6. Like eukaryotic genes, archaebacterial genes may contain introns, which are absent in eubacterial genes.

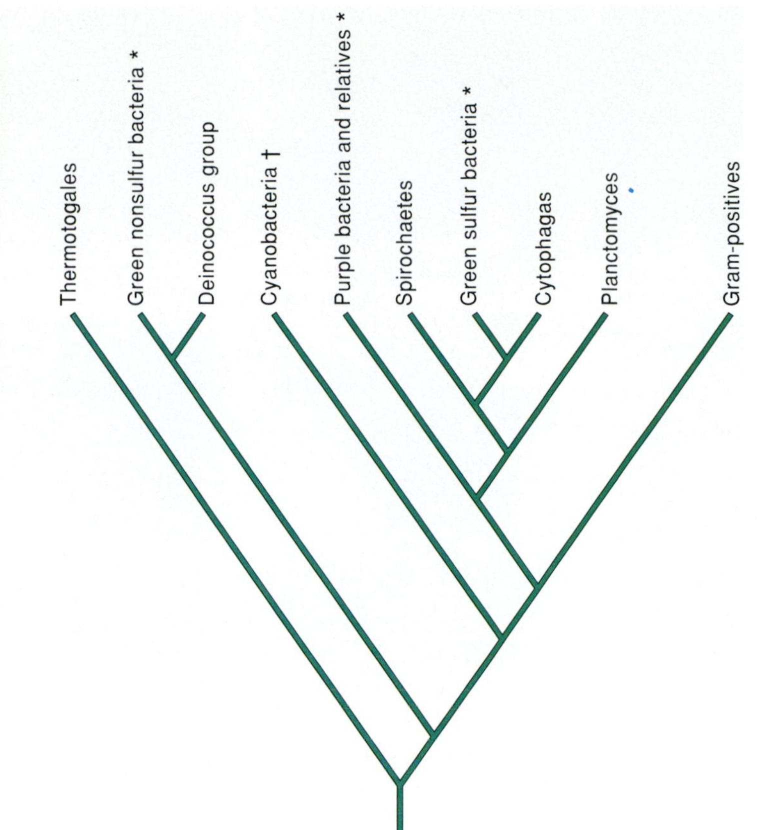

FIGURE 25.7

A phylogeny for eubacteria.

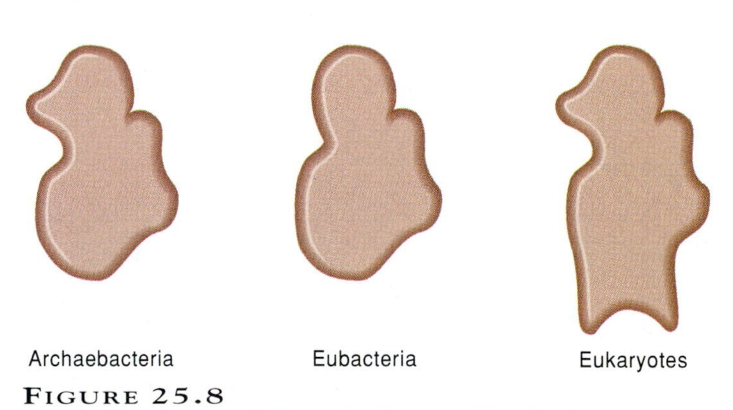

FIGURE 25.8

The different shapes of the 30S subunit of ribosomes in eubacteria, archaebacteria, and eukaryotes.

Even a brief summary of examples from all groups of bacteria is beyond the scope of this textbook, but two groups are of special interest here because they photosynthesize much as plants do. These groups are the Cyanobacteria and the Chloroxybacteria. They are classified together in Class Oxyphotobacteria in *Bergey's Manual*, but many botanists view them as separate classes or divisions.

FIGURE 25.9

Travertine surrounding a hot spring at Yellowstone National Park. The colors are produced by cyanobacteria.

Cyanobacteria

The cyanobacteria are named for the blue-green pigment **phycocyanin.** Phycocyanin, along with chlorophyll *a*, gives some cyanobacteria a blue-green appearance, which led to their being called *blue-green algae* before the Kingdom Monera was recognized. Phycocyanin and its red counterpart, **phycoerythrin,** are proteins called **phycobilins.** These proteins, which occur only in the cyanobacteria and red algae, function as photosynthetic pigments in photosystem II. This contrasts with photosystem II in plants, which relies on chlorophyll *b*.

Cyanobacteria live in a wide variety of habitats, including in snow fields, in frozen lakes of Antarctica, in extremely acidic, basic, salty, or pure (i.e., nutrient poor) water, in deserts, inside rocks, and in hot springs where water temperatures approach 85°C. In Yellowstone National Park, a different species of cyanobacterium lives in each temperature range of hot springs. In such habitats the cyanobacteria precipitate chalky, carbonate deposits, which become a rocklike substance called *travertine*. The travertine is often marked with brilliantly colored streaks of cyanobacteria (fig. 25.9). Cyanobacteria are usually the first photosynthetic organisms to appear on bare lava after a volcanic eruption. They also live on the shells of turtles and snails, as symbionts in lichens (which are discussed later in this chapter), in protozoans, amoebas, aquatic ferns, the roots of tropical plants, sea anemones, and a variety of other hosts. Cyanobacteria that live symbiotically in other organisms often lack a cell wall, and function essentially as chloroplasts inside their hosts.

The more than 1,500 known species of cyanobacteria are diverse in form (fig. 25.10); they range from single-celled spheres to filaments and colonies. Many of the cyanobacteria have a gelatinous sheath around their cells, which protects them and binds trace elements for metabolism. In some parts of the open ocean, cyanobacteria sometimes reach densities of 10 million cells per liter. By themselves, cyanobacteria account for 20% of the primary production in the seas, and provide fodder for the rest of the marine food chain.

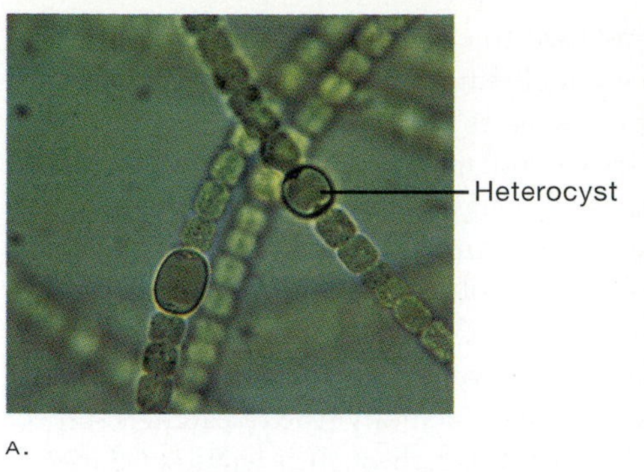

A.

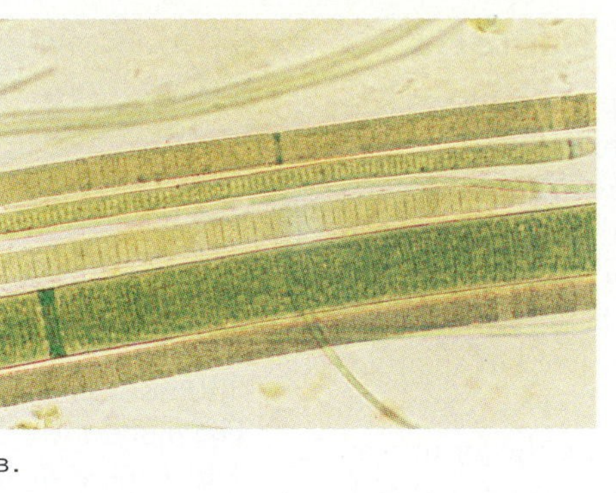

B.

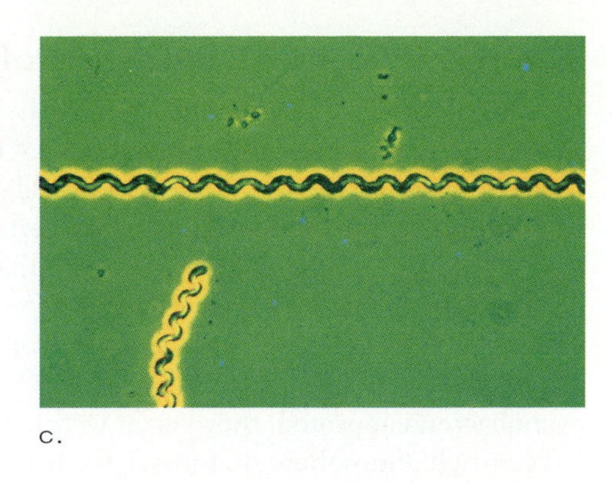

C.

Figure 25.10

Light micrographs of cyanobacteria: (a) *Anabaena*, ×850, (b) *Oscillatoria*, ×400, (c) *Spirulina*, ×300.

If the substrate in which they are growing is low in nitrogen, most cyanobacteria produce **heterocysts,** which are larger, thicker-walled cells that can fix nitrogen (fig. 25.11). Cyanobacteria may also produce thick-walled **akinetes** that resist desiccation and freezing. Akinetes can lie dormant during long, dry periods and then germinate to form a new chain of cells when water becomes available.

Cyanobacteria reproduce by fission, but they can also reproduce by fragmentation of filaments, by budding, or by multiple fission. Budding begins when a large swelling forms on a cell and enlarges until it becomes a mature cell. Multiple fission occurs when a cell enlarges and its contents divide several times. The wall of the original cell then breaks down and the smaller cells are released.

Cyanobacteria lack flagella, but filamentous forms like *Oscillatoria* exhibit curious gliding movements. Adjacent filaments glide up and down while touching each other, or they appear to rotate on their axes. These movements are apparently caused by the twisting of fibrils in the cell walls. *Synechococcus*, a nonfilamentous genus, includes marine forms whose cells can move at speeds of up to 25 µm (about the diameter of a pollen grain) per second. The mechanism responsible for this movement is unknown.

Chloroxybacteria

In 1976 Ralph A. Lewin of the Scripps Institute of Oceanography announced the discovery of a unicellular, prokaryotic organism living on or in sea squirts in the intertidal region of Baja California, Mexico. This prokaryote, although similar to cyanobacteria in structure and chemistry, possesses chlorophylls *a* and *b*, like higher plants, yet lacks phycobilins. Lewin named the new organism *Prochloron*. In 1984 a group of Dutch scientists discovered a similar organism in the shallow Loosdrecht lakes in the Netherlands. This organism, which they named *Prochlorothrix*, also has chlorophylls *a* and *b* and lacks phycobilins, but differs from *Prochloron* in being filamentous and free-living. *Prochloron* and *Prochlorothrix* are now classified together, sometimes as cyanobacteria and sometimes as the only two members of their own class or division, the Chloroxybacteria.

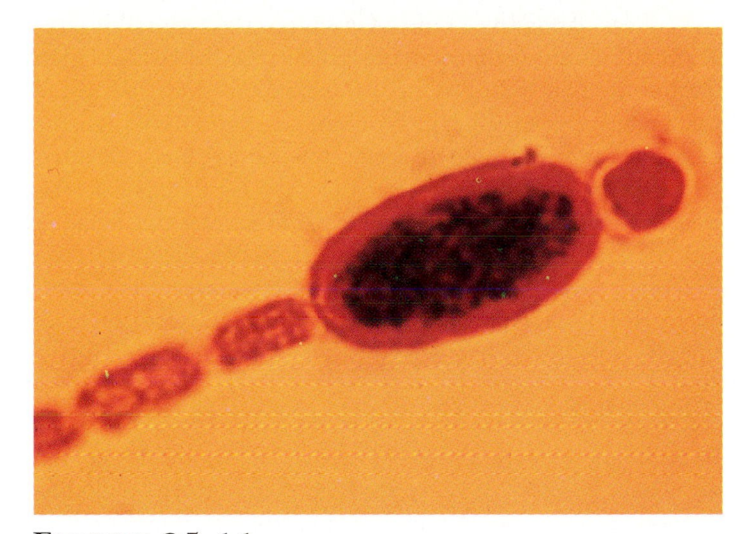

Figure 25.11

Light micrograph of *Cylindrospermum* with terminal heterocyst and subterminal akinete, ×500.

The chloroxybacteria have paired thylakoid membranes, unlike the unstacked membrane of cyanobacteria, but like the stacked thylakoids of green algae and plants. Their stacked thylakoids, their use of chlorophyll *b* in photosystem II, and their lack of phycobilins have led scientists to speculate that chloroxybacteria descended from the same ancestral prokaryotes as the chloroplasts of plants.

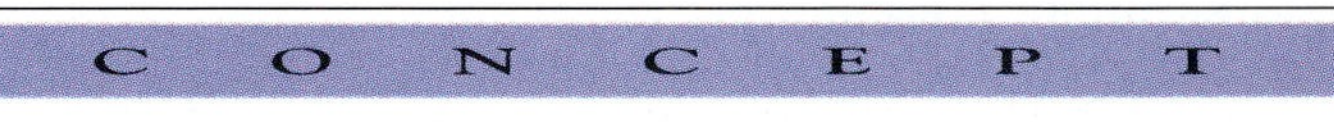

Biologists classify bacteria into two major groups, the Eubacteria and the Archaebacteria. The Eubacteria include the Cyanobacteria, which use chlorophyll *a* and phycobilins for photosynthesis, and the Chloroxybacteria, which resemble plants in their use of chlorophylls *a* and *b* for photosynthesis. The Archaebacteria are so distinctive that some biologists classify them as a separate kingdom. Archaebacteria seem to be more closely related to the eukaryotes than to the Eubacteria.

The Origin and Evolution of Bacteria

Bacterialike cells were probably the first living organisms. Fossil evidence for bacteria goes back about 3.4 billion years, which is much older than the oldest fossil evidence for eukaryotes (1.5 billion years). Moreover, cyanobacterialike fossils lived at least 2 billion years ago, which is evidence that oxygen-producing photosynthesis might be that old. Scientists believe that before the appearance of cyanobacteria, prokaryotes lived in an atmosphere that was probably no more than 1% oxygen. Once the cyanobacteria appeared, they began to transform the air into an oxygen-rich atmosphere that may have reached as much as 30% oxygen (the present level is about 21% oxygen). This transformation set the evolutionary stage for the diversification of aerobic bacteria and the origin of oxygen-respiring eukaryotes.

Ideas about the origin of bacteria are essentially ideas about the origin of life, since bacteria seem to be the oldest form of life that is still living. Scientists have long speculated that prebacteria arose by the spontaneous aggregation of simple molecules into polymers. Recent discoveries about the catalytic properties of RNA have led to the idea that prebacteria were probably regulated by small RNA molecules (see Chapter 11). According to this idea, the first bacteria evolved from RNA-based ancestors. All extant organisms, however, are DNA-based, so molecular phylogenies comparing the RNA-based relatives of modern-day bacteria cannot be made.

The Economic Importance of Bacteria

Bacteria affect every aspect of our lives. They are the pivotal organisms in many diseases of plants and animals, including humans, and they are the microscopic laborers in many multimillion-dollar industries. Thus, we can think of the economic importance of bacteria as either positive or negative. The following paragraphs discuss a few examples of each.

The Positive Aspects

The cyanobacterium *Spirulina* is cultivated for human consumption. This organism, which is common in saline lakes, has a dry-weight protein content of about 70%. It has been harvested for food in central Africa for several centuries. *Spirulina* is currently harvested in Mexico and is commercially cultivated for human consumption in both Mexico and Israel. In the United States, *Spirulina* is sold as a nutritional food supplement in most health food stores.

The genus *Bacillus* includes many species important to humans. For example, three species have been approved by the United States Department of Agriculture for biological control of pests. *Bacillus thuringiensis* (BT) reproduces only in the intestinal tracts of caterpillars, which are killed by a toxin from the bacterium. This toxin is harmless to all other living organisms. BT is currently being used to control more than one hundred species of plant pests, including tomato hornworms and corn borers. A variety of this species (var. *israelensis*) is being used to control mosquito larvae, and another variety (var. *nigeriensis*), which is even more effective as a mosquito control, may soon be approved for use in the United States.

The discovery in 1994 of a mutant of *Bacillus stearothermophilus* may eventually enable us to ferment ethanol from agricultural wastes as cheaply as we refine gasoline from crude oil. The mutant was discovered accidentally (the researchers suspect that a spore floated into the lab through a window) during a study of rapid anaerobic growth by bacteria at high temperatures. To everyone's surprise, the bacterium fermented wheat, corn, and sugarbeet wastes at 75°C , producing ethanol in the process. The mutant has a big advantage over yeast (the most commonly used fermenting agent) because it can digest hemicelluloses that yeast ignores. This could dramatically increase the proportion of plant waste that can be made into fuel. In Brazil, where more than two-thirds of cars are powered by ethanol from sugarcane, only the juice of the plants is converted into fuel. If the leftover hemicelluloses from sugarcane were also fermented, the yield of ethanol would be more than doubled.

Several species of *Lactobacillus* are mainstays of the dairy, beverage, and baking industries. Lactobacilli are used to make acidophilus milk, kefir, cheeses, yogurt, and related foods. These bacteria are also used in the production of wine, beer, sourdough bread, sauerkraut, and many other commercial products.

Antibiotic chemicals from actinomycete bacteria (i.e., nonphotosynthetic bacteria with branched filaments) account for about two-thirds of the more than 4,000 antibiotics that have been discovered. Actinomycetes were the original sources of such well-known antibiotics as tetracycline, neomycin, aureomycin, erythromycin, and streptomycin. Streptomycin is produced by *Streptomyces,* which is common in soil and is the largest genus of actinomycetes. The various species of *Streptomyces* play a significant role in breaking down and recycling plant and animal products such as lignin, latex, and chitin. They also produce the characteristic odor of damp soil. *Frankia* species are nitrogen-fixing actinomycetes that, like other kinds of nitrogen-fixers, form nodules on the roots of legumes and other higher plants. *Thermoactinomyces* is common and active in compost piles, haystacks, silos, and sod, where it thrives at temperatures between 45°C and 60°C.

The Negative Aspects

In water supplies, cyanobacteria frequently clog filters, corrode steel and concrete, soften water, and produce undesirable odors or coloration in the water. Cyanobacteria also produce toxins. Fish that are immune to the toxins produced by certain cyanobacteria become toxic to their predators after eating these bacteria. In summer months various cyanobacteria and algae often become abundant in bodies of fresh water, especially if the water is polluted. When this happens, a floating mat called a *bloom* may form. Such blooms can cover more than 2,000 square kilometers. Domestic animals and fowl have occasionally died as a result of drinking water that contains such a bloom. While the cyanobacteria of a bloom are alive, the oxygen content of the surrounding water may temporarily increase. When they die, however, the other bacteria that decompose the remains may so deplete the oxygen that fish and other aquatic organisms in the vicinity die from lack of oxygen. Many communities control cyanobacteria in reservoirs with dilute concentrations of copper sulfate.

Unlike their helpful counterparts mentioned, some bacilli cause diseases such as diphtheria, periodontal disease in humans, and anthrax in cattle. The most dangerous of the bacilluslike bacteria may be the clostridia: species of *Clostridium* cause tetanus, gas gangrene, and botulism. *Clostridium botulinum*, which causes botulism, makes the most powerful toxin known to humankind: just one gram of botulism toxin can kill 14 million people, and half a kilogram could kill the entire human population. Moreover, because clostridia produce spores that resist heat, they can survive a certain amount of boiling during the preparation of canned foods. Botulism in adults often results from eating home-canned, nonacidic food that was not kept at high temperatures long enough to destroy the bacterial spores.

FUNGI

Botanists still study fungi, sometimes mistakenly refer to them as plants in idle conversation, and continue to classify and name them according to the International Code of Botanical Nomenclature. Furthermore, botanists still occasionally slip up by calling the haploid phase of a fungus a "gameto*phyte*," and college botany departments throughout the world still have mycologists (biologists who study fungi) on their faculties. Because biologists in all disciplines now generally agree that fungi are not plants, these two groups of organisms are now usually separated into their own kingdoms. Moreover, recent molecular phylogenies suggest that fungi are more closely related to animals than to plants (see Chapter 24). Nevertheless, the past and current confusion between fungi and plants has a long tradition. Your appreciation for this tradition should develop as you note the plantlike features of fungi while reading the remainder of this chapter. More important are the unique features of fungi that finally caused taxonomists to classify them as a separate kingdom.

Features of Fungi

The fungi are usually filamentous, eukaryotic, spore-producing organisms that lack chlorophyll. They have cell walls made of **chitin** combined with other complex carbohydrates, including cellulose (chitin is also the main component of the exoskeletons of insects, spiders, and crustaceans). The main storage carbohydrate of fungi is **glycogen,** which is also the main storage carbohydrate of animals (but not plants). All species of fungi are either saprobes or **symbionts** (i.e., live with other organisms). As symbionts, they may be parasitic, they may provide a benefit to their host, or they may be parasitized by their host.

The main vegetative feature of fungi is a tubular, threadlike, whitish or colorless filament called a **hypha** (plural, **hyphae**). Hyphae, which vary in thickness between 0.5 and 100 µm, grow only at their tips and can grow indefinitely in favorable conditions. The hyphae of most fungi branch repeatedly, intertwine, or fuse with other hyphae, forming a mass known as a **mycelium** (plural, **mycelia**). Mycelia can become quite compact, and parts may take on the form of parenchyma tissue, in which individual hyphae are indistinguishable from one another. Such compact mycelia form mushrooms and similar spore-bearing structures that have traditionally been referred to as *fruiting bodies* (fig. 25.12). When growth is undisturbed, a mycelium tends to grow more or less equally in all directions from its point of origin, sometimes for hundreds of years. Recent size estimates of single-mycelium mats in the states of Washington and Michigan, each of whose mycologists claim world records for the largest fungus, put them at several metric tons for a single mycelium.

The hyphae of most fungi are partitioned into cells at regular intervals by crosswalls, or **septa** (singular, **septum**), but other groups are nonseptate. The cells of septate hyphae may be uninucleate, binucleate (**dikaryotic**), or multinucleate (**coenocytic**). Nonseptate hyphae are all coenocytic. Unlike mitosis in plants, mitosis in fungi occurs almost completely within a persistent nuclear envelope.

Fungi have many plantlike features, but they lack chlorophyll and their cell walls usually contain chitin. The main body of a fungus is a hypha or an aggregate of hyphae called a *mycelium*. Hyphae are either uninucleate, dikaryotic, or coenocytic. Mitosis in fungi is unique because it occurs within a persistent nuclear envelope.

Figure 25.12

The diversity of fruiting bodies in fungi. (a) Oyster mushrooms (*Pleurotus ostreatus*). (b) A jelly fungus (*Tremella mesenterica*). (c) A bird's nest fungus (*Cyathus striatus*). (d) A shelf or bracket fungus (*Grifola sulphurea*). (e) Puffballs (*Lycoperdon* sp.). Note the spores being released. (f) A common stinkhorn fungus (*Phallus impudicus*).

Writing to Learn Botany

What features of fungi prompted early botanists to classify them as plants? Why are fungi no longer considered to be plants?

The Generalized Life Cycle of Fungi

Reproduction in fungi can be either asexual or sexual; most fungi exhibit both forms. Asexual reproduction usually occurs by mitosis and cell divisions or by budding. Budding is typical of yeasts and other unicellular fungi; it produces new individuals by small outgrowths that pinch off from parent cells. Other forms of asexual reproduction include mycelial fragmentation and the mitotic production of spores. Such spores form either within a specialized container (**sporangium**) or in rows at the tips of specialized hyphae (fig. 25.13).

The sexual life cycles of many fungi involve a dominant haploid phase and a diploid phase consisting of just the zygote. The haploid phase consists of two parts in some fungi: one part derives from spore germination, and one part forms after the cells, but not the nuclei, of different mating strains fuse. The resulting dikaryotic cells can then grow into a dikaryotic mycelium. Fertilization eventually occurs in some of the dikaryotic cells, which form zygotes that undergo meiosis. These meiotic products develop into different kinds of spores, depending on what type of fruiting body the fungus produces. However, at no time in the life cycle are there any flagellated cells. Further details of fungal reproduction are presented in the discussions of particular groups of fungi later in this chapter.

Where Fungi Grow

The distribution and ecology of fungi resemble those of bacteria. Saprobic fungi and bacteria both decompose and recycle plant and animal remains, and have probably done so for as long as 2 billion years. Fungi attack virtually all organic materials. They can also etch the lenses of cameras, telescopes, and binoculars. In lichens, fungi can even degrade rocks.

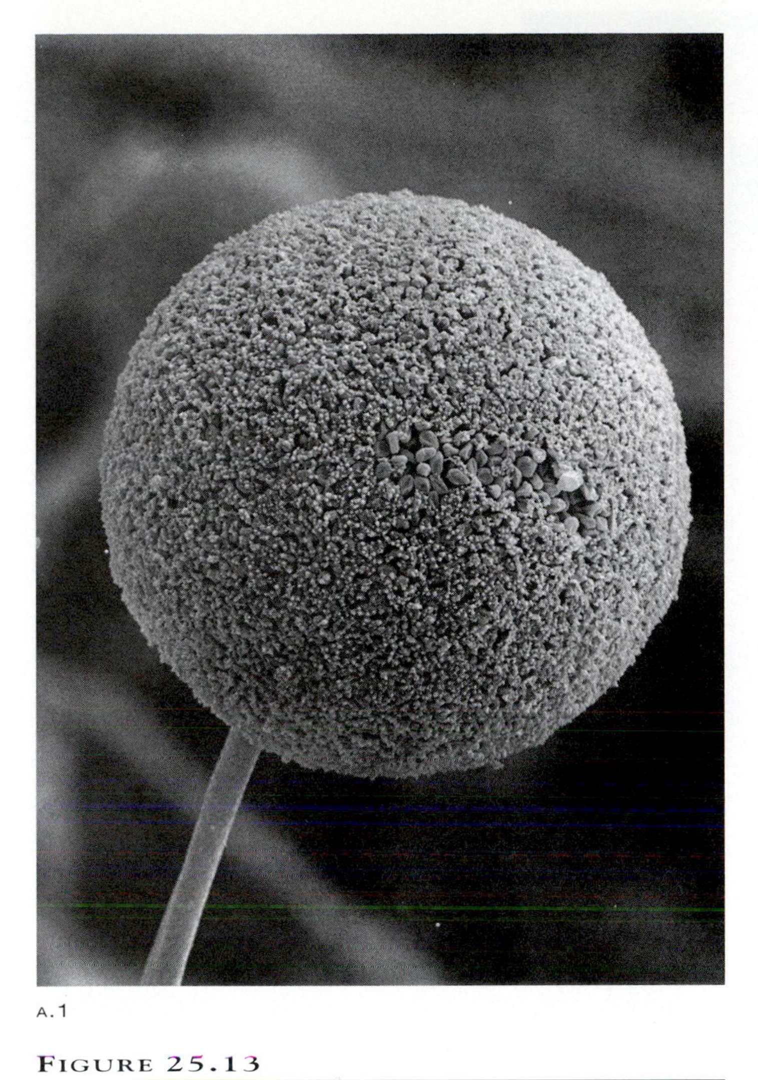

A.1

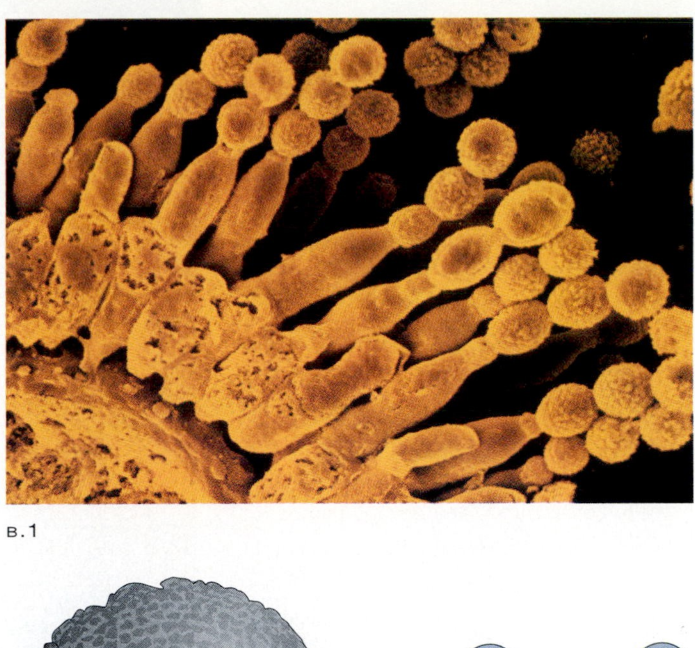

B.1

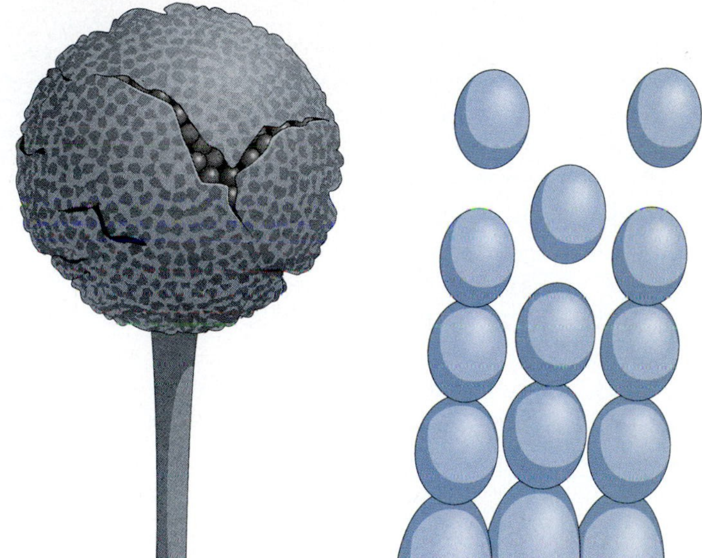

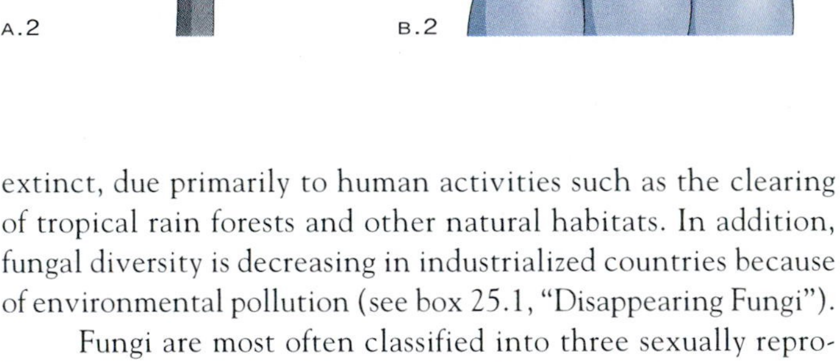

A.2

B.2

FIGURE 25.13

Fungal spores and sporangia. (a.1) A sporangium of *Rhizopus stolonifer*, ×500. (b.1) Spores being pinched off of hyphal tips of *Aspergillus*, ×1,600.

Most diseases of living plants are caused by fungi. Rusts and other fungi frequently attack crops and stored foods, causing millions of dollars of losses annually. Fungi also cause several diseases in humans, especially diseases of the skin and lungs. Although most fungi are not poisonous, the few exceptions include mushrooms that are relatively common and widespread. Consumption of such fungi has often proven fatal; no effective antidote for human poisoning by mushrooms has yet been found.

The Diversity of Fungi

Nearly 100,000 species of fungi are known, and descriptions of more than 1,000 new species are published each year. Moreover, biologists believe that more than half of all existing fungi have yet to be described. Mycologists also suspect that unknown numbers of undiscovered species have already become extinct, due primarily to human activities such as the clearing of tropical rain forests and other natural habitats. In addition, fungal diversity is decreasing in industrialized countries because of environmental pollution (see box 25.1, "Disappearing Fungi").

Fungi are most often classified into three sexually reproducing groups, variously treated as divisions, subdivisions, or classes. Regardless of their taxonomic level, the common names of these three groups persist as the **zygomycetes,** the **ascomycetes,** and the **basidiomycetes.** Furthermore, fungi that apparently cannot reproduce sexually, which are referred to as **imperfect fungi** or **deuteromycetes,** are often put into their own division. Most deuteromycetes, however, are believed to be ascomycetes. The distinguishing features of each of these groups primarily involve differences in reproduction. These differences are described in the next few paragraphs, along with general information about each group.

Boxed Reading 25.1

Disappearing Fungi

European gourmets with a taste for the subtle flavors of fresh wild mushrooms are finding these delicacies increasingly harder to find. A few years ago it was easy to pick a basket of the most prized fungus of all, the apricot-scented chanterelle. However, not only are these mushrooms becoming scarce, they are also getting smaller: it took fifty times as many chanterelles to make up a kilogram in 1975 as it did in 1958. Other fungi are also becoming rare. For example, the average number of fungal species in Holland has dropped from 37 to 12 per 1,000 square meters. These and many similar observations by other mycologists suggest that a mass extinction of mushrooms is happening all over Europe. But why?

Ecologists have recently begun to notice a negative correlation between the abundance and diversity of fungi and the amount of pollution. This negative correlation is stronger for fungi than for plants or other kinds of organisms. Apparently, fungi are more sensitive to air pollution than are plants because fungi have no protective covering, whereas the aerial parts of plants are protected by cuticles and bark. This distinction is a functional one: fungi absorb water directly from air, along with whatever else is in the air, but plants get water from soil through their roots. It seems, therefore, that fungi are being driven to extinction by bad air.*

* Colonies of luminescent fungi are often maintained aboard NASA's spaceflights. Sensitive to escaped fuel or other noxious fumes, the fungi stop luminescing when exposed to as little as 0.02 parts per million of fuel. Thus, the fungi serve as an early warning system for noxious fumes, much as the death of canaries warned miners of the lack of oxygen or the presence of dangerous gases such as methane.

Box Figure 25.2

A prized chanterelle mushroom.

Although the loss of a few gourmet treats does not seem to be important, the loss of fungi *is* harmful to forests. Hyphae associated with roots absorb some minerals more readily than do plants, and then transport them into the plants. This association normally changes as a tree gets older; one species of fungus gives way to another in a steady progression. However, this progression is speeding up in European forests; that is, "old-age" fungi are more frequently associated with "middle-age" trees. Unfortunately, trees associated with the right fungi but at the wrong age tend to die early. Changes in the patterns of fungus/plant associations and the loss of fungal diversity are early warning signals of problems for trees. This may explain why forests are dying faster where the heaviest losses of fungal diversity occur.

Similar disappearances of fungi have not been noticed in the United States, but they have probably occurred. Mycologists hypothesize that many species of fungi in the United States have not yet been described, and since people in the United States do not have a long history of collecting and eating wild fungi, there are few historical data from mushroom collectors about fungal diversity. This means that there is simply little or no information on the decline of edible wild fungi in the United States; it does not mean that such decline has not occurred. Suggestions that fungi are on the decline in the United States are nevertheless supported by collection records for lichens. These records show that most species that were native to areas such as the Los Angeles basin can no longer be found there.

A.

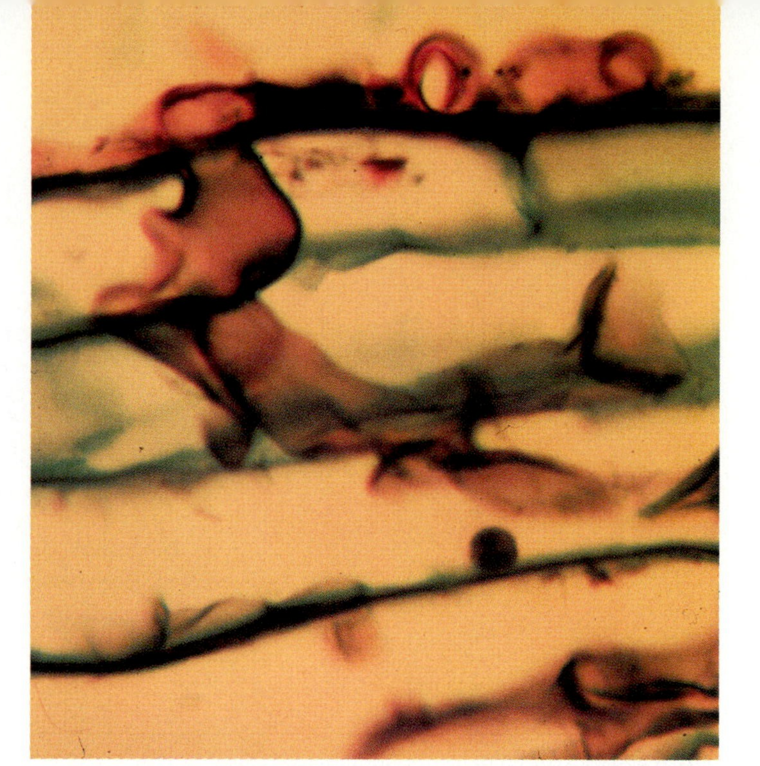

B.

Figure 25.14

Zygomycete fungi. (a) *Entomophthora* growing on a housefly; (b) *Gigaspora*, a mycorrhizal fungus penetrating a root of cotton plant.

The Zygomycetes

Zygomycetes have mostly coenocytic mycelia, with septa occurring in the hyphae of some species. The name *zygomycete* refers to the thick-walled, sometimes elaborately ornamented spore container, the **zygosporangium** (plural, **zygosporangia**), that zygomycetes form during sexual reproduction.

More than 750 species of zygomycetes are known, examples of which are the hat-throwing fungi (*Pilobolus;* see the discussion in the section on "The Physiology of Fungi"), which grow on horse dung, the fly fungi (*Entomophthora*), which are most commonly seen growing from dead flies on window panes, and most of the fungi that are symbiotic with plant roots (for example, *Gigaspora*) (fig. 25.14). One of the best-known and most widespread of the zygomycetes is *Rhizopus stolonifer,* a common bread mold that is also the chief cause of "leak," a disease of strawberries that appears while they are being transported to market. Because *Rhizopus* is so common, the following discussion of zygomycete reproduction uses this genus as representative of the entire group. Keep in mind, however, that reproduction in other zygomycete genera differs from that in *Rhizopus*.

The mycelia of *Rhizopus* occur in two morphologically identical but reproductively different mating strains, usually designated as plus (+) or minus (–). Sexual reproduction begins when mature hyphae of two different mating strains come close to each other and start to develop ovoid swellings (fig. 25.15). Once these swellings touch, a septum forms in each swelling a short distance behind the point of contact. The two multinucleate cells that form by this process are the **gametangia** (gamete containers). These gametangia (singular, **gametangium**) soon fuse, the wall separating them disintegrates, and each + nucleus unites with a – nucleus to form diploid nuclei. Any unfertilized nuclei disintegrate. The cell containing the diploid nuclei is a multinucleate zygote, which enlarges into the zygosporangium. After a resting period of a few months, the zygosporangium splits open, thereby allowing one or more filaments to grow out of it. Although meiosis has not been observed in *Rhizopus*, mycologists believe that it occurs just before these filaments germinate. As the filaments grow, the tip of each one swells, and the (probably) haploid nuclei from the zygosporangium migrate into it. A septum then forms between the swelling and the filament, and a cell wall forms around the cytoplasm that immediately surrounds each nucleus. At this point, each walled cell is a **spore,** and the swelling that contains the spores is the sporangium. Eventually, when the sporangium wall disintegrates, the spores are released into the environment, where they germinate and form new hyphae.

In addition to sexual reproduction by zygosporangia, each hypha can also sprout branches that form swellings at their tips. These swellings become sporangia filled with mitotically derived nuclei. When a cell wall forms around the cytoplasm that immediately surrounds each nucleus, it becomes a spore. These are asexual spores because neither fertilization nor meiosis is involved in their production. Because they are produced asexually, these spores form clones of the parent hypha. Most reproduction in *Rhizopus* and other zygomycetes is from asexual spores.

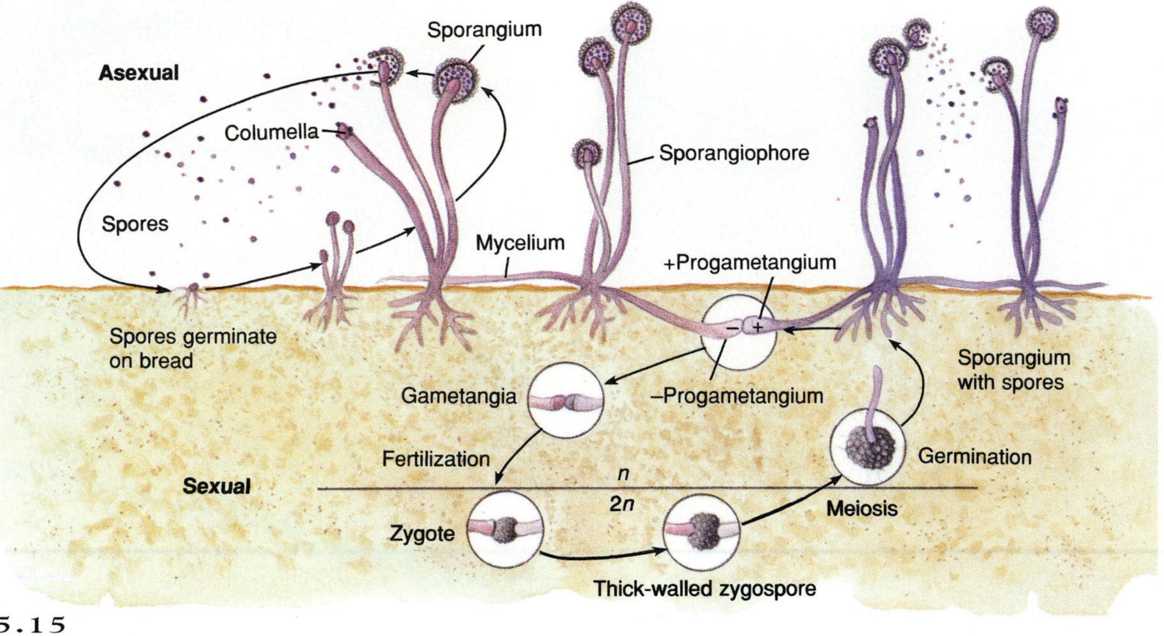

Figure 25.15

Sexual and asexual reproduction in *Rhizopus stolonifer*.

The Ascomycetes

Asco means "sac," which refers to the saclike structures where spores form in ascomycetes. The spore sacs are called **asci** (singular, **ascus**). There are about 30,000 known species of ascomycetes, the most famous of which are brewer's or baker's yeast (*Saccharomyces cerevisiae*), bread mold (*Neurospora crassa*), truffles (*Tuber melanosporum*), and morels (*Morchella esculenta*) (fig. 25.16). Yeast is probably the most important ascomycete commercially; bread mold is famous primarily as a research organism. People have used yeast to make beer since the fourth millenium b.c. Truffles and morels are gourmet delicacies, primarily in Europe.[2] Ascomycetes are also of interest because they cause plant diseases such as chestnut blight, Dutch elm disease, apple scab rot, apple bitter rot, stem rot of strawberries, powdery mildew, and brown rot of peaches, plums, and apricots. Ascomycetes are the most common type of fungi that occur in lichens.

The simplest ascomycetes are yeasts, which are unicellular. Other ascomycetes are filamentous with septate hyphae. Hyphal cells can be either uninucleate or multinucleate, but nuclei and cytoplasm are freely exchanged between adjacent cells through septal pores (fig. 25.17).

Sexual reproduction in the ascomycetes begins when uninucleate hyphae of opposite mating strains touch each other (fig. 25.18). Before contact, however, a hypha will form either male or female gametangia, depending on which mating strain it is; each gametangium forms around several nuclei at the tip of a short hyphal branch. Each female gametangium, called the **ascogonium** (plural, **ascogonia**), then sprouts slender outgrowths, called **trichogynes,** that grow toward the male gametangia, which are called **antheridia** (singular, **antheridium**).[3] Once the trichogyne touches the antheridium, nuclei migrate through it from the antheridium to the ascogonium.

The intermingling of nuclei from different mating strains is not followed immediately by fertilization. Instead, the ascogonium sprouts new hyphae (**ascogneous hyphae**) whose cells are dikaryotic, with one nuclear type from each of the mating strains in each cell. Cell division occurs in such a way that every cell contains one of each type of nucleus. These dikaryotic hyphae may grow extensively among the uninucleate hyphae of the parent mating strains, often becoming compacted into different kinds of fruiting bodies. As the fruiting bodies begin to form, the tip of each dikaryotic hypha forms a hook. Once the hook is in place, the cell just below the hyphal tip enlarges, and

2. Morels, which sell for as much as $40 per pound ($88 per kg), are an important part of the $50-million-per-year wild mushroom industry. Commercial mushroom pickers in the northwest earn $70 to $500 per day.

3. Maleness and femaleness are artificial here. Males and females are so designated simply because their gametes either move, as male gametes do, or are stationary, as female gametes are.

A.

B.

FIGURE 25.16

Fruiting bodies of ascomycetes: (a) truffles; (b) morel.

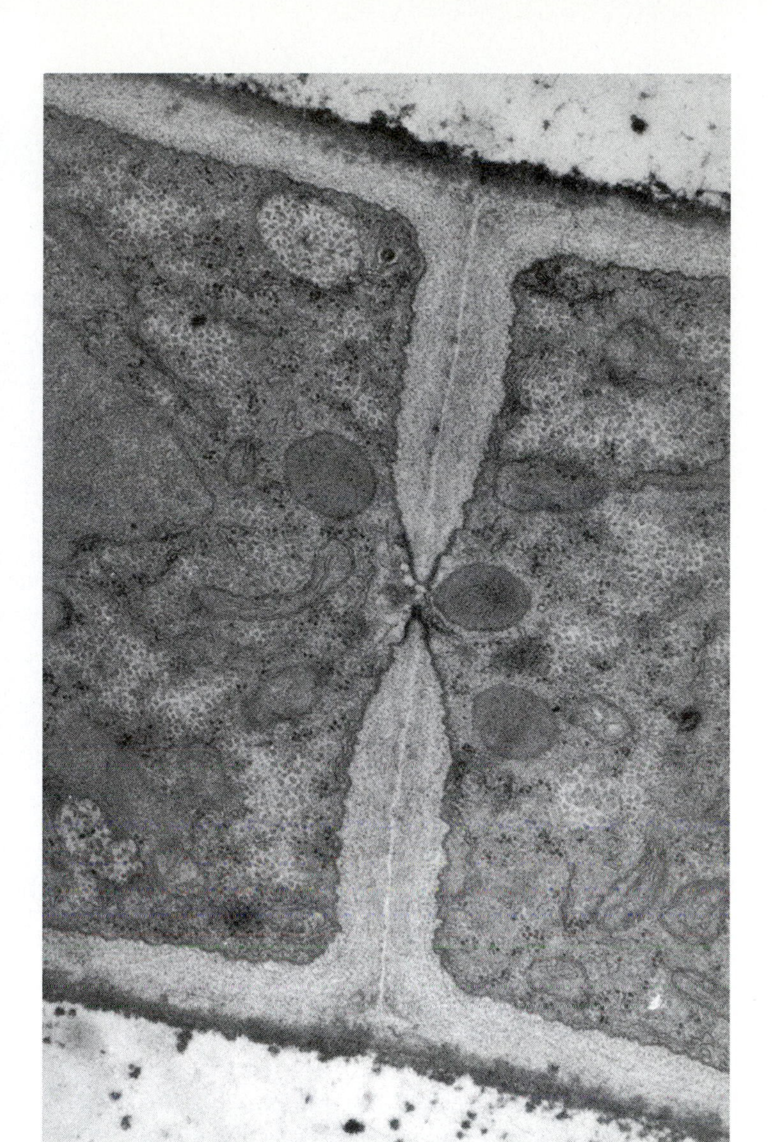

FIGURE 25.17

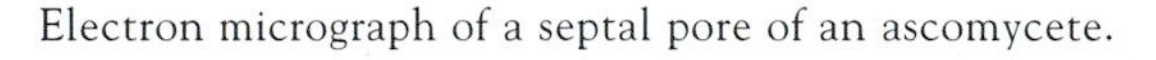

Electron micrograph of a septal pore of an ascomycete.

the two nuclei in it fuse to form a zygote nucleus, the only diploid nucleus in the life cycle. This zygote nucleus immediately undergoes meiosis. Mitosis follows meiosis in such a way that the resulting eight nuclei are arranged in a row. Each nucleus becomes a spore when a cell wall forms around it and the immediately surrounding cytoplasm. By this time the cell in which fertilization occurred has enlarged into an ascus that contains eight spores.

The fruiting body of ascomycetes is called an **ascocarp,** and the spores released from it are the **ascospores.** There are three main types of ascocarps: (1) a **cleistothecium** is a completely closed,

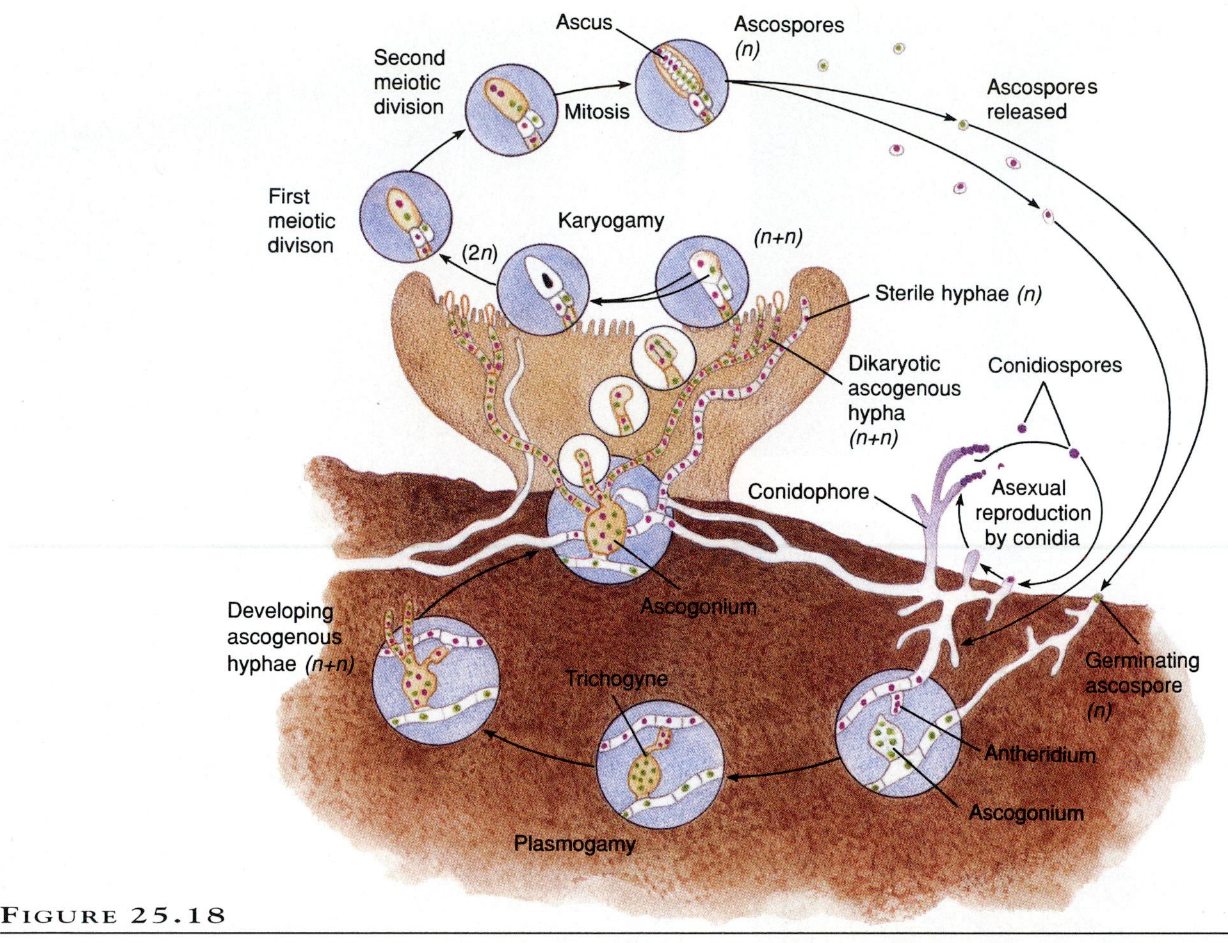

FIGURE 25.18

The sexual life cycle of an ascomycete.

almost spherical ascocarp; (2) a **perithecium** is a usually flask-shaped ascocarp that opens by a single pore; (3) an **apothecium** is shaped like an open cup.

Unlike the zygomycetes, the ascomycetes do not form spores within sporangia during asexual reproduction. Instead, the spores are pinched off in neat rows at the tips of exposed hyphae. Asexual spores that are formed by pinching off hyphal tips are called **conidia,** and the hyphae that bear them are called **conidiophores** (fig. 25.19).

The Deuteromycetes

The deuteromycetes, also called the Imperfect Fungi, form an artificial group, since it is defined by a single feature: the absence of sexual reproduction. The approximately 17,000 species of imperfect fungi reproduce almost exclusively by conidia. Because their asexual reproduction usually resembles that of the ascomycetes, most of the deuteromycetes probably descended from an ascomycete ancestor that lost the ability to reproduce sexually. This suggestion is supported by the observation that whenever sexual reproduction is discovered in a deuteromycete, it is usually of the ascomycete type. However, the sexual reproduction of a few deuteromycetes resembles that of the basidiomycetes (discussed later in this chapter), which means that a small proportion of deuteromycetes are of basidiomycete origin.

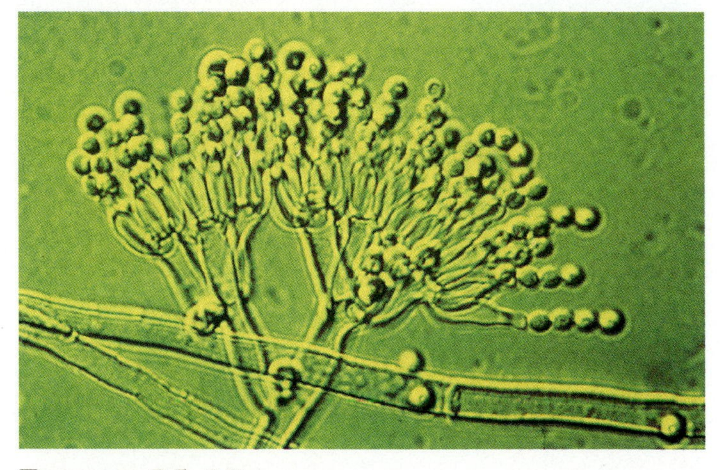

FIGURE 25.19

Light micrograph of the conidiophores of *Penicillium*, ×790.

Once sexual reproduction is found in a deuteromycete, the species is then reassigned to an appropriate genus of sexual

Boxed Reading 25.2

The Yeast Genome

Yeasts have long been a favorite experimental organism of biologists. These ascomycetes are single-celled and simple, and their genome—which is only one-fiftieth the size of the human genome—is considered to be a basic, no-frills set of instructions for maintaining the cells of higher organisms, including those of humans. Baker's yeast has only sixteen chromosomes, number three of which is the third smallest. Although this chromosome is relatively small and simple, it took a huge amount of work—147 biologists working in 35 labs in 17 countries—to spell out its DNA sequence. The chromosome consists of 315,356 base-pairs, making it the longest stretch of DNA ever sequenced and the first chromosome of any organism to be sequenced end-to-end.

When the sequencing project began in 1989, biologists had mapped just 34 genes on chromosome 3. However, the completion of its sequencing in 1992 showed that the chromosome has 182 genes. This huge number surprised biologists, who are now trying to figure out the function of each of the genes. The sequencing has produced other surprises: for example, only 10% of the new genes bear any resemblance to known genes of other organisms. One of the genes, strangely enough, codes for a protein normally used by some bacteria to fix nitrogen (see Chapter 20). Although yeasts do not fix nitrogen, they still need the gene. Indeed, when the gene was deleted, the yeast died, suggesting that the gene may be needed for other aspects of general metabolism.

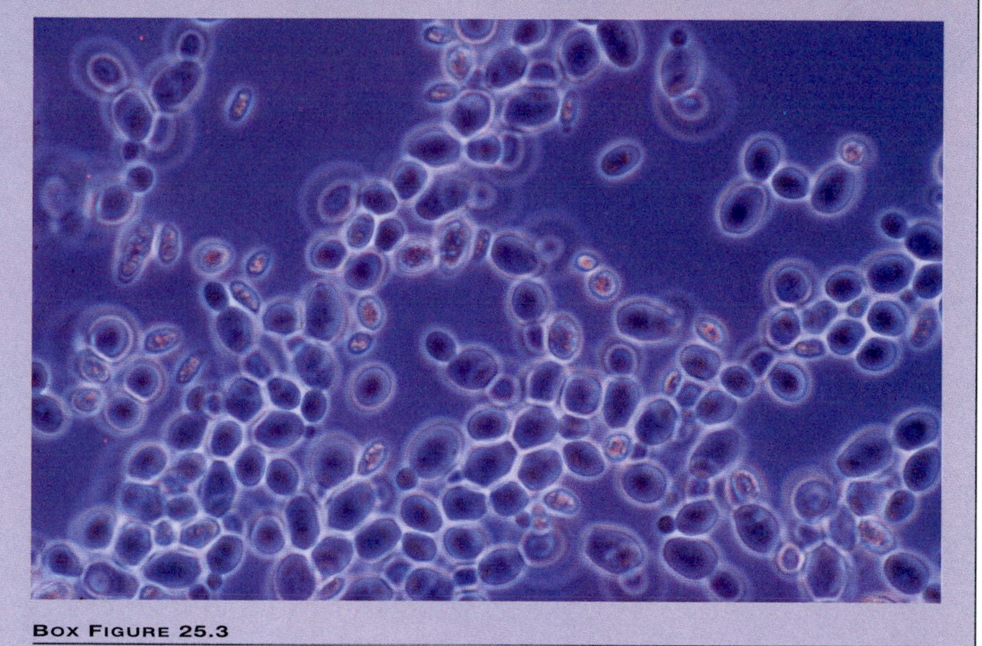

Box Figure 25.3
Light micrograph of yeast.

fungi. The taxonomic reassignment involves replacing the old deuteromycete genus name but keeping the second part of the binomial. The new genus name can be either new or one that already exists, depending on the similarity of the species in question to other fungi. For example, when the deuteromycete species *Aspergillus chevalieri* was discovered to produce ascocarps, it was reclassified as *Eurotium chevalieri*, an ascomycete. Nevertheless, mycologists still use both names: *A. chevalieri* for the conidial stage and *E. chevalieri* for the ascocarp stage. Classification of sexual and asexual fungi into separate groups results in the same species being classified in two genera in two divisions. Furthermore, species of *Aspergillus* that have no known sexual counterpart are probably more closely related to *Eurotium* than to other deuteromycetes.

Deuteromycetes are mostly free-living and terrestrial, but some are pathogenic. The most well-known of the pathogenic deuteromycetes include the causal agents of a respiratory disease called aspergillosus (*Aspergillus niger*), athlete's foot (*Epidermophyton floccosum*), ringworm (*Tinea* species), and candida "yeast" infections (*Candida albicans*). The most famous deuteromycetes are species in the genus *Penicillium: P. notatum* for its role in the discovery of penicillin, *P. chrysogenum* for the commercial production of penicillin, *P. griseofulvum* for the production of griseofulvin (the only effective antibiotic against ringworm and athlete's foot), and *P. roquefortii* and *P. camembertii*, which are used to make Roquefort and Camembert cheeses, respectively (fig. 25.20).

Lichens

Lichens are symbiotic relationships consisting of a fungus and a green alga, a fungus and a cyanobacterium, or a fungus with both, in a spongy body called a **thallus** (plural, **thalli**). This symbiosis seems to be a parasitic one, since the fungus gets carbohydrates from algae and cyanobacteria and fixed nitrogen from cyanobacteria, but the photosynthetic organisms derive no apparent benefit from the fungus.

The lichen thallus generally consists of several layers of cells or hyphae of an ascomycete, although about twenty

FIGURE 25.20

Commercial products from *Penicillium* fungi: (a) bottle of penicillin; (b) Roquefort cheese.

lichenized basidiomycetes are known (fig. 25.21). At the surface of the thallus is a protective layer where hyphae are so compressed that they resemble parenchyma cells. Below this upper layer is the algal or cyanobacterial layer, where the photosynthetic organisms are scattered among hyphae. Next is a layer of loosely packed hyphae that occupies at least half the volume of the thallus. Finally, a bottom layer of tightly packed hyphae is frequently but not always present. This layer is often accompanied by anchoring strands of hyphae.

Each of the approximately 20,000 known species of lichens is tacitly assumed to have its own unique fungal component, but the photosynthetic symbionts in about 90% of lichens come from five genera: the green algae *Trebouxia*, *Pseudotrebouxia*, and *Trentepohlia*, and the cyanobacteria *Nostoc* and *Anabaena*. Although a fungus, a green alga, and a cyanobacterium in a lichen might have their own Latin binomials, the names of lichens are independent of the species names of their symbionts. Nevertheless, their names and classifications are based on the features of fungi. Moreover, although lichens are classified as fungi, lichen species are separated into families, orders, and classes apart from nonlichenized fungi.

Individual symbionts of lichens can be isolated and cultured separately, but in pure culture the fungi often assume indefinite, compact shapes, and the algae and cyanobacteria grow faster than they do when they are part of a lichen. The

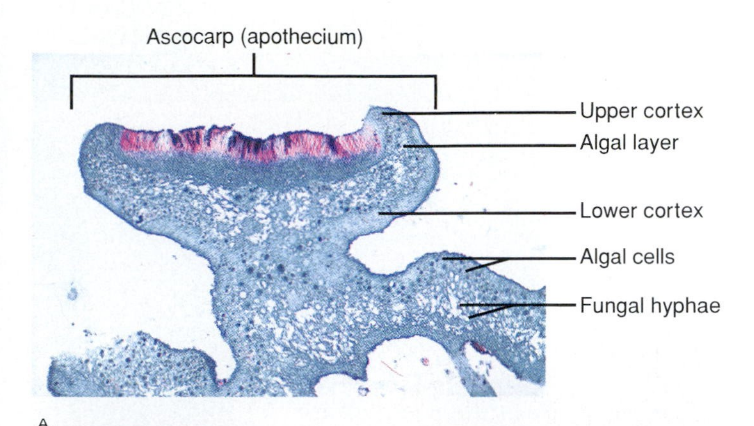

FIGURE 25.21

Lichens. (a) Section through part of a foliose lichen thallus (*Physcia* sp.). (b) A fruticose lichen (*Cladonia deformis*). The red structures are immature apothecia.

A.

B.

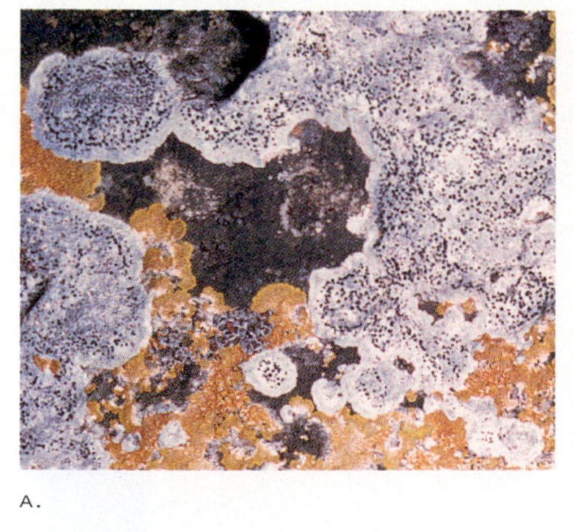

C.

Figure 25.22

Types of lichens. (a) Crustose lichens. (b) A foliose lichen. (c) A fruticose lichen with apothecia.

lichen-forming fungi rarely grow by themselves in nature, but the algal and cyanobacterial components of lichens often do. These observations are further evidence that lichen fungi parasitize algae and cyanobacteria.

Lichen thalli grow slowly, sometimes up to 1 cm per year or as slow as 0.1 mm per year. On the basis of such growth rates, lichenologists estimate that some lichens have lived at least 4,500 years. Furthermore, lichens tolerate environmental conditions that are too extreme for most other forms of life. Indeed, lichens live on bare rocks in the blazing sun or bitter cold in deserts, in both arctic and antarctic regions, on trees, and just below the permanent snow line of high mountains where nothing else will grow. One species grows completely submerged on ocean rocks. They even attach themselves to artificial substances such as glass, concrete, and asbestos.

Although lichens can withstand many environmental extremes, they are, like other fungi, sensitive to air pollution. Indeed, lichens are effective monitors of air pollution: lichens in or near large cities throughout the world have disappeared rapidly during the twentieth century. Similarly, they have recolonized areas in which air quality has improved.

Lichens are dispersed in nature primarily by asexual means. Rain, wind, running water, and animals disperse lichen fragments that grow into new thalli. The fungi reproduce sexually, as evidenced by the brightly colored apothecia often produced by lichens (figs. 25.21b; 25.22). Sexual reproduction is a bit of a problem though, because spores, or hyphae from them, must contact algae or cyanobacteria before the lichen can develop. No one has ever observed such an initiation of a new thallus in nature.

CONCEPT

Lichens are symbiotic relationships consisting of a fungus and a green alga, a fungus and a cyanobacterium, or a fungus with both. The classification of lichens is independent of the other fungal groups.

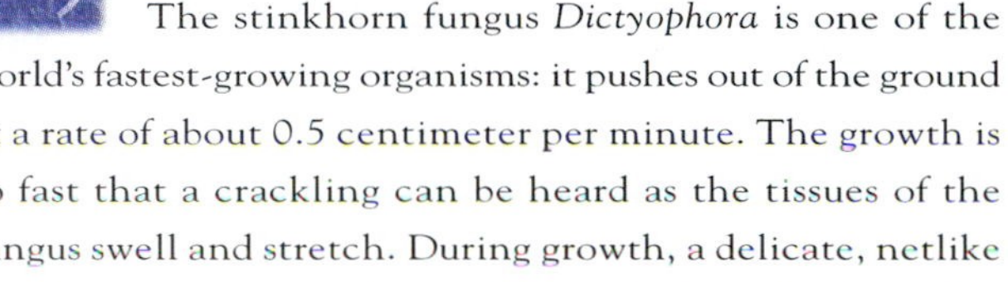

The Lore of Plants

The stinkhorn fungus *Dictyophora* is one of the world's fastest-growing organisms: it pushes out of the ground at a rate of about 0.5 centimeter per minute. The growth is so fast that a crackling can be heard as the tissues of the fungus swell and stretch. During growth, a delicate, netlike veil forms around the fungus (this is the basis for the other common name of the fungus, "the lady of the veil"). The fungus then decomposes and, in the process, produces a strong odor of decaying flesh. This odor attracts flies, which crawl over the fungus and collect spores on their feet, thereby ensuring that the spores are carried to new areas.

The Basidiomycetes

Like ascomycetes, basidiomycetes are filamentous with uninucleate or multinucleate (i.e., dikaryotic) hyphae. Basidiomycetes are so named because they form spores on structures called **basidia** (singular, **basidium**), which are hyphal swellings that bear spores on tiny pegs. The fruiting body is a **basidiocarp,** which is best known in the form of mushrooms (fig. 25.23), bracket fungi, jelly fungi, puffballs, earth stars, coral fungi, bird's nest fungi, and stinkhorns (see fig. 25.12). Basidiomycetes also include rusts and smuts, which cause plant diseases and do not form basidiocarps (fig. 25.24).

Except for the absence of basidiocarps in some species, the basidiomycetes have a common reproductive cycle. The discussion that follows regarding reproduction in the common edible mushroom (*Agaricus brunnescens*) (fig. 25.25), therefore, represents the entire basidiomycete group (fig. 25.25).

Reproduction by *A. brunnescens* begins when uninucleate hyphae of different mating types touch each other. Unlike the ascomycetes, the basidiomycetes have no specialized gametangia. Instead, a dikaryotic hypha grows from the fusion point

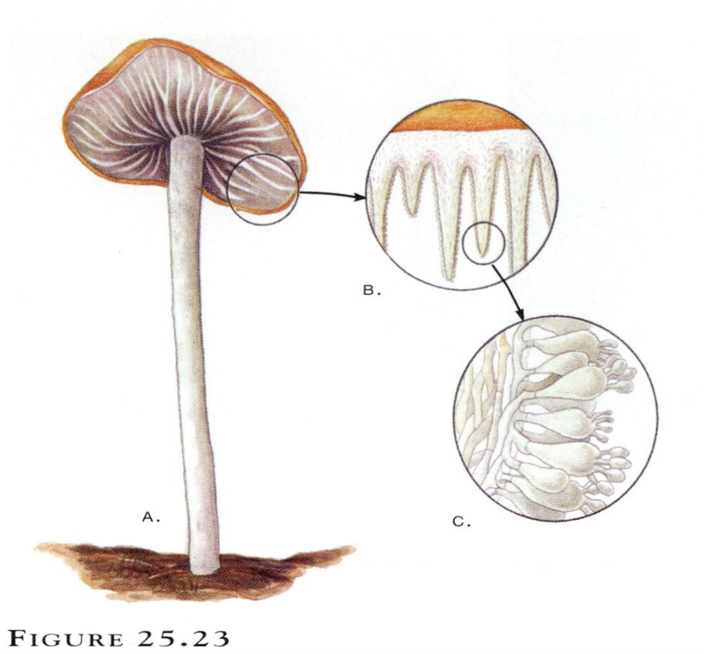

FIGURE 25.23

The structure of a mushroom.

of the two parent hyphae. As in the ascomycetes, the dikaryotic hyphae of basidiomycetes divide in such a way that each cell has one nucleus from each type of parental nucleus. The basidiomycetes are unique, however, because their dikaryotic hyphal growth involves the formation of a small bypass loop between cells. This loop is called a **clamp connection** because it looks like it clamps adjacent cells together (fig. 25.26).

As dikaryotic hyphae grow, they form a compact mycelium that eventually develops into a small, closed basidiocarp. As these mushroom-type basidiocarps develop, basidia form on fleshy plates called **gills,** in what will become the mushroom's cap (fig. 25.23). As basidia form, the nuclei in cells at the hyphal tips unite into diploid zygotes and then immediately undergo meiosis. The haploid nuclei from meiosis migrate into the pegs that extend from the basidia, and the tips of the pegs expand into spores.

Asexual reproduction in the basidiomycetes is primarily by means of conidia, budding, or fragmentation. Fragmentation involves the breakup of hyphae into single cells that develop into mycelia without the formation of thick walls or of internally produced spores.

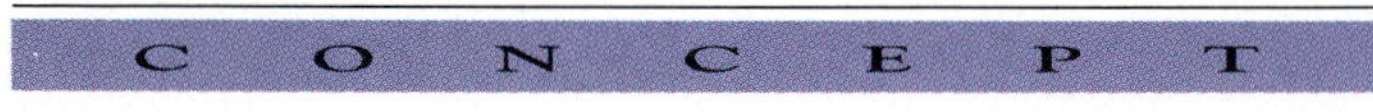

Fungi that have sexual life cycles are classified into three groups: zygomycetes, ascomycetes, and basidiomycetes. This classification is based primarily on features of the reproductive structures.

FIGURE 25.24

Wheat infected with wheat rust (*Puccinia graminis*).

Doing Botany Yourself

Walk around your neighborhood or a park and collect any fungi that you see. How would you classify each fungus? Explain your answer.

The Physiology of Fungi

Fungal metabolism is similar to plant metabolism, except that fungi are never photosynthetic, and they can metabolize a wider variety of carbon sources, including sugar, plastic, and jet fuel. Fungi may also show some of the same responses to light as do plants. For example, many fungi cannot reproduce without light, and others show positive phototropic responses. Fungal phototropism has been studied most thoroughly in the hat-throwing

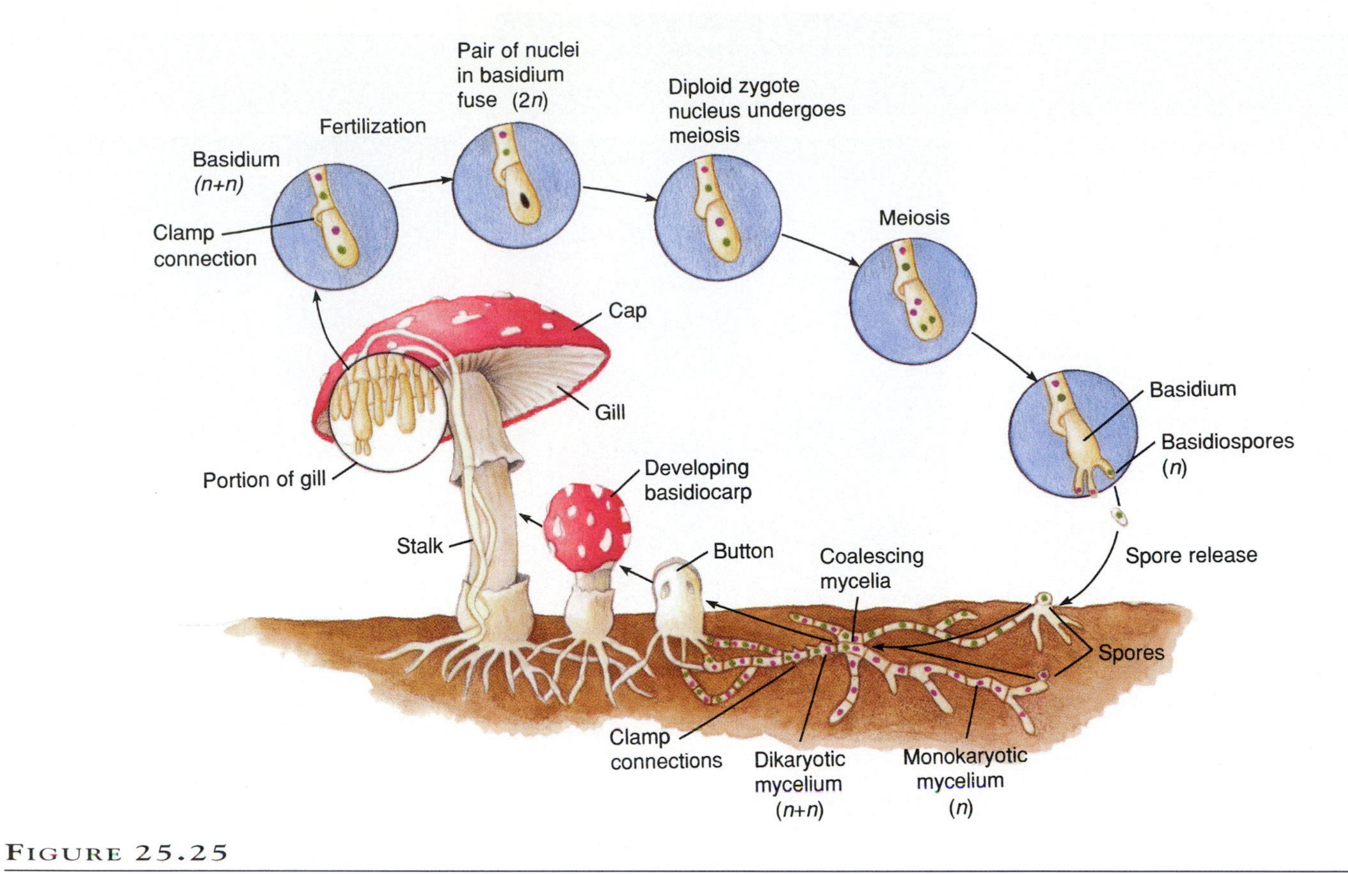

FIGURE 25.25

The sexual life cycle of a mushroom.

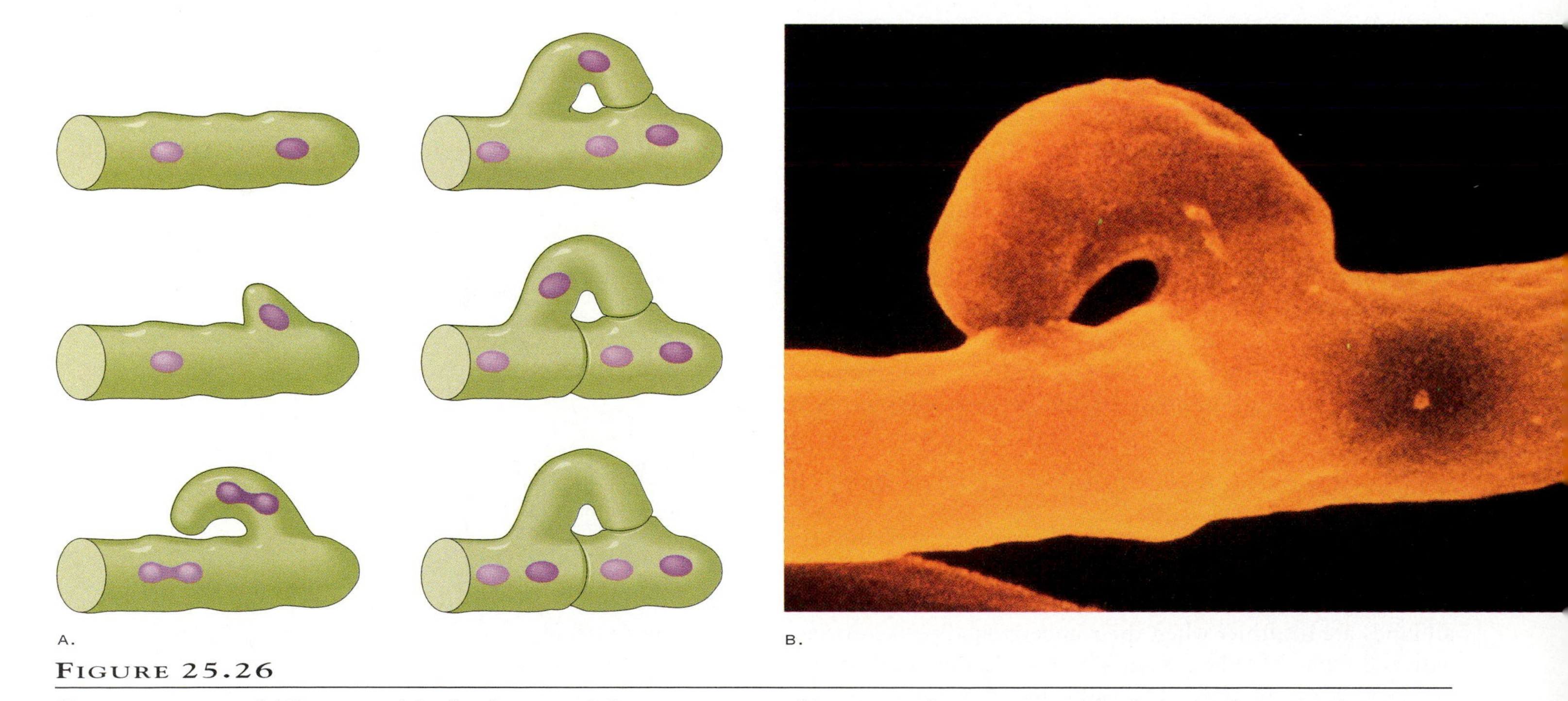

FIGURE 25.26

Clamp connections. (a) Drawings of the development of clamp connections; (b) scanning electron micrograph of a hypha showing a clamp connection, ×20,000.

Boxed Reading 25.3

Fungal Foolery

Rock cress (*Arabis holboellii*) usually produces delicate, pale purplish-pink flowers. However, plants infected by the rust fungus *Puccinia monoica* (a pathogen of several species of mustard) produce twice as many leaves and, in one species, a dense cluster of yellow leaves that make the plant look like a buttercup, both to many insects and botanists alike. The leaves appear yellow because they are covered with spermagonia that produce a sugary fluid that attracts pollinators such as flies, bees, and butterflies. This fluid contains 10–100 times more sugar than does the nectar of nearby flowers, and the pollinators stay up to five times longer at the pseudoflowers than they do at real flowers. As the pollinators shuttle from one counterfeit flower to another, they carry fungal gametes between mating types, thereby increasing the reproductive success of the fungus.

A. B.

Box Figure 25.4

(a) A typical rock cress (*Arabis* sp.) in flower; (b) the yellow, flowerlike leaves of a rock cress that has been invaded by a rust fungus. The fake flower exudes a fragrant, sugary fluid attractive to insects, which inadvertently redistribute fungal sex cells that adhere to their bodies.

fungus (*Pilobolus*). This zygomycete produces upright hyphae up to about 10 mm tall, primarily on horse dung. The sporangium-bearing hyphae grow toward light so that the light enters a lenslike swelling below the sporangium. Although the mechanism of this response to light is not known, some kind of photoreceptor probably influences the swelling to split, enabling the explosive release of turgor pressure in the cells of the swelling. The explosion of the swelling blasts its sporangium up to 2 m away, at a speed of almost 60 kilometers per hour, which is faster than the speed limit on most city streets. The sporangia stick to grass that may be eaten by a horse, through which the spores pass unharmed and germinate on the next round of dung.

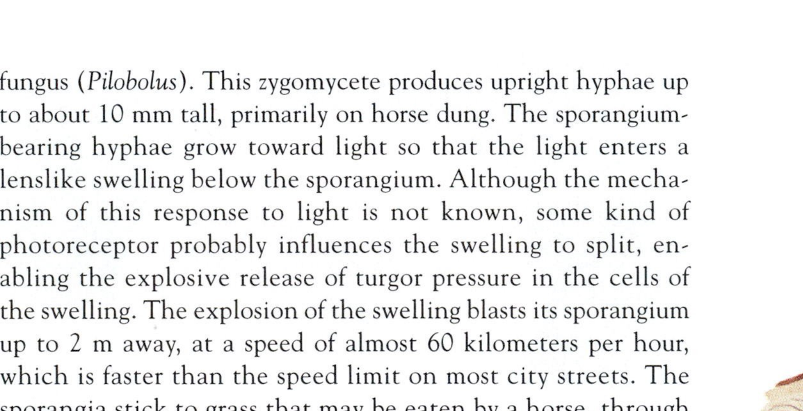

Figure 25.27

How light influences the release of sporangia of *Pilobolus*.

The Ecology of Fungi

The ecology of fungi is especially important for plants. For example, tiny orchid seeds cannot germinate until they are invaded by hyphae of the soil fungus *Rhizoctonia*, and plants of all kinds are healthier when their underground parts associate with soil fungi. Members of all 400 or so families of flowering plants, with the exception of possibly fewer than a dozen, form **mycorrhizae** (see pp. 348, 478–479). A mycorrhiza, meaning "fungus-root," is an association between a fungus and the underground parts of a plant. The association is a mutually beneficial symbiosis: the plants provide a source of carbon for the fungus, and the fungus absorbs phosphorus or other minerals that the plant cannot otherwise get easily from the soil.

Mycorrhizae are divided into two groups, depending on whether the fungal component penetrates the plant (**endomycorrhizae**) or forms only an external mantle around the plant's roots (**ectomycorrhizae**) (fig. 25.28). Endomycorrhizal fungi are mostly zygomycetes. They occur in about 80% of all vascular plants, usually forming balloonlike structures (vesicles) or treelike structures (arbuscles). These structures have led to the name **vesicular-arbuscular (V-A)** mycorrhizae for associations that have these fungal components (fig. 25.28a). Endomycorrhizae are especially important to tropical plants, because the fungi can help plants obtain phosphorus from the phosphate-poor soils that characterize tropical habitats. Ectomycorrhizal fungi are primarily associated with the roots of trees and shrubs in temperate regions. Ectomycorrhizal fungi apparently replace root hairs, which may be absent in ectomycorrhizal roots (fig. 25.28b). Most ectomycorrhizal fungi are basidiomycetes, but some are ascomycetes.

The most significant impact of fungi on plants is the formation of mycorrhizae. The fungal symbiont in a mycorrhiza gets carbohydrates from the plant host, and the plant gets minerals from the fungus. Most plants depend on mycorrhizae.

What are Botanists Doing?

Although most woodland mushrooms are either saprobic or mycorrhizal, the saprobic oyster fungus shown in figure 25.12a is also predatory. Predatory fungi lure, trap, lasso, paralyze, colonize, or enzymatically dissolve their prey. Go to the library and read a recent article about this unusual life-style. Summarize the main points of the paper. What are some questions you have about these fungi? What experiments could you do to answer these questions?

The Economic Importance of Fungi

Fungi have greatly benefited human societies as sources of industrial chemicals, antibiotics, medicines, and vitamins. They are the mainstay of the brewing and baking industries, and are also important for making certain dairy foods, including gourmet cheeses. Fungi also cause many plant and animal diseases.

Fungi produce gallic acid, which is used in photographic developers, dyes, and indelible black ink, and in the production of artificial flavoring and perfumes, chlorine, alcohols, and several acids. Fungi are also used to make plastics, toothpaste, and soap, and in the silvering of mirrors. In Japan, almost 500,000 metric tons of fungus-fermented soybean curd (tofu, miso) are consumed annually.

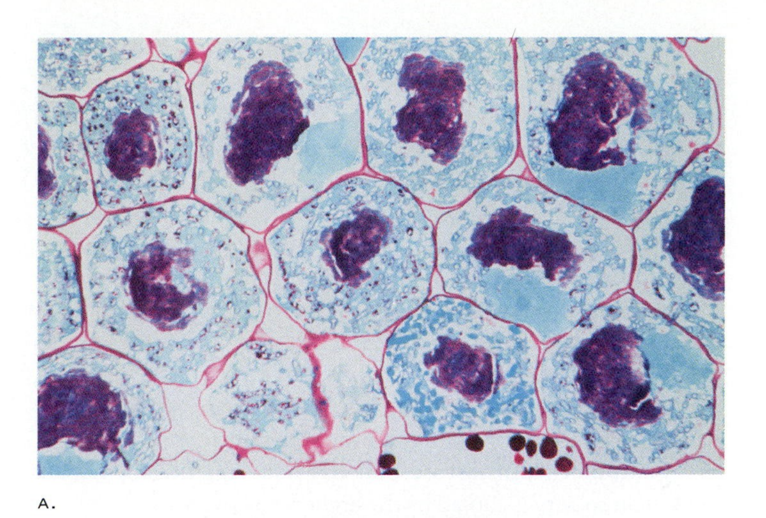

A.

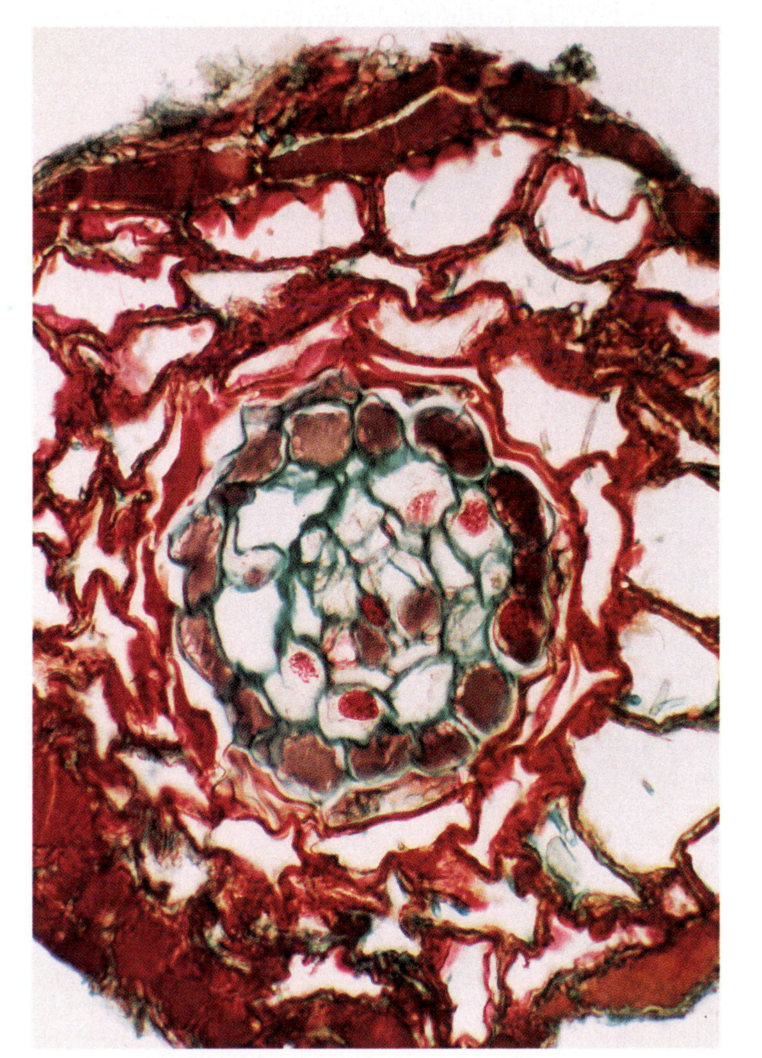

B.

Figure 25.28

Mycorrhizae. (a) Light micrograph of V-A endomycorrhizal fungus in plant root cells, ×400. (b) Transverse section of a plant root surrounded by ectomycorrhizal fungus.

Different strains of the rust fungus, *Puccinia graminis*, cause billions of dollars of damage annually to food and timber crops throughout the world (fig. 25.24). Plant breeders are constantly faced with the challenge of developing rust-resistant varieties of crops. As a result of this effort, for example, Donald Winkelmann of the International Maize and Wheat Improvement Center in Mexico City recently announced that the defeat of wheat-rust is imminent. Scientists at the center have found a Brazilian-grown wheat plant that somehow controls the disease by slowing its growth. The "slow-rust" genes of this plant have been incorporated into cultivated wheat by hybridization experiments, and the new rust-resistant wheat is now being grown in more than one hundred countries.

Another plant disease that has had a significant impact on human society is caused by the ergot fungus, *Claviceps purpurea* (fig. 25.29). This fungus infects the inflorescences of rye and other grain crops. Ergot seldom damages a crop significantly, but it produces several powerful drugs in the maturing grain. If infected rye is harvested and milled, a disease known as *ergotism* may occur in those who eat the bread made from the contaminated flour. The disease can affect the central nervous system, often causing hysteria, convulsions, and even death. Another form of ergotism causes gangrene of the limbs, and cattle that eat infected grass often abort their calves. Regardless, ergot-derived drugs have been used since the sixteenth century to hasten childbirth by stimulating uterine contraction. Because other ergot drugs constrict blood vessels, they are used to stop the bleeding often associated with childbirth. Ergot drugs have also been used to treat migraine headaches, heart palpitations, nervous stomach, menopausal disorders, and several other medical problems.

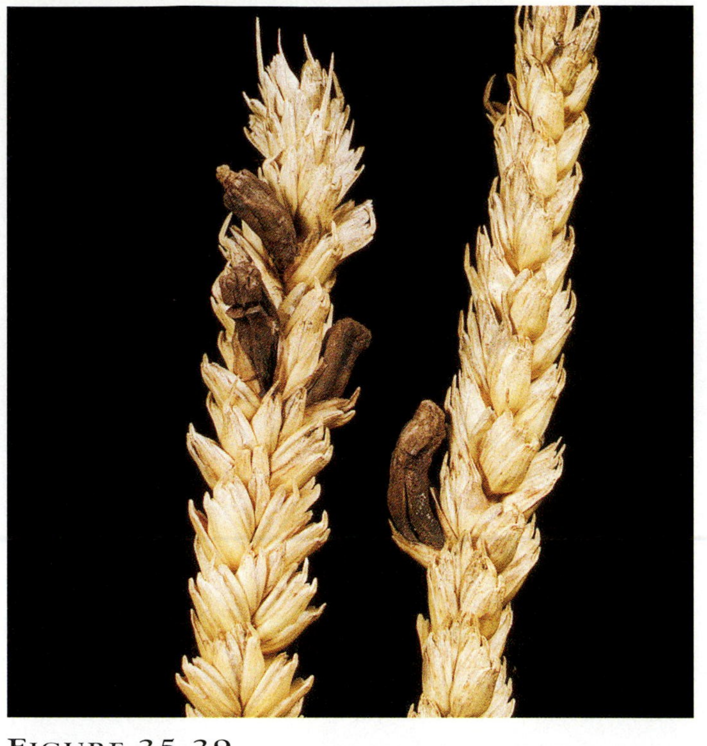

FIGURE 25.29

Plant infected with ergot (*Claviceps purpurea*).

The Relationships of Fungi

The fossil record of fungi is more fragmentary than that of any other group of multicellular eukaryotes (see box 25.4, "Fossil Mushrooms" on p. 616). Nevertheless, mycologists have used phylogenetic systematics to infer the evolutionary relationships of fungi. After dispensing with the deuteromycetes because of their artificiality as a group, one of the main questions about fungal relationships involves the zygomycetes, ascomycetes, and basidiomycetes. Traditionally, the zygomycetes have been thought of as "lower" fungi and the ascomycetes and basidiomycetes as "higher" fungi. This terminology reveals the traditional view of mycologists that the zygomycetes are more primitive, and the ascomycetes and basidiomycetes are more advanced and closely related to each other. Based on a cladistic analysis of fungi, Anders Tehler of the University of Stockholm has proposed a phylogeny that bears out this traditional view (fig. 25.30). He also supports the traditional view that flagellated molds, which are discussed in the next section of this chapter, should be included in Kingdom Fungi. These organisms are now more often classified as protoctists, a view supported by molecular phylogenies that show little relationship between flagellated molds and fungi. In addition to suggesting evolutionary relatedness, Tehler took the next step for phylogenetic classification by designating taxonomic categories based on relationships among monophyletic groups. These categories, as well as some of the character state changes that support the relationships suggested by Tehler, are also shown in figure 25.30. Note that one of the surprises in this cladogram is the close relationship between the basidiomycetes and *Saccharomyces*, which is traditionally classified as an ascomycete. One of the features that supports this idea is the shared origin of **meio-blastospores.** This is the name for spores that arise by budding from haploid, meiotically produced spores.

C O N C E P T

The zygomycetes are probably the closest representatives of the ancestor of true fungi. The ascomycetes and basidiomycetes are more closely related to each other than either group is to the zygomycetes.

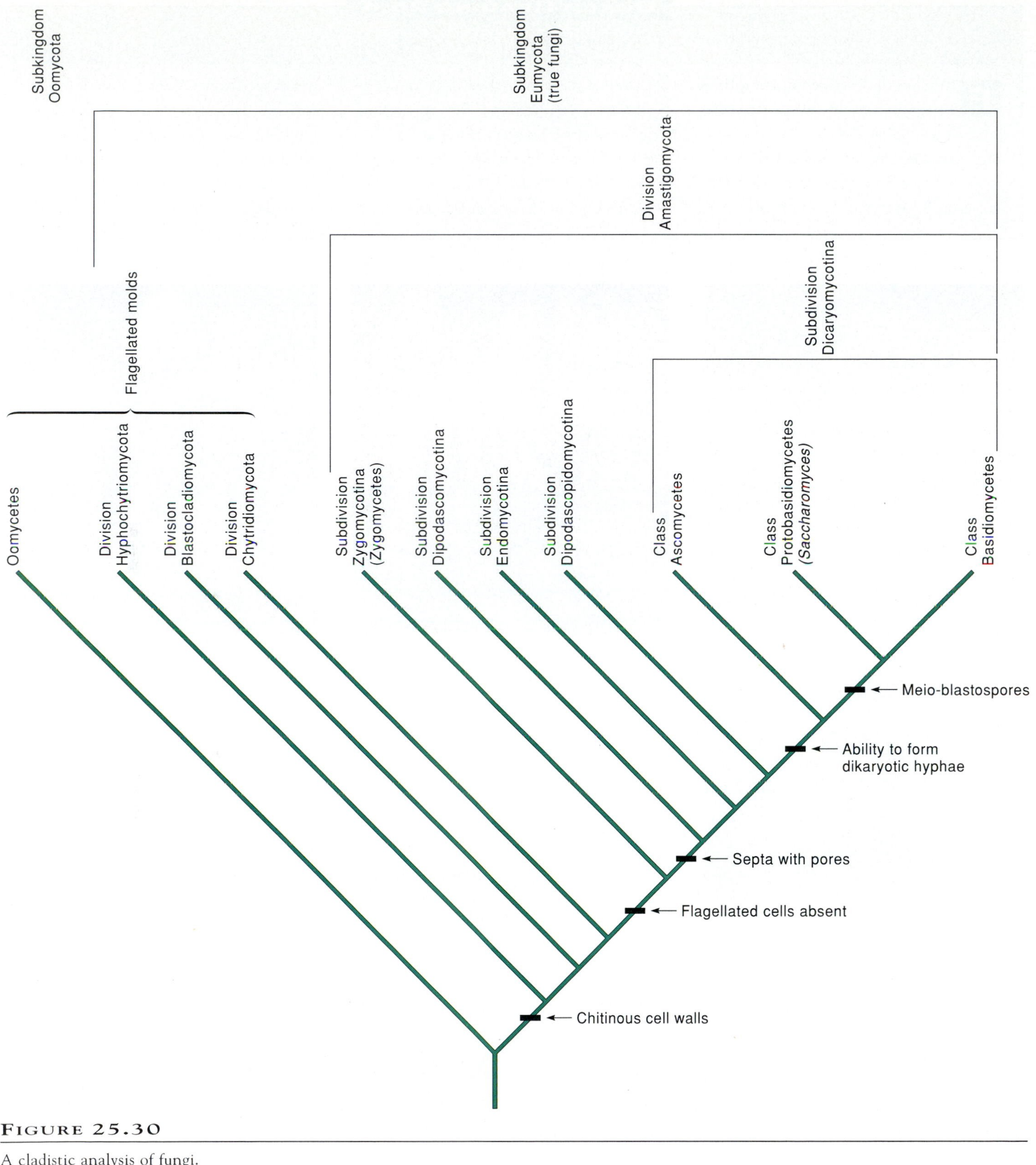

FIGURE 25.30

A cladistic analysis of fungi.

Boxed Reading 25.4

Fossil Mushrooms

Fungi have a long record of fossilization as spores and hyphae. However, because of the almost complete absence of any substance that can be fossilized, mushrooms are rare in the fossil record. In 1990, however, G. O. Poinar, Jr., of the University of California at Berkeley, and Rolf Singer, of the Field Museum of Natural History in Chicago, reported what is believed to be the earliest fleshy gilled mushroom and the only fossil mushroom known from the tropics. The mushroom was well-preserved in Dominican Republic amber, which is between 35 and 40 million years old. Because a microscopic examination of the fossil mushroom suggests that it is related to modern inky cap mushrooms (*Coprinus* species); its discoverers named it *Coprinites dominicana.*

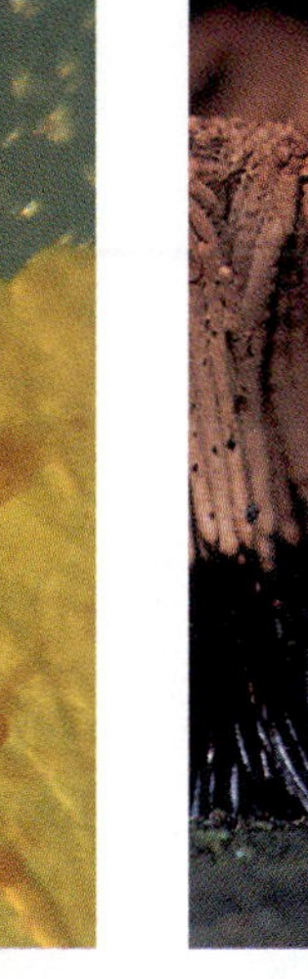

A.

B.

C.

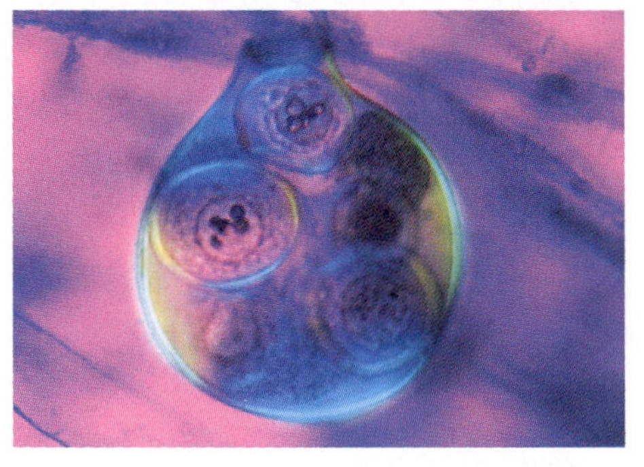

D.

Figure 25.31

Slime molds and water molds. (a) Plasmodium of the common slime mold *Physarum polycephalum*. (b) Sporangia of the slime mold *Stemonitis splendens*. (c) An immature sporangium of *Dictyostelium*, a cellular slime mold. (d) An oogonium of the water mold *Saprolegnia*.

Slime Molds and Water Molds

Two groups of organisms that resemble fungi in their general appearance, but which form motile cells, are informally referred to as *slime molds* and *water molds*. Slime molds include the plasmodial slime molds and the cellular slime molds (fig. 25.31a–c); water molds are grouped into those with uniflagellate motile cells and those with biflagellate motile cells (fig. 25.31d). Some classifications still treat slime molds and water molds as fungi, but they are not closely related to fungi. Many biologists now regard these molds as members of Kingdom Protoctista. As protoctists, they are usually classified into four divisions, whose main characteristics are summarized in table 25.2.

Slime molds and water molds resemble fungi in that they are heterotrophic, store glycogen, and have cell walls that contain cellulose and chitin. They differ from fungi in that they form swimming cells that have flagella or that are amoeboid. The slime molds also differ from true fungi by being **phagotrophic;** that is, they ingest solid food particles. In contrast, fungi obtain their nutrition by absorbing dissolved nutrients. Furthermore, some of the water molds store a carbohydrate (**mycolaminarin**) that is more similar to that of brown algae (see Chapter 26) than it is to the storage carbohydrates of fungi, plants, or animals.

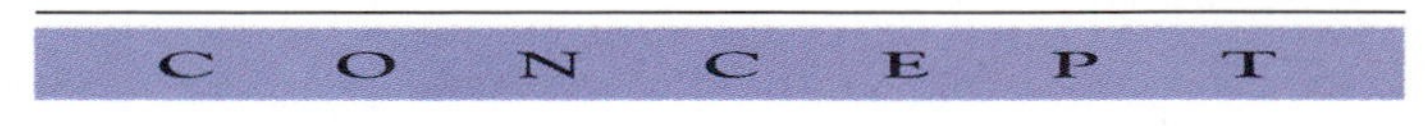

C O N C E P T

Slime molds and water molds, which were traditionally considered as fungi, are more appropriately classified in Kingdom Protoctista.

Considerable scientific attention has been paid to the cellular slime mold *Dictyostelium discoideum* (fig. 25.31c). Normally, this organism lives as single, amoeboid cells. When food gets scarce, however, the cells stream together into a moving

Table 25.2

The Main Characteristics of Slime Molds and Water Molds

Division	No. of Species	Mode of Nutrition	Storage Carbohydrate	Motility	Cell-Wall Composition	Main Habitat
Acrasiomycota (cellular slime molds)	70	Phagotrophic	Glycogen	Amoeboid	Cellulose	Terrestrial
Myxomycota (plasmodial slime molds)	500	Phagotrophic	Glycogen	2 flagella (whiplash)	None	Terrestrial
Chytridiomycota (uniflagellate water molds)	575	Absorptive	Glycogen	1 flagellum (whiplash)	Chitin, glucan*	Fresh water or marine
Oomycota (biflagellate water molds)	580	Absorptive	Glycogen or mycolaminarin	2 flagella (1 whiplash, 1 tinsel)	Cellulose, chitin, glucan*	Fresh water

* Glucan refers to structural polymers of glucose other than cellulose.

slug, which ultimately settles down and differentiates into a stalk with a spore-bearing fruiting body at its top (fig. 25.31c). The development from identical, free-living cells to a multicellular organism in *Dictyostelium* simulates many of the properties of cells that form embryos in much more complicated organisms, including mammals. For this reason, *Dictyostelium* has been studied for decades as a model for the developmental biology of complex organisms (see p. 69). More recently, a new technique has been devised for tagging and isolating the estimated 300 developmental genes in this slime mold. Using this technique, scientists expect to obtain the DNA sequences of almost all of these genes by about 1996. In addition to its significance for the genetics of slime molds, this work will also be important for the Human Genome Project: matches between developmental genes of the slime mold and unknown genes in the human genome will help identify developmental genes in humans.

Viruses

Viruses are parasites that cannot replicate on their own outside of their host, and they completely lack cellular structure. Viruses, therefore, are often referred to as *particles* instead of organisms. In addition, viruses do not grow by increasing in size or by dividing, nor do they respond to external stimuli. They have few (if any) enzymes, they cannot carry on independent metabolism, and they cannot move on their own. Nevertheless, viruses are incredibly numerous. In 1989, marine biologists at the University of Bergen in Norway discovered that a single teaspoon of seawater typically contains more than one billion viruses.

Viruses contain DNA or RNA (but not both), which is either single-stranded or double-stranded. The DNA or RNA is surrounded by a protein coat that may be attached to more complicated structures (fig. 25.32). With few exceptions, plant viruses have only RNA and a simple coat made of one or a few kinds of proteins. Tobacco mosaic virus, for example, has a core of RNA that is surrounded by 2,200 copies of the same protein. The genome has four genes, one for the coat protein, two for replicase enzymes, and the fourth for a protein that probably enables the virus to spread from cell to cell in the plant.

C O N C E P T

Viruses are nonliving, parasitic particles that consist of nucleic acids surrounded by a protein coat.

A.

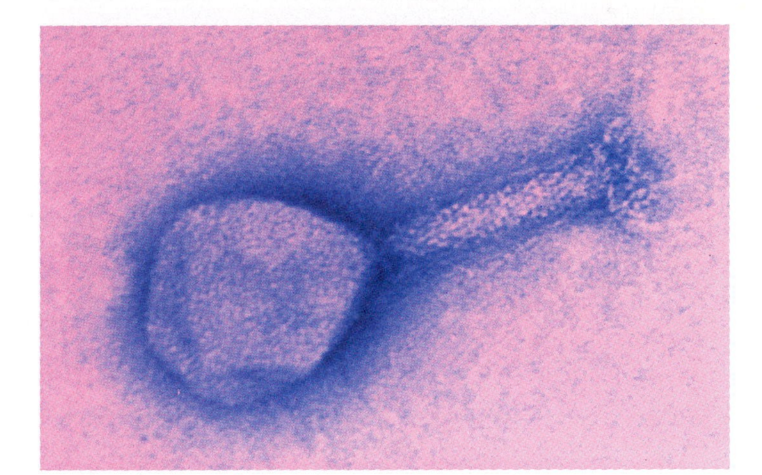

B.

Figure 25.32

Examples of viruses: (a) tobacco mosaic viruses, ×95,000; (b) a bacteriophage, ×85,000.

Viruses have been classified either according to their hosts and the types of tissues or organs they affect, or according to their type of nucleic acid (DNA or RNA), their size and shape, the nature of their protein coats, the number of nucleic acid molecules in their cores, and the type of host they can invade. In spite of the widely held views that viruses are not organisms, animal and bacterial virologists have agreed to use a classification system that is modeled after that of Linnaeus. In this system, viruses are grouped into families and genera, which are comprised of **serotypes** instead of species. A serotype is a protein that is a unique antigen; it induces and binds to antibodies that are specific to it alone. For example, the common cold is caused by as many as 113 serotypes of the genus *Rhinovirus*, which has an additional two serotypes that invade cows. The genus *Rhinovirus* is classified in the family Picornaviridae, which also includes polio and hepatitis-A viruses (serotypes of *Enterovirus*) and foot-and-mouth disease viruses (serotypes of *Aphthovirus*).

In opposition to the above system, plant virologists have so far refused to classify plant viruses in a Linnaean-type system. Plant viruses are classified in groups based on the same kinds of features that are used to classify other viruses, but the names of plant viruses are usually derived from a prototype for the group. For example, viruses that are similar to the tobacco mosaic virus are members of the **tobamo** group. As you might expect, the opposing taxonomic philosophies of plant virologists and other virologists can lead to confusion. Many plant viruses, for example, are similar enough to animal viruses to be classified side by side in the same families or genera. Rice dwarf virus, for instance, resembles animal viruses in the family Reoviridae (fig. 25.33). Animal virologists think it should be called *Phytoreovirus* (*phyto* to indicate a plant virus), which is how it appears in some virology textbooks (at least those written by animal virologists). To confuse things further, alternate hosts for *Phytoreovirus* include leafhoppers, which are insects.

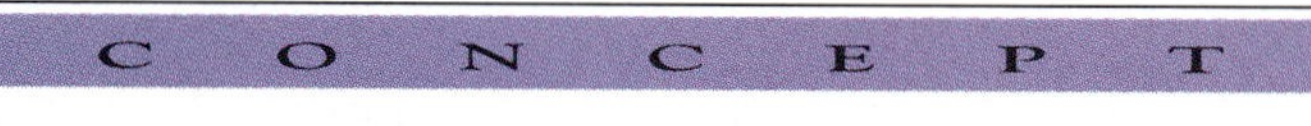

The classification of viruses is based mostly on composition, size, and shape. Viral taxonomy differs among plant virologists and other virologists.

Although viruses have a deservedly negative reputation, they can also be put to good use in controlling other kinds of pests and in making vaccines and other medically important products. Pest-control viruses, for example, include **Abby,** which is a virus that is found only in the caterpillars of gypsy moths. Abby, which is an abbreviation of Abington, is one of nineteen strains of a virus that forms particles in polyhedral crystals. These viruses are characterized by double-stranded DNA cores surrounded by protein coats that protect the viruses against chemicals, heat, and low pH. Gypsy moth caterpillars have been particularly destructive in forests of both North America and Europe, and it is hoped that this destruction may be greatly reduced by the dissemination of Abby in the forests under attack. Related viruses are receiving considerable attention as potential biological control agents for several other insect pests. For example, the Environmental Protection Agency has approved the use of viruses to control alfalfa looper, cotton bollworm, European pine sawfly, Douglas fir tussock moth, and other pests.

FIGURE 25.33

A rice plant with symptoms of infection by rice dwarf virus (left). The plant on the right is not infected with the virus.

The Origin and Evolution of Viruses

There are no fossil remains of the oldest viruses. However, clues about the possible origin of viruses come from the similarity between viral genomes and their host genomes. For example, DNA sequences of several hundred nucleotides from cancer-causing viruses are almost the same as host DNA sequences. This could mean that viruses arose from fragments of DNA (or RNA) that became self-replicating in the host cell.

Viruses that insert their DNA into the host genome, then recover it for making new particles after replication, may mimic this origin repeatedly.

Multiple origins for viruses from different hosts mean that viruses are more closely related to their hosts than they are to each other. Therefore, any phylogenetic classification should account for the different origins of plant and animal viruses. Accordingly, the view by plant virologists that rice dwarf virus, for example, should not be classified with animal viruses in the family Reoviridae has merit. This also means that classifications that rely on physical features such as size, shape, and number of nucleic acid molecules are artificial, because these features do not indicate phylogenetic relationships. Instead, because of genetic similarities to their hosts, plant reoviruses are more closely related to plants than they are to animal reoviruses.

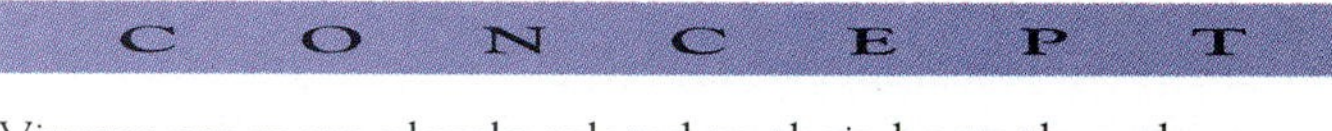

Viruses are more closely related to their hosts than they are to each other. Plant viruses, therefore, have more nucleotide-sequence similarity with plant hosts than they do with other viruses.

Can viruses evolve? They can in the sense that viruses are like genetic programs that are subject to change. As such, they can evolve faster than any organism. Evidence for such quick change comes from comparative studies of influenza viruses. Samples of these viruses were put into storage in the 1940s and used for genetic comparisons with the same kinds of viruses that were isolated every ten years through the 1980s. These comparisons showed that the genetic difference between influenza viruses in the 1940s and 1980s was equivalent to the difference between species. This means that "speciation" in viruses can occur in forty years or less. Different "species" of viruses may even arise from year to year, which may explain why new vaccines have to be developed almost constantly to combat new strains of viruses as they appear. This is far faster than evolution in eukaryotic organisms, which may require hundreds of thousands or, more likely, millions of years for the formation of a new species.

Viroids and Virusoids

If it seems that nature has innumerable agents of disease, it may be because viruses are only one type of infectious particle that plagues living organisms. In plants (but not animals), such additional particles include **viroids** and **virusoids,** so named because they have some viruslike features. Like viruses, for example, viroids have a single molecule of single-stranded RNA. Unlike viruses, the RNA molecule of a viroid is circular and is not surrounded by a protein coat. Viroids are known primarily from cultivated plants.

Virusoids are like viroids but are located inside the protein coat of a true virus. Virusoid RNA can be either circular or linear. Unlike viroids, virusoids are not infectious by themselves because they are replicated only in the presence of their host viruses. In some cases, the dependency is mutual; the virus cannot replicate in the absence of its virusoid.

Chapter Summary

Bacteria are the most metabolically diverse group of organisms because of the wide variety of metabolic pathways that occur in the Kingdom Monera. Because they are prokaryotic, bacteria cannot undergo true sexual reproduction.

Biologists divide bacteria into two groups, the Eubacteria and the Archaebacteria, which some systematists now classify as two separate kingdoms. The Eubacteria include organisms that have photosystems like those of plants. These are the cyanobacteria, formerly called the *blue-green algae,* and the chloroxybacteria. The cyanobacteria have chlorophyll *a* and phycobilins as their main photosynthetic pigments. In contrast, like plants, the chloroxybacteria have chlorophylls *a* and *b*. The Archaebacteria and Eubacteria differ in their lipid linkages, cell-wall components, RNA polymerase types, ribosome shapes, energy sources, habitats, and gene structures.

The ability of bacteria to incorporate foreign DNA into their plasmid genomes and then replicate and express it in new cells is an important feature of bacteria for genetic engineering. This feature allows scientists to modify bacteria to control pests and diseases and to produce medically or agriculturally important biochemicals.

Bacteria are probably more closely related to the ancestor of life than any other group of organisms. Like all other organisms, however, bacteria are based on DNA as a genetic code that is transcribed into RNA and translated into proteins. In contrast, the first life-forms probably originated as RNA-based organisms.

Fungi are the most outwardly plantlike of the groups presented in this chapter. They have multicellular or coenocytic thalli or filaments that look plantlike, and they often have cellulose in their cell walls. Unlike plants, though, fungi have only one diploid cell in their life cycle, they can have chitin in their cell walls, they store glycogen, and they can reproduce asexually by spores. All fungi are heterotrophic, including saprobic and parasitic forms. Most plant diseases are caused by fungi. The two most complicated kinds of fungi, the ascomycetes and the basidiomycetes, are useful for making many commercial products. Mushrooms and truffles, which are basidiomycetes and ascomycetes, respectively, include species that are eaten as delicacies.

Fungi have a fragmentary fossil history. The two groups of "higher" fungi (ascomycetes and basidiomycetes) are believed to be more recent and closely related to each other than either one is to the zygomycetes. The zygomycetes are the group that is most representative of the ancestor of true fungi.

Slime molds and water molds were classified as fungi by early botanists. These molds have funguslike, animallike and plantlike features, but they are now classified in Kingdom Protoctista.

Viruses are on the edge of life, since they are based on proteins and either DNA or RNA. However, because viruses lack many features associated with life, they are instead referred to as particles.

The structures of viruses provide the main characters for their classification. Viruses vary in whether they are based on DNA or RNA, in how many kinds of proteins make up their coats, and in their size. However, classification based on nucleic acid sequences is probably a more accurate reflection of the relationships of viruses. According to such comparisons, viruses are more closely related to their hosts than they are to each other. This suggests that viruses originated as genetic fragments from a host genome, and that their origin has occurred repeatedly.

Questions for Further Thought and Study

1. If bacteria can only reproduce asexually, how can new forms arise?
2. Since many of the diseases of plants and animals are caused by bacteria, explain how it would (or would not) benefit humans if a virus that would eliminate all bacteria could be developed through genetic engineering.
3. If you could develop the most useful bacterium known to humans from various existing bacteria, what features would you combine in your new bacterium?
4. If a single puffball can produce a trillion spores, each of which can germinate and develop, why are we not overrun with puffballs and other fungi that also produce prodigious numbers of spores?
5. Pollution, loss of habitat, and other factors contribute to the extinction of many species of animals and plants. Fungi are particularly sensitive to pollution, and many species are threatened or have disappeared. Should we be as concerned about the loss of fungal species as we are about the loss of, for example, Africa's big game animals? Explain.
6. For many years the U.S. Forest Service had a program to eradicate gooseberry bushes, which are the host for one phase of the life cycle of the fungus that produces white pine blister rust. The idea was that if gooseberry bushes could be eliminated, then the fungus could not reproduce and form spores that infect white pines. This program was eventually abandoned in favor of growing rust-resistant, nonnative pines instead. Can you think of any possible alternative measures to control the fungus?
7. Conifer seeds planted in fertilized, sterilized soil may germinate when watered but do not grow nearly as well at first as their counterparts in the forest. What might explain this phenomenon?
8. If viruses do not respond to external stimuli, lack enzymes, have no means of locomotion, and cannot reproduce outside of a living cell, should they be classified as lifeless, inanimate objects? Explain.
9. The morphology of cyanobacteria has changed relatively little during the past billion or so years; indeed, cyanobacteria have persisted 2 to 10 times longer than other "living fossils" (e.g., crocodiles). How would you explain this observation?
10. Recent evidence suggests that Archaebacteria have diverged less from the common ancestor they share with the eukaryotes than the eukaryotes have. How would you test this hypothesis?
11. In 1994, researchers discovered that spores of some fungi germinate when exposed to ethylene. Of what significance is this?

Suggested Readings

Articles

Barinaga, M. 1994. Molecular evolution: Archae and eukaryotes grow closer. *Science* 264:1251.

Batra, L. R., ed. 1967. Insect-fungus symbiosis: Nutrition, mutualism and commensalism. *Scientific American* 217 (5):112–120.

Butler, P. J. G., and A. Klug. 1978. The assembly of a virus. *Scientific American* 239 (5):62–69.

Carmichael, W. W. 1994. The toxins of cyanobacteria. *Scientific American* 270:64–72.

Chang, S. T., and P. G. Miles. 1984. A new look at cultivated mushrooms. *BioScience* 34:358–362.

Lewis, R. 1994. A new place for fungi? Molecular evolution studies suggest fungi should be taxonomically transposed. *Bioscience* 44:389–391.

Litten, W. 1975. The most poisonous mushrooms. *Scientific American* 232:14–22.

Marx, J. 1992. *Dictyostelium* researchers expect gene bonanza. *Science* 258:402–403.

Pennisi, E. 1994. Static evolution: Is pond scum the same now as billions of years ago? *Science News* 145:168–169.

Seaward, M. R. D. 1989. Lichens as monitors of recent changes in air pollution. *Plants Today* March–April:64–69.

Woese, C. R. 1981. Archaebacteria. *Scientific American* 244 (6):98–122.

Wright, K. 1990. Bad news bacteria. *Science* 249:22–24.

Books

Birge, E. A. 1992. *Modern Microbiology: Principles and Applications*. Dubuque, IA: Wm. C. Brown.

Carlile, M., and S. Watkinson. 1994. *The Fungi*. San Diego: Academic Press.

Carr, N. G., and B. A. Whitton, eds. 1982. *The Biology of Cyanobacteria*. Berkeley, CA: University of California Press.

Gibbs, A. J., and B. D. Harrison. 1979. *Plant Virology: The Principles*. New York: John Wiley.

Hale, M. E. 1983. *The Biology of Lichens*. 3d ed. Baltimore: University Park Press.

Harley, J. L., and S. E. Smith. 1983. *Mycorrhizal Symbiosis*. New York: Academic Press.

Ingram, D. S., and A. Hudson. 1994. *Shape and Form in Plants and Fungi*. San Diego: Academic Press.

Joklik, W. 1988. *Virology*. 3d ed. East Norwalk, CT: Appleton and Lange.

Pirozynski, K. A., and D. L. Hawksworth, eds. 1988. *Coevolution of Fungi with Plants and Animals*. San Diego: Academic Press.

Smith, A. H., and N. Weber. 1980. *The Mushroom Hunter's Field Guide*. Enlarged ed. Ann Arbor, MI: University of Michigan Press.

A kelp from along the coast of California. Kelps are brown algae; giant kelps can grow to more than 60 meters long.

Algae

CHAPTER

26

Chapter Outline

Chapter Overview

Algae include some of the smallest and largest eukaryotes; they can be unicellular, colonial, filamentous, coenocytic, or multicellular. Sexual life cycles of algae can be dominated by a haploid phase or a diploid phase, or they can be split evenly between the two. Although algae live in a diversity of habitats that is rivaled only by bacteria and fungi, most algae live in water. The great diversity of algae has led to many conflicting ideas about how they should be classified, but three main groups are consistently recognized as their own divisions: the green algae, the brown algae, and the red algae. Other divisions, consisting mostly of unicellular forms, are variously aligned with these three or are classified with protozoans or other single-celled protoctists. The unifying feature of algae that has traditionally led botanists to classify them with plants is their ability to photosynthesize using chlorophyll *a*. Certain groups of algae also share with plants the presence of chlorophyll *b*, starch for energy storage, and cellulose in their cell walls.

Introduction

People often think of algae only as pond scum or seaweeds, which they are, but algae are more diverse in their growth forms and habitats than we commonly think. Many species also grow in soil, snow, or clouds, or symbiotically with plants, animals, and fungi. Some algae live as epiphytes on aquatic plants, on tropical plants or on other algae, and still others live on the fur of animals (fig. 26.1).

The wide variety of habitats occupied by algae is matched by their morphological diversity. Algae range in size from single-celled green algae, which may reach a maximum of 50 μm in diameter, to giant kelps whose length may exceed 60 m (see photo that opens this chapter). Algae include species that either swim by flagella or just produce certain cells that do so. Because some marine and aquatic algae have no flagellated cells, their gametes must float or glide to their targets.

Many botanists still consider all or some of the algae to be plants; others exclude them all. This means that there are many interesting (and often heated) arguments about the best classification of algae. For example, all groups of algae—like plants—have chlorophyll *a* and plantlike photosynthesis. A few, such as the green algae, also have chlorophyll *b* and photosynthesize exactly as plants do; indeed, the pathway for carbon fixation in plants was discovered in *Chlorella*, a green alga (see Chapter 7). Conversely, algae exhibit a wide variety of life cycles that are unknown in plants, and different groups of algae have cell walls, carbohydrate reserves, and pigments that are unlike those of plants.

In this overview of algal diversity, we emphasize the three groups that have the greatest variety of forms: the green algae, the brown algae, and the red algae. Other groups, such as the diatoms, may include more species and be distributed more widely, but the green, brown, and red algae include representatives of almost all morphological and reproductive types that occur in algae. We present the key features of the other groups of algae later in this chapter, but in less detail.

General Features of Algae

Algae are an informally defined group of eukaryotes that are usually classified in seven divisions. The names and main features of each of these divisions are given in table 26.1. These divisions are at least partially distinguished by their pigments, their energy-storage polymers, their cell-wall components, and the number and types of their flagella. In addition, the organisms within each division have a variety of life cycles; representative life cycles are presented in this chapter.

Overview of Vegetative Organization

Although the simplest algae are unicellular (fig. 26.2), the most complex algae rival the giant redwoods as the largest of all photosynthetic organisms. Most algae are somewhere between these two extremes. **Colonial algae** are those with groups of cells that are loosely attached to each other and sometimes surrounded by a slimy sheath. Filamentous algae are either branched or unbranched, and have either uninucleate or multinucleate cells. Some filamentous algae are coenocytic because they have no cross walls (fig. 26.3).

The kelps and other macroscopic algae have organs that resemble leaves, stems, and roots. The blades of the leaflike structures consist of parenchyma cells. The stemlike organs, called **stipes,** have many cell types, including sieve cells (fig. 26.4). Sieve cells of kelps occur in sieve tubes, like the phloem of vascular plants. However, botanists hyphothesize that the sieve cells of algae arose separately from plant cells, that is, that algal and plant sieve cells arose by evolutionary convergence.

The divisions of algae are sufficiently different from one another that botanists hyphothesize the algae evolved from several different ancestors. If this is true, then algae are not a monophyletic group. This conclusion is currently supported by molecular phylogenies based on comparisons of ribosomal RNA sequences (see fig. 24.14).

A.

B.

Figure 26.1

Diverse habitats of algae: (a) *Pleurococcus* growing on a yew tree and stone; (b) algae growing on the fur of a three-toed sloth in Costa Rica.

Table 26.1

Comparison of Main Features of Algal Divisions

Division	Habitat	Photosynthetic Pigments	Cell Wall Components	Carbohydrate Storage	Flagella
Chlorophyta (green algae)	Mostly fresh water, some marine, terrestrial, or airborne	Chlorophylls *a* and *b*, carotenoids	Polysaccharides, including cellulose	Starch	None, 1–8, or dozens; whiplash
Phaeophyta (brown algae)	Almost all marine, rarely fresh water	Chlorophylls *a* and *c*, fucoxanthin and other carotenoids	Cellulose, alginic acid, and sulfated polysaccharides	Laminarin, mannitol	2, lateral; forward tinsel, rearward whiplash
Rhodophyta (red algae)	Mostly marine, some fresh water	Chlorophyll *a*, carotenoids, phycobilins	Callulose, pectin, calcium salts	Floridean starch	None
Chrysophyta (diatoms, yellow-green & golden-brown algae)	Marine and fresh water, some terrestrial or airborne	Chlorophylls *a* and *c*, fucoxanthin and other carotenoids	Cellulose or silica; sometimes absent	Chrysolaminarin	None, 1, or 2; whiplash or tinsel
Euglenophyta (euglenoids)	Marine or fresh water, some airborne	Chlorophylls *a* and *b*, carotenoids	Absent	Paramylon	1–3; tinsel
Pyrrhophyta (dinoflagellates)	Marine and fresh water, some airborne	Chlorophylls *a* and *c*, peridinin and other carotenoids	Cellulose; sometimes absent	Starch	None or 2; tinsel
Cryptophyta (cryptomonads)	Mostly fresh water, some marine	Chlorophylls *a* and *c*, carotenoids, phycobilins	Absent	Starch	2; tinsel

FIGURE 26.2

Algae vary greatly in size. *Chlamydomonas*, shown here, is a single-celled green alga, that is less than 100 μm long. Compare the microscopic size of *Chlamydomonas* with the kelps shown in the opening photograph of this chapter, ×150.

A.

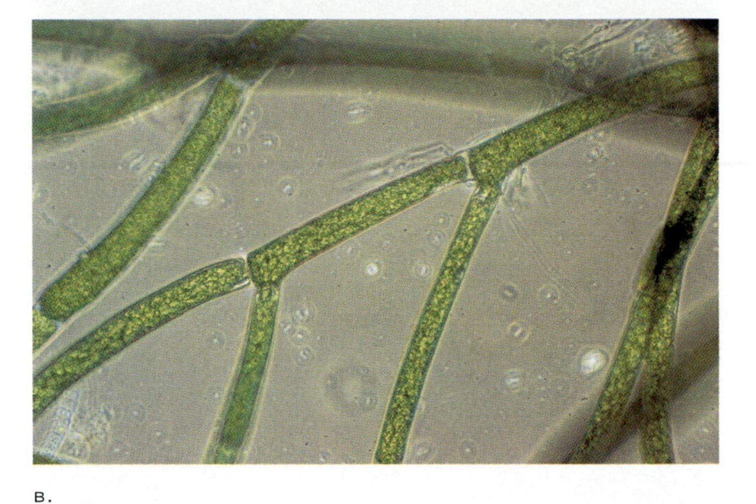

B.

FIGURE 26.3

Examples of other common types of vegetative organization of algae. (a) *Volvox* is a colonial green alga, ×50. (b) *Cladophora* forms branched filaments that consist of multinucleate cells, ×100.

Generalized Life Cycles

As in plants and fungi, sexual reproduction in algae entails an alternation between diploid and haploid phases, which alternate between fertilization and meiosis. The diploid phase produces the haploid phase by meiosis; the haploid gametes then fuse to make a zygote that starts another diploid phase. Unlike those of plants, however, the diploid and haploid phases of algae are usually free-living; that is, neither phase is attached to the other phase at maturity. Moreover, unlike that of fungi, the diploid phase in many algae is plantlike in being multicellular.

Three distinct versions of a generalized life cycle occur among algae. One version resembles the life cycle of plants (see Chapters 27–30) because certain cells of a multicellular diploid phase undergo meiosis to make spores (fig. 26.5a). The diploid phase is therefore a sporophyte. Diploid sporangia produce haploid spores that can be either motile **(zoospores)** or nonmotile **(aplanospores).** Spores grow by mitosis, which may or may not be followed by cytokinesis, into the haploid phase of the life cycle. This haploid phase produces gametes and is therefore a gametophyte. The gametes can be either motile or nonmotile. Fertilization restores the diploid phase. This type of life cycle is based on **sporic meiosis,** because meiosis produces spores. Sporic meiosis is common among all divisions of algae that have multicellular forms.

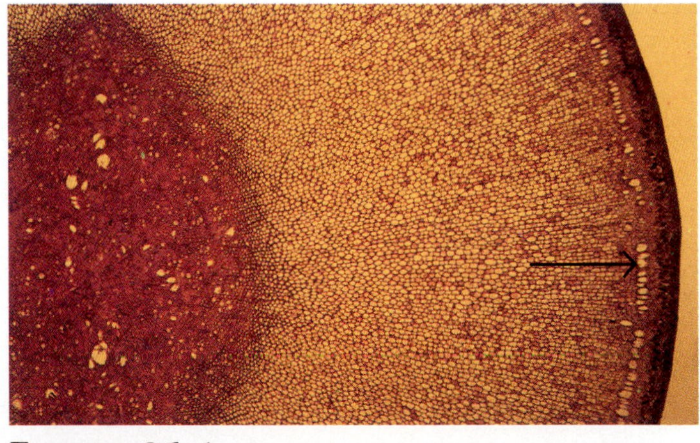

FIGURE 26.4

Cross section of a stipe of *Laminaria cloustonii*. The arrow points to a region of the outer cortex that has many sieve cells, which are common in kelps.

The second type of life cycle resembles that of animals. Certain cells of a multicellular diploid phase undergo meiosis to make gametes, not spores (fig. 26.5b). Because meiosis produces gametes directly, it is called **gametic meiosis.** Gametic meiosis is rare in the algae.

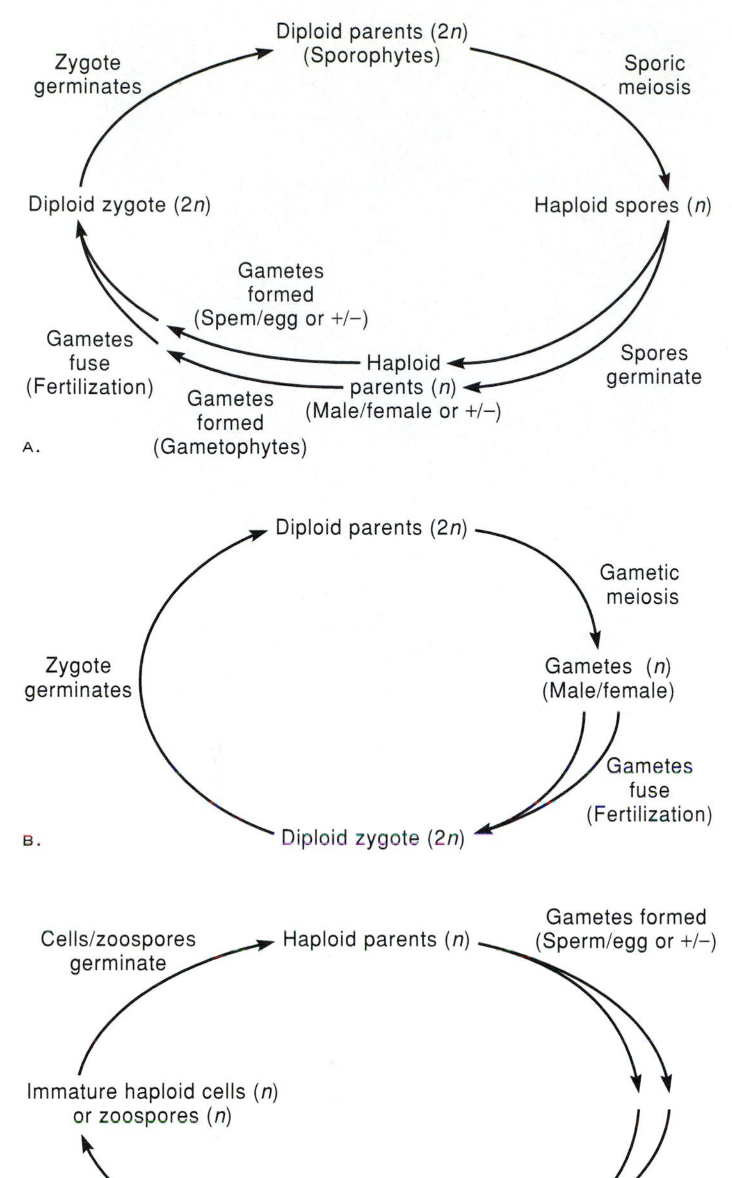

FIGURE 26.5

The three types of life cycles in algae. (a) Diploid parents form spores by sporic meiosis; spores produce haploid parents, which produce gametes; (b) diploid parents produce gametes directly by gametic meiosis, thereby bypassing a haploid parental phase; (c) haploid parents produce gametes; gametes fuse into a zygote, which is the only diploid cell in this type of life cycle.

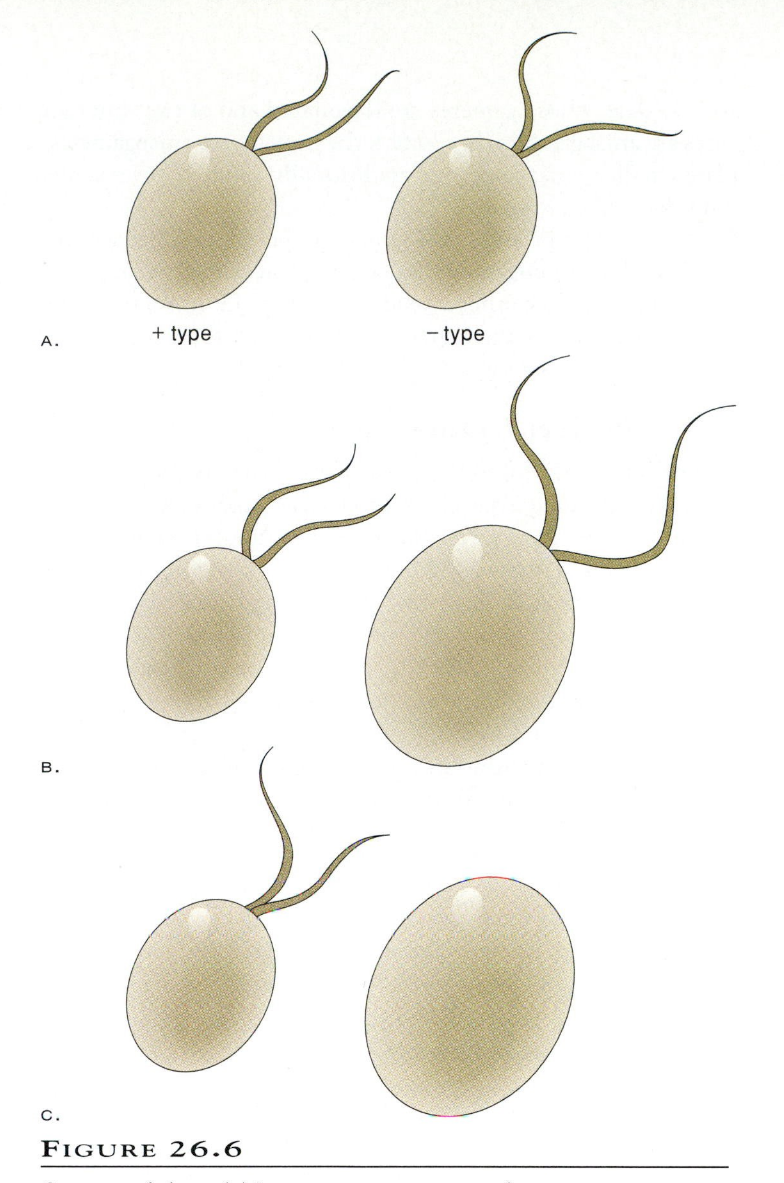

FIGURE 26.6

Gametes of algae. (a) Isogametes are gametes of opposite mating types that look alike; they are generally designated as + and – mating strains. (b) Flagellated gametes of different sizes are called anisogametes. (c) An egg is a large, nonmotile gamete fertilized by a much smaller gamete that is motile or nonmotile.

The third type of life cycle resembles that of fungi, in that the only diploid cells are the zygotes (fig. 26.5c). This means that the dominant phase of the life cycle is haploid, regardless of whether the organism is unicellular or multicellular. The haploid phase forms when the zygote undergoes meiosis. Although the zygote may produce spores when it divides, this type of meiosis is called **zygotic meiosis** to distinguish it from the sporic meiosis of a multicellular diploid phase. Most of the green algae, including almost all of the unicellular forms, reproduce by zygotic meiosis.

The gametes of algae show more morphological diversity than any other group of organisms. With the exception of the red algae, at least some of the members of every algal group produce gametes that swim by one or more flagella. In addition, in some species the gametes are neither male nor female; all the gametes look alike (fig. 26.6a). Such gametes are called **isogametes,** and organisms that reproduce by isogametes are referred to as **isogamous.** Isogametes are not genetically identical, however; they belong to one of two mating strains, and therefore are different genetically. Mating strains are arbitrarily designated as + or – strains. Fertilization can occur only between gametes of different strains.

Algae whose gametes are flagellated and of two different sizes are **anisogamous** (fig. 26.6b); the gametes are **anisogametes.** The smaller anisogamete is traditionally referred to as male, and the larger as female.

The most pronounced differences between gametes occur in algae that are **oogamous.** In oogamy, one gamete is large and nonmotile, and the other is small and either motile or nonmotile (fig. 26.6c). The larger gamete is designated the egg, and the smaller one the sperm.

Asexual Reproduction

Unicellular algae reproduce asexually by mitosis and cell division. Multicellular algae also reproduce asexually, either by vegetative fragments, by vegetative propagation (i.e., growth of new individuals from rootlike structures), or by mitotically produced spores that form clones of the parent. Like spores produced by meiosis, mitotically derived spores are either motile or nonmotile. Mitotically derived spores are sometimes indistinguishable from meiotically derived spores.

Vegetative propagation and mitotic spore production occur almost continuously during a growing season. Because asexual reproduction is generally much faster than sexual reproduction, most algal populations consist of several clones.

C O N C E P T

Algae include organisms that are unicellular, colonial, filamentous, or large and multicellular. They reproduce sexually by fertilization and meiosis that is either sporic, gametic, or zygotic. Algae also reproduce asexually by different kinds of motile or nonmotile spores, or by different vegetative means.

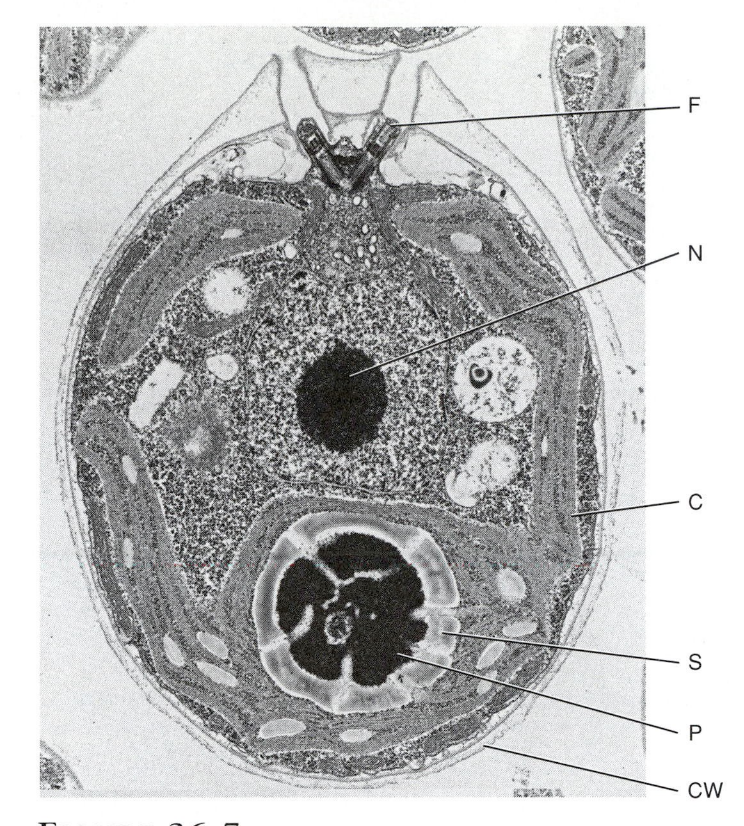

FIGURE 26.7

Transmission electron micrograph of a *Chlamydomonas* cell in median longitudinal section, ×10,500. C=chloroplast, CW=cell wall, F=flagellum, N=nucleus, P=pyrenoid, S=starch granule. A stigma was not preserved in this cell.

Division Chlorophyta: The Green Algae

The division Chlorophyta, which includes about 7,500 species, has a higher diversity of vegetative organization, life cycles, and habitats than any other group of algae. Most green algae live in fresh water, but different species also occur in marine habitats, clouds, snow banks, soil, or on the shady moist sides of trees, buildings, and fences. Green algae also live symbiotically with several different kinds of animals and with the fungi of lichens (see p. 609–611). Green algae and plants share many important characteristics which support the hypothesis that plants arose from a green-algalike ancestor (see box 26.1, "Evolutionary Relationships of Green Algae and Plants").

Unicellular Green Algae

There are two types of unicellular algae: motile and nonmotile. The most well-known example of a motile green alga is *Chlamydomonas* (see fig. 26.2). Cells of *Chlamydomonas* swim by means of two equal flagella at the anterior end of each cell. The dominant feature of each cell is a single, large chloroplast that encloses at least one **pyrenoid** and usually a **stigma** (fig. 26.7). Each pyrenoid contains the enzyme RuBP carboxylase-oxygenase (see Chapter 7) and is surrounded by starch granules. The red stigma contains the pigment rhodopsin, the same pigment that vertebrates use in vision. The stigma functions as a light-receptor for phototaxis.

Most cells in a population of *Chlamydomonas* are products of asexual reproduction by mitosis and cell division. However, environmental conditions such as nitrogen availability or daylength stimulate sexual reproduction (fig. 26.8). Sexual reproduction begins when flagella of cells from compatible mating strains touch. At first, cells aggregate into clusters; then a delicate cytoplasmic thread develops between pairs of cells. The cell walls dissolve in the region of this thread, and the cytoplasm of the two cells moves together and fuses. Fertilization occurs when the two nuclei fuse into a large, diploid nucleus. This diploid zygote is surrounded by a thick, spiny wall that resists heat and desiccation. The zygote then becomes dormant, after which it divides by meiosis and releases four haploid zoospores, two of each mating strain.

In stressful conditions, *Chlamydomonas* retracts its flagella and becomes dormant. When water becomes available

Boxed Reading 26.1

Evolutionary Relationships of Green Algae and Plants

Botanists have long believed that green algae are the closest relative of plants, and recent molecular phylogenies support this idea (see Chapter 24). However, botanists are not satisfied with general agreement on such an apparently easy question; the question now asked is, "Which green algae are *most* closely related to plants?" This is a more interesting question scientifically, because there is more than one possible answer, depending on the significance of different characters.

One possibility is presented in cladogram A, which shows the relationships between the land plants and the green algae that have been traditionally thought to be their closest relatives. Some of the characters that support these relationships are noted on the cladogram. The cladogram shows that the closest green alga to the land plants is *Coleochaete;* this relationship is supported by the shared derivation of the feature that zygotes are retained in the gametophyte. However, an equally likely possibility is presented in cladogram B, in which the closest green alga to the land plants is *Chara*. This relationship is supported by the evolution of flavonoid biosynthesis. Note that in both cladograms, however, one or the other derived state (i.e., zygote retention or flavonoid biosynthesis) is a parallelism: flavonoid biosynthesis in A and zygote retention in B. This means that flavonoid biosynthesis arose twice according to cladogram A, or that zygote retention arose twice according to cladogram B. Botanists who regard zygote retention as more significant than flavonoid biosynthesis accept *Coleochaete* as the green alga that is most closely related to land plants. Others consider flavonoid biosynthesis to be too complicated to have arisen twice in evolution, which leads them to choose *Chara* as the closest relative to land plants.

Molecular phylogenies may help resolve this issue. At the moment, overall similarities of ribosomal RNA sequences support cladogram B. However, recall from Chapter 24 that overall similarity may not be the same as evolutionary relatedness. Thus, the jury is still out on the details of a land plant/green algae phylogeny.

Perhaps the most significant and generally agreed upon result of the cladistic analysis of green algae and land plants is that the green algae are apparently a paraphyletic group. This has led plant systematists to consider classifying all of the organisms shown in these cladograms in one supergroup, the Streptophyta, and redefining the Chlorophyta to exclude the streptophytes. No decision has been made, however, about whether the streptophytes should be classified as a division, a superdivision, or any other taxonomic category.

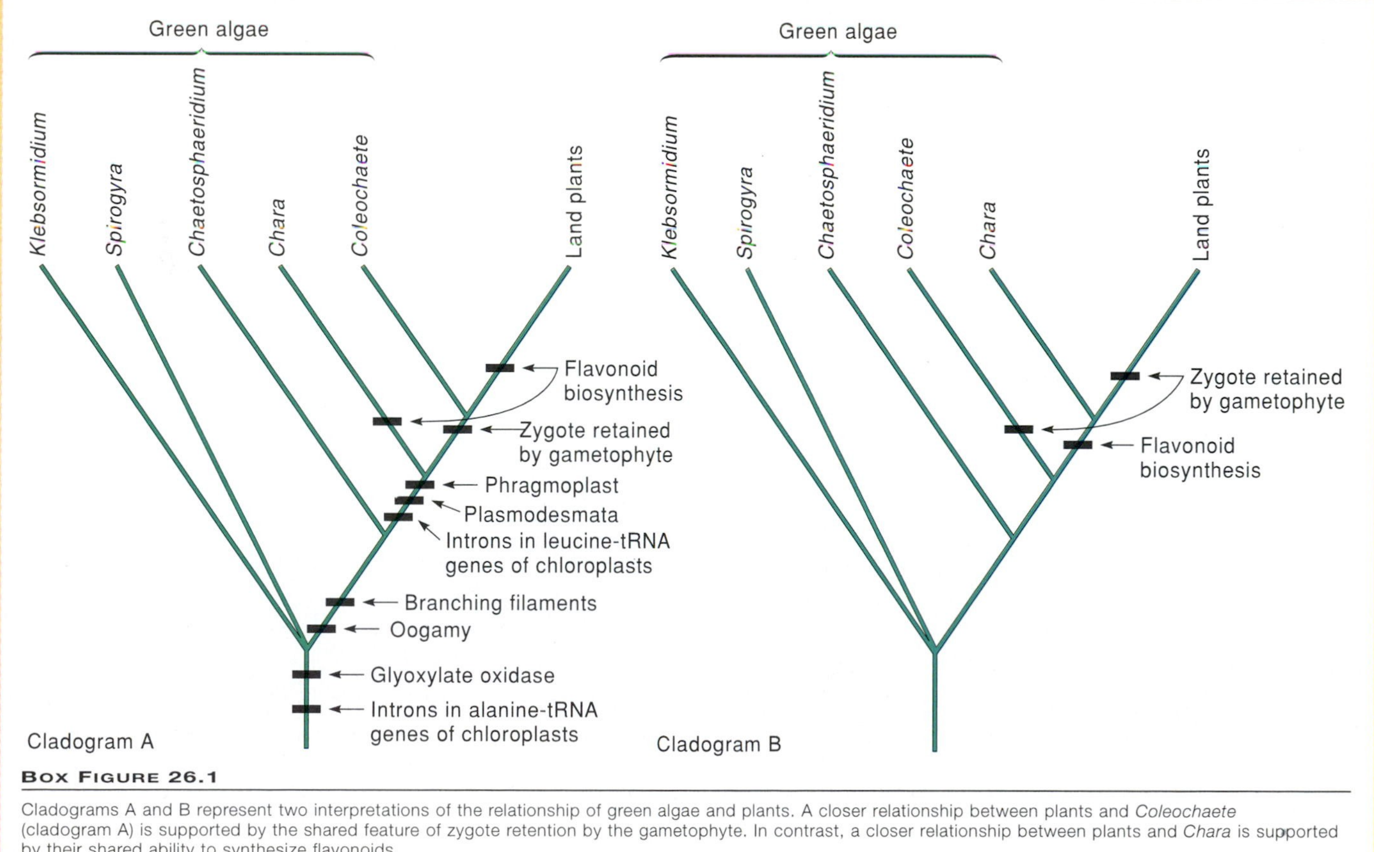

Box Figure 26.1

Cladograms A and B represent two interpretations of the relationship of green algae and plants. A closer relationship between plants and *Coleochaete* (cladogram A) is supported by the shared feature of zygote retention by the gametophyte. In contrast, a closer relationship between plants and *Chara* is supported by their shared ability to synthesize flavonoids.

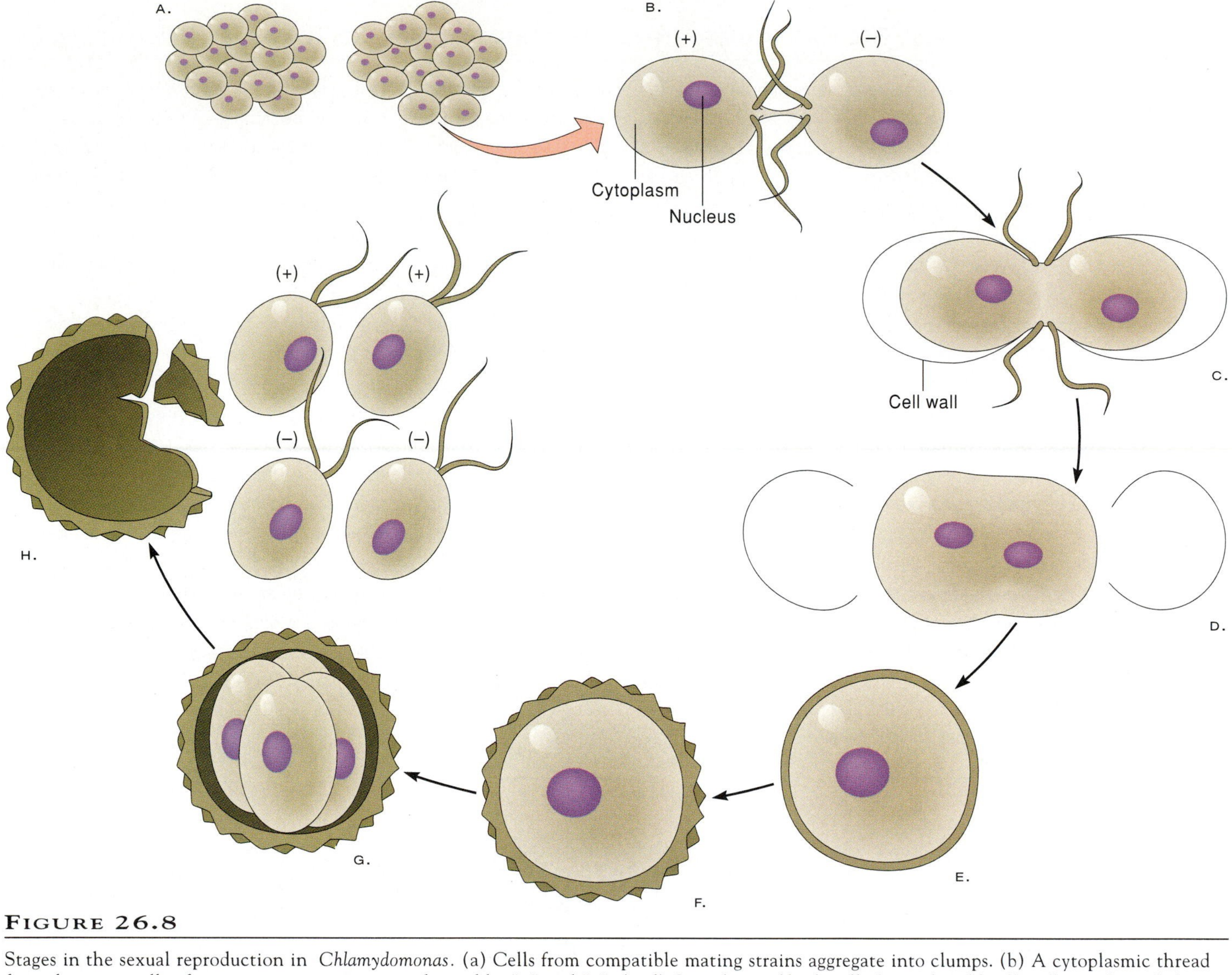

FIGURE 26.8

Stages in the sexual reproduction in *Chlamydomonas*. (a) Cells from compatible mating strains aggregate into clumps. (b) A cytoplasmic thread forms between cells of opposite mating strains, indicated by "+" and "–". (c–d) Cytoplasm of both cells fuses; the old cell walls are discarded. (e) Nuclei fuse (fertilization) and a new cell wall forms around the zygote. (f) The cell wall surrounding the zygote thickens and becomes spiny, becoming a zygospore. (g) Four haploid cells are produced by meiosis and cytokinesis of the zygote. (h) Two of the new cells after meiosis are of one mating strain and two are of the other mating strain.

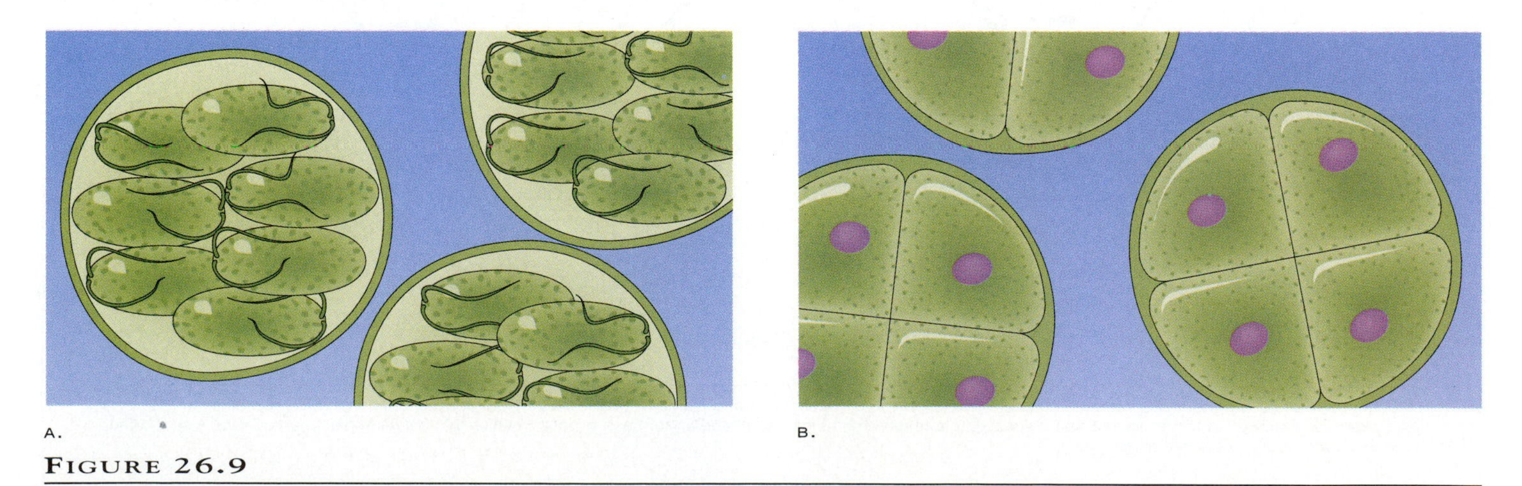

FIGURE 26.9

Tetracystis, a nonmotile green algae. (a) *Tetracystis* forms biflagellate zoospores whenever there is enough moisture for swimming; (b) when water is scarce, these algae often stick together in two-cell or four-cell complexes.

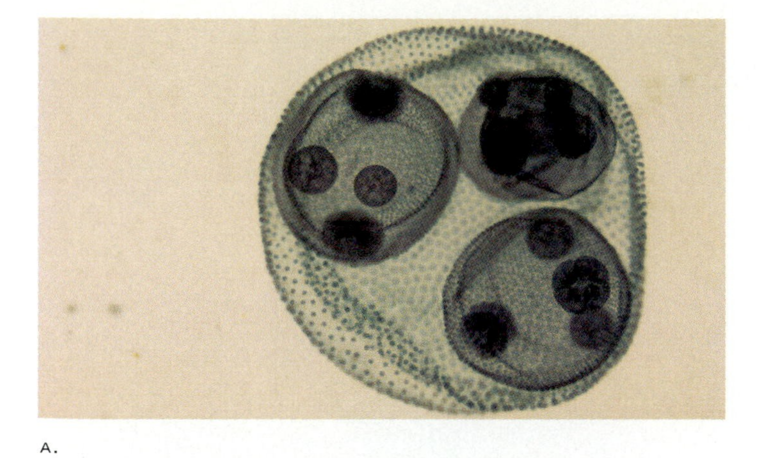

A.

B.

Figure 26.10

Volvox. (a) *Volvox,* a motile, colonial green alga containing daughter colonies, ×250. (b) Portion of a colony with spiny-coated zygotes, ×500.

again, *Chlamydomonas* regrows its flagella, enlarges, and reproduces. After a rainfall, *Chlamydomonas* cells that have been dormant grow quickly in puddles and drainage ditches.

Most unicellular green algae that are nonmotile can produce zoospores. One example is *Tetracystis,* a soil-inhabiting alga that releases zoospores from mitotic cell division whenever there is enough moisture for swimming. Each zoospore looks like a cell of *Chlamydomonas* (fig. 26.9a). Zoospores lose their flagella quickly, and the cells enlarge and divide by mitotic cell division. New vegetative cells often stick together in two-celled or four-celled complexes (fig. 26.9b). In some species of *Tetracystis,* the zoospores may function as gametes. By analogy with *Chlamydomonas,* meiosis in *Tetracystis* is thought to be zygotic, but this has not been observed.

Some unicellular green algae such as *Chlorella* seem to reproduce exclusively by asexual means. Zoospores and sexual reproduction have not been observed in *Chlorella,* so the rapid growth of a *Chlorella* population is apparently due to asexual reproduction alone.

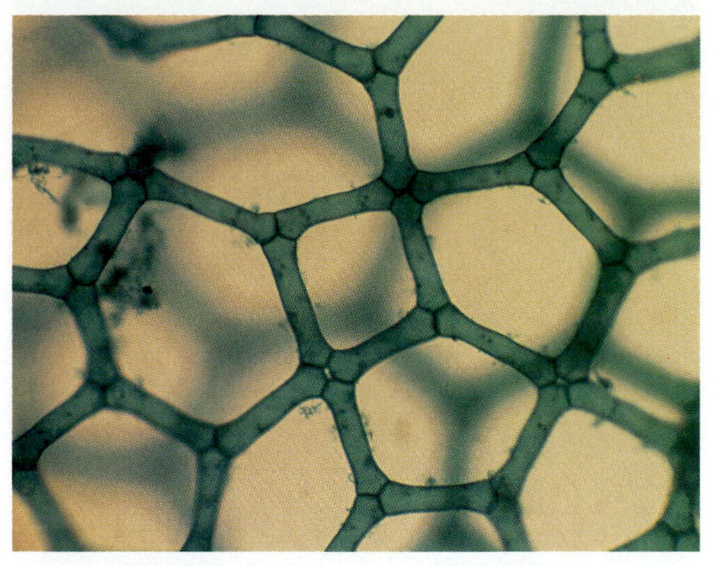

Figure 26.11

Hydrodictyon, a nonmotile, colonial green alga, ×100.

Colonial Green Algae

Like the unicellular green algae, the colonial green algae include both motile and nonmotile organisms. An example of colonial green algae is shown in figure 26.3a. The main features of representative genera of colonial green algae are discussed next.

The largest and most spectacular motile colonies of green algae occur in the genus *Volvox* (fig. 26.10). Each *Volvox* colony may contain thousands of *Chlamydomonas*-like cells arranged at the periphery of a hollow sphere. The cells are attached to each other by delicate cytoplasmic threads. Their flagella beat in a coordinated motion that rolls the colony like a ball. During asexual reproduction, certain cells enlarge and grow inward, where they pinch off and form **daughter colonies** (fig. 26.10a). These colonies are released when the parent colony disintegrates. During sexual reproduction, *Volvox* is oogamous; the same enlarged cells that grow into daughter colonies may instead become fertile eggs (fig. 26.10a). Packets of sperm are produced elsewhere in the same colony or in a different colony; that is, a colony is either **bisexual** or **unisexual.** The sperm cells swim to the eggs and fertilize them, thereby forming zygotes (fig. 26.10b). The zygotes go through a dormant period in the parent colony before they undergo meiosis and release haploid zoospores.

Mature colonies of *Hydrodictyon,* the "water net," consist of elongated cells joined at their ends to make polygonal shapes (fig. 26.11). The cells are coenocytic and contain many nuclei. The whole colony is a hollow cylinder that can grow up to 75 cm in length if it is not crowded by other colonies. Asexual zoospores develop when mature coenocytic cells cleave into numerous uninucleate segments, each of which produces a biflagellated zoospore. Miniature nets grow within the parent cell from zoospores that come together into preliminary polygonal shapes and lose their flagella. Young nets are released as

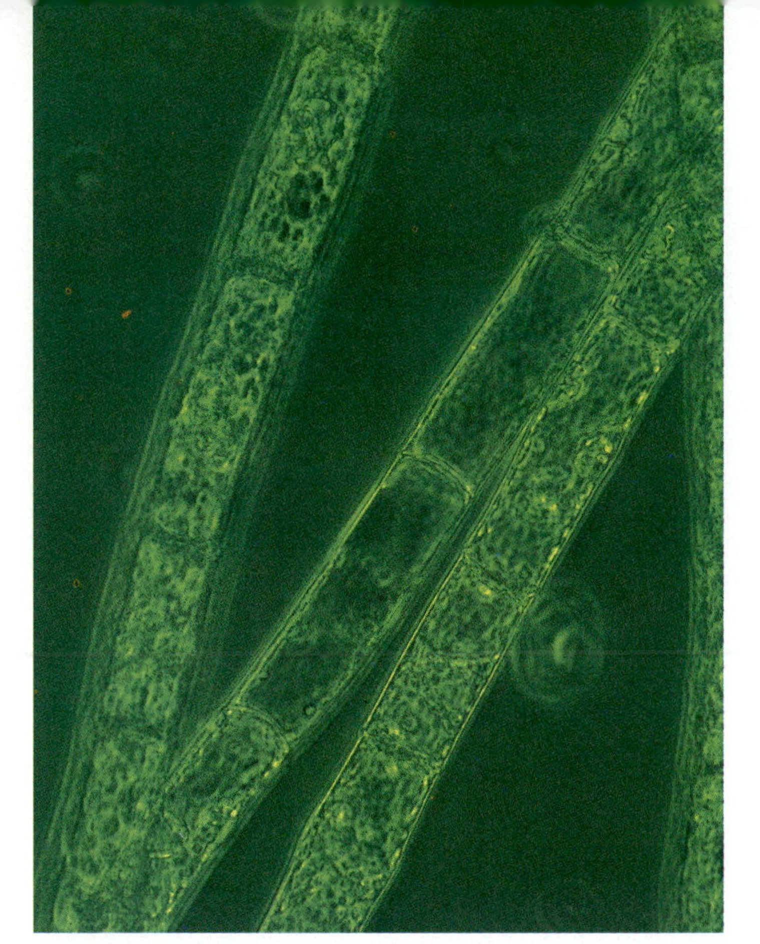

FIGURE 26.12

Filamentous green algae: light micrograph of unbranched filaments of *Oedogonium*, ×700.

their cells expand and rupture the parent cell. If biflagellated cells are discharged from the parent cell before they assemble into young nets, they function as isogametes. Zygotic meiosis produces four biflagellated zoospores, each of which enlarges and divides mitotically into a coenocytic cell in a polyhedral gelatinous matrix. This large coenocytic cell cleaves to another set of zoospores, which make small nets in the matrix, just as asexual zoospores do in cells of mature colonies.

Filamentous Green Algae

Filamentous green algae are microscopic and are either branched or unbranched (figs. 26.3b and 26.12). Branched filaments are usually coenocytic, but unbranched filaments consist mostly of uninucleate cells. Filamentous green algae often grow epiphytically on aquatic flowering plants; they also attach to rocks or other objects under water. Some filaments are free-floating.

Asexual reproduction in filamentous green algae is usually by quadriflagellate or biflagellate zoospores, but some organisms produce zoospores with up to 120 flagella (fig. 26.13). In sexual reproduction, meiosis is usually zygotic or sporic, while gametic meiosis is rare. In most filamentous green algae, gametes are either oogamous or isogamous. In oogamous filaments, such as those of *Oedogonium*, the egg forms in an enlarged container called the **oogonium** (fig. 26.14). Sperm arise in pairs

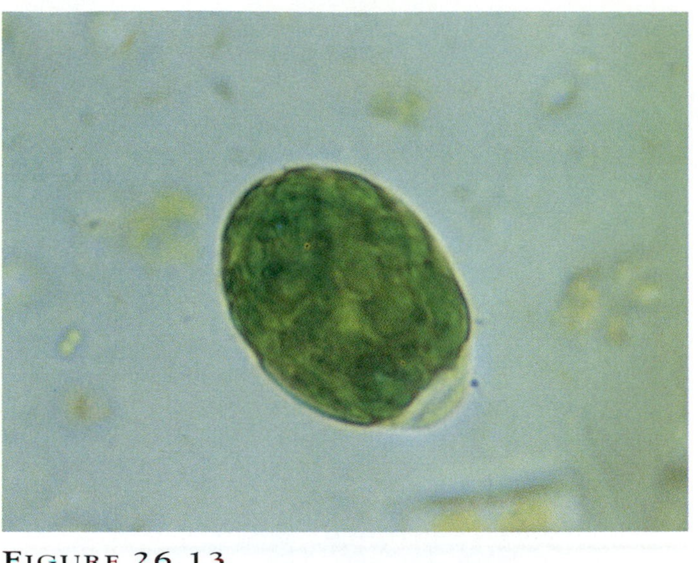

FIGURE 26.13

Zoospores of *Oedogonium* have up to 120 flagella, ×1,000.

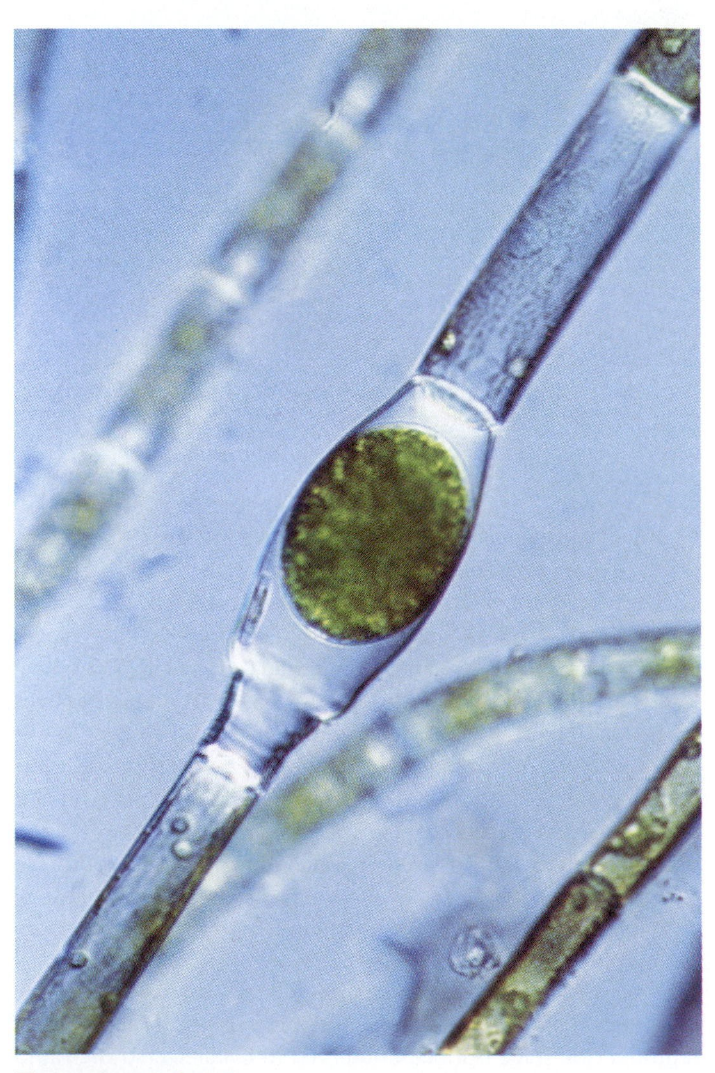

FIGURE 26.14

Light micrograph of an oogonium of *Oedogonium*, ×780.

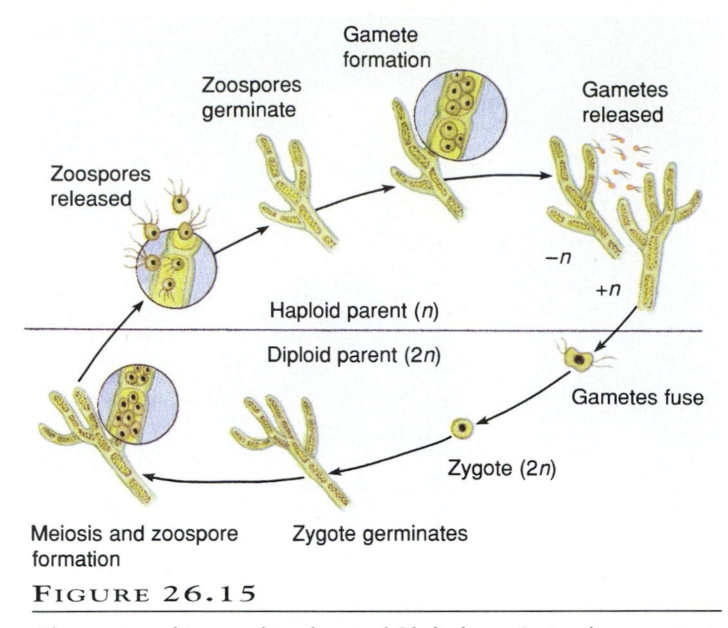

FIGURE 26.15

Alternation of isomorphic phases of *Cladophora*. As in plants, meiosis in *Cladophora* is sporic.

A.

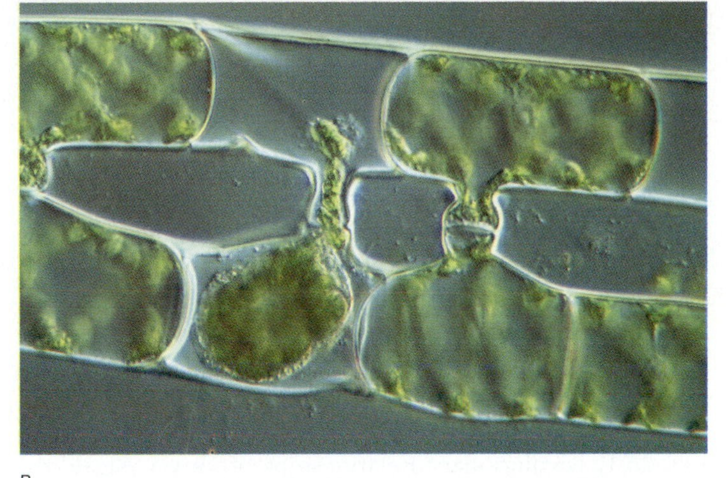

B.

FIGURE 26.16

Spirogyra. (a) Light micrograph of *Spirogyra* filaments showing spiral chloroplasts, ×250. (b) Light micrograph of *Spirogyra* filaments in a late stage of conjugation, ×150.

in each of several boxlike cells, together called the **antheridium.** Like zoospores, sperm have about 120 flagella in *Oedogonium*, but they are biflagellate or quadriflagellate in most other genera. Although meiosis is zygotic in oogamous organisms, many isogamous genera have sporic meiosis. In the genus *Cladophora*, for example, the zygote germinates into a diploid branched filament that looks like the haploid branched filaments (see fig. 26.3b). Although the sporophyte and the gametophyte are **isomorphic**—that is, vegetatively they look alike—they differ reproductively. After meiosis, the sporophyte releases quadriflagellate zoospores that grow into filamentous gametophytes. At maturity, these gametophytes produce biflagellate isogametes. The sporophytic and gametophytic phases may also produce asexual zoospores. Sexual reproduction in *Cladophora* is summarized in figure 26.15. The life cycle of *Cladophora* resembles that of plants because meiosis is sporic.

Some filamentous green algae reproduce differently than do *Oedogonium* and *Cladophora*. The most common representative of other types of reproduction occurs in the genus *Spirogyra*, which is named for the spiral chloroplasts in each cell (fig. 26.16a). Species of this genus grow as frothy or slimy green masses of unbranched filaments that float in the water of small ponds. Flagellated cells are absent in all species of *Spirogyra*, and asexual reproduction is restricted to fragments of filaments that form new filaments. Sexual reproduction in *Spirogyra* is called **conjugation.** During conjugation, filaments growing side by side in a dense mass from small protuberances, or **papillae,** that grow toward each other. Upon contact, the end-walls of the papillae dissolve to create a tube. The cell of one filament then moves through the tube, and the cytoplasm and nuclei of both cells fuse. Although each cell is apparently an isogamete, the migrant cell is considered to be the male. Conjugation occurs between almost every pair of cells along the length of the two filaments (fig. 26.16b). The diploid nucleus in each cell undergoes meiosis, but only one of the four haploid cells grows into a new filament. The other three nuclei disintegrate.

The most complex and plantlike green algae in overall appearance are the stoneworts (also known as tank mosses and brittleworts). The branched filaments of stoneworts, such as *Chara*, are whorled like those of some plants, and they have leaflike structures at their nodes (fig. 26.17). A single apical cell produces cells that form nodes, internodes, and leaflike structures. Indeed, *Chara* is often identified incorrectly as an aquatic vascular plant because of its cylindrical branches attached to nodes, its large size (up to a meter long), and its plantlike body. However, when *Chara* is collected with its distinctive reproductive structures present, there is no doubt that it is a stonewort.

The sexual organs of stoneworts are unique in the algae. Antheridia and oogonia are multicellular, and each is surrounded by a layer of sterile cells (fig. 26.18). Herein lies a botanical and taxonomic dilemma: in other algae, sex organs are either unicellular or, if multicellular, entirely fertile (i.e., not covered by

Figure 26.17

Chara is a stonewort that has branched filaments and whorls of leaflike organs. The brown structures are antheridia and oogonia.

sterile cells). If *Chara* is considered ancestral, the sterile jacket of cells that surrounds their reproductice organs is yet another shared characteristic.

Stoneworts, like plants, can also make flavonoids (see box 26.1, "Evolutionary Relationships of Green Algae and Plants"). No other algae have yet been found to do this.

Other Multicellular Forms

Seaweeds in the genus *Ulva,* the sea lettuce, are leafy green algae. These seaweeds are widely distributed on rocks, woodwork, and larger algae in shallow marine habitats. As in some filamentous green algae, the sporophytes and gametophytes of *Ulva* are isomorphic (fig. 26.19).

A population of sea lettuces includes three kinds of organisms: the sporophyte, the male gametophytes, and the female gametophytes. Sporophytes release large quadriflagellate zoospores produced by meiosis. Half of these zoospores grow into male gametophytes, and half grow into female gametophytes. The male gametophytes produce small biflagellate gametes, and the female gametophytes produce large biflagellate gametes. Sexual reproduction is bypassed when gametes that fail to unite grow directly into new haploid organisms.

Classification of the Green Algae

Less than two decades ago, the classification of green algae was based primarily on features such as vegetative organization (unicellular, filamentous, colonial), presence or absence of flagellated cells, and type of life cycle. Current classifications also use features determined from electron microscopy and comparative biochemistry. Because these subcellular characteristics are time-consuming to determine, most green algae have not been surveyed. Consequently, the classification of green algae is changing rapidly as more organisms are examined.

One recent classification divides green algae into five classes. Representative organisms from each of the three largest of these classes were discussed earlier in this chapter: the Chlorophyceae (*Chlamydomonas*, *Tetracystis*, *Chlorella*, *Volvox*, *Hydrodictyon*, and *Oedogonium*), the Charophyceae (*Spirogyra* and *Chara*), and the Ulvophyceae (*Ulva* and *Cladophora*). Several relatively new kinds of features are used for defining groups of green algae in current classifications, including types of cell division, details of flagellar ultrastructure, enzymes used

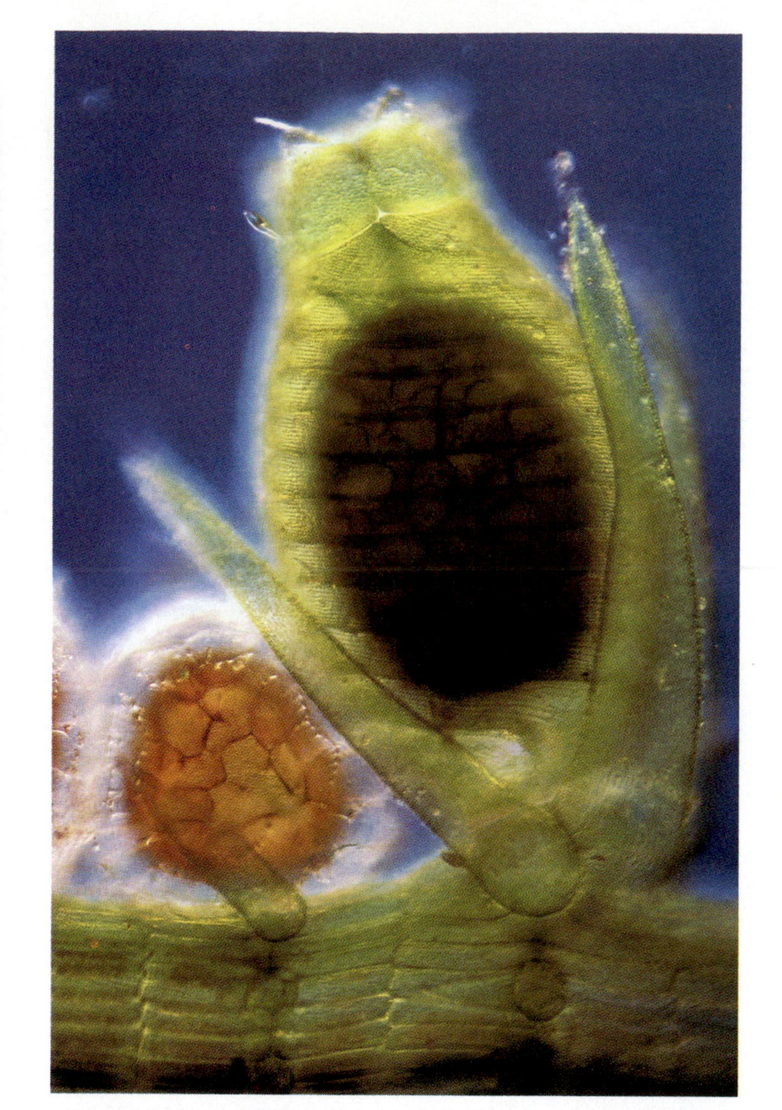

Figure 26.18

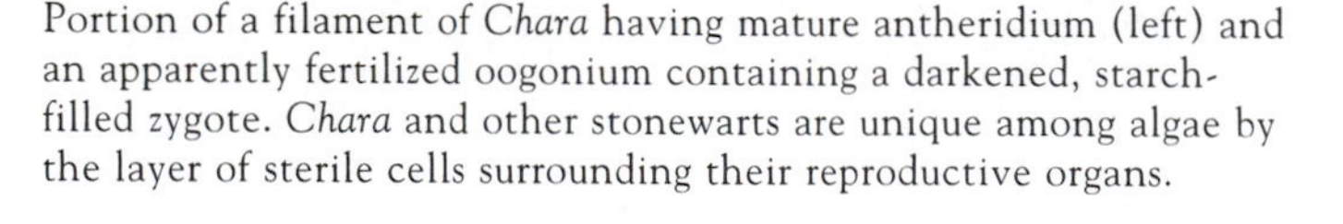

Portion of a filament of *Chara* having mature antheridium (left) and an apparently fertilized oogonium containing a darkened, starch-filled zygote. *Chara* and other stonewarts are unique among algae by the layer of sterile cells surrounding their reproductive organs.

in photorespiration, and biosynthetic capabilities (see box 26.1, "Evolutionary Relationships of Green Algae and Plants"). For example, cytokinesis in the Charophyceae includes a phragmoplast, which is also a feature of plants (fig. 26.20). The Chlorophyceae, instead, have a **phycoplast** to aid in the construction of the cell plate. In contrast to phragmoplasts, which are perpendicular to the spindle apparatus, phycoplasts arise parallel to spindle fibers. Finally, cytokinesis in the Ulvophyceae occurs by furrows that are made entirely without a cell plate.

The flagella of the Charophyceae attach to a flat band of microtubules called a **multilayered structure,** which resembles that in the flagella of a few plants (most plants do not produce flagellated cells at all). However, the Chlorophyceae and Ulvophyceae have flagella that attach to a cross-shaped system of microtubular "roots." Other green algae have either one of these types, or they have a star-shaped arrangement of microtubules in the region of flagellar attachment.

The green algae are the most diverse of algae. Green algae include examples of all the major types of vegetative organization, including organization into colonies, which is absent or rare in other divisions. Although the most common type of meiosis in green algae is zygotic, both sporic meiosis and gametic meiosis also occur. Gametes may be isogamous, anisogamous, or oogamous. Asexual reproduction is common and involves motile or nonmotile spores, or growth by vegetative means.

FIGURE 26.19

Ulva, a green alga that is commonly referred to as sea lettuce. Populations of *Ulva* consist of isomorphic sporophytes and gametophytes.

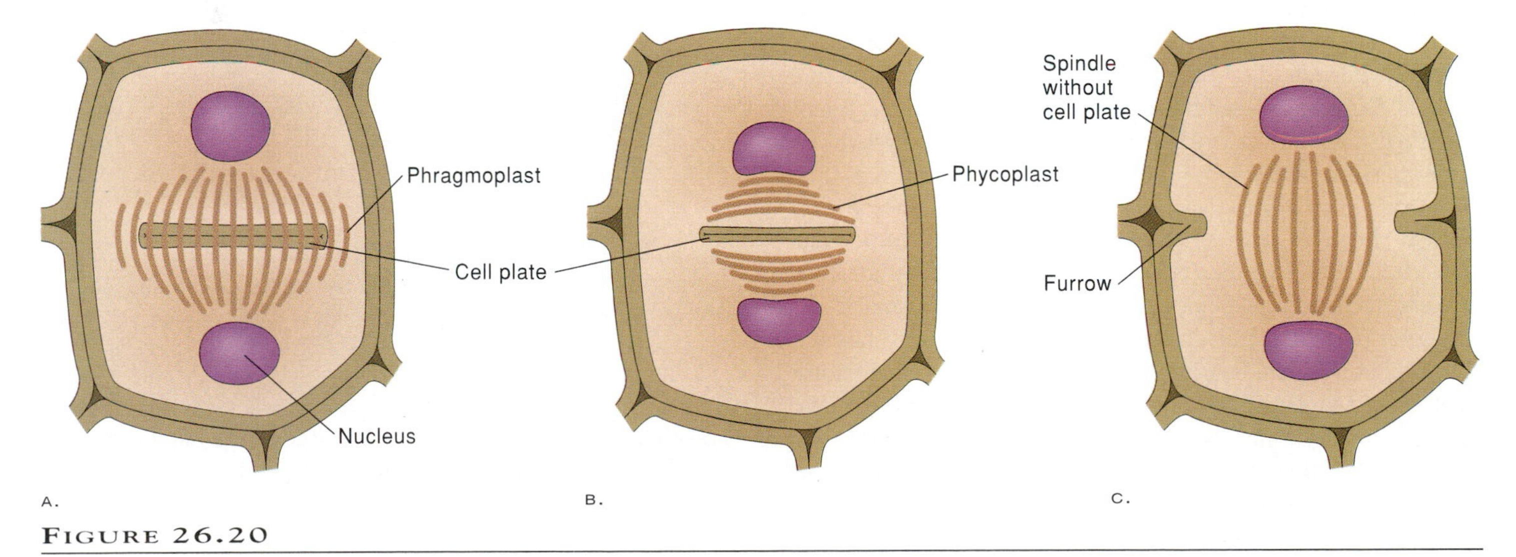

FIGURE 26.20

Cytokinesis occurs by (a) a phragmoplast in the Charophyceae, (b) a phycoplast in the Chlorophyceae, and (c) furrows in the Ulvophyceae.

What are Botanists Doing?

Go to the library and read an article about the classification of green algae. Summarize the main points of the article. Considering what you know about algae, were you convinced by the arguments presented by the author(s)? Explain your answer.

Division Phaeophyta: The Brown Algae

Almost all of the approximately 1,500 species of brown algae are marine organisms, but a few species live in fresh water. The division Phaeophyta includes the most complex of all the algae, the kelps (see the photo at the beginning of this chapter); the smallest brown algae are branched filaments (fig. 26.21). There are no unicellular, colonial, or unbranched filamentous organisms among the Phaeophyta.

The flagella of brown algae are characteristic of this division. Every motile cell has two flagella, which are usually on the side of the cell. The shorter of the two flagella is a smooth **whiplash** type that points toward the back of the cell. The longer flagellum points toward the front of the cell and is a **tinsel** type, meaning that it has several rows of small appendages projecting from it (fig. 26.22). Except for the order Dictyotales, all of whose members produce motile cells having a single tinsel flagellum, all brown algae release biflagellate motile cells at some time in their life history, either as gametes or as zoospores.

Filamentous Brown Algae

Ectocarpus, perhaps the best known of the filamentous brown algae, grows on rocks or on larger marine algae along ocean shores worldwide. The outward appearance and reproductive cycle of *Ectocarpus* resemble those of some green algae. Small branched filaments are either haploid or diploid. The haploid gametophyte and the diploid sporophyte are isomorphic, as they are in the green algae *Cladophora* and *Ulva*. In the sporophyte, cells at the ends of lateral branches enlarge into **unilocular** (one chamber) sporangia. After the diploid nucleus undergoes meiosis, the new haploid nuclei divide mitotically, forming 32–64 zoospores (fig. 26.23). Each zoospore germinates into a haploid gametophyte. When the gametophyte is mature, isogametes form in elongated multicellular organs called **plurilocular** (i.e., have many chambers) gametangia on lateral branches. Although they are isogamous, the cells of one mating strain settle on the ocean floor and secrete **ectocarpene,** which attracts cells of the other mating strain. The ectocarpene-secreting gametes are designated as female, and the gametes that are attracted to them are designated as male. The gametophytes of *Ectocarpus* are unisexual; that is, each is either male or female.

Figure 26.21

Light micrograph of *Ectocarpus* filaments, among the smallest of the brown algae.

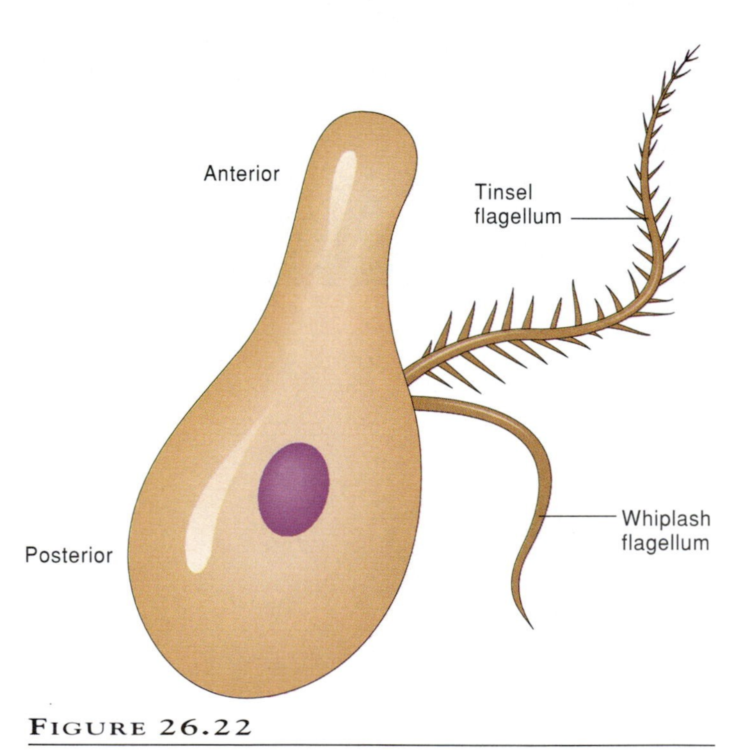

Figure 26.22

Motile cells of brown algae typically have two lateral flagella, one whiplash and one tinsel.

Asexual reproduction in *Ectocarpus* is versatile, because mitotic zoospores can be produced by diploid and haploid phases. Plurilocular organs on diploid individuals release diploid zoospores, which germinate into new sporophytes. Plurilocular organs of the diploid phase are morphologically indistinguishable from those of the haploid phase. Moreover, gametes that fail to unite may become zoospores and germinate into new haploid organisms.

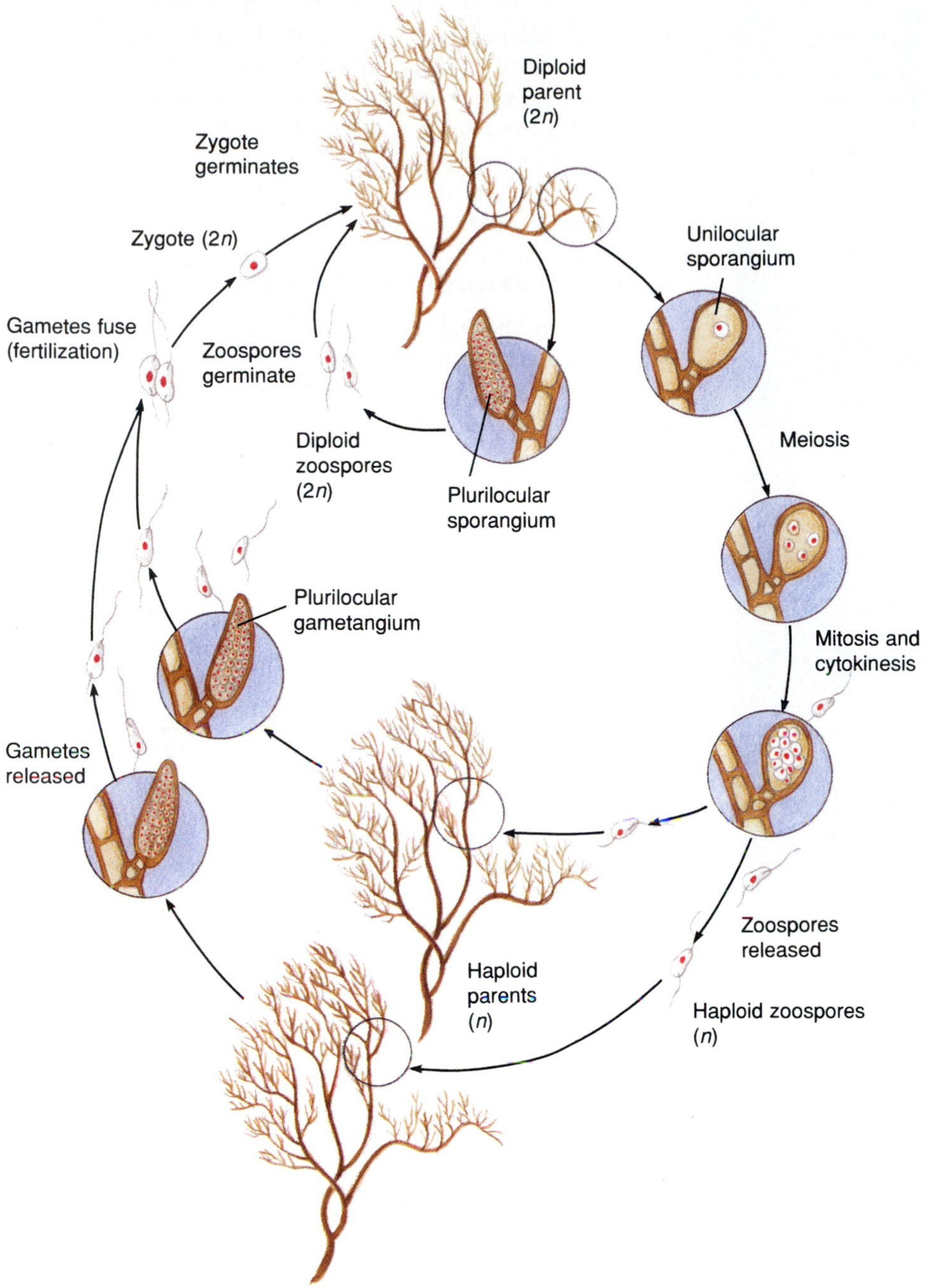

FIGURE 26.23

The life cycle of *Ectocarpus* alternates between isomorphic gametophytes and sporophytes, as in some green algae.

The full range of reproductive alternatives in a single species of *Ectocarpus* is known only from a single population of *E. siliculosus* near Naples, Italy. Other populations of this species, from the coasts of England and the northeastern United States, are not as versatile. No other species have been studied enough to know whether the *E. siliculosus* population near Naples is typical of the genus. However, reproduction in the Naples population shows that other populations and species of *Ectocarpus* may have the latent genetic capability to produce zoospores regardless of ploidy level.

The production of zoospores by haploid and diploid phases blurs the distinction between sporophytes and gametophytes. This is apparently a general phenomenon, since spore production by "gametophytes" also occurs in some plants. Such variations in basic life histories are evidence that chromosome number does not determine whether a "sporophyte" or a "gametophyte" develops when a specific cell germinates.

Kelps and Rockweeds

Kelps and rockweeds dominate shorelines and nearby offshore habitats in cool climates worldwide. Some kelps, such as certain species of *Sargassum*, are free-floating. They grow rapidly by vegetative propagation into massive populations in the open ocean in such places as the Sargasso Sea near the Caribbean Islands. Kelps and rockweeds are of interest because of their vegetative organization and because of the distinctiveness of their life cycle in comparison with those of green algae.

The life cycle of kelps is dominated by a large sporophyte. The huge sporophytes of *Macrocystis* (giant kelp), for example, consist of stipes, blades, and branching **holdfasts** that anchor them to the substrate. The leafy blades often have air bladders that keep the kelps afloat (fig. 26.24). Intercalary growth occurs in the stipe, so the oldest tissues are at the tips of the blades.

During sexual reproduction in kelps, certain cells on the surfaces of some blades enlarge into unilocular sporangia, each of which releases 32–64 biflagellate zoospores. Sporangia occur in large groups called **sori** (singular, **sorus**). The spores of some kelps have sex chromosomes of the X-Y type, which are common in animals but rare in algae and plants. Each spore germinates into a microscopic filament that is either a male or a female gametophyte. The female gametophytes have oogonia that extrude eggs, and the male gametophytes produce biflagellate sperm that swim to the eggs. After fertilization, the zygote germinates into a new sporophyte. The life cycle of kelps resembles those of *Ulva*, *Cladophora*, and *Ectocarpus* because they all have sporic meiosis. However, the kelps differ in the dominance of the sporophytic phase of the life cycle.

FIGURE 26.24

Giant kelp (*Macrocystis*), showing stipes, blades, holdfast, and air bladders.

Rockweeds are similar to the kelps. However, meiosis in rockweeds is gametic, as it is in animals. In the rockweed genus *Fucus*, the fertile swollen tips of branches are embedded with **conceptacles,** which are containers that bear oogonia and antheridia (fig. 26.25). Eggs and sperm are forced out of the conceptacles by a slimy substance that absorbs water and sets the gaametes free. Immediately after fertilization, the zygote germinates into another diploid rockweed.

Classification of the Brown Algae

The division Phaeophyta has just one class, the Phaeophyceae, which includes about thirteen orders. Classification of brown algae is based on the complexity of vegetative organization and the type of life cycle. There are three main orders of brown algae based on these kinds of features; these groups are represented by *Ectocarpus* (Order Ectocarpales), *Laminaria* (Order Laminariales), and *Fucus* (Order Fucales). The filamentous brown algae are assumed to be primitive because of the simplicity of their vegetative organization and the similarity of their life cycle to that of certain green algae. These and other assumptions about the evolutionary relationships within the Phaeophyta have yet to be examined in the framework of testable phylogenetic hypotheses.

Brown algae are exclusively filamentous or multicellular and complex. They include the largest of the algae, the giant kelps. Although sexual reproduction generally involves sporic meiosis and isogamous or anisogamous gametes, some brown algae are oogamous and have gametic meiosis. Spores and motile gametes generally have two lateral flagella. Asexual reproduction is mostly by zoospores.

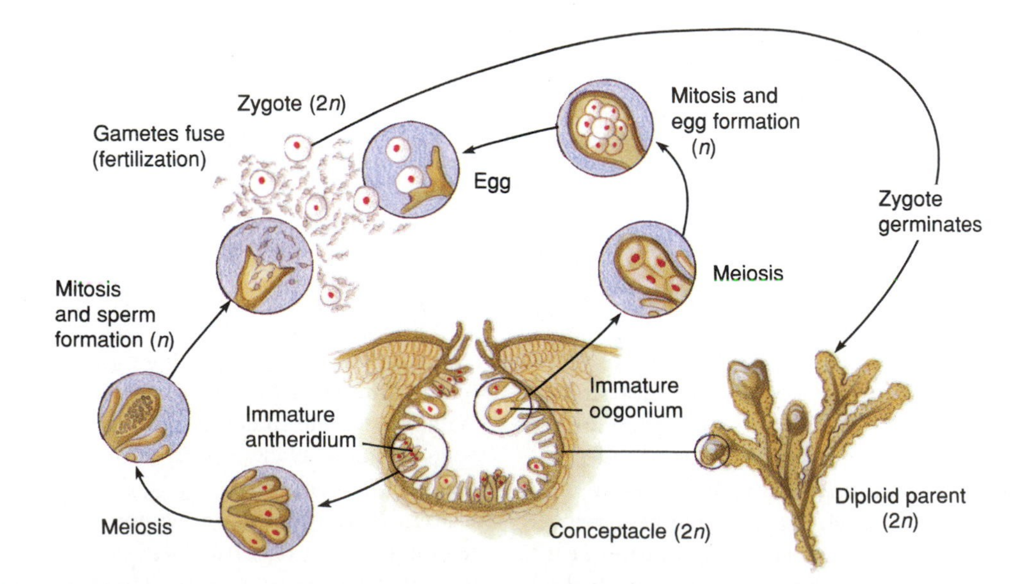

FIGURE 26.25

Conceptacle of *Fucus* are embedded containers that bear oogonia and antheridia.

Figure 26.26

Corallina, a macroscopic red alga. The hard, crusty cell walls of *Corallina* contain carbonates of calcium and magnesium.

Division Rhodophyta: The Red Algae

Like the brown algae, the red algae are mostly marine organisms that are either microscopic filaments or macroscopic leafy branches. One group, the coralline algae, have cell walls impregnated with carbonates of calcium and magnesium, which makes the walls hard and crusty (fig. 26.26). The coralline algae are an important part of coral reefs. The division Rhodophyta also includes unicellular species and species with coenocytic filaments, features that are common in the Chlorophyta but unknown in the Phaeophyta. The most significant features of the approximately 3,900 species of red algae are their phycobilins, their single thylakoid lamella per band, and their lack of flagellated cells.

Red algae are red because they have an abundance of phycoerythrin, a red phycobilin. Phycoerythrin absorbs blue light, which penetrates more deeply into water than do other colors of light. This means that red algae can photosynthesize at greater depths than other algae, and explains why red algae can grow at depths of more than 200 meters.

The gametes of red algae are all nonmotile. The female gametes are therefore eggs, so sexual reproduction in this division is oogamous. The nonmotile male gametes are called **spermatia** (singular, **spermatium**) to distinguish them from the motile sperm of other algae.

Unicellular Red Algae

The simplest red algae are represented by unicellular organisms in the genus *Porphyridium*. Each single-cell individual lacks a cell wall and has a prominent, star-shaped chloroplast (fig. 26.27).

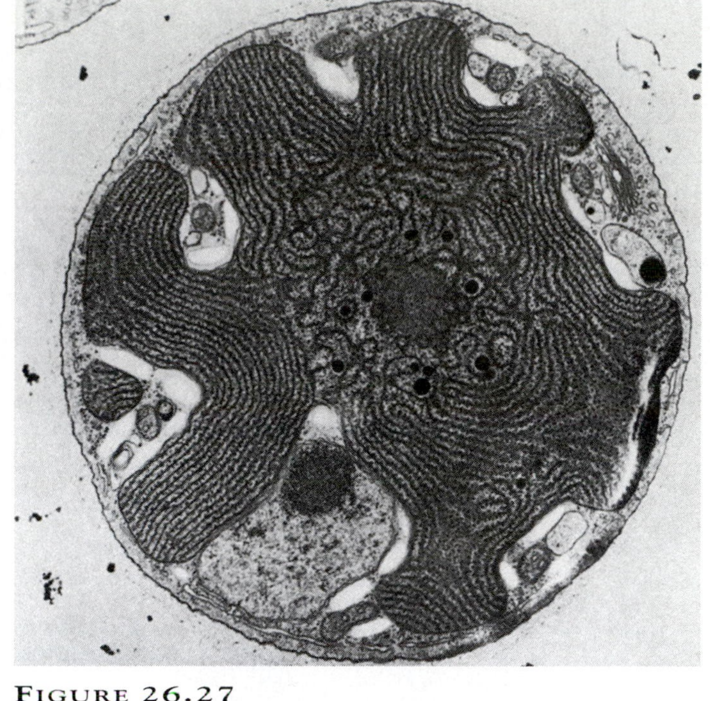

Figure 26.27

Electron micrograph of a star-shaped chloroplast of *Porphyridium*.

Porphyridium often grows on the surfaces of moist soils or pots in greenhouses; populations form shiny, blood-red or bluish-green patches, depending on the relative amounts of red or blue phycobilins present. The cells move by secreting a slimy substance from one end. Reproduction is entirely asexual by mitotic cell division.

Multicellular Red Algae

Red algae are mostly macroscopic and leafy or filamentous. Many of these larger forms have an unusual three-phase life cycle. In *Polysiphonia*, for example, the haploid phase consists of male and female gametophytes that look alike (fig. 26.28). These filamentous gametophytes are coenocytic. The males release free-floating spermatia that are carried by currents to the female gametophytes. The egg container has a special name, the **carpogonium,** because of its unique appearance and function in sexual reproduction. As it is formed, the carpogonium sprouts a sticky, hairlike projection, called a **trichogyne,** that spermatia stick to as they float around the female gametophyte. Once it gets stuck, each spermatium releases its nucleus into the trichogyne, after which the nucleus migrates toward the egg nucleus in the carpogonium. After the zygote is formed by fertilization between a spermatium and an egg, it moves to an **auxiliary cell** and divides mitotically into a multicellular sporophyte. This phase of the life cycle is the **carposporophyte,** and it releases diploid **carpospores** that are produced mitotically.

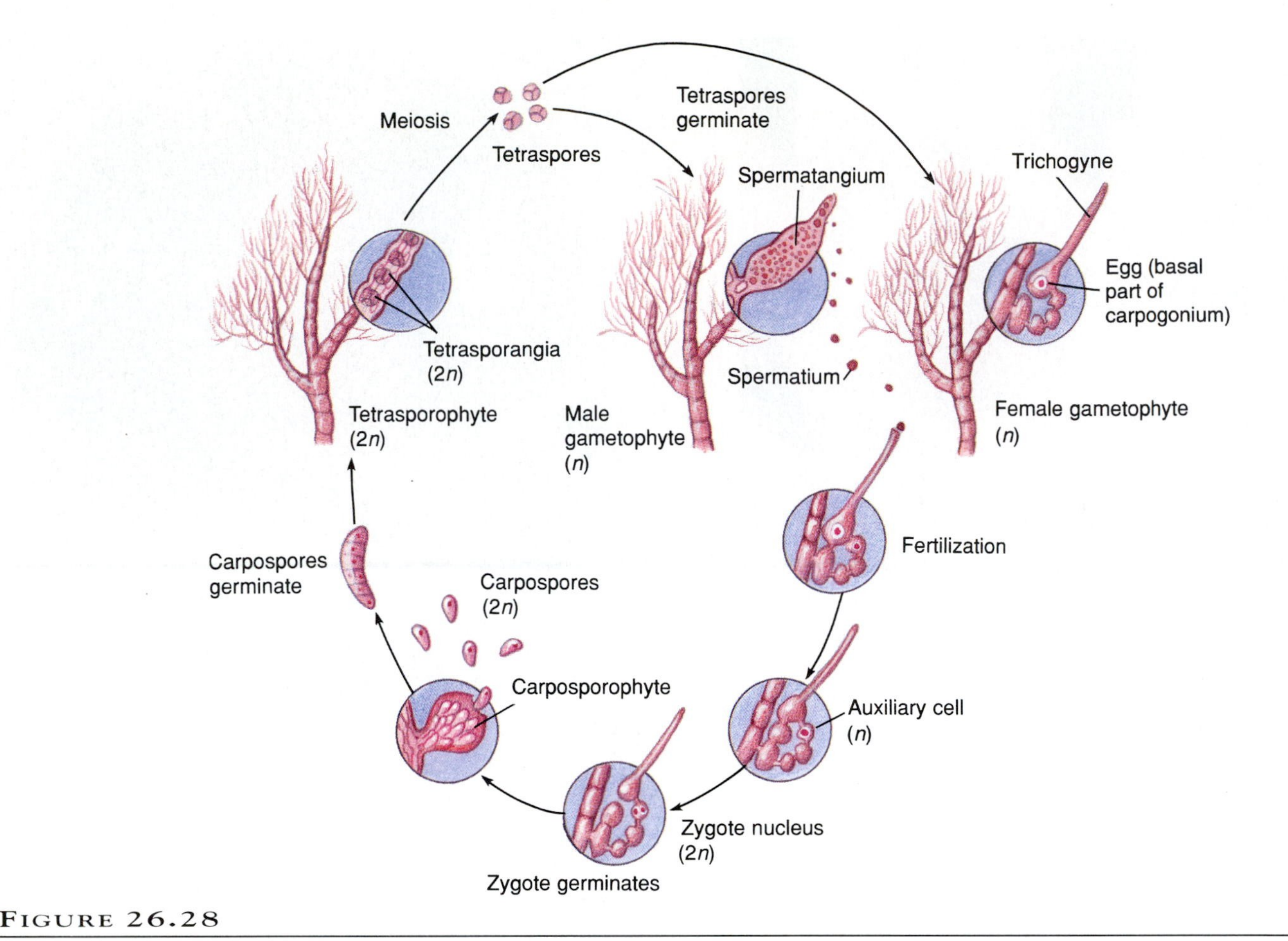

Figure 26.28

Life cycle of *Polysiphonia.* This genus represents a three-phase life cycle that characterizes some red algae; there are two sporophytic phases and one gametophytic phase.

Carpospores settle onto rocks or other substrates and germinate into diploid sporophytes, called the **tetrasporophytes.** Because the tetrasporophyte looks like the gametophytes, these two phases are isomorphic. Unlike the carposporophyte, which stays attached to the parent female gametophyte, the tetrasporophyte is free-living. Lateral branches on the tetrasporophyte differentiate into **tetrasporangia,** where meiosis occurs. Meiosis produces **tetraspores** that form male and female gametophytes. Although the gametophytes and tetrasporophytes are large and isomorphic in *Polysiphonia,* the microscopic tetrasporophytes of other red algae are much smaller than the gametophytes.

Classification of the Red Algae

Like the Phaeophyta, the division Rhodophyta has only one class, the Rhodophyceae. This class is divided into two groups, the subclass Bangiophycidae and the subclass Florideophycidae. The Bangiophycidae comprise about 1% of the red algae. They are unicellular or multicellular, and each cell has a single nucleus and a single, star-shaped chloroplast. Sexual reproduction in the Bangiophycidae is rare (see box 26.2, "The Strange Case of Bangia"). In contrast, the Florideophycidae are exclusively multicellular, and each cell has many disc-shaped chloroplasts. Many species are coenocytic. Sexual reproduction is well-developed in the Florideophycidae; most species have a three-phase life cycle like that of *Polysiphonia.*

Several phylogenies have been proposed for the red algae, but are all plagued by apparent parallelisms. Nevertheless, all eleven orders of the subclass Florideophycidae can be distinguished by two shared-derived character-states: the presence of tetrasporangia and the formation of **gonimoblasts,** which are unbranched filaments that bear sporangia in carposporophytes (fig. 26.29). Earlier suspicions that the Bangiophycidae are an artificial subclass have been borne out; they appear as a paraphyletic group in every cladistic analysis that has been done so far.

C O N C E P T

Red algae are unicellular, filamentous, or multicellular and complex. Unicellular and small filamentous members of this division rarely reproduce sexually. Multicellular red algae have a more complex life history than any other group of algae because their reproductive cycles have three phases: the gametophyte, the carposporophyte, and the tetrasporophyte. Red algae do not produce motile cells.

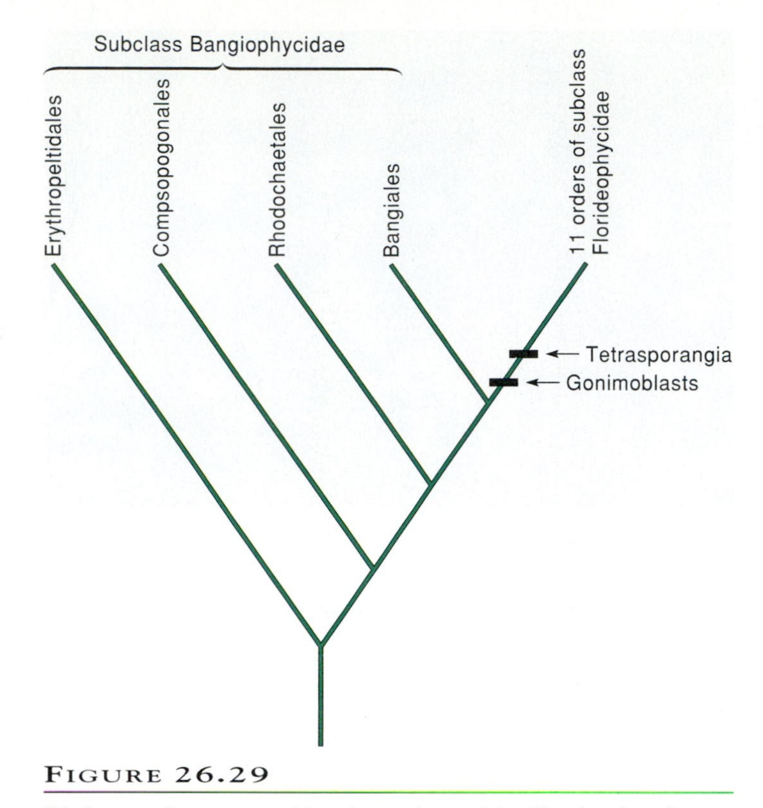

Figure 26.29

Cladogram showing possible relationships of the Florideophycidae.

Other Divisions of Algae

In addition to the three divisions already discussed, algae also include the division Chrysophyta and three divisions that have only unicellular species. Each of these four divisions has one or more features that distinguish it from other algae (table 26.1). Each division also shares important biochemical and ultrastructural characteristics with the green, brown, or red algae. Some of the unicellular species resemble protozoans because they lack chlorophyll and engulf their food. Although photosynthetic pigments and chloroplast structures probably reflect a common ancestry, the origin of other features is not clear. The possible origins of algae are discussed later in this chapter.

Division Chrysophyta: Diatoms and the Golden-Brown and Yellow-Green Algae

The Chrysophyta form the largest division of algae, with more than 11,000 species. About 10,000 species are diatoms in the class Bacillariophyceae, almost all of which are unicellular. Diatoms are unique because of their exquisitely ornamented glass cell walls (fig. 26.30). The golden-brown algae (Chrysophyceae) and yellow-green algae (Xanthophyceae) are also mostly unicellular, but some are colonial or filamentous and coenocytic (fig. 26.31). Chrysophytes occur both in fresh water and in salt water.

Sexual reproduction in chrysophytes includes isogamy, anisogamy, and oogamy. Meiosis is gametic in the diatoms but

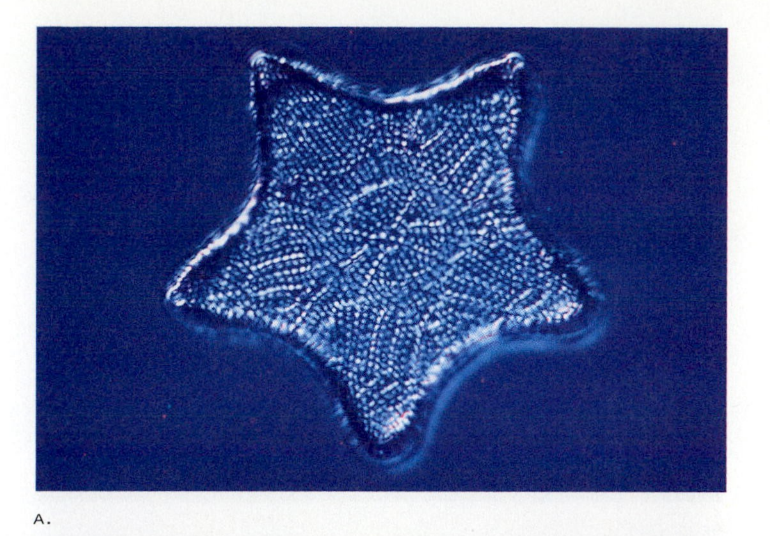

A.

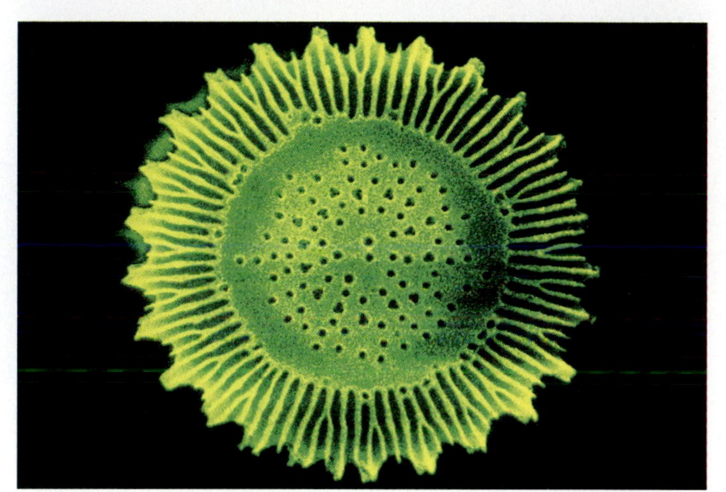

B.

Figure 26.30

(a) Light micrograph of diatom, ×1,000; (b) scanning electron micrograph of diatom, ×1,000.

zygotic in the other classes. Reproduction has been more thoroughly studied in diatoms, probably because of their large number and their great economic importance (see "The Economic Importance of Algae," later in this chapter). In asexual reproduction, diatom cells divide mitotically while they are still encased in their rigid glass walls. The walls consist of two parts, called **valves,** one of which fits tightly over the other like the lid of a petri dish (fig. 26.32). The new cells inside the rigid wall expand and force the valves to separate. New inner valves form inside the old valves so that one of the new cells is smaller than the parent cell and the other new cell is the same size. Division of the smaller cell again produces two new cells, one of which will be even smaller. Thus, in a population of diatoms, some cells are about half the size of others. When the lower size limit is reached, the smallest cells become sexual and produce new large cells.

Vegetative cells of diatoms are diploid and undergo meiosis to produce gametes. In certain diatoms, three of the four meiotic

A.

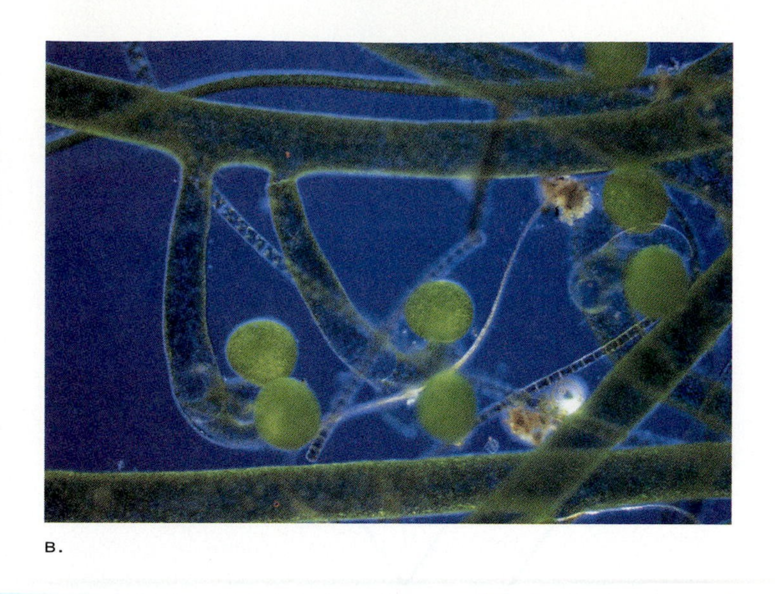

B.

Figure 26.31

Chrysophytes (a) *Dinobryon* is an example of a colonial chrysophyte whose cells live in urn-shaped shells that are attached in a branched series, ×150. (b) *Vaucheria* is a coenocytic, filamentous chrysophyte, ×100.

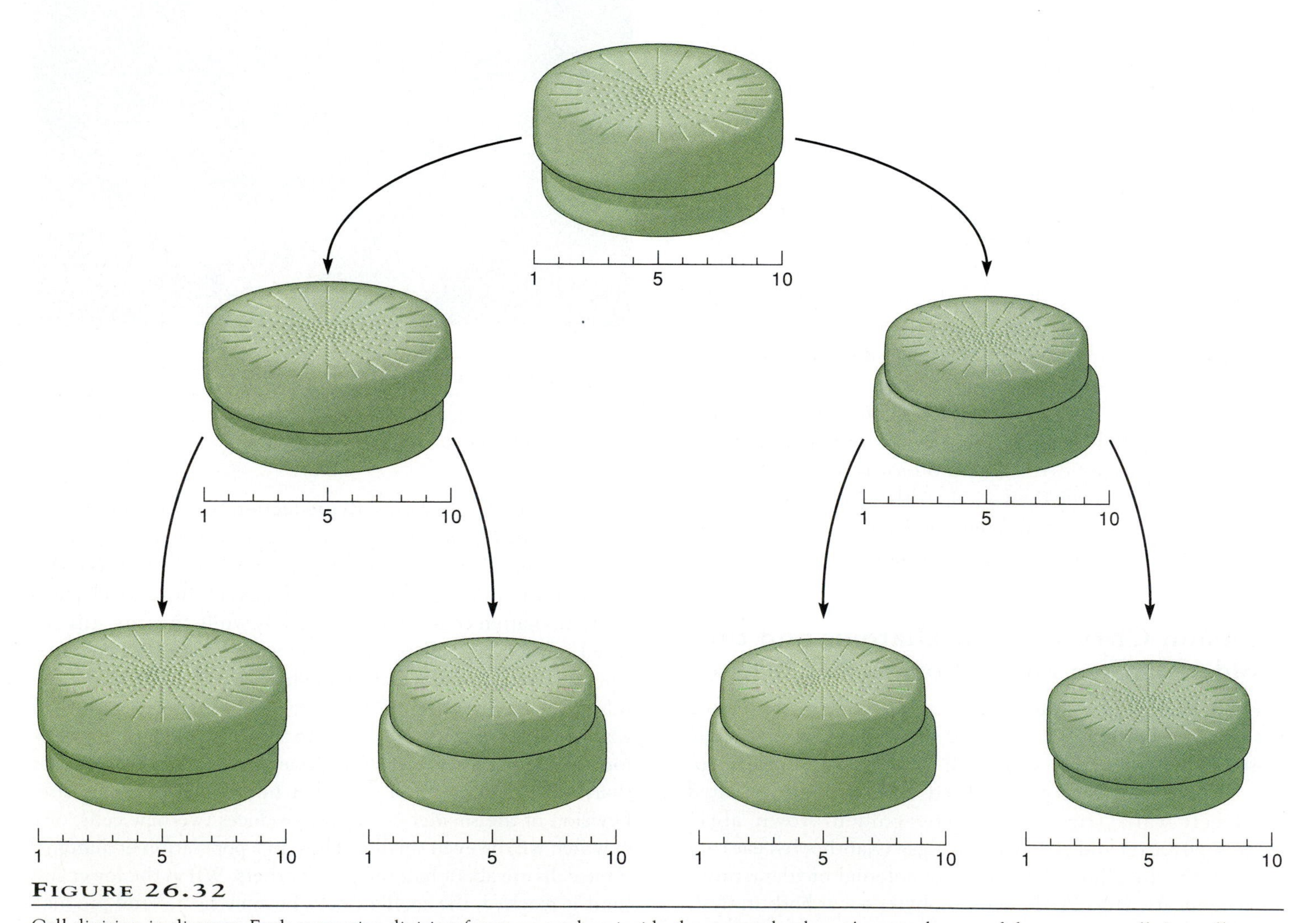

Figure 26.32

Cell division in diatoms. Each successive division forms new valves inside the parental valves. As a result, one of the two new cells is smaller than the parent cell and one is the same size.

products disintegrate, and the remaining cell becomes an egg. In others, all four products become uniflagellate sperm, which are attracted to the egg for fertilization.

In some other diatoms, two of the meiotic products disintegrate, and the remaining two cells function as isogametes, which fertilize another pair of isogametes to make two zygotes. The zygotes of diatoms, called **auxospores,** lack cell walls and expand considerably before making new glass walls.

The Chrysophyta share many features with the Phaeophyta. For example, except for the yellow-green algae, the Chrysophyta have chlorophylls *a* and *c* and the golden-brown pigment fucoxanthin. The color of the yellow-green algae comes from the dominance of yellow beta-carotene over green chlorophyll. Motile cells can have one to three whiplash or tinsel flagella. Some motile cells have lateral flagella and look like the motile cells of the brown algae. The photosynthetic storage carbohydrate of the Chrysophyta is **chrysolaminarin,** which is similar to the brown algal laminarin.

Classification of the Chrysophyta is controversial. Some botanists argue that the similarities between the Chrysophyta and Phaeophyta are significant enough to merge both groups into the same division. Others argue that the reproductive and vegetative features of *Vaucheria* and other Xanthophyceae are sufficiently distinctive to elevate this group to a division, the Xanthophyta. Regardless of their taxonomic status, the Chrysophyta have chloroplasts that were apparently derived from the same prokaryotic endosymbiont that gave rise to chloroplasts in the brown algae (see box 3.2, "Cellular Invasion: Origin of Chloroplasts and Mitochondria").

Divisions of Unicellular Algae

Three divisions of algae consist exclusively of flagellated unicellular organisms: the euglenoids (**Euglenophyta,** 800 species), the dinoflagellates (**Pyrrhophyta,** 3,000 species), and the cryptomonads (**Cryptophyta,** 100 species). These three divisions are traditionally studied by botanists and zoologists alike because of the plantlike and animallike features of different species. Each division includes species that are nonphotosynthetic and species that have chlorophyll *a* and beta-carotene. In addition, the euglenoids have chlorophyll *b*, and the dinoflagellates and cryptomonads have chlorophyll *c;* cryptomonads also have phycobilins. The chloroplasts in all three divisions are surrounded by an extra envelope of endoplasmic reticulum.

Euglenoids and cryptomonads have no cell walls. Instead, their cells are bounded by a flexible **periplast,** which is a plasma membrane that has extra inner layers of proteins and a grainy outer surface (fig. 26.33). A few dinoflagellates also lack cell walls, but most species have cellulose plates interior to the plasma membrane. These plates are grouped into armorlike arrangements that are useful for identifying different genera (fig. 26.34). The dinoflagellates have the most distinctive flagellar pattern among the unicellular algae. Both flagella are lateral, but one coils around the cell like a belt, while the other trails the cell as a rudder. The rhythmic beating of these flagella

FIGURE 26.33

Scanning electron micrograph of *Euglena*, ×2,800. *Euglena* cells are bounded by a flexible periplast instead of a cell wall.

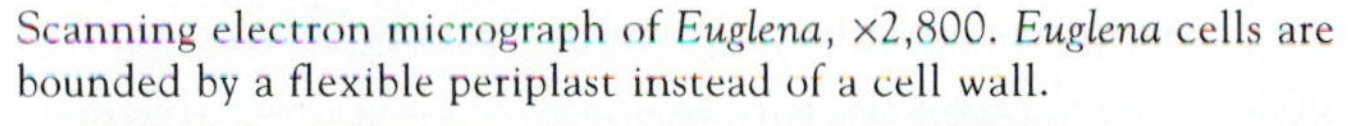

FIGURE 26.34

Dinoflagellates have cellulose-containing armor plates.

propels the dinoflagellate through the water like a spinning top. Most dinoflagellates live in saltwater habitats.

Sexual reproduction is unknown in the Euglenophyta and Cryptophyta; reproduction occurs only by mitotic cell division. In euglenoids, the nuclear envelope persists during mitosis. The cryptomonads have a nucleuslike organelle between the outer chloroplast membrane and the chloroplast ER. This organelle, called a **nucleomorph,** is not present in any other algae.

A few species of the Pyrrhophyta reproduce sexually, but most species are exclusively asexual. Sexual dinoflagellates are either isogamous or anisogamous, and meiosis is usually zygotic. Some dinoflagellates are apparently diploid, but their sexual cycles are not well understood. Nuclear division in the Pyrrhophyta is unusual, because dinoflagellate chromosomes are permanently condensed and lack histones. Like the euglenoids, the dinoflagellates have a persistent nuclear envelope during mitosis.

C O N C E P T

The largest group of algae are the diatoms, which are unicellular organisms. Regardless of their life histories, diatoms are usually classified with the golden-brown and yellow-green algae in the division Chrysophyta because of their similar photosynthetic accessory pigments. The relationships of the other divisions of unicellular algae are not so clear. Many organisms in these other divisions are nonphotosynthetic and are often classified as protozoans.

Writing to Learn Botany

Although plants probably evolved from green algae, they are classified in different kingdoms by some botanists. Do you agree with this classification? Explain your answer.

The Fossil History of Algae

The earliest photosynthetic organisms were undoubtedly prokaryotic; their fossils are more than three billion years old. The first photosynthetic eukaryotes may be represented by fossils of the late Precambrian era (1.2 to 1.4 billion years old) from Bitter Springs, Australia. These fossils have nucleuslike or pyrenoidlike bodies, but not all paleobotanists agree that they are truly eukaryotic organelles. Nevertheless, green algae certainly existed in the Paleozoic era, approximately 500 million years ago. They are common in the fossil record because they are so well-preserved as imprints in calcium carbonate that precipitated around them when they were alive. Most of the earliest green algae resemble living species that have branched, coenocytic filaments. Oogonia from *Chara*-like green algae also occur as calcium carbonate imprints; such representatives of the most complex green algae are more recent, about 360 million years old. Complex red algae probably lived slightly earlier than complex green algae. Ancestors of the coralline red algae are represented in calcium carbonate deposits that are about 400 million years old.

Figure 26.35

Huge deposits of diatomaceous earth at Lompoc, California.

The most common algae in the fossil record are diatoms. Their cell walls become "instant fossils" when a cell dies, because glass does not decompose. Diatoms have been identified in sediments deposited about 120 million years ago. Organisms that seem to be diatoms lived almost 200 million years ago. The fine, sculptured patterns of diatom cell walls makes them easily distinguishable from one another. Approximately 40,000 species of fossil diatoms, all of which are probably extinct, have been named from different cell-wall patterns. Diatom cell walls sometimes settle in large deposits, the largest of which is about 900 meters deep and 2 kilometers long, in Lompoc, California (fig. 26.35). Diatoms in this deposit are harvested commercially and used in many ways (see p. 648).

The most complex fossil algae are in the genus *Prototaxites*, which is about 400 million years old. Specimens of this genus have massive, trunklike stipes that are often misidentified as tree stumps (fig. 26.36). They also contain tubes of two kinds of cells laid end to end, which seem to resemble vascular tissue. The larger tubes are hollow, however, and the smaller ones have cross walls with pores that look like those of some algae and fungi. *Prototaxites* is probably related to the Phaeophyta, because no other algae have such large, complex stipes.

Unicellular organisms that resemble dinoflagellates also occur in the Bitter Springs deposit. Like the other organisms in that deposit, however, their resemblance to extant groups may be superficial. The oldest fossils that are confirmed as dinoflagellates are about 430 million years old.

A.

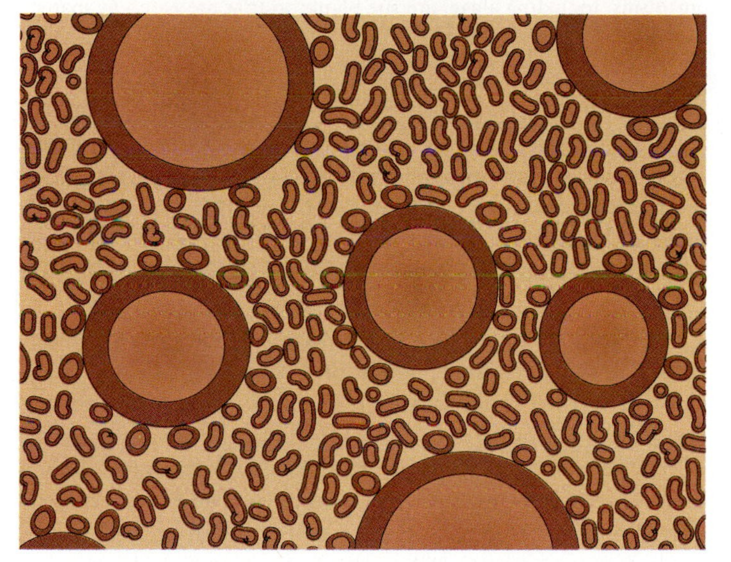

B.

Figure 26.36

The trunklike stripe that was named *Prototaxites southworthii* has conductive cells that resemble those of the brown algae. (a) Longitudinal section; (b) cross section.

The Origins of Algae

Botanists hypothesize that the chloroplasts of algae came from at least three kinds of photosynthetic prokaryotes. Chloroplasts of the green algae and euglenoids came from a *Prochloron*-like ancestor of the Chloroxybacteria that had chlorophylls *a* and *b*. Chloroplasts of the red algae and cryptomonads came from an *Anacystis*-like ancestor of the Cyanobacteria that had chlorophyll *a* and phycobilins. The genus *Heliobacterium* of the Eubacteria represents the most likely ancestor of the chloroplasts of the Phaeophyta, Chrysophyta, and Pyrrhophyta, which all have chlorophylls *a* and *c*.

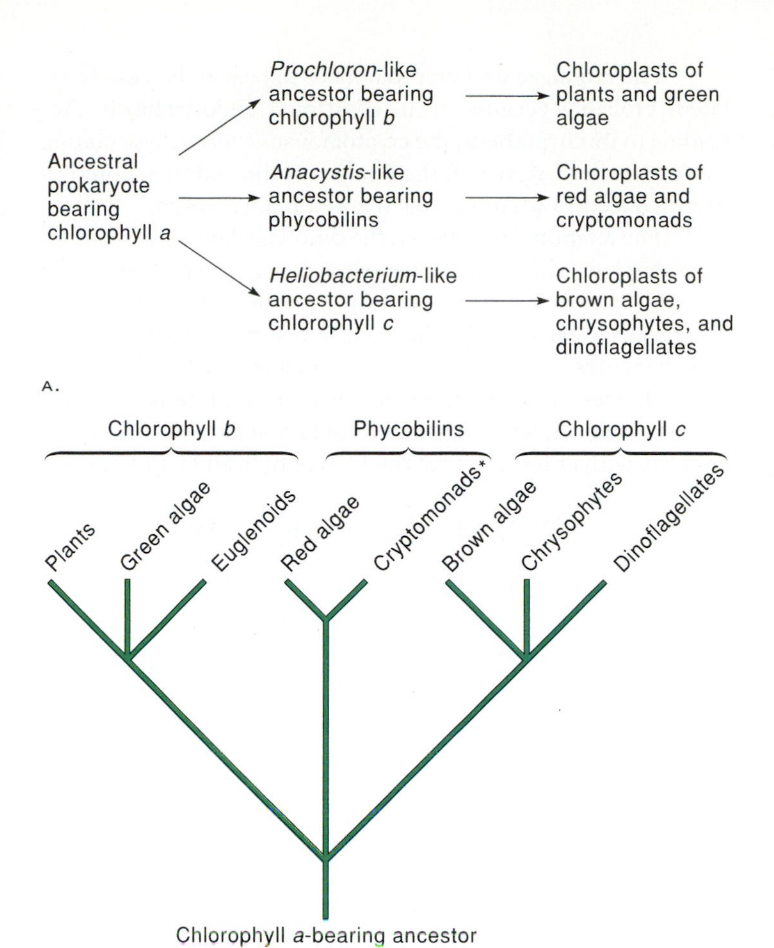

Figure 26.37

Evolution of the algae. (a) One possible evolutionary pathway for the origin of chloroplasts from prokaryotic ancestors based on the evolution of their main photosynthetic pigments; (b) example of a cladogram of the main groups of algae, showing unresolved trichotomy at the base and later divergence of different divisions on each of the three evolutionary branches.

One hypothesis about the evolutionary relationships of the three main chloroplast ancestors is shown in figure 26.37a. This hypothesis shows that chlorophyll *a* evolved once from a porphyrin-bearing ancestor. However, there is no clear candidate for the porphyrin-bearing ancestor of the photosynthetic prokaryotes, since porphyrins are also common in all organisms as components of cytochromes.

Aside from the evolution of their chloroplasts from prokaryotes, there are no good characters for evaluating the evolutionary relationships of the three main groups of algae. This means that a cladogram of these three groups must be three-parted at its base, as shown in figure 26.37b. Assuming that cellulosic cell walls evolved only once, then this character is a shared derivation for all divisions of algae. If so, we must assume that groups without cell walls lost the ability to make cellulose, even though their ancestors had it. The cladogram must also include other protoctists, such as water molds, because they have cellulosic cell walls, too.

The red algae and cryptomonads appear to be closely related, probably because of an additional endosymbiosis. According to this hypothesis, the cryptomonads evolved by acquiring a eukaryotic red algal cell; the nucleus of the endosymbiont has been reduced to what we now see as a nucleomorph.

The relationship between the dinoflagellates and the other algae that have chlorophyll *c* is more puzzling. The prokaryotelike chromosomes of dinoflagellates are so different from those of other eukaryotes that the dinoflagellates are sometimes called **mesokaryotes.** The mesokaryotic nucleus in dinoflagellates evolved independently of the eukaryotic nucleus in other algae. However, the dinoflagellate chloroplast apparently shares its ancestry with chloroplasts of the Phaeophyta and Chrysophyta.

C O N C E P T

The fossil history and evolutionary origins of algae show that there are three main evolutionary lines. These are best explained by the character of their chloroplasts. One line has chlorophylls *a* and *b* from a *Prochloron*-like ancestor, one has chlorophyll *a* and phycobilins from an *Anacystis*-like ancestor, and the third line has chlorophylls *a* and *c* from a *Heliobacterium*-like ancestor. The divergence of these three groups probably began at least 500 million years ago, but their origins may be more than one billion years old.

The Ecology of Algae

Algae are nearly ubiquitous and occupy a wide variety of habitats. They live primarily in marine and fresh water, either free-floating or attached to rocks, wood pilings, shells of shellfish, or to other algae. Many species are terrestrial on moist soil, rocks, stone roofs and walls, or tree bark. Some of their more unusual habitats include clouds and airborne dust, snow, and the fur of certain animals. Some unicellular species are the food-producing endosymbionts inside the cells of protozoans, sponges, sea slugs, sea anemones, and saltwater fish. Green algae live symbiotically with fungi and cyanobacteria in lichens.

The most significant ecological role of algae is as plankton (fig. 26.38). Algae and other unicellular organisms are eaten by small animals, which in turn are eaten by larger animals. Thus, algae are the primary producers that support life in marine and freshwater habitats. Commercial fish farmers often exploit this relationship by fertilizing tanks and ponds to enhance the growth of plankton. The oxygen that planktonic algae produce is equally important to life; perhaps 50%–70% of the earth's atmospheric oxygen comes from unicellular marine algae.

Brown and red algae form "forests" in intertidal zones and shallow coastal waters. These marine forests are habitats for teeming populations of many kinds of animals. Coral reefs are special habitats whose primary production comes from coralline red algae. Some red algae precipitate calcium carbonate around them or bond the calcium carbonate from marine sponges.

Figure 26.38

Light micrograph of a drop of pond water, ×40. Microscopic organisms, including algae, are the plankton that are eaten by many aquatic and marine animals.

These interactions of red algae, sponges, and other organisms form coral reefs. Some corals also harbor green algae or yellow-green algae as food-producing endosymbionts. By using food from algae, coral reefs can form extensive, highly complex ecosystems that support many species of marine animals. The most spectacular example of a coral reef is the Great Barrier Reef off the east coast of Queensland, Australia.

Planktonic and other microscopic algae are especially conspicuous when their populations expand rapidly into large **blooms.** These blooms may be induced by warm temperatures, phosphate pollutants, or other factors that enhance algal growth. In fresh water, algal blooms sometimes get so large that when they ultimately begin to decay, oxygen is used by the respiration of decomposing bacteria faster than it can be replenished by algal photosynthesis. When this happens, fish and other animals die from lack of oxygen, and their decaying bodies make the problems worse.

Blooms of dinoflagellates are called *red tides* because of the abundance of these red-pigmented organisms. Unlike other algal blooms, the cause of red tides is unknown. Red tides are especially common along the Gulf coast of Florida and the coast of central California. One of the most interesting features of red tides is that they are bioluminescent—they glow in the dark just like fireflies. Red tides kill millions of fish each year because of a nerve toxin secreted by dinoflagellates. Shellfish that eat dinoflagellates appear to be immune to these toxins, but humans or other vertebrates that eat these shellfish may suffer poisoning and death.

The Lore of Plants

During a bloom, algae can be so abundant that only 0.03 cubic meters (about 1 cubic foot) of water can contain more than thirteen million organisms.

Boxed Reading 26.2

The Strange Case of *Bangia*

Bangia atropurpurea is a macroscopic, filamentous red alga that grows attached to rocks or woodwork in marine and fresh water. It rarely reproduces sexually. Algae like *Bangia* are often brought into the laboratory to be studied in culture because it is difficult to observe all of the components of their life cycles in nature. Botanists who collect these algae often do not know what phase of the algal life cycle they have collected. For example, the red alga *Bangia atropurpurea* had only one known phase for a long time, until it was brought into laboratory culture. At first, when it was grown in no more than 12 hours of light per day, *Bangia* produced spores that formed more *Bangia*. However, when the daylength was increased to more than 12 hours, *Bangia* produced spores that grew into another genus, *Conchocelis,* whose alternate phase had also been previously unknown. Furthermore, when cultures were switched back to less than 12 hours of light, they produced spores that grew back into *Bangia.* These two genera—that is *Bangia atropurpurea* and *Conchocelis*—turned out to be alternate phases of the same organism. Because *Bangia* was named first, the genus name for the two phases is *Bangia*. Nevertheless, biologists still refer to this organism as having a *Bangia* phase alternating with a *Conchocelis* phase in its life cycle.

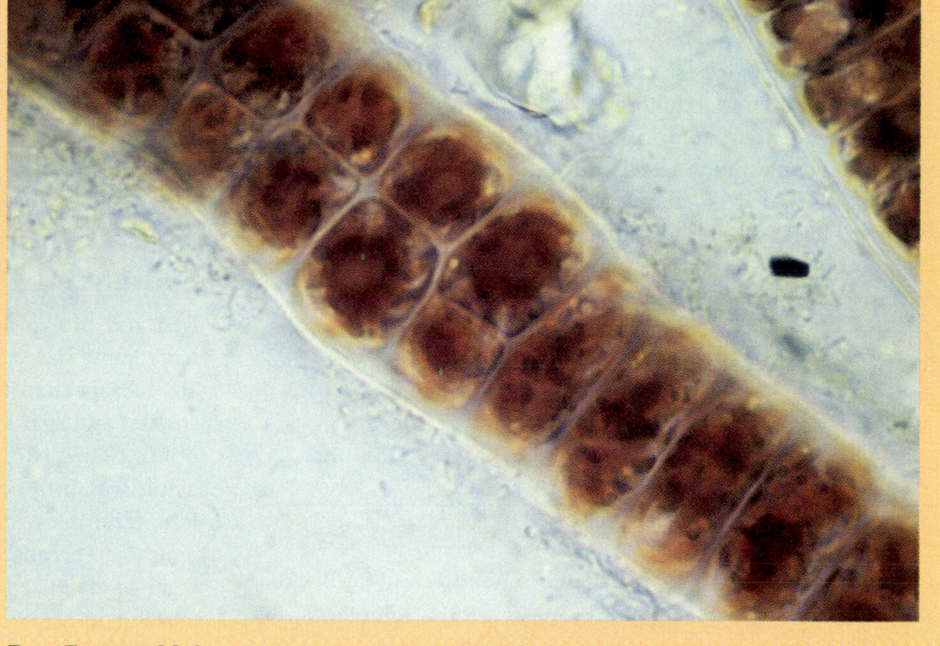

Box Figure 26.2

Bangia atropurpurea.

The puzzle about the relationship between *Bangia* and *Conchocelis* is still not fully resolved, because the sexual cycle is poorly understood. Both phases have the same chromosome number, either $n = 3$ or $n = 10$; therefore, if fertilization occurs at all, then meiosis must be zygotic.

The Physiology of Algae

The Influence of Light and Temperature on Reproduction

Reproduction in many multicellular algae is influenced by daylength and temperature (see box 26.2, "The Strange Case of Bangia" above). For example, gametophytes of the brown alga *Sphacelaria* produce gametes at 4°C and 12°C under a 12-hour photoperiod. At 20°C, they reproduce only asexually. Reproduction is also asexual when temperatures are lower but the photoperiod is longer.

Euglena cells point their eyespots toward light, so that the cells are parallel to the direction of the light. In this way, *Euglena* cells are oriented for swimming toward the light. Conversely, in multicellular algae, zoospores with eyespots often swim away from the light toward the seabed, where they can germinate. Regardless of the response, cells with eyespots have light receptors. Light receptors are usually associated with the absorption of blue light by carotenoids, but in *Euglena* the receptor pigment may be riboflavin.

Osmotic Regulation

The unicellular, wall-less green alga *Dunaliella* has a remarkable range of tolerance for different salt concentrations: it can live in water with NaCl concentrations between 0.05 M and 5 M (seawater is about 0.03 M). *Dunaliella* maintains cytoplasmic concentrations of Na^+ at 50–100 times lower than those of the external environment. Because it has no cell wall and it is easy to grow in laboratory culture, *Dunaliella* is used as a model for studying membrane functions. Much of our information on osmotic regulation and control of membrane potential comes from this organism.

Silicon Metabolism

Silicon is an essential element for the development of diatoms, certain seedless vascular plants, and many flowering plants. In living organisms, silicon is only known in the form of silicates, such as SiO_2, which polymerize to form glass. The physiology of silicate metabolism is best known in diatoms, because they can be easily manipulated in laboratory cultures. Silicon is absorbed by diatoms as silicic acid, $Si(OH)_4$. In general, growing cells absorb silicic acid much faster in the light than in the dark.

Diatoms seem to have an ecological and evolutionary advantage because they can make cell walls out of silicates. Silicate cell walls are more economical to make, since their synthesis requires up to ten times less metabolic energy than does the synthesis of cellulose or other cell-wall materials. Moreover, silicates are highly adsorptive, which means that nutrients stick to them. This adsorption may help diatoms to compete for nutrients, especially when nutrients are at low concentrations.

Algae are the major primary producers in marine and freshwater habitats. In the form of plankton, algae are the foundation of food webs that support animal life in water. The rapid vegetative growth of kelps and other macroscopic algae is also important in "marine forest" habitats along coastlines. Algal reproduction and growth are sensitive to light and temperature.

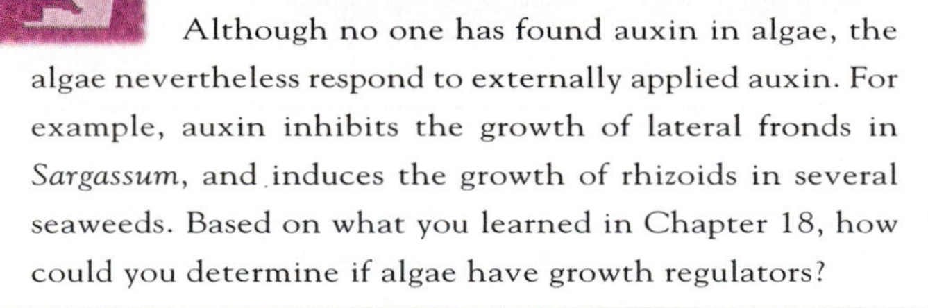

Doing Botany Yourself

Although no one has found auxin in algae, the algae nevertheless respond to externally applied auxin. For example, auxin inhibits the growth of lateral fronds in *Sargassum*, and induces the growth of rhizoids in several seaweeds. Based on what you learned in Chapter 18, how could you determine if algae have growth regulators?

The Economic Importance of Algae

Diatoms

As a group, the diatoms are perhaps the most economically important algae. Indeed, they make the fishing industry possible because of their role as food for freshwater and marine animals. Diatom cell walls are also harvested from large deposits as *diatomaceous earth* (see fig. 26.35), which is used in many industries. For example, diatomaceous earth is an abrasive in metal polishes and a few brands of toothpaste; it is also used in swimming-pool filters and in filters for clarifying beer and wine. Reflective paint on highways, road signs, and license plates also contains cell walls of diatoms. One of the recipes for dynamite includes diatomaceous earth as an inert absorbent mixed with nitroglycerin. Because much of the earth's fossil oil originates from diatoms, the presence of diatoms in sample cores of the earth is often a useful indicator of oil deposits.

Industrial Polymers

Red and brown algae produce cell-wall polysaccharides that have many industrial uses. The main red algal polysaccharides are **carrageenan** from seaweeds such as Irish moss (*Chondrus crispus*) and **agar** from several other seaweeds, including species of *Gracilaria* (fig. 26.39). Carrageenan is used to stabilize or emulsify paints, cosmetics, cream-containing foods, and chocolate. The main use of agar is in preparing culture media to grow bacteria and other microorganisms for laboratory research. Agar is also used as a gel for canning fish and meat and for making desserts. Some medicines contain agar as an inert carrier. The most useful polysaccharide from brown algae is **algin,** which is obtained from species of *Macrocystis* that are cultivated in large kelp beds (see photo at the beginning of this chapter). The uses of algin are similar to those of carrageenan and agar.

One of the more promising uses of algal polymers is in medicine. Polysaccharides extracted from certain unicellular and colonial green algae stimulate the immune systems of animals, which may enhance resistance to disease. Although several patents for these potential medicines have been awarded in Japan, algal polysaccharides are not yet approved for human medicinal use.

Algae as Food

Many algae have been used traditionally as food, although most algae are not particularly nutritious. Some seaweeds, however, contain a lot of iodine, which is an important mineral in the thyroid gland. Furthermore, although much of the cell-wall material of algae cannot be digested by humans, some seaweeds contain useful amounts of protein. These include nori (*Porphyra tenera*), a red alga that is used mostly as a flavoring in soups and salads and as a wrapping around small rolls of rice (*maki-sushi*). Species of *Ulva*, *Laminaria*, and other seaweeds are also eaten fresh or pickled.

A.

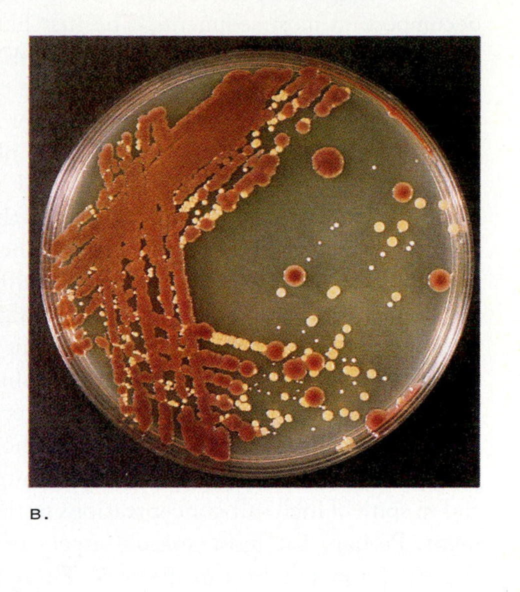

B.

Figure 26.39

(a) Irish moss (*Chondrus crispus*), is a red algae that is commercially important as a source of carrageenan. (b) Microbiologists grow a variety of organisms on media solidified with agar (shown here), which is extracted from seaweeds such as *Gracilaria*.

Chapter Summary

The vegetative organization of algae includes microscopic unicellular, colonial, and filamentous organisms and complex macroscopic seaweeds. Of the seven divisions of algae, three include macroscopic forms, and four consist almost entirely of unicellular organisms. Most algae reproduce both sexually and asexually. In sexual reproduction, gametes may look alike (isogamy), be of two different sizes (anisogamy), or consist of sperm and eggs (oogamy). In most green algae, zygotes undergo meiosis, so the only diploid cell in the life cycle is the zygote itself. In other green algae and in other multicellular algae, meiosis is either sporic or gametic. Only the red algae lack motile cells in all phases of the life cycle. In filamentous algae, the sexual alternation between two phases may occur between dominant diploid and reduced haploid phases, or the reverse. Some algae alternate between isomorphic sporophytes and gametophytes; larger seaweeds, such as kelps and rockweeds, have dominant diploid phases that alternate with microscopic haploid filaments or with gametes as the only haploid cells in the life cycle.

The evolutionary relationships of algae show that green algae are the ancestors of plants. Within the green algae, genera of the Charophyceae are apparently the closest relatives of land plants. Some of the most significant features shared by green algae and plants include the development of a phragmoplast during cytokinesis, the retention of the zygote on the gametophyte, and the ability to make flavonoids.

Relationships among algae are shown by the features of their chloroplasts; however, these features do not show which two of the three main evolutionary lines of algae are more closely related to each other. The cryptomonads (Cryptophyta) are closest to the red algae; the diatoms and other Chrysophyta are closest to the brown algae. Dinoflagellates (Pyrrhophyta) may also be closely related to the brown algae, but dinoflagellate nuclei are mesokaryotic and apparently very primitive. The euglenoids (Euglenophyta) share their main photosynthetic features with the green algae.

The fossil record of algae contains representatives that are at least 500 million years old. Some fossils that may be algae are up to 1.4 billion years old, but there is no general agreement that these fossils are truly eukaryotic. The most common fossil algae are diatoms, because their glass cell walls do not decompose in most sediments. The first likely brown algae are represented by giant stipes that are about 400 million years old.

The most important ecological role of algae is as plankton. In this role, algae are the primary producers in the food webs of marine and freshwater habitats; plankton feeders are then eaten by other animals. Certain red algae cause calcium carbonate to precipitate around them, or they cement the calcium carbonate from marine sponges. The interaction of red algae with sponges, corals, and other sea life forms coral reefs.

The growth and reproduction of algae are sensitive to photoperiod and temperature. Some kelps reproduce sexually only at lower temperatures, whereas certain red algae alternate phases only when daylength is increased or decreased. Unicellular marine algae, especially those without cell walls, have evolved effective mechanisms for maintaining osmotic potential in spite of high salt concentrations in the external environment. Perhaps the most unusual aspect of algal physiology is the silicon metabolism of diatoms. These algae absorb silicic acid and convert it into silicate-based cell-wall materials.

Diatoms are probably the most economically important group of algae because they produce large deposits of discarded cell walls. These fine glass particles are used in many commercial, nonfood products. Polysaccharides from kelps and red algae are used to stabilize and emulsify many foods and nonfood products. Some algae are used directly as human food.

Questions for Further Thought and Study

1. What is the evidence that green algae and plants had a common ancestor that is not shared with either the brown algae or the red algae?
2. How can you tell if isogamy, anisogamy, or oogamy are the results of parallel evolution?
3. Why can red algae grow at greater depths than other algae?
4. Why are only some algae well preserved in the fossil record?
5. Why are algal polysaccharides so important as emulsifiers in many commercial products?
6. What features of algae are plantlike? What features of algae are not plantlike?
7. Botanists generally agree that the chloroplasts of green algae, brown algae, and red algae are derived from three different prokaryotic endosymbionts. How do the other divisions of algae fit into this three-part pattern of evolution? What features of these other algal divisions do not seem to fit this pattern?
8. Both *Ectocarpus* and *Ulva* have life cycles that alternate between isomorphic phases. *Ectocarpus* populations have two kinds of individuals, and *Ulva* populations have three kinds of individuals. Why are they different? What might be the adaptive significance of this difference?
9. What features of coralline red algae are important for their role in making coral reefs? Why is the growth pattern of coral reefs determined by algae, even though their main components are corals?
10. What features of charophytes distinguish them from other algae? What is the potential evolutionary significance of these distinguishing features?
11. Figure 26.31b shows pores in the glass shells of diatoms. Of what function might these pores be?
12. When grown in darkness, euglenoids such as *Euglena* lose their chloroplasts and live as heterotrophs. Of what selective advantage is this?

Suggested Readings

Articles

Anderson, D. M. 1994. *Scientific American* 271:62–68.

Bremer, K. 1985. A summary of green plant phylogeny and classification. *Cladistics* 1:369–385.

Corliss, J. O. 1984. The Protista kingdom and its 45 phyla. *BioSystems* 17:87–126.

Graham, L. E. 1984. *Coleochaete* and the origin of land plants. *American Journal of Botany* 71:603–608.

Manhart, J. R., and J. D. Palmer. 1990. The gain of two chloroplast tRNA introns marks the green algal ancestors of land plants. *Nature* 345:268–270.

Maxwell, Christine D. 1991. A seaweed buffet. *The American Biology Teacher* 53:159–161.

Saffo, M. B. 1987. New light on seaweeds. *BioScience* 37:654–664.

Books

Bold, H. C., and J. W. LaClaire, II. 1987. *The Plant Kingdom*. 5th ed. Englewood Cliffs, NJ: Prentice-Hall.

Bold, H. C., and M. J. Wynne. 1985. *Introduction to the Algae*. 2d ed. Englewood Cliffs, NJ: Prentice-Hall.

Margulis, L. 1980. *Symbiosis in Cell Evolution*. New York: W. H. Freeman.

"Leafy" mosses growing among large, thalloid lichens near Glacier Bay in Alaska.

CHAPTER 27

Bryophytes

Chapter Outline

Chapter Overview

Plants are multicellular eukaryotes with cellulose-rich cell walls, chloroplasts containing chlorophylls *a* and *b* and carotenoids, and starch as their primary food reserve. They evolved from green algae, and are currently divided into two main groups: bryophytes and vascular plants. Vascular plants include ferns, conifers, flowering plants, and related groups, and will be considered in subsequent chapters. Bryophytes include mosses, liverworts, and hornworts. Although bryophytes are most noticeable when they grow in dense mats, they can grow just about everywhere—including on bark and exposed rocks, where other plants cannot grow. Bryophytes are especially common in moist areas, and lack the specialized vascular tissues that characterize other groups of plants. Gametophytes dominate the life cycle of bryophytes. Sporophytes are short-lived and obtain their food from the gametophyte to which they are attached.

Introduction

Although most biologists agree that early photosynthetic organisms evolved in the water, true plants are nearly absent in the oceans, and those that occupy freshwater habitats seem to be highly specialized. Aquatic habitats offer tremendous stability when compared to land. There is always plenty of water, and the temperature range is small. Yet this watery medium has its drawbacks. Light does not penetrate far into the water, and most stationary plants cannot grow at depths greater than a few meters. Furthermore, limited amounts of CO_2 (for photosynthesis), nutrients, and O_2 for their roots can sometimes make fresh water a very inhospitable environment. Clearly, a movement to land would open vast new opportunities to obtain gases and light, if the adventurous plant could solve its water problems.

If we speculate about those first organisms to survive upon the land, we would probably consider them to be simple organisms with no organized vascular systems, perhaps like the **bryophytes** (mosses and liverworts). After all, there was no selection pressure for any wasteful vascular tissue while these organisms were living in the water. When we consider these organisms, we are tempted to think about wet habitats where the mosses are close to water, basking in the sun of a bog, or cooling off in the spray of a waterfall. Certainly these are habitats where bryophytes are common. But keep thinking. What about those rocks on the cliff or the sand of the dunes? In fact, can you think of a habitat that has plants where it is impossible to find mosses? There aren't many, and if you visualize some of the rocky habitats in your mind, certainly these organisms undergo tremendous changes in moisture and temperature, even within a single day, occupying habitats where no vascular plants can survive.

Even with so many diverse habitats occupied by plants today, we still consider the move from water to land to have been a major one. The greatest challenge to plants colonizing land was to keep their cells wet. Land plants responded to this challenge in two ways. Some, the ones we call **vascular plants,** acquired lignin, developed a complex water-transport system, and encased themselves in a waxy, waterproof **cuticle.** These adaptations enabled land plants to maintain a watery environment in their leaves, despite the fact that they were suspended in desertlike air. Other plants, the bryophytes, developed other strategies that we are only beginning to understand. They lack lignin, and as a result they lack the complex transport system of other plants. They have only a thin cuticle, if any, and their leaves are only one cell thick, making it easy to lose water. Yet, there are about 15,000 species, more than any other group of plants besides flowering plants.

Add to these simple survival problems the problem of transferring gametes from a male organ to a female organ when the male gamete, the sperm, requires free water to swim! It seems that one of the best "solutions" was to produce gametes only when water was available, but that requires developing the gametangia well in advance to be ready on time. Something has to trigger the plants to stop using all their energy for growth and put some of it into gametangia. That is, a method of receiving and responding to environmental signals was necessary. Nevertheless, even the bryophytes have been successful at organizing their life cycles in a way best suited to their individual environments.

Features of Bryophytes

Bryophytes have the following general features:

1. Most bryophytes are small, compact, green plants (fig. 27.1). Like green algae, they produce chlorophylls *a* and *b*, starch, cellulose cell walls, and motile sperm. Bryophytes usually grow very slowly.
2. Bryophytes lack well-developed vascular tissues and lignified tissues. As a result, they grow low to the ground and absorb water by capillarity. However, not all bryophytes are strictly nonvascular. For example, some

Figure 27.1

(a) A hornwort (*Anthoceros*) growing in a highland rain forest; (b) a liverwort (*Conocephalum*) growing near the entrance of a cave; (c) a moss (*Dawsonia*) growing on Mount Kinabalu, Borneo; (d) another moss (*Hypnum*, also known as feather moss) growing in a rain forest in Olympic National Park, Washington.

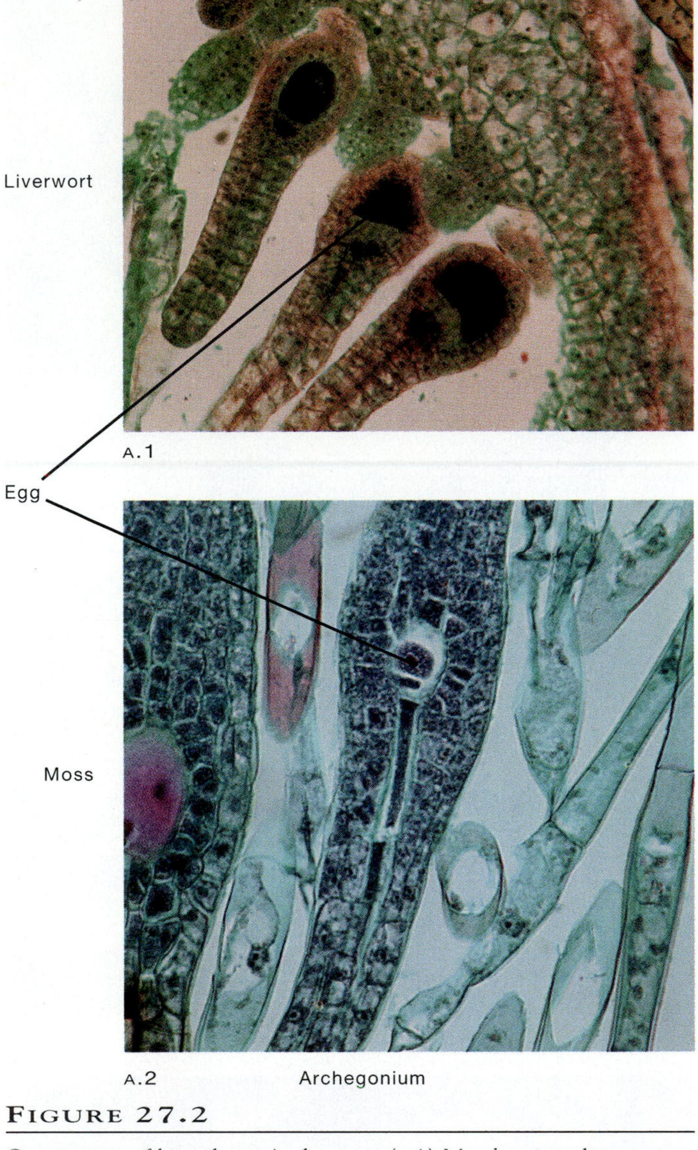

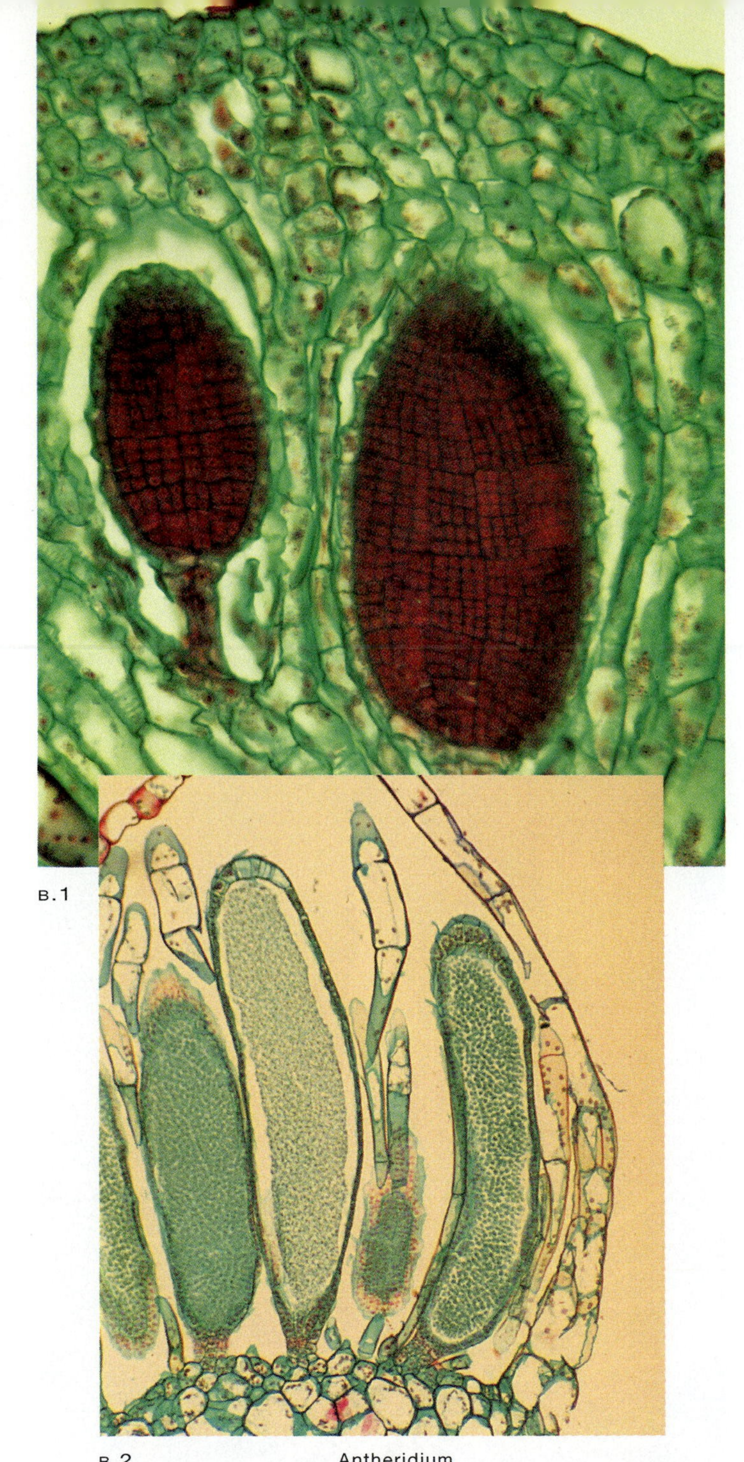

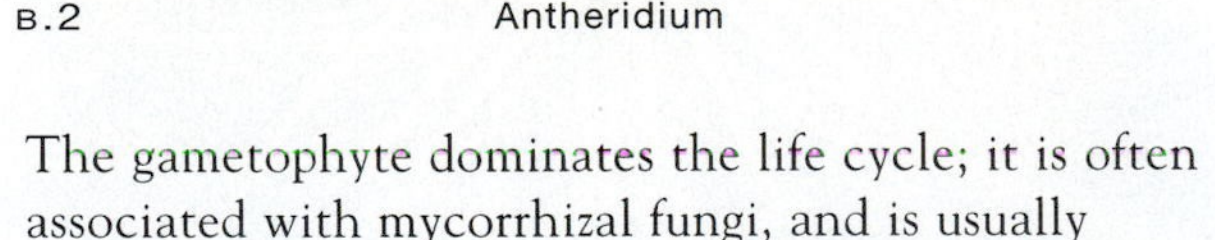

FIGURE 27.2

Gametangia of bryophtes: Archegonia (a.1) *Marchantia*, a liverwort, (a.2) *Mnium*, a moss. Antheridia of (b.1) *Marchantia*, (b.2) *Mnium*.

mosses have a central strand of conducting cells that are functionally equivalent to xylem and phloem.

3. Organs such as leaves and roots of vascular plants are defined by the arrangement of their vascular tissues. Since bryophytes lack true vascular tissues, they also lack true leaves and roots. However, many bryophytes have structures that are similar and functionally equivalent to leaves, so they are often referred to as such.
4. Bryophytes get their nutrients from dust, rainwater, and substances dissolved in water at the soil's surface. Tiny **rhizoids** (hairlike extensions of epidermal cells) along their lower surface anchor the plants but do not absorb water or minerals. Water and dissolved minerals move by capillarity over the surface of bryophytes.
5. The gametophyte dominates the life cycle; it is often associated with mycorrhizal fungi, and is usually perennial. Gametes form by mitosis in multicellular gametangia called **antheridia** (male) and **archegonia** (female) (fig. 27.2). Each flask-shaped archegonium produces one egg, and each saclike antheridium produces many sperm. Gametangia are surrounded by a protective, sterile sheath of cells.
6. The sporophyte is short-lived and unbranched, and produces a terminal sporangium. Although photosynthetic for most of its life, the sporophyte is permanently attached to and partially dependent on

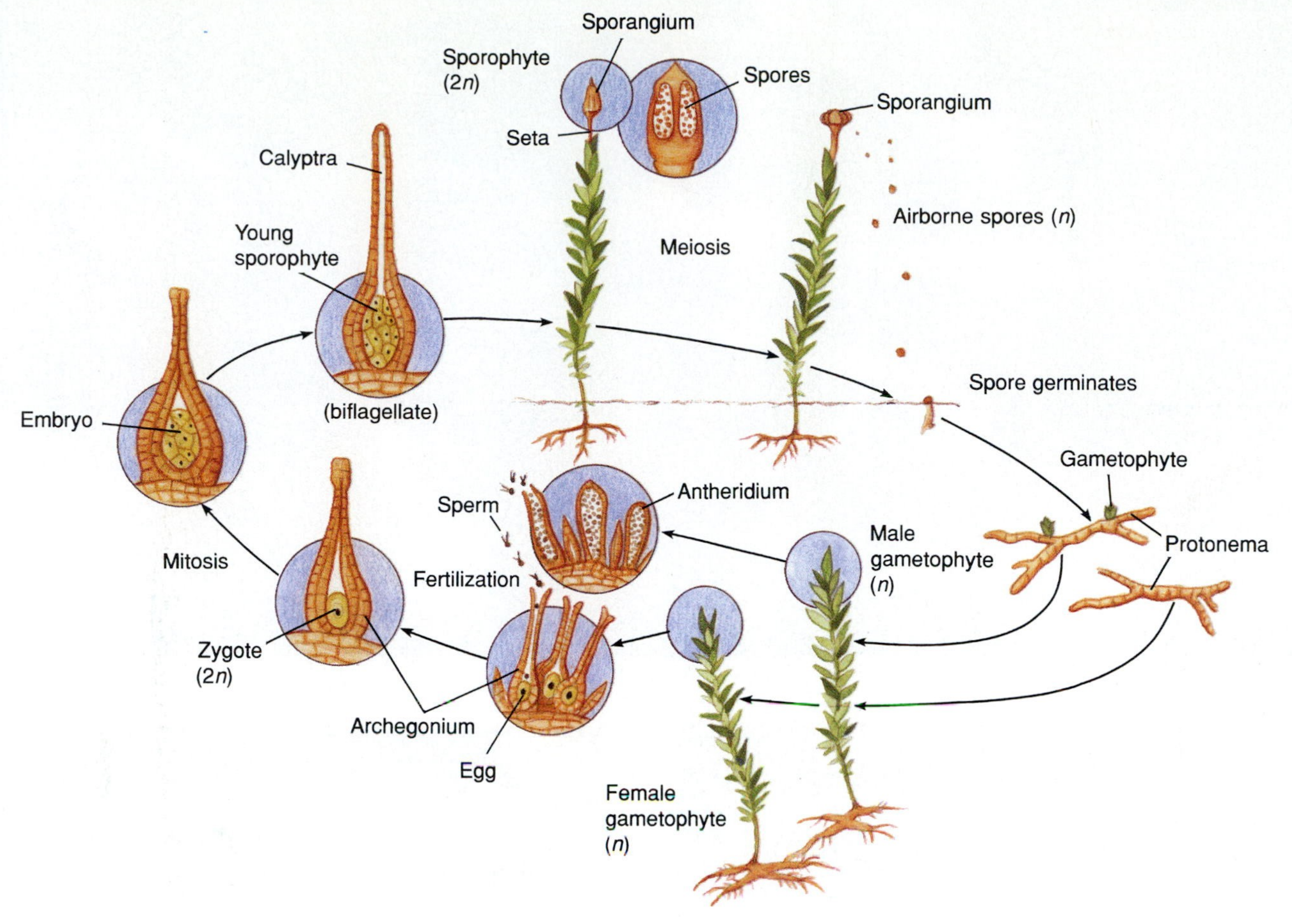

FIGURE 27.3

The general life cycle of a moss.

the gametophyte. This contrasts with the situation in green algae such as *Ulva* and *Ulothrix*, in which the sporophyte and gametophyte live independently of each other. The sporophytes of bryophytes have no direct connection to the ground. Spores have a cutinized coat and are usually dispersed by wind.

7. Biflagellate sperm swim through water to eggs. Thus, bryophytes need free water for sexual reproduction. In this regard they are like amphibians, since they cannot live too far from water. Unlike those of nonplants, fertilized eggs develop in protective sex organs. Most reproduction in bryophytes is asexual.

Although small and inconspicuous, bryophytes are studied intensely. For example, much of the early work on sex determination in plants involved bryophytes. In unisexual species, spores that produce female gametophytes contain a large X chromosome, while those that produce male gametophytes contain a corresponding and smaller Y chromosome. This topic remains actively studied, as do the evolutionary relationships of bryophytes.

A Generalized Life Cycle of Bryophytes

Unlike many green algae, bryophytes have heteromorphic alternation of generations; that is, the sporophyte and gametophyte are distinctly different (fig. 27.3). Haploid spores formed by meiosis begin the gametophyte generation. The spore germinates to form a gametophyte, which is more diverse in bryophytes than in any other group of plants. Rhizoids attach the gametophyte to its substrate. The gametophyte grows from an apical cell and produces antheridia and archegonia. Antheridia (singular, antheridium) produce sperm, and archegonia (singular, archegonium) produce eggs. Sexual reproduction in bryophytes requires free water, because the sperm must swim to the egg. Since sperm cannot swim far, antheridia and archegonia must be close to each other (usually within a few centimeters) for sexual reproduction to occur. Sperm released from an antheridium do not swim randomly; rather, they are attracted by a gradient of a still-unidentified substance(s) produced by an archegonium. A sperm fertilizes an egg and forms a diploid zygote, thus beginning the sporophyte generation of the life cycle.

Figure 27.4

The diversity of mosses. (a) Mosses often dominate wet places, such as this waterfall; (b) moss (*Rhacomitrium*) colonizing a lava bed in Iceland; (c) cushion moss growing in Antarctica, where daily summer temperatures range from −10° to −30°C; (d) *Splachnum luteum*, a moss that resembles a small flowering plant. The colors of the moss, and the chemicals it produces, attract insects that disseminate its spores; (e) *Grimmia*, a rock moss that survives on bare rocks, often in scorching sun.

The developing sporophyte depends on the gametophyte for nutrition and water. Although sporophytes develop differently in mosses, liverworts, and hornworts, they all ultimately produce sporangia containing sporogenous (spore-producing) tissue. This tissue undergoes meiosis to produce spores, which are released to the environment. If the spore lands in a dry area, it can remain dormant, often for several decades. When water becomes available, the spore germinates (sometimes within a few hours) and forms the gametophyte, thus completing the sexual life cycle.

C O N C E P T

Bryophytes are small plants lacking complex vascular tissues and supporting tissues. Their life cycle is dominated by free-living, photosynthetic gametophytes that produce archegonia and antheridia, each surrounded by a sterile, protective sheath of cells. Bryophytes require free water for sexual reproduction. Egg and sperm fuse to form a zygote that begins the sporophyte generation. The sporophyte, which is attached to and dependent on the gametophyte for nourishment, produces spores that form gametophytes, thus completing the life cycle.

Some bryophytes produce only antheridia or archegonia, so that extensive colonies of bryophytes may have only archegonia or antheridia. Furthermore, the sporophytes of some mosses have never been found. These bryophytes apparently reproduce only asexually.

The Diversity of Bryophytes

The three major groups of bryophytes are variously treated as classes of a single division, Bryophyta, or as three separate divisions. In the latter case, they are Division Bryophyta (mosses), Division Hepatophyta (liverworts), and Division Anthocerotophyta (hornworts). The classification of all bryophytes in a single division implies a monophyletic origin for mosses, liverworts, and hornworts. As you will see later in this chapter, this idea is in dispute (see "The History and Relationships of Bryophytes").

Mosses

Mosses are remarkably successful plants that thrive alongside more conspicuous vascular plants. The approximately 12,000 species of mosses make up the largest and most familiar group of bryophytes.

Moss morphology is very diverse (fig. 27.4). The smallest moss may be the pygmy moss, which is only 1–2 mm tall and completes its life cycle within a few weeks. Also, luminous mosses such as *Schistostega* (cave moss) and *Mittenia* often grow near entrances to caves and glow an eerie golden green. The upper surface of these mosses is made of curved, lenslike cells that concentrate the cave's dim light onto chloroplasts for photosynthesis.

FIGURE 27.5

Gametophytes of *Polytrichum*, the haircap moss, showing antheridial heads. Note the radial symmetry and differentiation into "leaves" and "stems."

The Lore of Plants

Cave moss (*Schistostega*) has long fascinated the Japanese. Its eerie glow has been the subject of countless books, newspaper and magazine articles, television shows, and even an opera. There is a national monument to *Schistostega* near the coast of Hokkaido, where it grows near a small cave.

Contrary to what you may think, not all organisms referred to as *mosses* are true mosses. For example, Irish moss and other sea mosses are red algae, Iceland moss and reindeer moss are lichens, club mosses (i.e., a type of lycopods) are seedless vascular plants, and Spanish moss is a flowering plant in the pineapple family.

Mosses have several features that distinguish them from other bryophytes.

1. Gametophytes of most mosses are radially symmetrical and often have protective scales along their lower surface (fig. 27.5). Their rhizoids are multicellular and lack chlorophyll.
2. Their spores form a filamentous **protonema** (plural, **protonemata,** meaning "first thread") that is an extensive, branched phase of the life cycle (fig. 27.6). Unlike the cross walls of algae, the cross walls of protonemata are oblique.
3. Their gametophytes are always leafy, and the leaves are usually arranged radially in three rows. Leaves are not lobed, and usually have a midrib. Furthermore, leaves are usually only one cell thick and have no mesophyll, stomata, or petiole. Chloroplasts are lens-shaped.

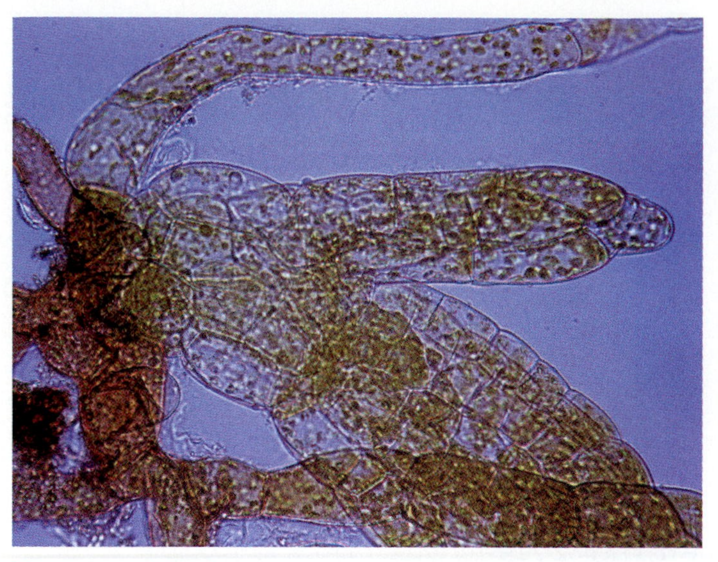

FIGURE 27.6

Moss protonema with a developing bud, ×450.

4. Intermixed with archegonia and antheridia are sterile, absorptive filaments called **paraphyses** (singular, **paraphysis**).
5. The sterile jacket surrounding the sporangium has stomata and opens by means of an apical lid called an **operculum** that dries and falls off of the capsule. This exposes the tooth-shaped segments of the peristome (fig. 27.7).
6. Their spores are dispersed over relatively large distances.

Keep these characteristics in mind as we examine the life cycle of a typical moss.

Gametophyte

A moss spore germinates to form a green protonema, which is strikingly similar to a filamentous green alga (fig. 27.6). The protonema grows rapidly from its tip—often more than 40 cm in only a few months—and branches into a tangled mass over the substrate. As the protonema grows, it forms specialized cells having fewer chloroplasts and oblique cross walls. At this stage, the protonema accumulates hormones (e.g., cytokinins) and forms buds that become leafy gametophytes with stemlike axes.

Protonemata can form from many kinds of cells. For example, fragments of gametophytes can usually form a protonema (fig. 27.8). This type of fragmentation is an important means of asexual reproduction in mosses. The Japanese use tiny fragments of *Polytrichum* to create their lavish moss-gardens. In nature, the fragmentation of mosses is so extensive that a cubic meter of arctic snow can contain more than five hundred fragments of moss gametophytes, each of which can form a protonema.

In most mosses, archegonia and antheridia form on separate gametophytes, which means the gametophytes are **unisexual** (i.e., monoecious). Other mosses are **bisexual** (i.e., dioecious), meaning that antheridia and archegonia form on

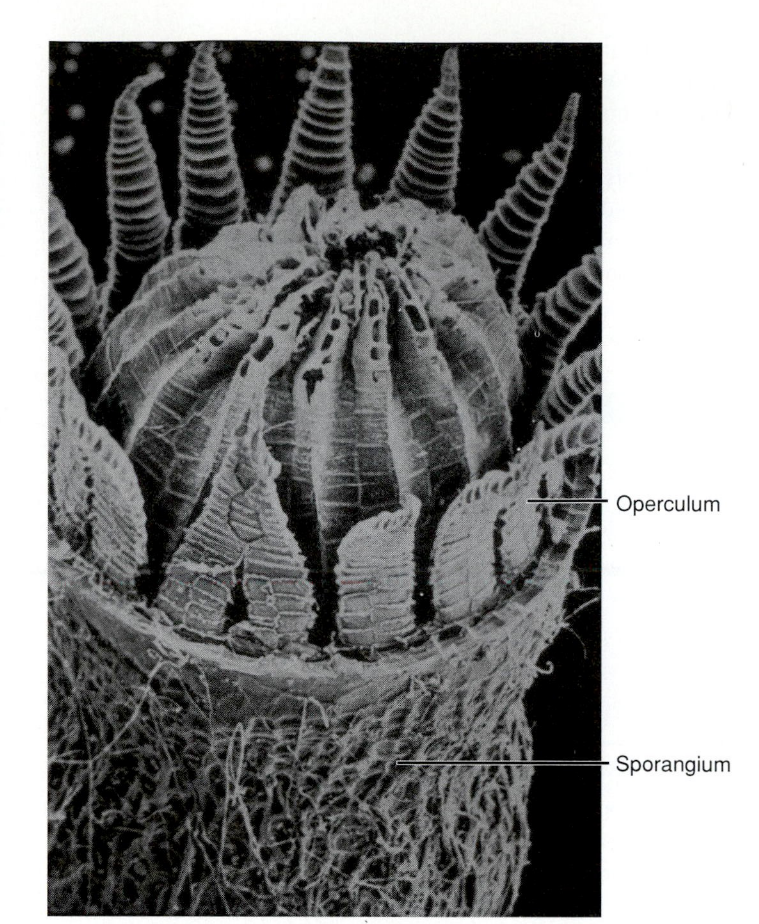

FIGURE 27.7

Scanning electron micrograph of the spore capsule of a moss, ×125.

FIGURE 27.8

A leaf tip of the moss *Bryum capillare*. Protonemata are growing from a cell on the lower surface of the leaf, while a new shoot is developing from the upper surface.

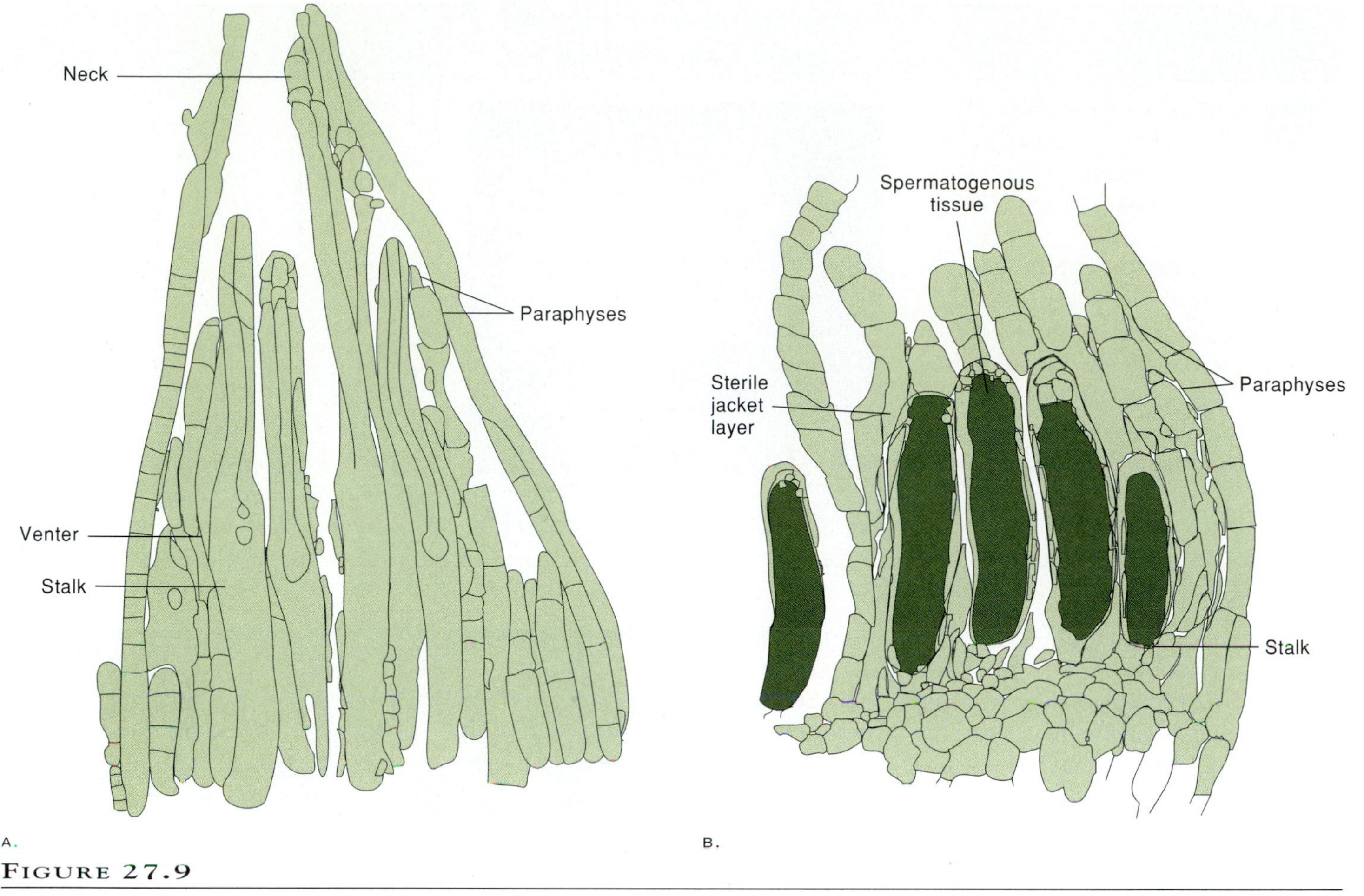

FIGURE 27.9

(a) Longitudinal section through archegonial head, showing archegonia surrounded by sterile structures called paraphyses. (b) Longitudinal section through antheridial head, showing antheridia surrounded by paraphyses.

the same gametophyte. Gametangia form at the tips of shoots or on separate branches. Archegonia terminate the main shoot or form on reduced lateral branches (fig. 27.9a). The swollen base of the archegonium is called the **venter** and contains a single egg. The slender region above the venter is the **neck.** Cells in the center of the neck disintegrate and produce a liquid that attracts sperm and provides a fluid-filled canal through which sperm swim to reach the egg. Archegonia are often separated by rows of paraphyses that absorb water.

Round or sausage-shaped antheridia form atop short stalks (fig. 27.9b). Inside each antheridium is a mass of tissue that forms coiled sperm. Water-absorbing paraphyses swell the antheridium and help expel the sperm, often in a cottony, mucilaginous mass. Once released, the mass of sperm disperses as individual cells. These sperm are attracted to archegonia by compounds such as sugars, proteins, and acids released by the archegonia. Sperm swim through the fluid-filled canal to the venter, where one sperm fertilizes the egg. The resulting zygote begins the sporophyte generation.

The gametophytes of some mosses (e.g., *Dawsonia, Polytrichum, Atrichum, Mnium*) have a central strand of conducting tissues made of specialized cells called **leptoids** and **hydroids.** Leptoids resemble sieve tubes in vascular plants because they transport sugars, lack nuclei, have many plasmodesmata, contain much callose, and are associated with parenchyma. Hydroids are tracheidlike cells that are dead and empty at maturity and transport water and dissolved minerals. Unlike tracheids, however, hydroids lack lignified secondary cell walls, and therefore contribute little support to the gametophyte.

Sporophyte

The sporophyte generation in a moss life cycle begins when an egg and sperm fuse to form a diploid zygote. The spindle-shaped embryo begins developing in the archegonium; cells of the venter divide to accommodate its growth. Cells beneath the embryo also divide and contribute to a protective sheath called the **calyptra.** Continued growth soon tears the base of the calyptra from the gametophyte, leaving the remains of the calyptra atop the venter like a tiny pixie cap (fig. 27.10). The calyptra has different shapes in different mosses. For example, it looks like a tiny candle-snuffer in extinguisher mosses. In hair cap mosses, such

Boxed Reading 27.1

Shooting Spores

The *Sphagnum* growing in peat bogs uses a battery of natural air guns to spread its tiny spores. As they mature, the spore capsules shrink to about a fourth of their original size.

This shrinkage, which compresses the air in the capsule, also shapes the capsule into a tiny gun barrel, with its own airtight cap (i.e., operculum). Eventually the cap is blown off the capsule with an audible pop, firing the spores as far as 2 m.

Box Figure 27.1

Sphagnum spores being fired from a capsule.

as *Polytrichum*, the calyptra prevents desiccation of immature cells of the sporophyte. If the calyptra is removed too early, the sporophyte stops growing, and no sporangium forms. Stomata on the sporangium have doughnut-shaped guard cells that remain open unless the surrounding air becomes excessively dry. Once the sporangium is mature, the stomata close permanently.

The mature sporophyte has three parts: a **foot,** a **seta,** and a **capsule** (fig. 27.11a). The foot, which penetrates the base of the venter and grows into the gametophyte, absorbs water, minerals, and nutrients from the gametophyte. The wiry seta elongates and raises the capsule as much as 15 cm above the gametophyte. In the process, the top of the exposed archegonium is torn from the gametophyte and forms the calyptra.

The capsule is a sporangium that begins developing while still within the protective calyptra. The outer cells form a sterile, protective jacket, and the innermost cells elongate to form another sterile tissue called the **columella.** Sporogenous cells form between the jacket and the columella and undergo meiosis to form as many as 50 million haploid spores per capsule. When these spores are mature, the calyptra falls off, thereby exposing the operculum (i.e., lid of the capsule). Contraction forces associated with drying often then burst off the operculum, exposing tooth-shaped segments of the **peristome** (figs. 27.7, 27.11b). In some mosses the peristome is cone-shaped and has pores through which spores are released. The mature sporophyte is usually yellow or brown, and its spores are dispersed by wind (however, see box 27.1 above, "Shooting Spores"). Spores that land in suitable environments form protonemata, thus completing the life cycle.

Figure 27.10

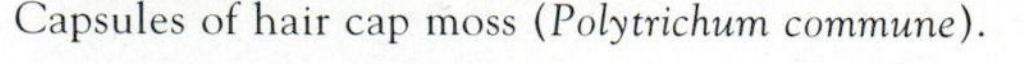

Capsules of hair cap moss (*Polytrichum commune*).

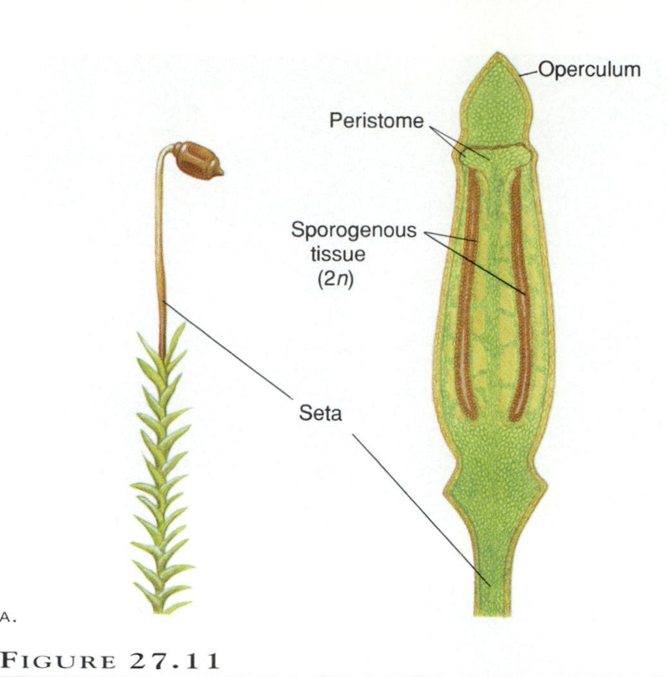

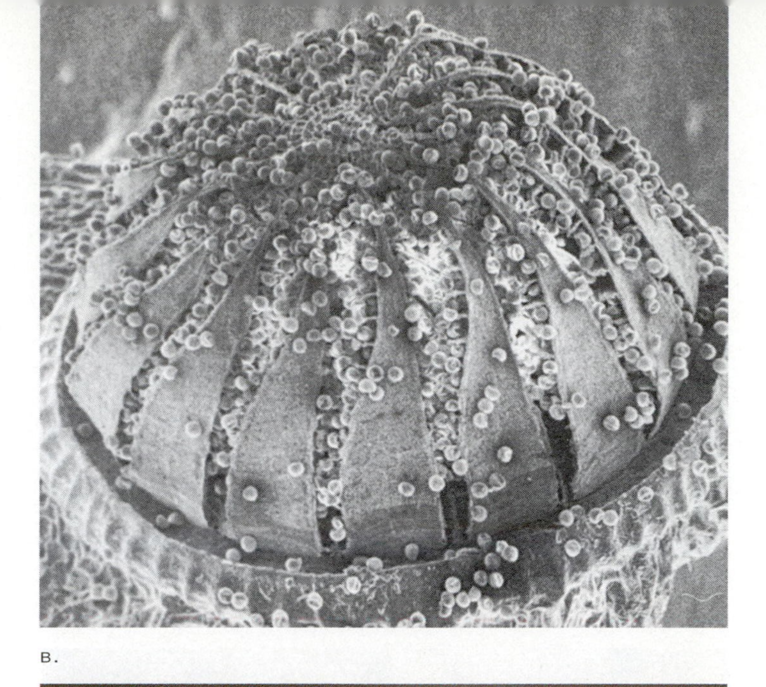

Figure 27.11

Moss sporophyte. (a) Diagram of parts of a mature moss sporophyte; (b) photograph of a moss capsule showing peristome teeth; (c) moss sporophytes. How does the capsule shown in 27.11b differ from the one shown in fig. 27.7?

C O N C E P T

Mosses are the largest and most familiar group of bryophytes. Their gametophytes are leafy and usually radially symmetrical. Sperm swim through the neck of an archegonium to reach an egg held in the venter. Spores are released from sporangia through the peristome. Spores germinate and form a filamentous protonema, which grows into a gametophyte. The gametophytes of some mosses contain special conducting tissues that resemble sieve tubes and tracheids.

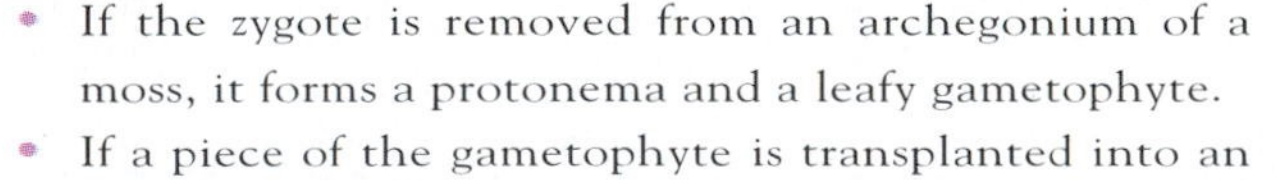

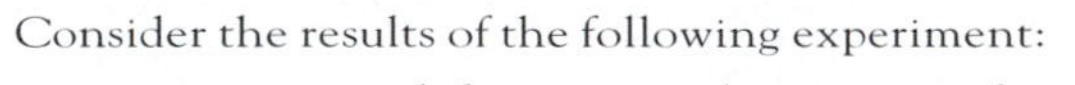

Writing to Learn Botany

Consider the results of the following experiment:

- If the zygote is removed from an archegonium of a moss, it forms a protonema and a leafy gametophyte.
- If a piece of the gametophyte is transplanted into an archegonium, it will form a sporophyte.

What do you conclude from these results?

Liverworts

Liverworts were named during medieval times, when herbalists followed the Doctrine of Signatures (see Chapter 24). A few liverworts are lobed and thus have a fanciful resemblance to the human liver, so the word *liver* was combined with *wort* (herb) to form the name *liverwort*. Although we now know that the Doctrine of Signatures is invalid and that eating liverworts will not help an ailing liver, the name *liverwort* has endured. Almost 8,500 species of liverworts have been named.

Liverworts range in size from tiny, leafy filaments less than 0.5 mm in diameter to a thallus more than 20 cm wide (fig. 27.12). All liverworts have a prominent gametophyte, which sometimes has a cuticle. Also, the spores of liverworts have thick walls. These features are believed to be important adaptations that enable liverworts to live on land. Liverworts also have the following distinguishing features.

1. Their rhizoids are always unicellular.
2. Their gametophytes are leafy or thallose, and are often lobed and bilaterally symmetrical. They lack a midrib. The upper side of the thallus is photosynthetic; the lower side is achlorophyllous and used for storage.

Figure 27.12

The diversity of liverworts: (a) Leafy liverworts growing on the leaf of an evergreen tree in the Amazon Basin in Brazil; (b) *Calypogeia muelleriana*, a leafy liverwort having unlobed leaves; (c) *Bazzania trilobata*, a leafy liverwort having dichotomous forking; (d) *Fossombronia cristula*, a thallose liverwort; (e) an aquatic thallose liverwort (*Ricciocarpus natans*) floating on the shallows of a pond; (f) *Conocephalum conicum*, a thallose liverwort. The conspicuous white spots are pores.

A.

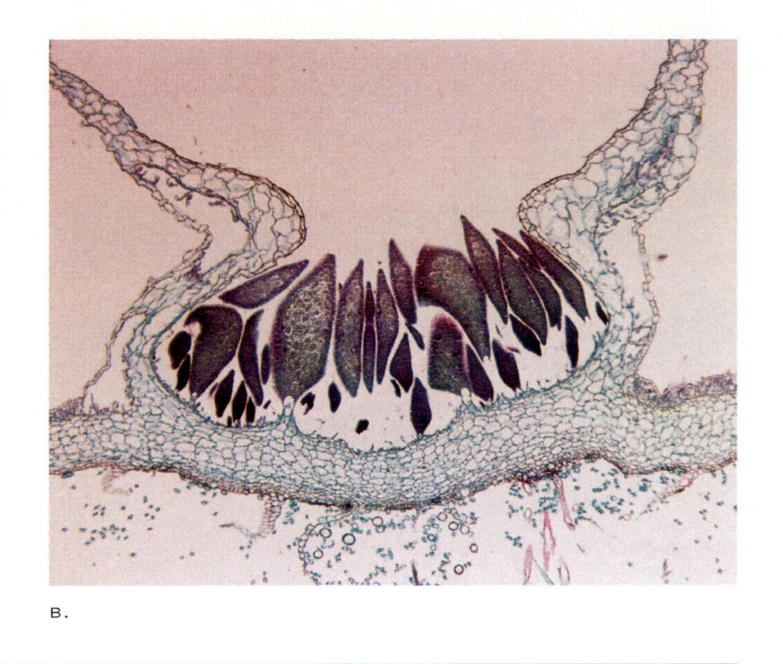

B.

FIGURE 27.13

(a) Gemmae cups ("splash cups") containing gemmae on the gametophytes of *Lunularia*. Gemmae are splashed out by raindrops and can then grow into new gametophytes, each identical to the parent plant that produced it by mitosis. (b) Longitudinal section of a gemmae cup, ×10.

3. Their sporangia are often unstalked.
4. They shed spores from sporangia for a relatively short time.

Liverworts frequently reproduce asexually. One way they do this involves the death of older parts of the plant. When this occurs, the growing areas are left isolated from the parent plant. A second way is by producing ovoid, star-shaped, or lens-shaped pieces of tissue called **gemmae** (singular, **gemma**). Gemmae form in small "splash cups" on the upper surface of the gametophyte or in clusters at the base of thalli (fig. 27.13). Gemmae become detached from their cups by falling raindrops, and are often splashed up to a meter from the parent plant. Gemmae grow into gametophytes. Pieces of gametophytes that are broken or torn from the parent plant can also regenerate entire plants.

Gametophyte: Leafy Liverworts

The gametophytes of liverworts have two shapes: leafy and thallose (see fig. 27.12). Approximately 80% of liverworts are leafy; they usually grow in wetter areas than mosses and are abundant in tropical jungles and fog belts. Their gametophytes have a leafy stem and resemble mosses, except that leafy liverworts appear bilaterally symmetrical while most mosses appear radially symmetrical. Leafy liverworts have three ranks of sessile leaves, two that project laterally and one (often colorless) rank along their lower side. These leaves are usually one cell thick and lack a midrib. The overall appearance of the gametophyte is flattened.

Gametophyte: Thallose Liverworts

Thallose liverworts have a flat, ribbonlike gametophyte. *Marchantia* is the most intensively studied liverwort. As a thallose type it is most specialized and, consequently, is an atypical representative.

The gametophytes of *Marchantia* are perennial and branch dichotomously. Each branch grows from an apical cell in the notch at the tip of a lengthwise groove. The flat gametophyte is 2–20 mm wide, and its surface consists of small, diamond-shaped plates (fig. 27.14). Each plate is covered by cutin and delimits an underlying chamber. These chambers contain filaments of photosynthetic cells arranged like plants in a cactus garden. Each chamber is connected to the atmosphere by a chimneylike pore surrounded by barrel-shaped cells. This pore is analogous to a stoma. Colorless rhizoids and scales form along the lower side of the gametophyte.

Many liverworts such as *Marchantia* are unisexual (see fig. 27.15). In *Marchantia*, archegonia form on the underside of rays protruding from the umbrella-shaped caps of **archegoniophores** (archegonium bearers), and antheridia form in furrows along the

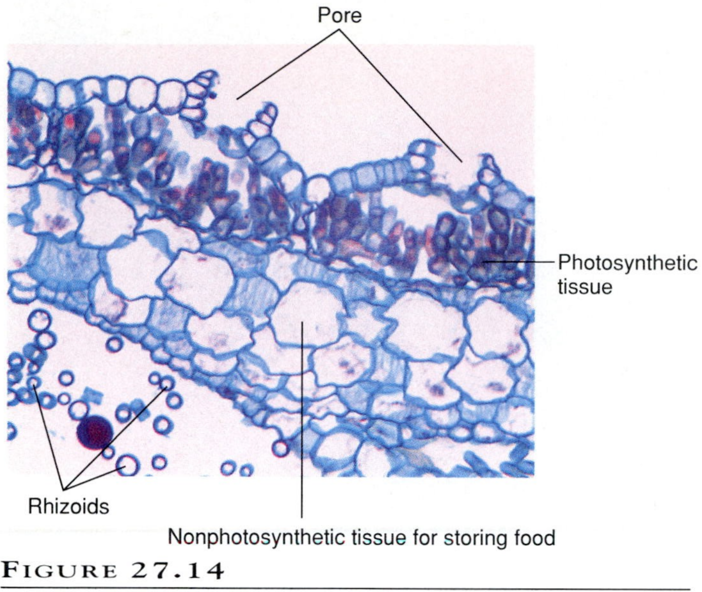

FIGURE 27.14

Marchantia. Section of gametophyte showing internal structure. p.: pore; a.c.: air chamber of photosynthetic zone; f.s.: nonphotosynthetic food storage zone; s.: scale; r.: rhizoid.

upper side of disk-shaped caps of **antheridiophores** (antheridium bearers) (fig. 27.16). Other thallose liverworts are structurally simpler than *Marchantia*. For example, the gametophytes of *Pellia* lack chambers and pores, and tiny *Riccia* can produce archegonia and antheridia on the same plant.

Sporophyte

The sporophyte of liverworts lacks stomata and is anchored in the gametophyte by a knoblike foot (fig. 27.17). The sporophytes of many liverworts are spherical, unstalked, and held within the gametophyte until they shed their spores. In other liverworts, a stalklike seta grows upward from a foot. The capsule forms atop the seta and is covered by a calyptra. Inside the capsule are sporogenous cells that undergo meiosis to form spores. Among these spores are long, dead cells called **elaters** (fig. 27.17e). Because elaters are hygroscopic, they twist violently when they dry, thus dispersing spores.

C O N C E P T

The gametophytes of liverworts are either leafy or thallose, and they often reproduce asexually by pieces of tissue called gemmae. Antheridia and archegonia in some liverworts are borne on antheridiophores and archegoniophores, respectively. Hygroscopic elaters help disperse spores.

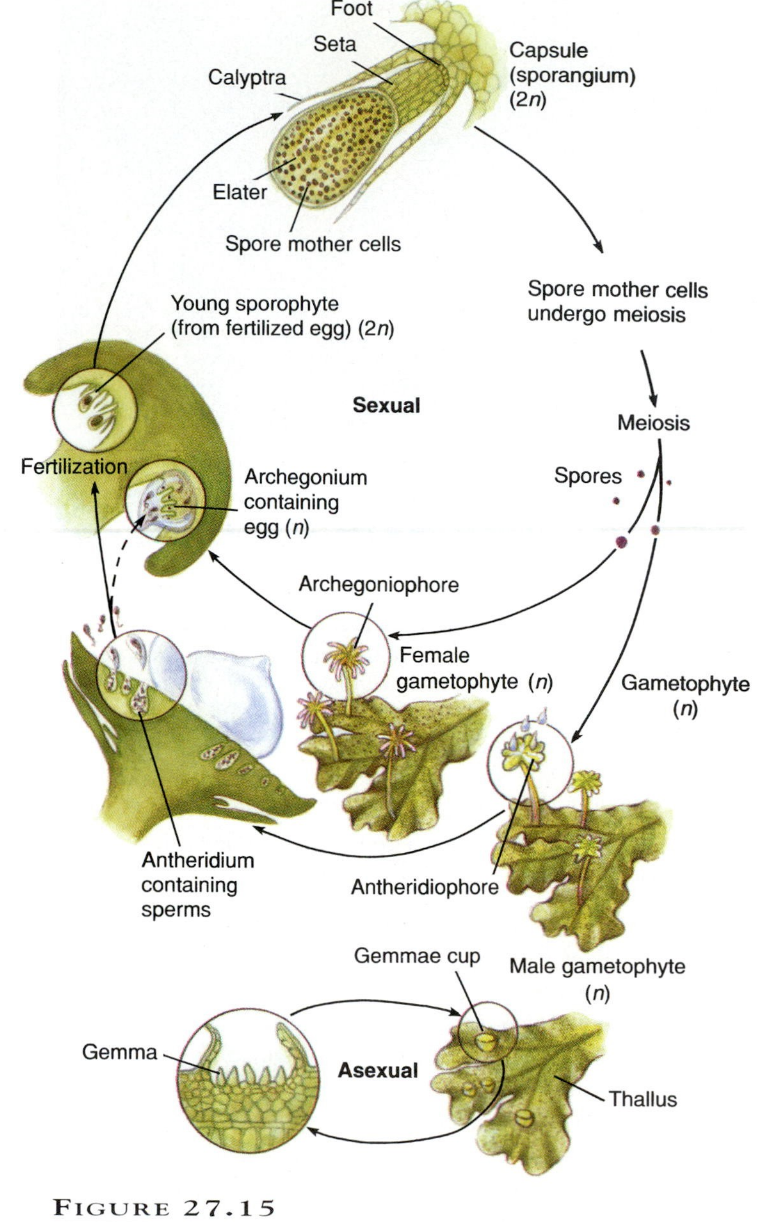

FIGURE 27.15

Life cycle of *Marchantia* a thalloid liverwort. During sexual reproduction, spores produced in the capsule germinate to form independent male and female gametophytes. The archegoniophore produce archegonia, each of which contain an egg; the antheridiophore produces antheridia, each of which produce many sperm. After fertilization, the sporophyte develops within the archegonium and produces a capsule with spores. *Marchantia* reproduce asexually by fragmentation and gemmae.

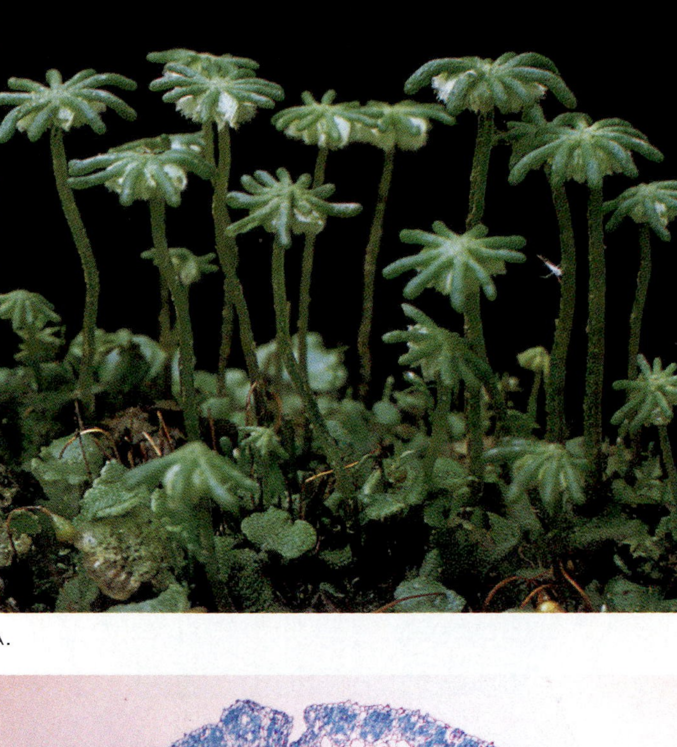

A.

B.

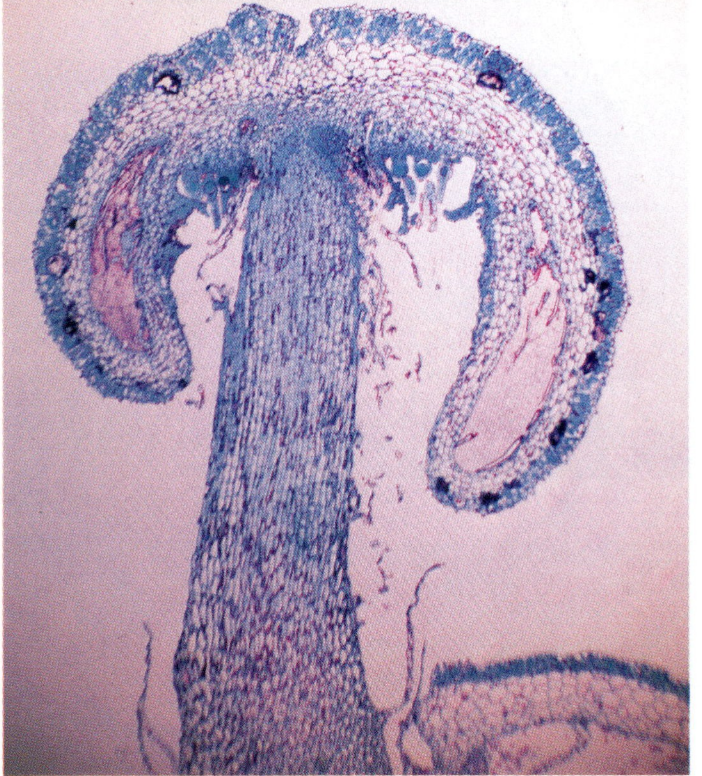

C.

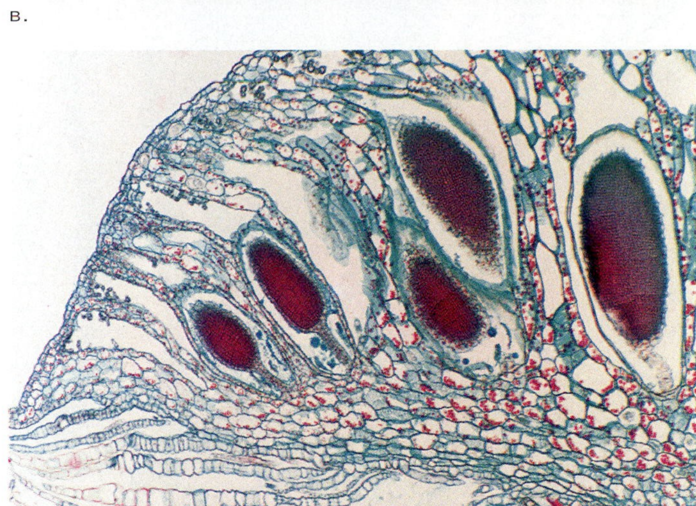

D.

FIGURE 27.16

Reproductive structures of *Marchantia:* (a) gametophyte with archegoniophores (female); (b) gametophyte with antheridiophores (male); note the archegoniophores in the upper right; (c) *Marchantia polymorpha* section through an archegoniophore, ×10; (d) Section through an antheridiophore, showing individual antheridia, ×100.

A.

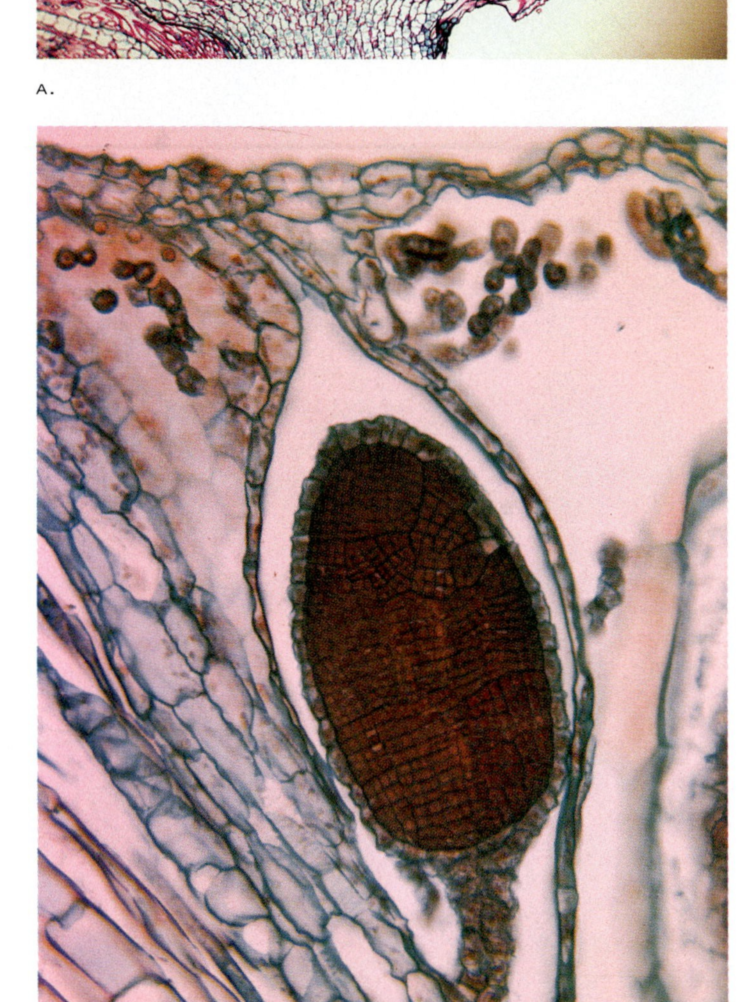

B.

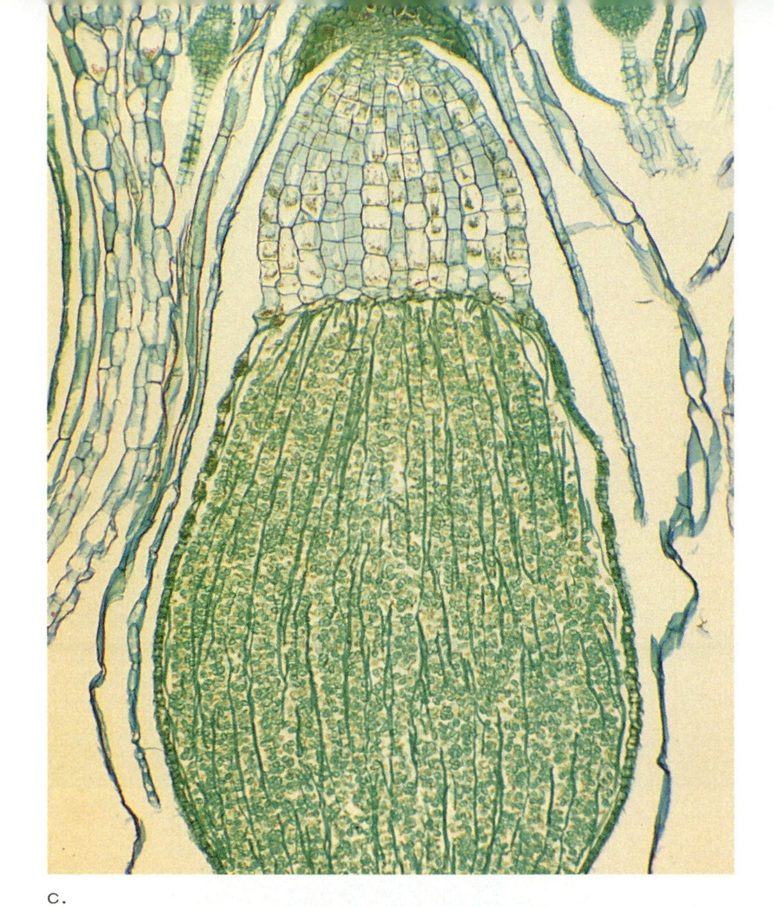

C.

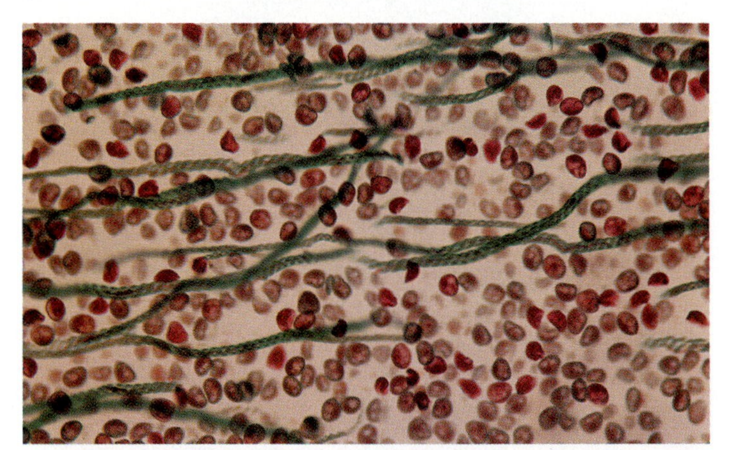

D.

E.

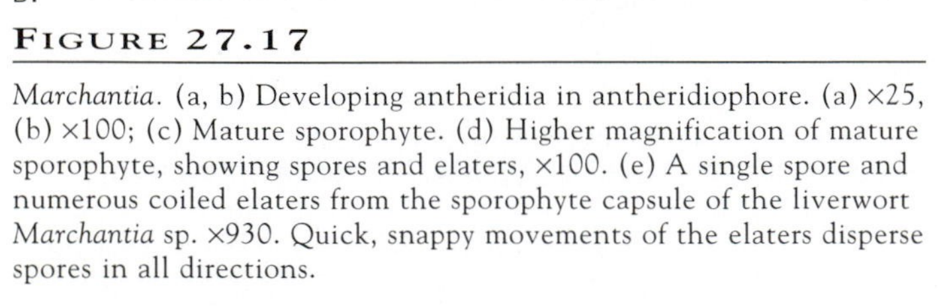

Figure 27.17

Marchantia. (a, b) Developing antheridia in antheridiophore. (a) ×25, (b) ×100; (c) Mature sporophyte. (d) Higher magnification of mature sporophyte, showing spores and elaters, ×100. (e) A single spore and numerous coiled elaters from the sporophyte capsule of the liverwort *Marchantia* sp. ×930. Quick, snappy movements of the elaters disperse spores in all directions.

FIGURE 27.18

Hornworts.

Hornworts

The hornworts are the smallest group of bryophytes; there are only about 100 species in six genera. The most familiar hornwort is *Anthoceros*, a temperate genus (fig. 27.18). Hornworts have several features that distinguish them from most other bryophytes.

1. The sporophyte is shaped like a tapered horn (fig. 27.19).
2. Each photosynthetic cell contains only one chloroplast. Each chloroplast is associated with a starch-storing pyrenoid, as are the chloroplasts of green algae and *Isoetes*, a vascular plant.
3. The sporophyte has an intercalary meristem at its base above the foot. As a result, the sporophyte grows indeterminately.
4. Archegonia are not discrete organs. Rather, they are embedded in the thallus and are in contact with the surrounding vegetative cells (fig. 27.20).
5. The thallus of the gametophyte has stomalike structures. Stomata (even nonfunctional ones) are absent from the gametophytes of all other plants.
6. The cavities of hornwort gametophytes are usually filled with mucilage.

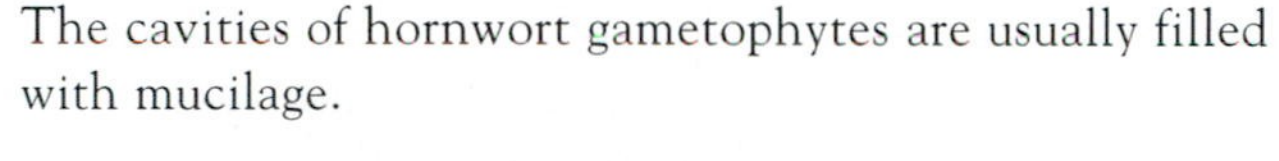

A.

B.

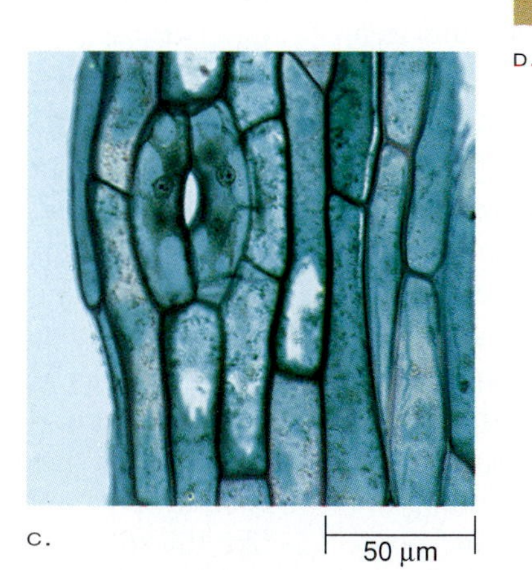

C.

Spores

D.

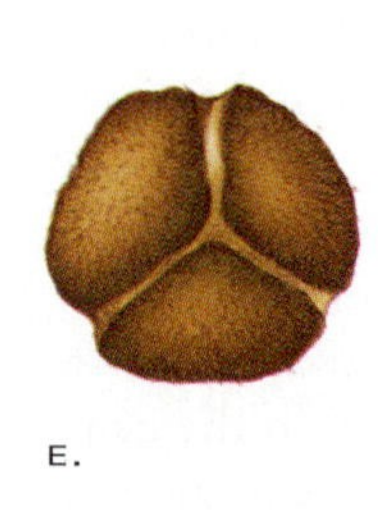

E.

FIGURE 27.19

Anthoceros, a hornwort. (a, b) *Anthoceros* with mature sporangia. (c) Stomatalike structure on the gametophyte of *Anthoceros*; such structures are absent from the gametophytes of all other plants. (d) Spores form along the central axis of the sporophyte; (e) mature spores.

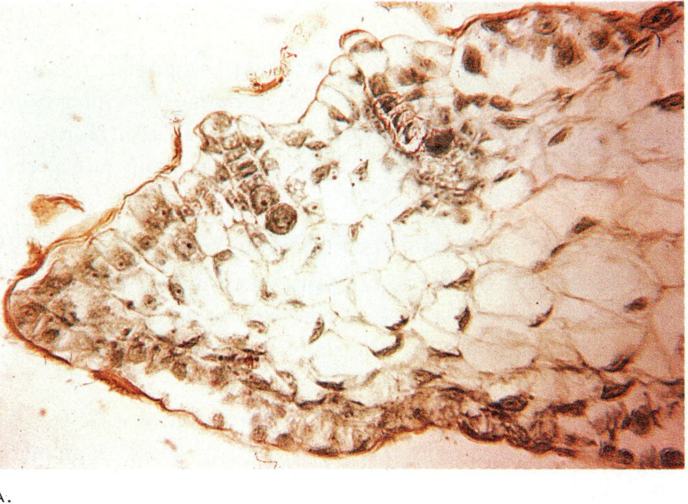

A.

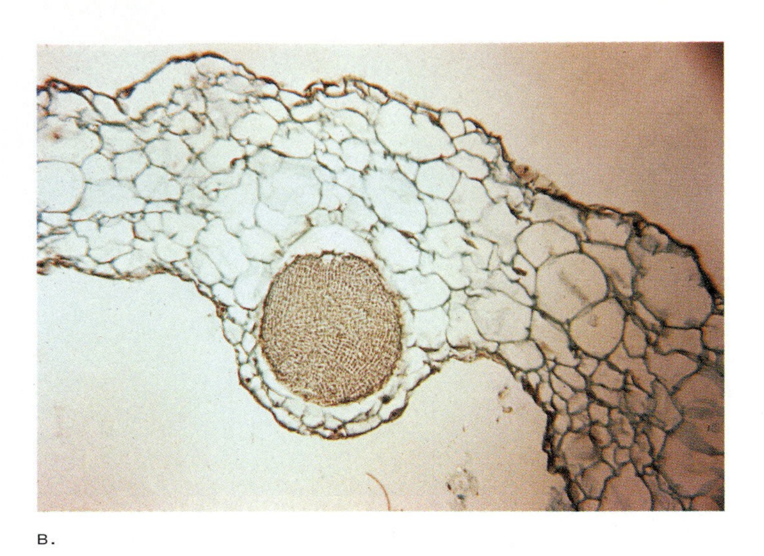

B.

Figure 27.20

Anthoceros: (a) section through two embedded archegonia; (b) section through an antheridial cavity with a single antheridium. (Antheridia are usually clustered.)

Gametophyte

The flat, dark green gametophytes of hornworts are structurally simpler than those of other bryophytes. They are dorsiventrally flattened and superficially resemble thallose liverworts such as *Marchantia.* The gametophyte is round or slightly oblong, and there is little internal differentiation. Hornwort gametophytes are annual or perennial, and are anchored to the substrate by rhizoids.

Most hornworts are unisexual. Sex organs form on the upper surface of the thallus. One or more antheridia resembling those of liverworts form in roofed chambers in the upper portion of the thallus, and archegonia form in rows beneath the surface (fig. 27.20). Hornworts reproduce asexually by fragmentation.

The pores and cavities of hornwort gametophytes are filled with mucilage instead of air. Nitrogen-fixing cyanobacteria such as *Nostoc* live in this mucilage and release their excess nitrogen compounds to the hornwort.

Sporophyte

The sporophytes of hornworts differ remarkably from those of other bryophytes. Hornwort sporophytes are long, green spindles (1–4 cm long) having tapered tips (fig. 27.18). They have a distinct epidermis and stomalike structures, but lack setae. Sporophytes begin forming beneath the surface of the thallus. During development, their tips pierce the upper surface of the thallus and appear as miniature horns. The base of the green capsule remains embedded in the gametophyte. Spores form along the central axis of the capsule from sporogenous cells. The sporophyte remains photosynthetic and can live for several months until spores are released. This semi-independence is viewed by many botanists as an evolutionary step toward the independent sporophytes that characterize vascular plants. Indeed, many botanists no longer consider hornworts to be bryophytes; they consider them to be more closely related to ferns.

C O N C E P T

Hornworts comprise the smallest group of bryophytes. Their sporophytes are shaped like tapered horns, and their photosynthetic cells each contain one chloroplast associated with a pyrenoid. The sporophyte has an intercalary meristem, and archegonia are not discrete organs.

The Ecology and Distribution of Bryophytes

Bryophytes live in almost all places that plants can grow and in some places where vascular plants cannot grow (see fig. 27.4). For example, bryophytes grow on moist soil, rooftops, the faces of cliffs, tombstones, and birds' nests; they carpet forest floors, dangle like drapery from branches, and sheathe the trunks of trees in rain forests. The greatest diversity of bryophytes, especially of hornworts, occurs in tropical habitats. Examples of the extremes of bryophyte habitats include exposed rocks and volcanically heated soil (up to 55°C), and they grow in Antarctica where summer temperatures seldom exceed −10°C. However, the most unusual habitat for bryophytes is reserved for *Splachnum*, the mammal dung moss (see fig. 27.4d). This moss produces a colored stalk and releases a putrid odor that attracts flies. Unlike the spores of other mosses, those of *Splachnum* are sticky and adhere to the visiting flies. The spores are disseminated when the flies move from the moss to piles of dung.

Bryophytes, which are often the first plants to invade an area after a fire, grow at elevations ranging from sea level to 5,500 m. There are no marine bryophytes, but some, such as dune mosses, grow near the seashore. Each group of bryophytes

Figure 27.21

Andreaea rothii, a rock moss.

Figure 27.22

Scopelophila cataractor growing on copper-rich soil.

has aquatic species. Bryophytes dominate the vegetation in peatlands. Mosses are especially abundant in the arctic and antarctic, where they far outnumber vascular plants.

Some bryophytes can withstand many years of dehydration and often grow in deserts. Other bryophytes can withstand prolonged periods of dark and freezing, which explains why they are the most abundant plants in Antarctica. Although generally widespread, some bryophytes grow only in specific habitats, such as on the bones and antlers of dead reindeer or on animal dung.

The ecology of bryophytes is determined by their gametophytes. Bryophytes, along with lichens, are among the first organisms to colonize bare rocks and volcanic upheavals. For example, *Andreaea* is a black to reddish brown moss that grows on exposed rocks (fig. 27.21). These organisms convert rock to soil, thus paving the way for colonization by other organisms.

Many bryophytes, such as the genus *Hypnum*, are notoriously sensitive to pollution, especially sulfur dioxide. As a result, most bryophytes are rare in cities and industrialized areas; for example, twenty-three species of bryophytes that grew in Amsterdam in 1900 no longer grow in that area. Other mosses, such as *Ceratodon* and *Bryum*, thrive in urban and polluted areas. Unlike pollution-sensitive mosses, *Ceratodon* and *Bryum* have short-lived protonemata and produce gametophytes rapidly.

Bryophytes increase the humus in soil and often indicate the presence or absence of particular salts, acids, and minerals. For example, the liverwort *Carrpos* grows only on gypsum-rich "salt pans," and *Mielichhoferia* and *Scopelophila* are mosses that grow only on copper-rich substrates (fig. 27.22). Other bryophytes concentrate elements such as barium, lead, strontium, and zinc—sometimes at more than two hundred times their levels in the soil. Bryophytes typically accumulate twice as much radioactivity as flowering plants on a weight-for-weight basis.

Figure 27.23

Sphagnum: (a) Carpet of peat moss; (b) close-up of photosynthetic (i.e., chloroplast-containing) and water-retaining cells of *Sphagnum*.

Bryophytes usually grow very slowly. However, peat moss (*Sphagnum*) accumulates at rates exceeding 12 metric tons per hectare annually (fig. 27.23; for comparison, yields of corn and rice average about 6 and 5.5 metric tons per hectare, respectively). Some mosses in Antarctica also grow rapidly. It is interesting that the mosses of the northern hemisphere often grow more slowly than do those in Antarctica.

A.

C.

B.

FIGURE 27.24

Peat and peat bogs. (a) A peat bog showing the diversity of plants that grow on peat moss; (b) A family in Ireland stacking peat sods to dry. This peat will later be used as the fuel to heat their home; (c) The bucolic image of peat fires smoldering on small hearths misrepresents peat's use. Indeed, more than 95% of the world's peat harvest is burned to generate electricity (note the power plant in the background). Peat-fired power plants are preferred over coal-fired power plants because of peat's low sulfur and ash content.

Carpets of bryophytes are excellent seedbeds for vascular plants (fig. 27.24). Peat moss can absorb 20–25 times its weight in water; it is therefore an excellent soil conditioner that helps to prevent flooding and minimizes erosion. Peat moss produces large amounts of acids and antiseptics that kill decomposers. As a result, dead peat moss accumulates and forms large deposits called **peat bogs.** The pH of these bogs often approaches 3.0 (i.e., about the same pH as vinegar) and thus eliminates all but the most acid-tolerant plants, such as cranberry, a few carnivorous plants, blueberry, and larch, a conifer that is often a climax species in acidic communities. The acidity of peat bogs also makes these communities very stable; some peat bogs are estimated to be more than 50,000 years old. Peat bogs cover approximately 1% of the earth's surface, an area that is equivalent to half of the United States.

The acidity and anaerobic environment of peat bogs also preserves dead plants and animals. Paleobotanists have used pollen preserved in peat to determine which plants grew in a particular area and how they have changed over time. The preservative effects of peat bogs have also yielded horrifying secrets about past civilizations (see box 27.2, "Secrets of the Bog").

Doing Botany Yourself

The preserving effects of peat bogs (see box 27.2, "Secrets of the Bog") correlate with (1) low pH, and (2) anaerobic conditions. How could you determine which of these factors is most important for preserving organisms and materials in a bog?

Boxed Reading 27.2

Secrets of the Bog

In the 1950s, several countries in northern Europe began mining peat as fuel. This peat from the bogs not only provided valuable fuel but also yielded some 2,000-year-old secrets that were as grotesque as they were valuable. Horrified workers uncovered several hundred human bodies. The bodies had been tanned by the bog's acids (i.e., much as we use acids to tan leather), and many were in such good condition that the workers suspected that they had been recently dumped into the bogs. Furthermore, many of these bodies were mutilated; they had slit throats, severed vertebrae, and nooses around their necks.

Further study showed that the bodies were actually 2,000–3,000 years old. The acidity and anaerobic environment of the bogs had preserved them by inhibiting the growth of bacteria and fungi that normally decompose dead organic material. Archaeologists gave the bodies that were pulled from the peat bogs names such as Lindow man and Tollund man, names that represented the geographical locations of the bogs that became graves. The most intensively studied of these bog people is Lindow man, a fellow who lived about the time of Aristotle. The preserved stomach-contents of Lindow man showed that his last meal was barley and linseed gruel. Thanks to the preservative effects of the bog, archaeologists have also learned about musical instruments and household items of the 2,000-year-old civilizations and determined that bogs were the sites of human sacrifice in religious ceremonies.

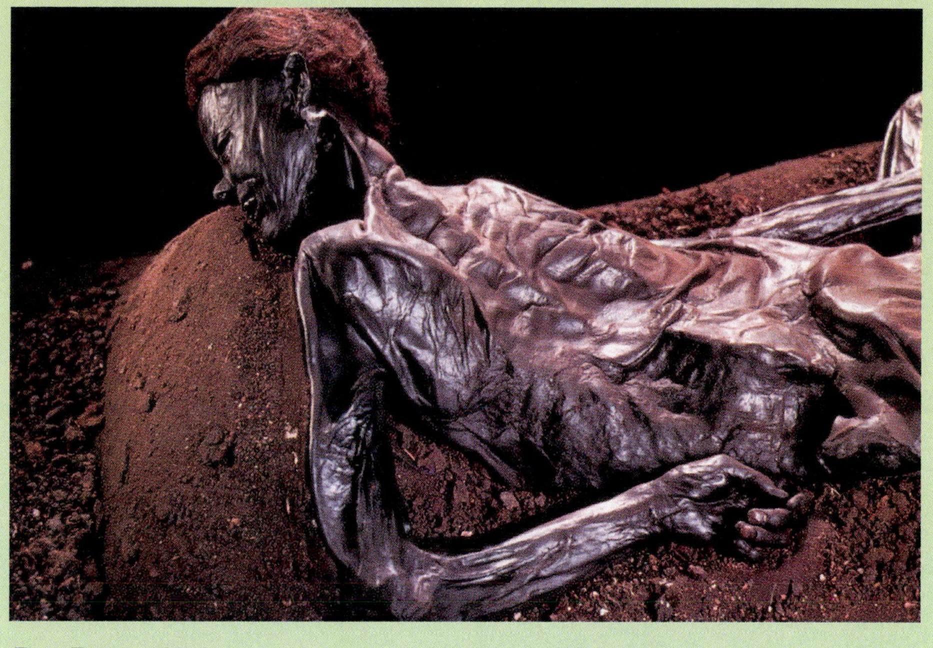

Box Figure 27.2

The Physiology of Bryophytes

The Influence of Light

Several bryophytes produce gametangia in response to daylength. For example, most leafy liverworts, such as *Porella*, produce antheridia and archegonia only during long days; other bryophytes, such as *Riccia*, *Anthoceros*, and *Sphagnum*, produce gametangia only on short days. Bryophytes such as *Pogonatum* are day-neutral.

Light also affects how protonemata develop. For example, consider these experimental observations (see fig. 27.25):

1. The protonemata of *Funaria* grow in darkness if supplied with nutrients, but produce no leafy, upright shoots. Similar growth occurs in blue light.
2. Changing the light from blue to red induces the formation of buds.
3. Buds form in the light or dark within two days if the protonema is supplied with the hormone cytokinin.

These results suggest that light may affect growth and development by altering the production of cytokinins. Budding is inhibited by auxin, another hormone.

Hormones and Apical Dominance

Protonemata require auxin for growth, a hormone that is made in the apex of the thallus. The protonemata of many mosses exhibit a strong apical dominance, meaning that buds farthest from the apex develop earlier than buds closer to the apex. Removing the apex releases this apical dominance. Applying auxin to a detipped protonema restores apical dominance, suggesting that auxin or a similar compound may control apical dominance in the plant. Auxin also influences the formation of gemmae and rhizoids.

Interactions with Other Organisms

Soil bacteria such as *Agrobacterium* can induce formation of buds on protonemata of mosses such as *Pylaisiella*. Interestingly, species of *Agrobacterium* that induce formation of roots in vascular plants also induce formation of rhizoids in many bryophytes.

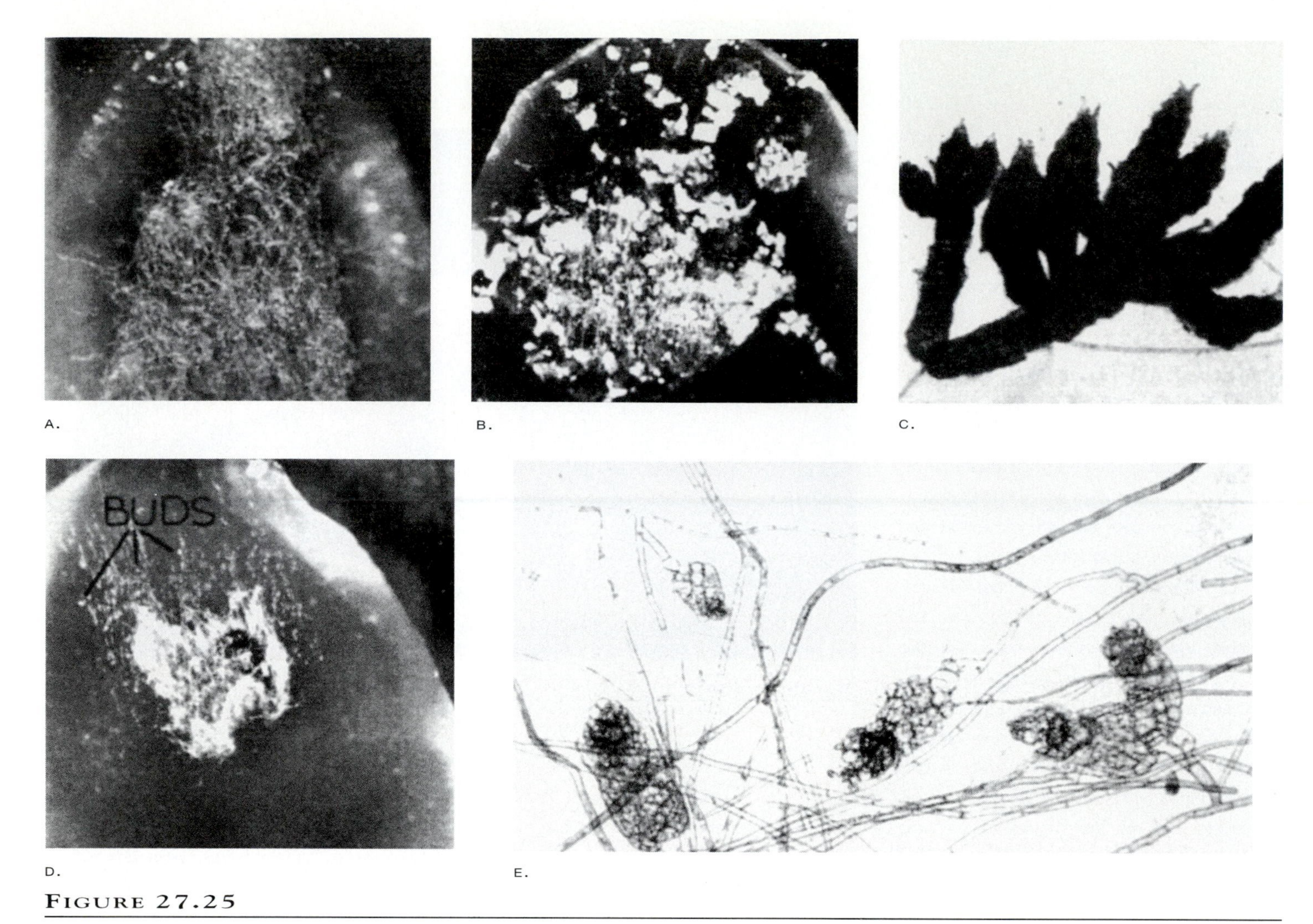

Figure 27.25

The influence of light and dark on the formation of buds in *Funaria hygrometrica*. (a) In cultures raised from spores in the dark on a medium containing 2% sucrose, buds did not form; (b) on cultures containing 2% sucrose and 1 part per million kinetin, buds (cottony white pinheads) formed; (c) in cultures containing 2% sucrose and 0.5 parts per million kinetin, branched buds formed; (d, e) on a medium lacking sucrose, but containing kinetin (1 part per million), protonema with tiny buds formed. All photographs are of five-month cultures.

The Formation of Gametangia

In bryophytes such as *Riccia,* cytokinin and auxin induce the formation of archegonia, whereas in mosses such as *Barbula* and *Bryum*, auxin induces the formation of antheridia. In *Ricciocarpus* and other liverworts, the formation of archegonia is delayed, and young plants have only antheridia.

Water Relations

Bryophytes absorb most of their water from the atmosphere and the surface of their substrate. Their water content varies from less than 50% to more than 2,500% of their fresh weight. Many bryophytes are extremely tolerant of desiccation. For example, star moss (*Tortula ruralis;* also called *dune moss*) becomes photosynthetic within a few hours of being moistened, even after being air-dried for almost a year (see box 27.3 "Quick Recovery"). When dry, *Tortula* can tolerate temperatures ranging from over 100° C down to that of liquid nitrogen (−196° C), a temperature range of about 300° C. Some bryophytes are photosynthetic at temperatures below freezing.

Sporophyte Physiology

We know relatively little about the physiology of bryophyte sporophytes. The sporophytes of most bryophytes (e.g., *Mnium* and *Pleuridium*) receive 80%–90% of their water and carbohydrates from their gametophytes, and it typically takes 1–4 days for solutes to move from the gametophyte to the sporophyte.

Light and hormones influence several aspects of bryophyte growth and development, including the production of gametangia, the branching of protonemata, and apical dominance. Bryophytes absorb water and dissolved minerals from the atmosphere or from the surface of their substrate. Some bryophytes are extremely tolerant of desiccation.

Boxed Reading 27.3

Quick Recovery

Botanists trying to find better ways to engineer drought-resistant plants may have found an unlikely star in the star moss (*Tortula ruralis*). *T. ruralis* grows in forests throughout the United States. During drought, it dries up and looks dead for periods of up to several years. When given just a few drops of water, however, the moss springs to life: within ten hours, all of its systems are back to normal. It is interesting that this recovery actually begins as the moss dries. At that point, the moss makes mRNA that directs the synthesis of recovery proteins as the moss later rehydrates. These proteins, which have apparently been lost by higher plants, enable the moss to repair the massive cellular damage caused by the desiccation.

Botanists are now using DNA probes to locate the genes that produce the recovery proteins. They will then try to clone those genes and insert them into crop plants such as cotton, hoping that the engineered plants will begin to produce their own recovery proteins. If they succeed, we may be able to grow crops on millions of hectares of arid land that now yield crops only with heavy irrigation.

What are Botanists Doing?

Go to the library and read a recent article in a popular magazine (e.g., *Natural History*) about mosses. Summarize the main points of the article. How does the information in the article relate to what you have learned from this chapter?

The History and Relationships of Bryophytes

The fossil record of bryophytes is poor and shows relatively little diversity. Fossils dating from the Carboniferous period (286–360 million years ago) have been interpreted as being mosses. The most ancient liverworts (*Pallaviciniites devonicus*) date to the Devonian period (360–408 million years ago), and other fossils that resemble liverworts appear in the Carboniferous and succeeding periods. Many of these fossil bryophytes have vague relationships and are classified as *Muscites*, a catch-all genus for problematic mosslike fossils, or *Thallites*, an equivalent all-purpose genus for liverwortlike fossils. The gametophytes of most fossilized bryophytes are rarely associated with sporophytes, which contributes to the uncertainty of their identity. Although fossils of genera such as *Sporogonites* and *Torticaulis* dating from the Devonian and late Silurian periods (408–438 million years ago) may be mosses, these plants are preserved so poorly that clear interpretation is difficult. The problem is further complicated because the sporangia of some bryophytes are almost indistinguishable from those of ancient lignified plants.

In spite of the meager fossil record, botanists suspect that bryophytes diverged from an ancestor common to vascular plants more than 430 million years ago (i.e., some time during the Silurian period). There is no general agreement as to what the common ancestor was, or even whether bryophytes diverged as a monophyletic group. Key features in identifying a possible ancestor may be the flagellated sperm, photosynthetic pigments, similar cell walls, and storage of starch that are common among bryophytes, vascular plants, and green algae. Among the green algae, the most likely ancestral group may be the charophytes (e.g., *Chara, Coleochaete*), because they are the only group of algae with a flavonoid biosynthetic pathway, a feature of plants (see box 26.1, "Evolutionary Relationships of Green Algae and Plants"). Another possibility is that the ancestor is represented by the green alga *Coleochaete*, which retains the zygote on the gametophyte, as do plants, but does not produce flavonoids (fig. 27.26). *Coleochaete* also has a thalluslike body that resembles a thalloid liverwort.

The bryophytes as a group seem to be paraphyletic, based on two shared-derived characters that are present in liverworts, mosses, and plants, but not hornworts: the occurrence of stomata and the enzymatic ability to distinguish d-methionine from l-methionine. Also, if leptoids and hydroids are interpreted as vascular tissue like that of plants, then mosses are closer to land plants than to other bryophytes. These comparisons confirm the traditional view that bryophytes should not be classified as a single division.

Disputes and unanswered questions about bryophyte evolution are also being approached by molecular phylogenetic studies. This work shows promise but also raises some new questions. For example, the evolutionary relationships shown in figure 27.27, which are based on a cladistic analysis of ribosomal RNA sequences, confirm most of the suggestions based on other features. However, the cladogram also indicates that thallose liverworts are more closely related to vascular plants than they are to leafy liverworts. Keep in mind, though, that this evolutionary tree represents just one of several hypotheses about the relationships of bryophytes, and it will undoubtedly be tested, modified, and improved as botanists pursue more kinds of molecular and morphological data.

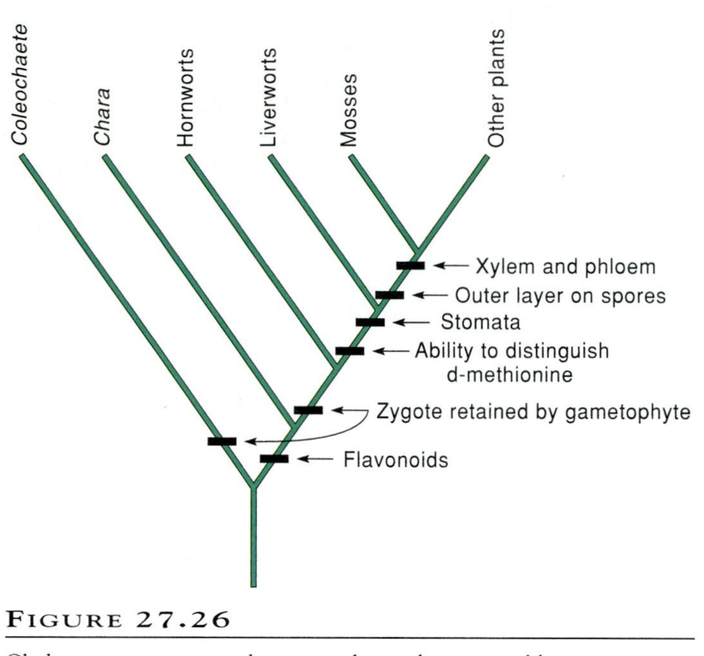

FIGURE 27.26

Cladograms comparing characters that indicate possible common ancestry of green algae, bryophytes, and vascular plants.

THE ECONOMIC IMPORTANCE OF BRYOPHYTES

Bryophytes are generally not edible, although liverworts were often soaked in wine and eaten in the 1500s. Today, the earthy aroma of Scotch whiskey is partly due to the fact that the malted barley is dried over peat fires.

Mosses are used as stuffing in furniture, as a soil conditioner, as an absorbent in oil spills, and for cushioning. For example, florists use peat moss as a damp cushion when shipping plants. *Sphagnum* has also been used by aboriginal people for diapers and as a disinfectant. Because of its acidity, peat moss is an ideal dressing for wounds. Indeed, the British used more than one million such dressings per month during World War I, and the Red Cross refers to *Sphagnum* as a wound dressing in its publications. North American Indians used *Mnium* and *Bryum* to treat burns. *Dicranoweisia* has been used to waterproof roofs in Europe.

The rapid growth of peat moss suggests that peat bogs may be an important source of renewable energy. Peat and lignite (a soft brown type of coal) have several properties that make them good sources of fuel. For example, peat has a caloric value of 3,300 calories per gram, a value that is greater than that of wood (but only half that of coal). Furthermore, peat is abundant. The United States (excluding Alaska) has more than 60 billion tons of peat, an amount of fuel equivalent to approximately 240 billion barrels of oil. Countries of the former Soviet Union have an annual harvest of more than 200 million tons of peat, which is used as fuel for nearly eighty power plants. Ireland obtains more than 20% of its energy from peat, and the United States annually harvests more than a million tons of peat in twenty-two states.

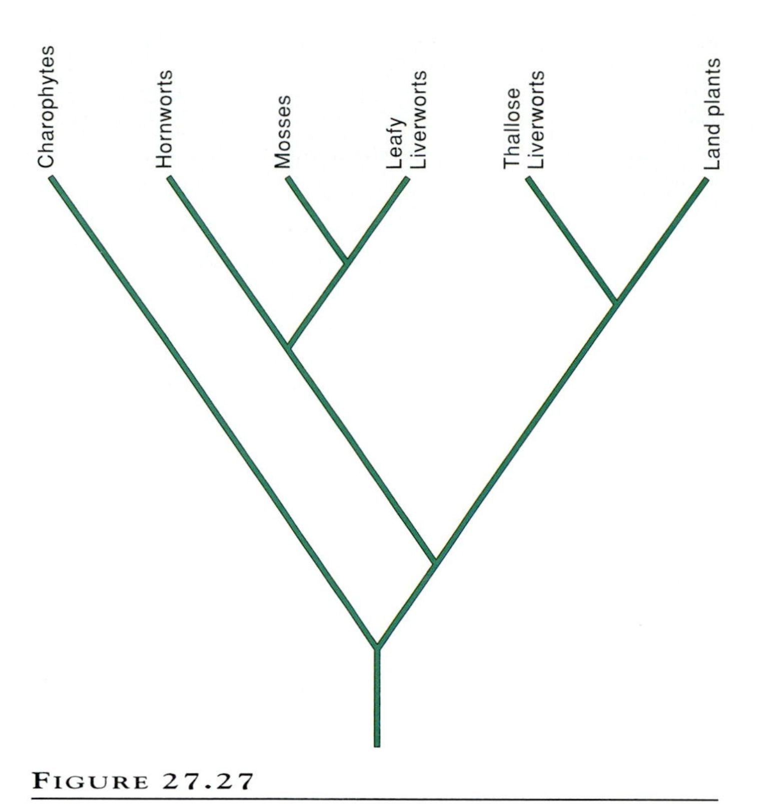

FIGURE 27.27

The evolutionary history of bryophytes, based on a cladistic analysis of ribosomal RNA sequences.

Chapter Summary

Bryophytes, which include mosses, liverworts, and hornworts, are small plants that produce chlorophylls *a* and *b*, starch, cellulosic cell walls, and motile sperm. They lack complex vascular tissues and supporting tissue, and absorb water by capillarity. Bryophytes have a heteromorphic alternation of generations dominated by a photosynthetic and independent gametophyte. Eggs and sperm form in gametangia called archegonia and antheridia, respectively, both of which are surrounded by a protective jacket of sterile cells.

Bryophytes require free water for sexual reproduction, so that sperm can swim to eggs. The fertilized egg begins the diploid sporophyte generation, which is attached to and nutritionally dependent on the gametophyte. Haploid spores formed by meiosis are released from sporangia and are usually dispersed by wind. Spores germinate and form gametophytes, thus completing the life cycle. Bryophytes have modest growth requirements and grow almost everywhere that other plants grow. Most reproduction in bryophytes is asexual.

Mosses are the largest and most familiar group of bryophytes. They produce multicellular rhizoids and have leafy gametophytes that are radially symmetrical. Spores form in sporangia that open by means of a lidlike operculum. A spore produces a filamentous and branched protonema, which develops into the gametophyte. Most mosses are unisexual. Sperm released from antheridia swim to archegonia and through the neck canal to the egg located in the venter. The zygote, which begins the sporophyte generation, is covered by a protective calyptra. The sporophyte consists of a foot, seta, and capsule (sporangium). The foot penetrates the gametophyte and absorbs food and water for the sporophyte. The stalklike seta raises the spore-producing capsule above the gametophyte.

Liverworts have leafy or thallose gametophytes that are either lobed or bilaterally symmetrical. They have unicellular rhizoids and frequently reproduce asexually by pieces of tissue called *gemmae*. Most liverworts are leafy and unisexual. The gametangia of some species form on upright, stalked structures called antheridiophores and archegoniophores. Hygroscopic elaters help to disperse spores from sporangia.

Hornworts are the smallest group of bryophytes. Their sporophytes are shaped like tapered horns and have an intercalary meristem at their base. The thallus of the gametophyte has stomatalike structures, and each photosynthetic cell has a chloroplast associated with a pyrenoid. Archegonia in hornworts are not discrete organs. Some botanists do not consider hornworts to be bryophytes.

Light and hormones strongly affect several aspects of bryophyte growth and development, including the branching of protonemata, apical dominance, and the formation of gametangia. Some bryophytes are extremely tolerant of desiccation.

The ecology of bryophytes is determined by their gametophytes. Bryophytes reduce erosion, condition soil, and are often among the first organisms to invade disturbed areas. Many bryophytes grow in specific habitats and are sensitive to pollution.

The evolutionary history of bryophytes is vague. They probably diverged after plants invaded land. Botanists do not agree on the relationships of bryophytes to other groups of plants.

Questions for Further Thought and Study

1. Mosses and liverworts have been used extensively to monitor radioactive fallout from the Chernobyl reactor accident that occurred in April 1986. What features of these organisms make them ideal for such a use?
2. Much of the early work on sex chromosomes in plants was done with bryophytes. The spores of unisexual species contain either a large X chromosome or a smaller Y chromosome, depending on whether they will form female or male gametophytes, respectively. In light of these findings, consider the following experiment. The lower portion of the capsule of many mosses can be removed and cultured in the laboratory. Given the proper nutrients, this diploid tissue does not form a sporophyte but forms a diploid gametophyte by a process called **apospory** (without spore formation; i.e., without meiosis). If apospory occurs in a species that normally forms unisexual gametophytes, the diploid gametophytes are bisexual. How can you explain this phenomenon?
3. What is the evidence that bryophytes and vascular plants had a common ancestry? What does this evidence suggest about the common ancestor of bryophytes and other plants?
4. Although bryophytes require free water for sexual reproduction, several bryophytes grow in deserts. How do you think these bryophytes reproduce in spite of their mostly dry environment?
5. What might be the adaptive significance of unisexual versus bisexual gametophytes?
6. Why are there so few fossils of bryophytes?
7. What does the fossil record show regarding the relationships of bryophytes to other plants?
8. Why is it important that pollution-tolerant bryophytes have short-lived protonema?
9. When are bryophytes most vulnerable?

Suggested Readings

Articles

Levanthes, L. E. 1987. Mysteries of the bog. *National Geographic* 171(3):396–420.

Miller, N. G. 1980. Bogs, bales, and BTUs: A primer on peat. *Horticulture* 59:38–45.

Stebbins, G. L., and G. J. C. Hill. 1980. Did multicellular plants invade land? *American Naturalist* 115:342–353.

Thieret, J. W. 1955. Bryophytes as economic plants. *Economic Botany* 10:75–91.

Wang, T. L., and D. J. Cove. 1989. Mosses—lower plants with high potential. *Plants Today* (March–April), pp. 44–50.

Books

Bates, J. W., and A. M. Farmer (eds). 1992. *Bryophytes and Lichens in a Changing Environment*. Oxford: Oxford University Press.

Chopra, R. N., and P. K. Kumra. 1988. *Biology of Bryophytes*. New York: John Wiley.

Conard, H. S., and P. L. Redfearn, Jr. 1979. *How to Know the Mosses and Liverworts*. New York: Academic Press.

Dyer, A. F., and J. G. Duckett. 1984. *The Experimental Biology of Bryophytes*. London: Academic Press.

Smith, A. J. E. 1982. *Bryophyte Ecology*. London: Chapman and Hall.

Certain species of *Selaginella*, a seedless vascular plant, have iridescent blue leaves, as shown here.

CHAPTER 28

Seedless Vascular Plants

Chapter Outline

Chapter Overview

Seedless vascular plants share features with bryophytes, including the same types of pigments, the basic life cycle, and the storage of starch as their primary food reserve. However, the evolution of vascular tissue enabled vascular plants to invade and dominate the drier habitats on land more effectively than could nonvascular plants. The sporophyte of seedless vascular plants dominates the life cycle; the gametophyte is always smaller, sometimes microscopic, and nutritionally independent of the parent sporophyte. The ancestors of modern seedless vascular plants dominated the earth's vegetation for more than 250 million years, eventually giving way to seed plants. Most seedless vascular plants are true ferns, but they also include horsetails, whisk ferns, and club mosses. Most varieties of these plants live in tropical areas.

Introduction

Vascular plants, like bryophytes, evolved from aquatic, nonvascular ancestors. In contrast to the bryophytes, however, vascular plants became the dominant organisms on land, primarily because of their diversity and large size. Early vascular plants developed upright stems with large leaves to absorb sunlight for photosynthesis. Roots anchored and supported the stems and absorbed water and minerals from the soil.

Structural support for large vascular plants came primarily from the lignin that strengthened their secondary cell walls. Conducting tissues enabled them to transport water and nutrients rapidly and efficiently throughout the larger plant body. A waterproof cuticle and functional stomata controlled gas exchange and minimized water loss.

Although all of the earliest vascular plants are extinct, their successful strategies for adapting to land habitats persist in their descendants. There are more than 250,000 species of vascular plants living today. These plants are conveniently divided into two groups: those that produce seeds and those that do not. In this chapter, we discuss the seedless vascular plants, which are divided into four divisions of extant plants (fig. 28.1). The seedless vascular plants include about 13,000 species, most which are ferns (Division Pteridophyta, with about 12,000 species). The other three divisions consist of the whisk ferns (Division Psilotophyta), the club mosses (Division Lycopodophyta), and the horsetails (Division Equisetophyta). The seedless vascular plants are distinguished from seed plants by their vascular organization and their lack of seeds; they reproduce by spores in much the same way that bryophytes do.

Features of Seedless Vascular Plants

Vascular plants are not watertight, nor do they use or conserve water efficiently. Water escapes when stomata open and exchange gases for photosynthesis. However, vascular plants can continuously replace water that is lost and keep the entire plant body moist; water lost by transpiration is replaced by absorption into the roots. When water loss increases, water is absorbed and transported faster to compensate for the loss. When the demand for water uptake exceeds the ability of the roots to absorb it, then stomata close and prevent further loss. Bryophytes lack such control over their use of water.

Although many features of seedless vascular plants are not found in the bryophytes, these two groups do have many characters in common. Some of these features are shared with a few algae as well. For comparison with the algae and bryophytes, the general features of seedless vascular plants are summarized as follows:

1. The life cycle of seedless vascular plants is similar to that of bryophytes and algae that exhibit sporic meiosis (fig. 28.2). The diploid sporophyte produces haploid spores by meiosis. Each spore germinates and grows into a gametophyte that produces gametes by mitosis. The gametes (eggs and sperm) fuse to form diploid zygotes.
2. Eggs are produced in archegonia, and sperm are produced in antheridia.
3. The zygote germinates to produce a multicellular embryo that depends on the gametophyte for its nutrition. To complete the life cycle, the embryo grows into a mature sporophyte. A multicellular embryo also characterizes the bryophyte life cycle, but it is absent in algae.
4. Seedless vascular plants produce chlorophylls *a* and *b*, carotenoids, starch, cellulose cell walls, and motile sperm. These features are also shared with bryophytes and many algae.
5. As in bryophytes and a few green algae, a cell plate separates new cells during cytokinesis in seedless vascular plants.
6. Each sporangium is protected by a multicellular jacket of nonreproductive cells. Spores are dispersed from sporangia by wind and are cutinized to resist desiccation.
7. Seedless vascular plants have a well-developed cuticle to minimize water loss. They also have stomata to allow gas exchange for photosynthesis. Although hornworts and mosses have stomata, they are not well developed and do not function efficiently to prevent water loss. Most bryophytes lack a cuticle.
8. Flagellated sperm swim through water to eggs. Like bryophytes, seedless vascular plants require free water for sexual reproduction.

Figure 28.1

The ferns, represented here by bracken fern (*Pteridium aquilinum*), are the largest group of seedless vascular plants.

9. Seedless vascular plants have well-developed vascular tissues. Xylem transports water and dissolved minerals great distances from the soil. Carbohydrates are transported up and down throughout the plant.
10. Many seedless vascular plants have lignin and cellulose in their secondary cell walls. Lignin strengthens cellulosic microfibrils, thereby enabling plants to stand upright to much greater heights than bryophytes.
11. Sporophytes and gametophytes of seedless vascular plants are nutritionally independent of each other. Sporophytes are photosynthetic, and gametophytes are

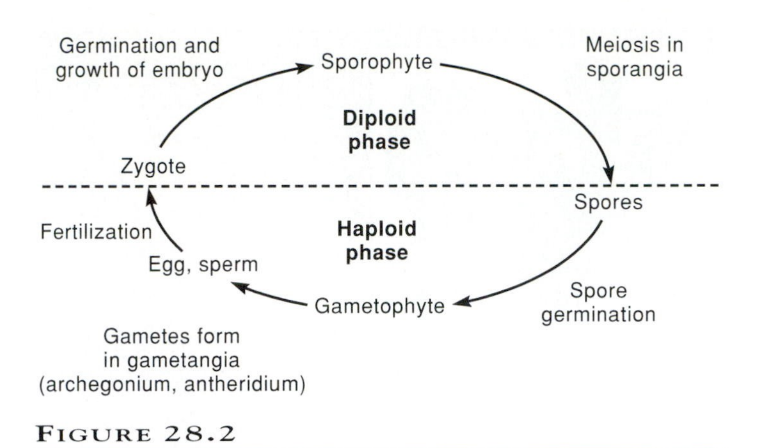

Figure 28.2

The generalized life cycle of seedless vascular plants.

either photosynthetic or **saprophytic** (i.e., they obtain nutrition from dead and decaying organic matter).

12. The sporophyte of seedless vascular plants, which dominates the life cycle, is long-lived and often highly branched.

CONCEPT

Seedless vascular plants share many of their reproductive and vegetative features with bryophytes. However, the evolution of a resistant cuticle, complex stomata, vascular tissue, and lignin enabled these plants to dominate habitats on land.

A Generalized Life Cycle of Seedless Vascular Plants

As in bryophytes, sexual reproduction in seedless vascular plants entails alternation between heteromorphic diploid and haploid phases. Meiosis produces haploid spores in the sporangia of diploid sporophytes. Spores then germinate and grow into gametophytes that produce eggs in archegonia and sperm in antheridia. Sperm must have water to swim to archegonia, where fertilization occurs to produce a diploid zygote. The zygote starts another diploid phase (fig. 28.2).

The Sporophyte

The life cycle of seedless vascular plants is dominated by the sporophyte, which is the "plant" that everyone thinks of when they think of plants. Sporophytes become nutritionally independent soon after the zygote grows out of the archegonium. In many genera, the sporophytes are perennial; new growth sprouts from underground rhizomes year after year. For example, the branching rhizomes of horsetails form extensive underground networks, and in ferns new leaves grow from the same rhizome every year. It is common for sporophytes to reproduce asexually, producing populations consisting entirely of clones.

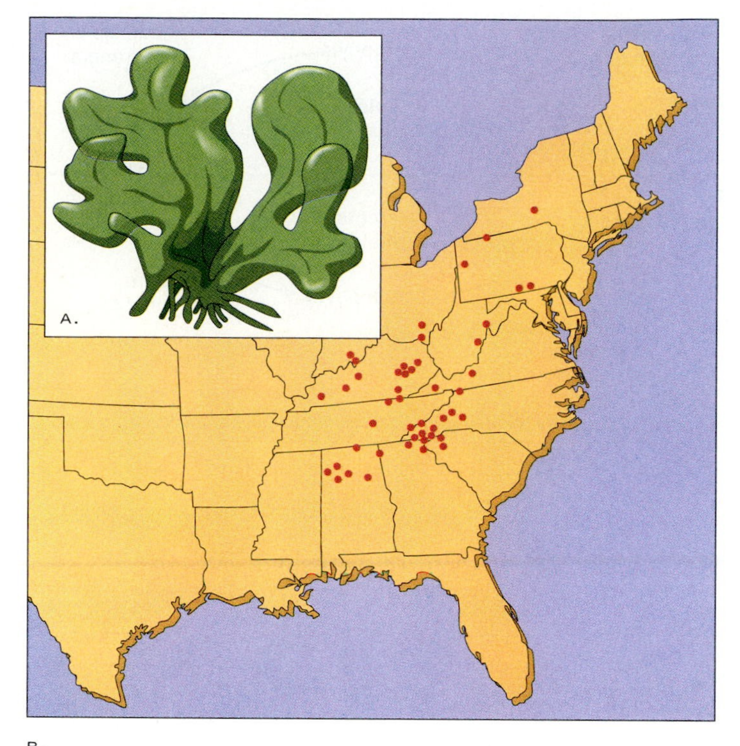

B.

FIGURE 28.3

(a) The Appalachian gametophyte of *Vittaria*. (b) Map showing the distribution of the Appalachian gametophyte and of the sporophyte of its nearest relative, *Vittaria lineata*.

The Gametophyte

The gametophyte of seedless vascular plants is short-lived and small, sometimes microscopic. With few exceptions, it is nutritionally independent of the sporophyte as soon as the spore leaves the sporangium. The gametophytes of whisk ferns, club mosses, and many ferns are saprophytic; they obtain their nutrition from decaying organic matter. In horsetails and most ferns, the gametophytes are photosynthetic. The gametophytes of many ferns look somewhat like thallose liverworts; they sometimes reproduce asexually and, like many sporophytes, can produce extensive clonal populations where the sporophytes are absent or rare. One example that has fascinated botanists for many years is the "Appalachian gametophyte" of the genus *Vittaria*, whose sporophyte is thought to be extinct. The gametophyte is distributed over several hundred square miles in the eastern half of the United States, but it is not known to produce a sporophyte (fig. 28.3).

ORGANIZATION OF REPRODUCTIVE STRUCTURES

Organs that produce gametes and spores in seedless vascular plants are similar in several ways to those of bryophytes (fig. 28.4).

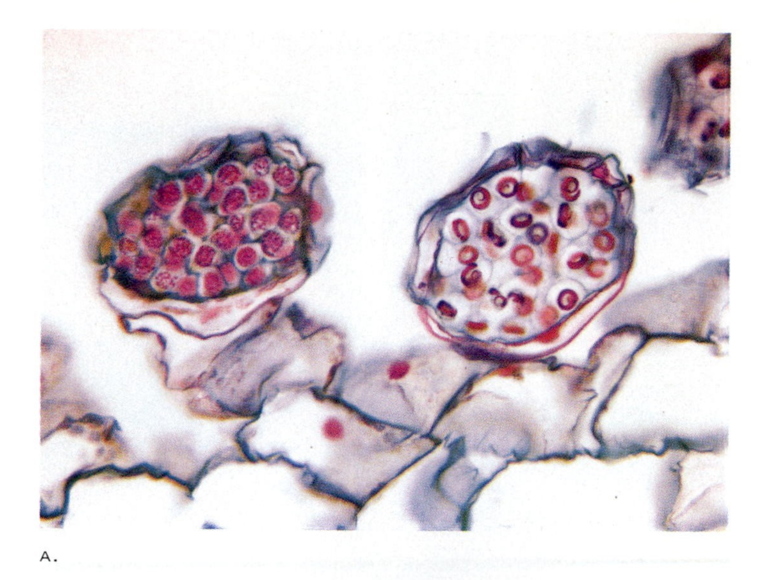

A.

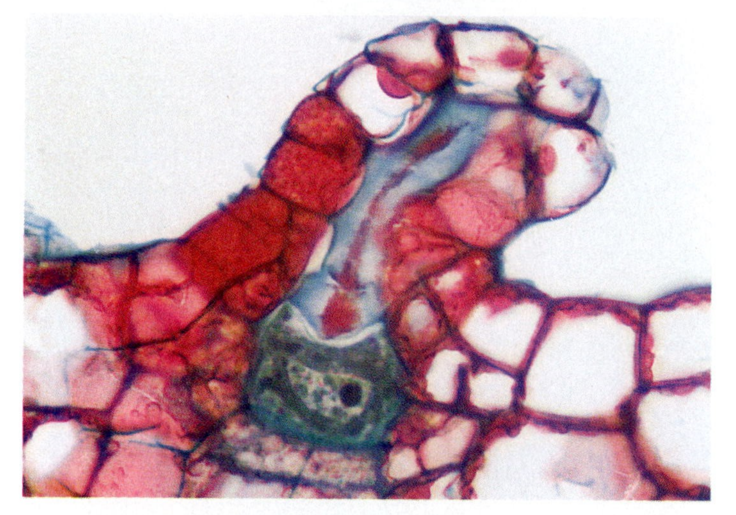

B.

FIGURE 28.4

The sexual organs of ferns are similar to those of bryophytes: antheridia and archegonia are multicellular and include a protective layer of sterile cells. (a) Antheridia are nearly spherical, each antheridium producing many flagellated sperm cells. (b) Archegonia are flask-shaped, with a narrow neck on top of an expanded region containing a single nonmotile egg.

Antheridia and archegonia are multicellular and protected by a layer of sterile cells. Each antheridium produces many sperm cells, but each archegonium makes only one egg.

The most significant advancement in the reproductive structures of vascular plants is the aggregation of sporangia into cones, called **strobili** (singular, **strobilus**). A strobilus is essentially a stem tip with several closely packed leaves or branches that bear sporangia (fig. 28.5). This advancement does not consistently distinguish vascular from nonvascular plants, however, since it occurs primarily in club mosses, spike mosses, and horsetails, but not in ferns. Nevertheless, the strobilus is one of the most significant developments in reproductive organization

Figure 28.5

Strobili, or cones, are aggregations of closely packed sporangium-bearing branches or leaves. Shown here are strobili of *Lycopodium obscurum*, a club moss.

among plants. The flowering plants, considered to be the most advanced group of plants, bear sporangia exclusively in flowers that are highly modified strobili.

Seedless vascular plants reproduce by means of spores; the sporophyte, unlike those of the bryophytes, is the dominant phase of the life cycle. Gametes form archegonia and antheridia. Spores are produced in sporangia that in some plants are aggregated into cones (strobili). Spores are usually dispersed by wind.

Organization of Vascular Tissues

Vascular tissues in stems and roots have two main functions: conduction and support. Nevertheless, the organization of vascular tissues is variable. For example, not all stems have a central pith, and phloem is not always outside of xylem, but may be on the inside as well. Regardless of the organization, each arrangement of vascular tissues is referred to as a **stele.** A stele is the combined vascular cylinder of xylem and phloem in roots and stems, along with the associated tissues in pith, if any. The basic terminology of stem and root structure was presented in Chapters 15 and 16. Those chapters focused on seed plants, but the basic cell and tissue types are the same in the seedless vascular plants.

Simple Steles

The simplest type of stele is a solid or nearly solid core of xylem and phloem. This basic stele type is called a **protostele.** Because they occurred in the earliest vascular plants, protosteles are considered to be the most primitive type of stele. Among the seedless vascular plants, protostelic stems occur only in the psilotophytes and a few lycopods and pteridophytes. With the exception of one class of flowering plants (i.e., monocots), protostelic roots are common in all groups of vascular plants.

Complex Steles

Vascular tissues in complex steles are interspersed with nonvascular, parenchymatous tissues that comprise the pith. Like the tissues in the cortex, the pith is derived from the ground meristem during primary growth (see Chapter 14). Among seedless vascular plants, most ferns have a pith that is surrounded by a tube of vascular tissues. This arrangement of pith and a vascular cylinder is referred to as a **siphonostele** (fig. 28.6a). The xylem is often surrounded by phloem in fern siphonosteles, but in many species the phloem is restricted to the outside of the xylem.

The configuration of siphonosteles, in contrast to that of protosteles, is often influenced by leaf traces. As a trace branches from the siphonostele, it interrupts the vascular tissue and creates what is called a **leaf gap** (fig. 28.7). A single leaf trace can form a wide leaf gap, or several nearby leaf traces can cause the siphonostele to appear highly dissected (fig. 28.6b). Such a siphonostele, which is called a **dictyostele,** is common in ferns.

The second major type of complex stele is a **eustele.** A eustele has vascular bundles that, like those of a siphonostele, are arranged in a circle around the pith (fig. 28.8). The bundles appear discrete in cross section, but they are actually interconnected longitudinally. Among seedless vascular plants, eusteles are common among horsetails and ferns.

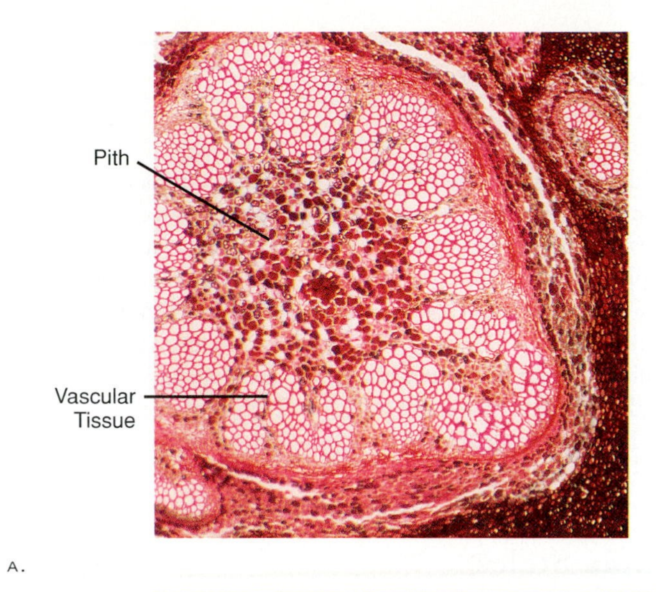

A.

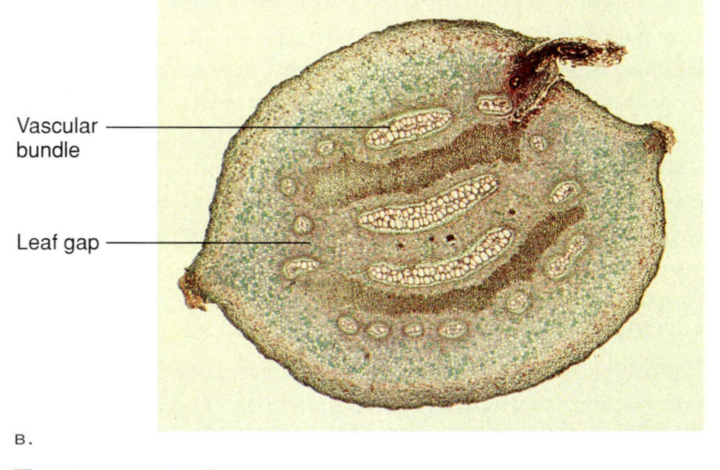

B.

FIGURE 28.6

Steles in seedless vascular plants. (a) The stems of most ferns are characterized by a siphonostele, which is a cylinder of vascular tissue that surrounds a central pith. The example here is from a species of *Osmunda*. (b) In other ferns, such as *Pteridium*, the vascular cylinder is interrupted by leaf gaps, thereby forming a dictyostele.

FIGURE 28.7

Longitudinal and cross-sectional views of a leaf gap.

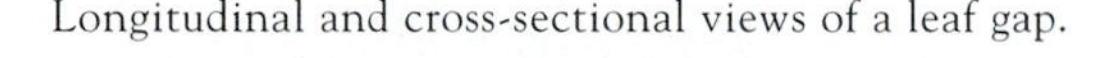

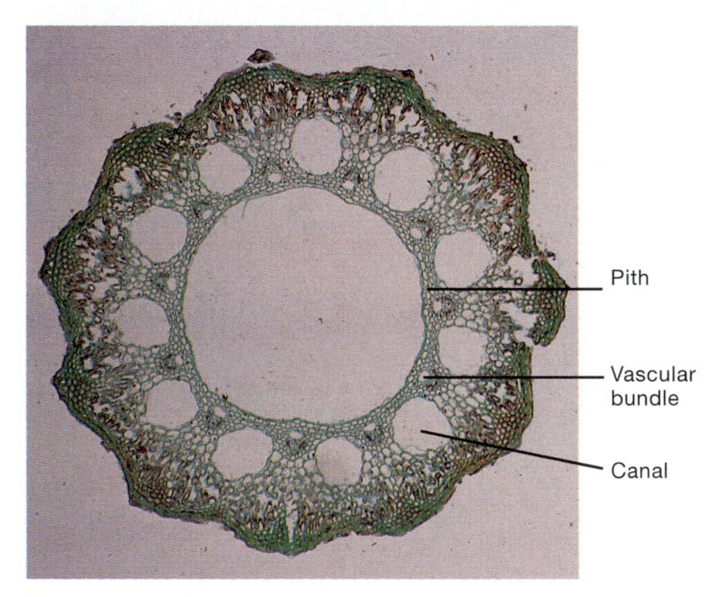

FIGURE 28.8

This cross section of an *Equisetum* stem shows an example of a eustele, which is the arrangement of discrete vascular bundles in a circle around the pith. *Equisetum* stems also have different kinds of canals, which appear in cross section as circular clear areas.

Microphylls and Megaphylls

Two kinds of leaves occur in the seedless vascular plants: **megaphylls** and **microphylls.** Megaphylls are what you usually think of as leaves and are defined by their evolutionary origin: they are believed to have arisen from modified branches. Conversely, microphylls first appeared from **enations,** which are flaps of tissue that protrude from stems. Megaphylls are identifiable by their multiple, branched veins, whereas microphylls have a single, unbranched vein.

Although the name megaphyll means "large leaf," megaphylls can be smaller than the microphylls of club mosses. Moreover, the microphylls ("small leaves") of some extinct relatives of the club mosses were more than a meter long.

C O N C E P T

Arrangements of vascular tissues vary from simple to complex among seedless vascular plants. Whisk ferns have the simplest steles, and ferns have the most complex. Seedless vascular plants have either microphylls, which first arose from enations, or megaphylls, which evolved from modified branches.

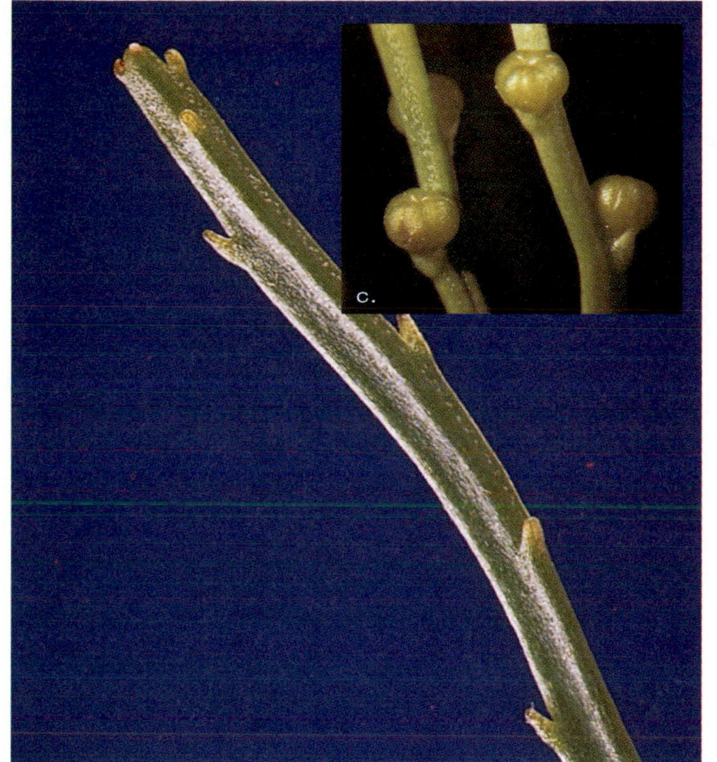

A.

B.

C.

FIGURE 28.9

(a) Whisk ferns (*Psilotum* sp.) are so called because their branching pattern gives the impression of a whisk broom. The stems bear nonvascularized, scalelike projections called prophylls (b) and lobed sporangia that are called synangia (c).

THE DIVERSITY OF SEEDLESS VASCULAR PLANTS

The four divisions of living seedless vascular plants are distinguished by both sporophytic and gametophytic features; significant variation exists among their branching patterns, leaf morphology, vascular organization, and underground absorptive organs. The types and arrangements of sporangia on the sporophyte can also be used in defining different divisions. Gametophytes vary in their origin from different spore types and in their sources of nutrition. No single characteristic defines each division; rather, classification depends on syndromes of features.

FIGURE 28.10

Tmesipteris with synangia. *Tmesipteris* grows on islands in the South Pacific.

Division Psilotophyta: Whisk Ferns

The Psilotophyta are the simplest vascular plants, primarily because they have no roots and because most species have no obvious leaves. Instead of roots with root hairs, the psilotophytes have rhizomes with absorptive rhizoids. The larger of the two genera in the division, *Psilotum*, has scalelike, nonvascularized projections called **prophylls** instead of leaves (fig. 28.9b). Botanists hypothesize that prophylls may be the reduced remnants of leaves, which were larger and more leaflike in the ancestors of the psilotophytes. The stems of *Psilotum* are green and photosynthetic.

The other genus in the division, *Tmesipteris*, has leaflike structures, each with a single vascular strand (fig. 28.10). Botanists are not sure whether these structures are flattened branchlets or leaves.

The name *whisk ferns* for this group comes from the highly branched stems of *Psilotum*, which give the plant the appearance of a whisk broom. *Psilotum* is widespread in subtropical regions of the southern United States and Asia, and it is also a popular and easily cultivated plant that is grown in greenhouses worldwide. *Tmesipteris*, on the other hand, is restricted to islands in the South Pacific, where it often grows as an epiphyte on the trunks of tree ferns. *Tmesipteris* is rarely cultivated.

The stems of *Psilotum* are protostelic and usually dichotomously branched. Lobed sporangia form at the tips of lateral branches, each of which grows in the axil of a forked projection on the stems. The occurrence of separate vascular strands to each sporangial lobe suggests that each lobe evolved from a separate sporangium. This type of fused sporangium is called a **synangium.** Meiosis occurs in the synangium, after which it splits open and releases its spores.

The spores of *Psilotum* germinate into microscopic, brownish gametophytes that live just below the soil surface. These gametophytes are usually cylindrical but sometimes have one dichotomous branch (fig. 28.11). Their cells are inhabited by a mycorrhizal fungus that absorbs nitrates, phosphates, and organic compounds used by the gametophyte. Sex organs develop from surface cells of monoecious (i.e., bisexual) gametophytes. Antheridia protrude slightly above the surface and contain a small number of coiled, multiflagellate sperm; archegonia are partially sunken and have necks that are shorter than those in bryophytes. The sperm cells, each having eight to ten flagella, must swim to the archegonium, often through thin layers of water on the surface of the gametophyte. Because the gametophytes are monoecious, fertilization can occur between egg and sperm either from different gametophytes (cross-fertilization) or from the same gametophyte (self-fertilization). Following fertilization, the zygote forms a foot and an embryonic stem. The primary stem is a branched rhizome that separates from the foot and becomes infected with the mycorrhizal fungus as it emerges from the gametophyte.

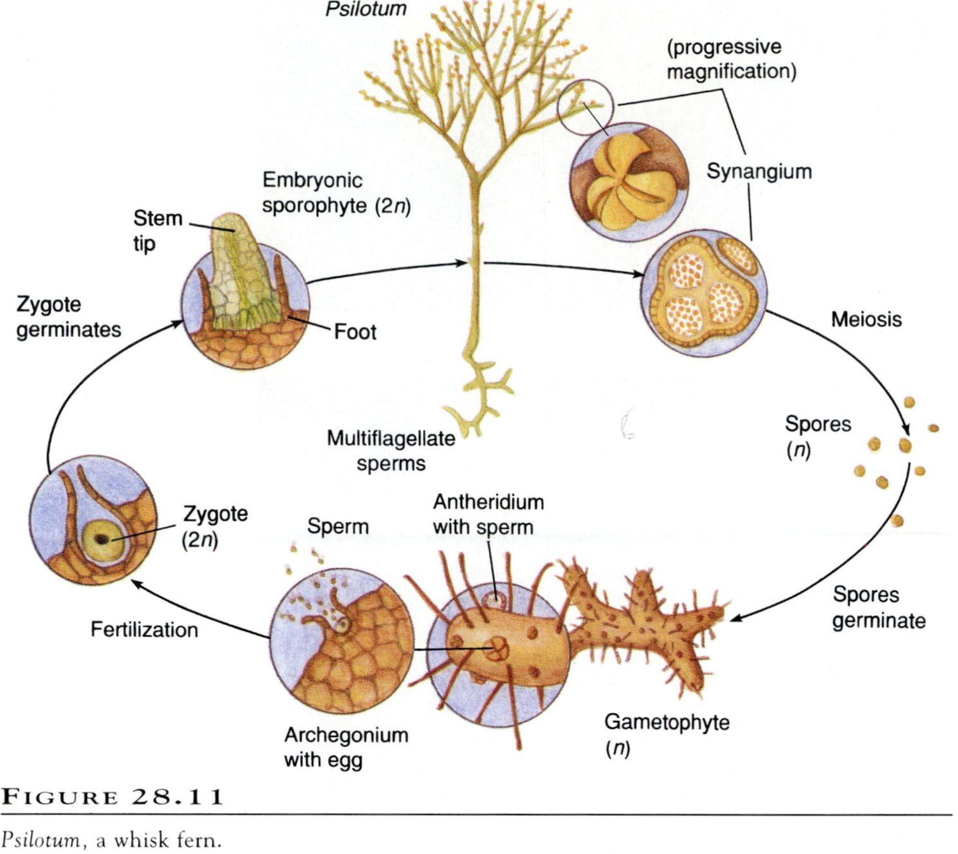

FIGURE 28.11

Psilotum, a whisk fern.

Division Lycopodophyta: Club Mosses

The division Lycopodophyta consists of more than 1,100 species that live in various habitats worldwide. They are primarily tropical but also form a conspicuous part of the flora in temperate regions. Most of the species are included in two genera, the club mosses (*Lycopodium*, about 400 species) and the spike mosses (*Selaginella*, about 700 species), both of which get their common names from their club-shaped or spike-shaped strobili (see fig. 28.5). Most species are terrestrial, but many are epiphytic. One species of *Selaginella*, called the *resurrection plant* (*S. lepidophylla*) because of its ability to defy drought conditions, occurs in the deserts of southwestern United States and Mexico. During periods of drought, this plant forms a tight, dried-up ball; when rain comes, its branches expand and come alive again with photosynthetic activity (fig. 28.12).

The sporophytes of club mosses are differentiated into leaves, stems, and roots. The roots branch from perennial rhizomes that sometimes grow outward from a central point to form "fairy rings" (as many mushrooms do). One such fairy ring of a *Lycopodium*, when measured for its size and annual growth rate, was calculated to have started growing in 1839.

In mature club mosses, some leaves bear a single sporangium on their upper surfaces. These fertile leaves are called **sporophylls** to distinguish them from sterile vegetative leaves (fig. 28.13). In *Selaginella* and in many species of *Lycopodium*, sporophylls are nonphotosynthetic and are packed together at the tops of stems into club-shaped or spike-shaped strobili. In other species of *Lycopodium*, the sporophylls are scattered among the sterile leaves; the two types of leaves are indistinguishable except for the presence of sporangia on the former.

There are two types of spore production in club mosses. In *Lycopodium*, spore production is homosporous (*homo* = same), meaning that all spores are morphologically indistinguishable. Whisk ferns and all bryophytes are also homosporous. In *Selaginella*, spore production is **heterosporous** (*hetero* = different) because there are two kinds of spores, a small type and a large type.

The small spores are called **microspores** and produce male gametophytes; the large spores are called **megaspores** and produce female gametophytes (fig. 28.14). Microspores and megaspores are produced in **microsporangia** and **megasporangia,** respectively; microsporangia are borne on **microsporophylls,** and megasporangia on **megasporophylls.** Both types of sporophylls usually occur in the same strobilus in *Selaginella* (fig. 28.14). In the megasporangium, one cell undergoes meiosis to produce four megaspores. In contrast, many cells divide meiotically in the microsporangium, producing hundreds or thousands of spores.

A.

B.

Figure 28.12

Selaginella lepidophylla, the resurrection plant, (a) forms a tight, dried-up ball during periods of drought, (b) but when rain comes its branches expand and once again become green and vibrant with photosynthetic activity.

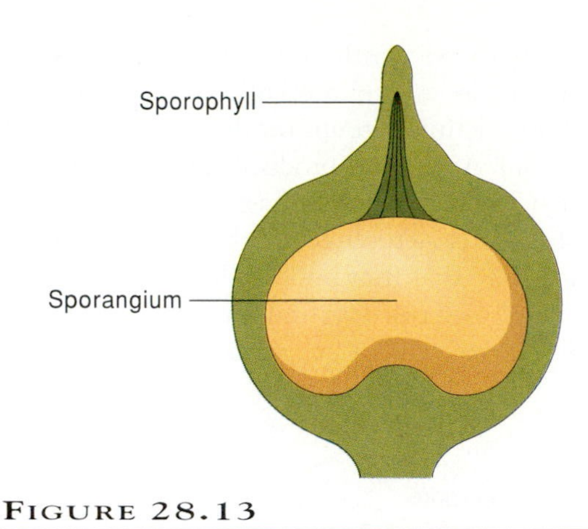

Figure 28.13

A sporophyll of *Lycopodium*, a club moss.

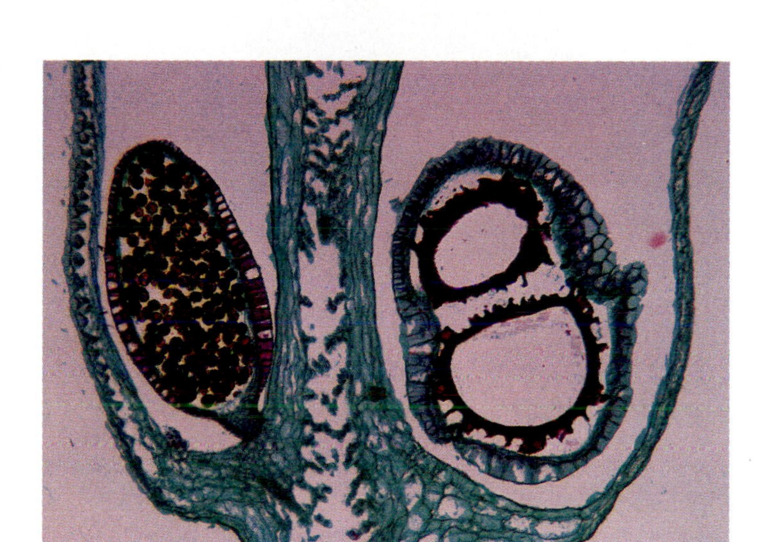

Figure 28.14

Longitudinal section through part of a strobilus of *Selaginella*, showing a microsporangium (left) and a megasporangium (right) on the same strobilus. The microsporangium produces many microspores; the megasporangium usually produces only four megaspores, two of which can be seen in this thin section.

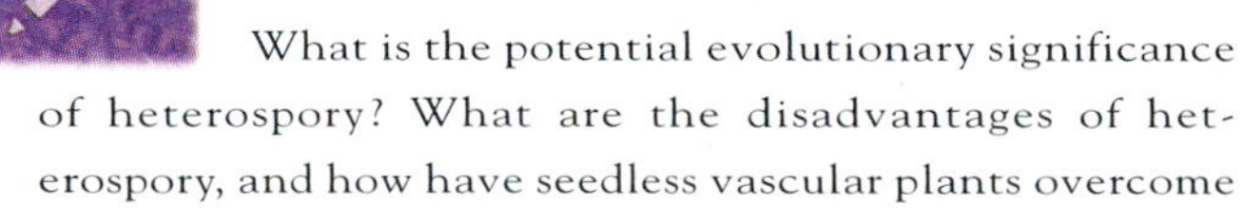

Writing to Learn Botany

What is the potential evolutionary significance of heterospory? What are the disadvantages of heterospory, and how have seedless vascular plants overcome these disadvantages?

The megaspores of *Selaginella* divide numerous times inside the spore walls, producing a multicellular female gametophyte within each spore. As the gametophyte grows and matures, the spore wall breaks open to expose the archegonia (fig. 28.15a). Likewise, a multicellular male gametophyte forms in each microspore. The male gametophyte consists almost entirely of a single antheridium, which protrudes from the spore just before releasing the biflagellate sperm (fig. 28.15b). The sperm must swim from the tiny male gametophyte to the large female gametophyte. Immediately after fertilization, the first cell division of the zygote produces a cell that will become the embryo and a cell that is called a **suspensor** (fig. 28.16a). This name is misleading, because nothing really hangs from the suspensor. Rather, the suspensor pushes the developing embryo into the gametophyte tissue where it can obtain sufficient nutrition during its early development. The embryo also develops a foot and an embryonic root, stem, and leaf (fig. 28.17b).

The gametophytes of *Lycopodium* are mycorrhizal and either photosynthetic and short-lived or saprophytic and long-lived (sometimes up to ten years). Antheridia and archegonia are intermingled among the lobes of photosynthetic gametophytes but are segregated into definite groups on the upper surface of saprophytes (fig. 28.17). Antheridia produce biflagellate, swimming sperm that may be attracted to archegonia by citric acid. Even though *Lycopodium* gametophytes are bisexual, studies of isozyme inheritance show that they undergo little self-fertilization. It is not known how cross-fertilization is promoted, but the gametophytes of some species are **protandrous,** meaning that they produce antheridia first, then archegonia. In a population of gametophytes of different ages, the young ones would all be male, thereby promoting outcrossing, if only temporarily. Ultimately, there can be many archegonia on a sexually mature individual, but generally only one zygote forms on each short-lived gametophyte. In contrast, long-lived gametophytes can support more than one young sporophyte in various stages of development. As in *Selaginella,* the zygote divides to produce a suspensor cell and an embryo that differentiates into a root, stem, and leaf.

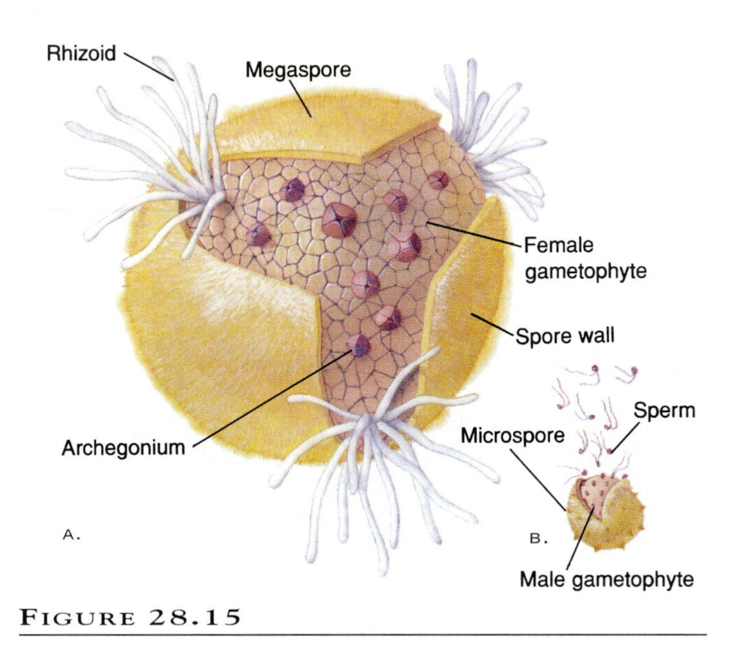

FIGURE 28.15

Selaginella spores containing gametophytes: (a) Megaspore with female gametophyte bearing archegonia. (b) Microspore with male gametophyte bearing a single antheridium that is releasing sperm.

The Lore of Plants

According to Druid mythology, Druid nuns living on the Isle of Sain in the Loire River valley of France had elaborate rituals to gather *Lycopodium,* which they displayed on their altars. The plants were considered to be sacred, and their display was believed to bring good fortune to the Druids.

The Lycopodophyta also includes the quillworts (*Isoetes*), so named because of their narrow, quill-like leaves (fig. 28.18). Quillworts, which are almost exclusively aquatic, live in freshwater habitats on almost every continent. Most of the leaves of quillworts are fertile and do not aggregate into strobili; some leaves produce sporangia that abort before they mature. Leaves grow from a corm instead of a rhizome. The corm has a peculiar cambium that adds both secondary phloem and secondary xylem to its interior, rather than to both sides, as in woody plants. This secondary growth in quillworts is unique among extant Lycopodophyta, although some extinct members of this division also had secondary tissue. Quillworts are also distinctive

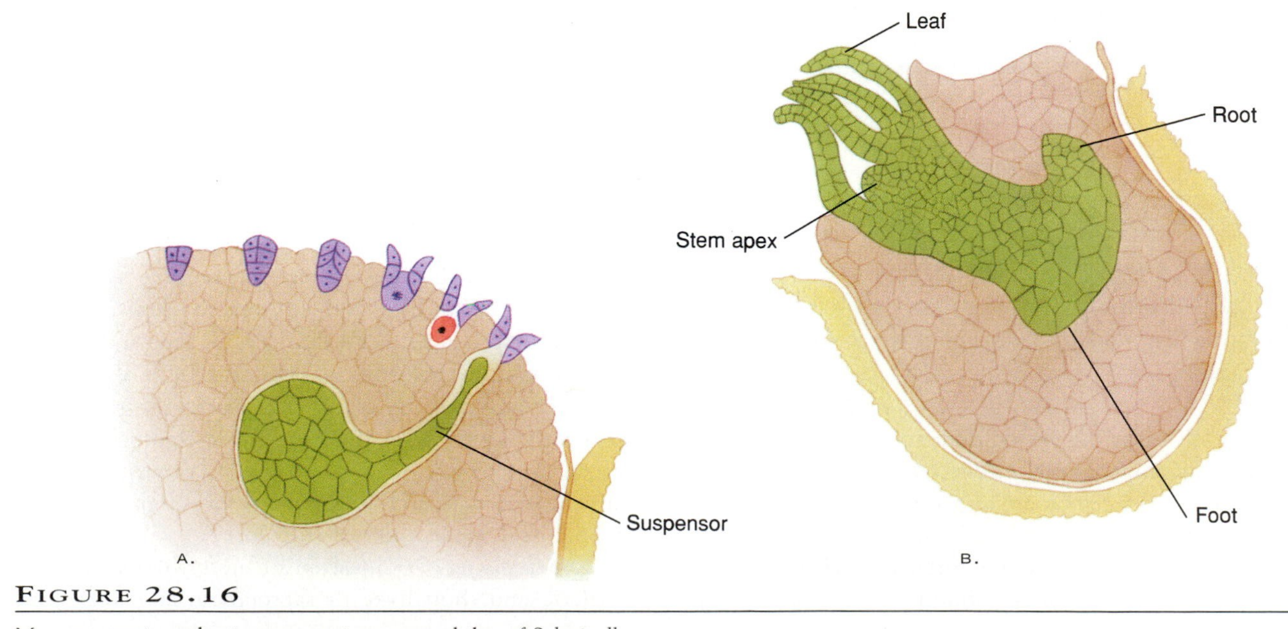

FIGURE 28.16

Megasporangia and microsporangia on a strobilus of *Selaginella.*

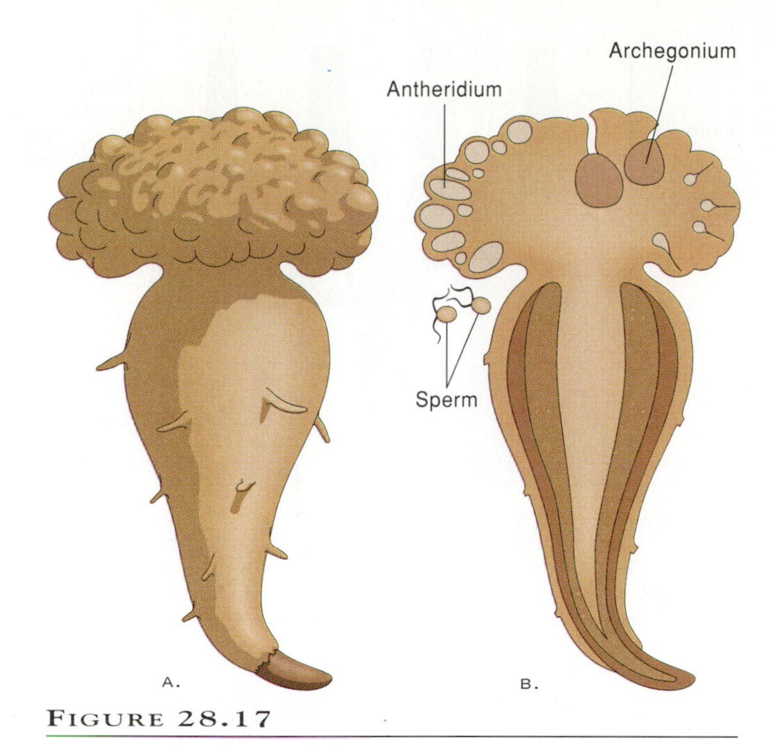

Figure 28.17

(a) Side view of an intact gametophyte of *Lycopodium;* such gametophytes are differentiated into a lobed above ground region and a taproot-like below ground region. (b) The distribution of antheridia and archegonia is shown in this longitudinal section.

Figure 28.18

Quillworts (*Isoetes* sp.) are so named because of their narrow, quill-like leaves.

in the division because they have CAM photosynthesis (see Chapter 7). Some lake-dwelling species in the Peruvian Andes (an extreme form) have no stomata in their leaves and obtain carbon dioxide for photosynthesis from the muddy substrate where they grow.

Division Equisetophyta: Horsetails

Equisetum is the only living genus of the Equisetophyta. Some of the fifteen species of *Equisetum* have branched stems that (with a good imagination) look like horses' tails (fig. 28.19). Both branching and nonbranching species are also called *scouring rushes* because their epidermal tissue contains abrasive particles of glass. Horsetails were used by American Indians to polish bows and arrows and by early colonists and pioneers to scrub their pots and pans. *Equisetum* occurs worldwide in moist habitats along streams or forest edges. Its rhizomes are highly branched and perennial. Because its rhizomes can grow rapidly, and its aerial stems are poisonous to livestock, *Equisetum* can be a serious problem for farmers and ranchers. Gardeners have often chopped up the rhizome while trying to remove it from the soil, only to have new plants arise from each of the fragments left behind.

Although *Equisetum* has true leaves, the stem is the dominant photosynthetic organ of the plant body. The most conspicuous feature of the stem is the series of "joints" made by whorls of small leaves (fig. 28.20b). The leaves are fused along most of their length, but their brown tips give the appearance

Figure 28.19

With some imagination, plants of *Equisetum* that have several whorls of branches look like horses' tails. For that reason, the entire genus is often referred to as the horsetails.

FIGURE 28.20

(a) With some imagination, plants of *Equisetum* that have several whorls of branches look like horses' tails. For that reason, the entire genus is often referred to as the horsetails. (b) The conspicuous "joints" of *Equisetum* stems consist of whorls of small leaves. These leaves, whose brown tips give the appearance of a collar just above the node, are fused into a sheath around the stem.

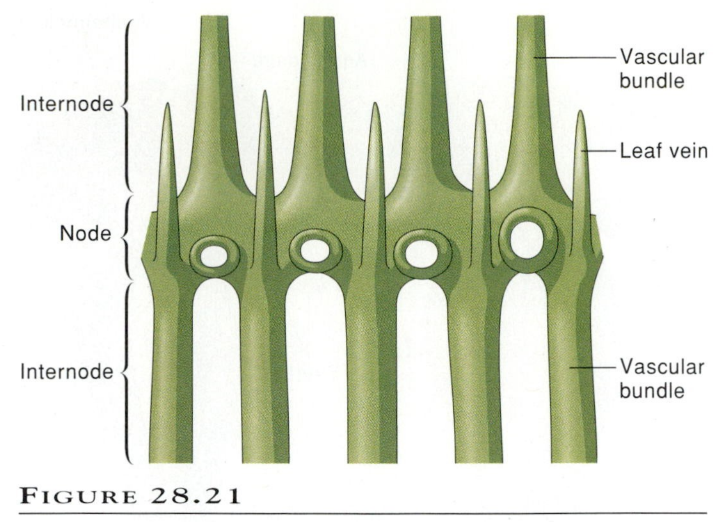

FIGURE 28.21

As shown in this longitudinal view of the vascular tissue of *Equisetum*, the discrete vascular bundles of eustelic internodes meet to form a continuous, siphonostelic pattern at the node.

of a collar around the stem just above each node. When the stems are pulled apart, they break easily at the nodes to yield pipelike internodal pieces. In addition, *Equisetum* stems are notable for a branching pattern that is unique among vascular plants. Instead of growing from axillary buds opposite the leaves, the lateral branches of horsetails sprout from between the leaf bases.

The internal structure of a horsetail stem fascinates botanists because of its complexity. As seen in cross section at an internode, the most prominent features of the stem are the epidermal ridges and the internal canal system (see fig. 28.8). The internodal stele is a eustele because of its regular, discrete vascular bundles; however, the vascular strands meet at the nodes in such a way that they form a siphonostele (fig. 28.21). Furthermore, at the node the central canal is interrupted with a short cylinder of pith.

Equisetum is homosporous, and sporangia occur in terminal strobili, which are either on vegetative stems or on nongreen fertile branches that develop from the rhizome. Rather than being on sporophylls, the sporangia of *Equisetum* are borne on branches called *sporangiophores* (fig. 28.22). Several fingerlike sporangia are attached to each sporangiophore, giving the overall appearance of tiny sausages hanging from the underside of a small umbrella. When sporangia are mature, the strobili elongate to separate the sporangiophores from each other, so that spores can be released into the air. The spores have an extra layer of wall material that peels off in four spoon-tipped strips called **elaters** (fig. 28.23). The elaters uncoil when they dry out and add to the buoyancy of the spores in the wind. The elaters coil again in humid air, thereby ensuring that the spores drop in a damp area. Although the term *elater* is also used in relation to liverworts, the elaters of *Equisetum* are not cellular like the elaters of *Marchantia*.

The gametophytes of *Equisetum* are photosynthetic, pincushion-shaped plants that can grow up to 1 cm in diameter. They are differentiated into a basal region with rhizoids and a branched region with bright green, platelike lobes (fig. 28.24). The sexuality of *Equisetum* gametophytes is not well understood because it is variable and appears to be related to environmental conditions (see the section titled "The Physiology of Seedless Vascular Plants").

All gametophytes are potentially monoecious, with antheridia borne on the upright lobes and archegonia embedded in cushions at the bases of the lobes. The antheridia release multiflagellate sperm that swim to the archegonia to fertilize the eggs.

As it grows, the zygote produces two regions of differentiation: (1) the stem and leaf sheath, and (2) the foot and root (fig. 28.25). There is no suspensor. The root immediately grows through the gametophyte and establishes the nutritional independence of the emergent sporophyte. Each gametophyte can bear many young sporophytes.

A.

B.

Figure 28.22

(a) The spores of *Equisetum* have an extra layer of wall material that forms four strips called elaters. The elaters entangle the spores together, thereby ensuring that male and female gametophytes will germinate close to one another. (b) Each sporangiophore consists of a stalk bearing a cap to which several sporangia are attached.

Figure 28.23

Equisetum spores, showing elaters.

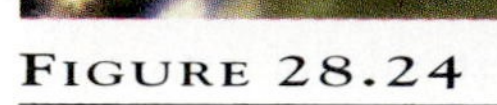

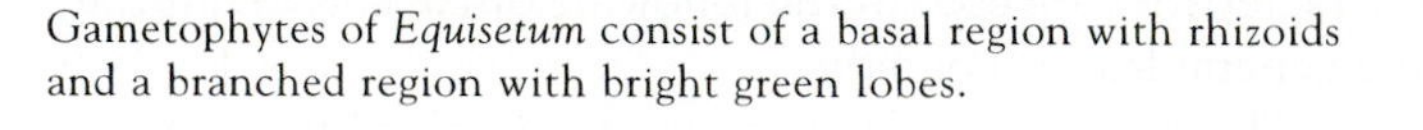

Figure 28.24

Gametophytes of *Equisetum* consist of a basal region with rhizoids and a branched region with bright green lobes.

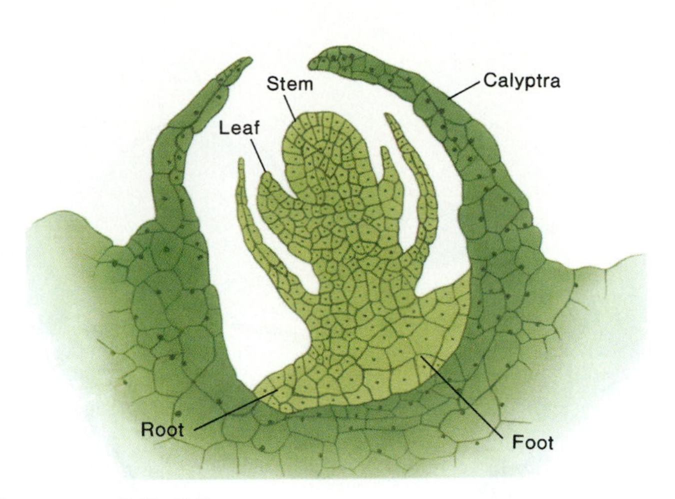

FIGURE 28.25

Emergent sporophyte showing two regions of differentiation.

FIGURE 28.26

The diversity of ferns. Ferns range from (a) *Marattia* and other tree ferns, some of whose leaves are the largest known among plants, to (b) tiny-leaved aquatic ferns such as *Azolla*.

Doing Botany Yourself

The gametophytes of homosporous plants, such as *Equisetum*, arise from only one type of spore and are therefore morphologically monoecious. Is self-fertilization common in these gametophytes? What experiments could you do to test this hypothesis?

Division Pteridophyta: Ferns

Ferns include approximately 12,000 living species, making this division by far the largest among the seedless vascular plants. Ferns are primarily tropical plants, but some species inhabit temperate regions, and some even live in deserts. North or south of the tropics, the number of species decreases because of decreasing moisture. In Guam, for example, about one-eighth of the species of vascular plants are ferns, but ferns constitute only about one-fiftieth of the total species in California.

Some genera of ferns have leaves that are the largest and most complex in the plant kingdom. For example, one species of tree fern in the genus *Marattia* has leaves that are up to 9 meters long and 4.5 meters wide, which is nearly the size of a two-car garage (fig. 28.26a). At the other extreme, the aquatic ferns *Salvinia* and *Azolla* have relatively tiny leaves (fig. 28.26b). For most people, however, a typical example of a fern is the bracken fern, *Pteridium aquilinum* (see fig. 28.1). The bracken fern, like most ferns, is classified in the order Filicales, which includes about 10,000 species. Because the bracken fern and its relatives are typical examples of the largest order of ferns, their characteristics are used in the following discussion to illustrate the general features of ferns.

The most conspicuous parts of a fern are its leaves, or **fronds.** New leaves grow from a fleshy, siphonostelic or eustelic rhizome each year. Early leaves are lopsided because they grow faster on their lower surface than on their upper surface. This growth pattern, which is called **circinate vernation,** produces young leaves that are coiled into "fiddleheads" (fig. 28.27). New fiddleheads arise close to the growing tip of the rhizome at the beginning of each growing season. The leaves of most ferns dieback each year, but the leaves of the walking fern (*Asplenium rhizophyllum*) can form new plants. Near the tip of each leaf, certain cells revert to meristems and grow into new roots, leaves, and a rhizome (fig. 28.28).

Fern leaves are usually fertile but do not form strobili. The leaves have dark spots on their lower surfaces, each of which is a collection of sporangia, together called a **sorus** (plural, **sori**). The sori of some species are covered by an outgrowth from the leaf surface called an **indusium,** while the sori of other species are either not covered or are enfolded by the edge of the leaf (fig. 28.29). With few exceptions, ferns are homosporous. However, there are significant differences in sporangium development between the primitive ferns and their specialized relatives.

Like most vascular plants, primitive ferns are **eusporangiate.** This means their sporangia are relatively large, with either massive stalks or no stalks at all, and contain many spores surrounded by a multilayered sporangial wall; the megasporangia of *Selaginella,* which often produce only four spores, are an exception. In contrast, the more specialized ferns, including most cultivated and wild species, are **leptosporangiate;** their sporangia are relatively small, with a delicate stalk and a thin sporangial wall (fig. 28.30). The small number of spores per leptosporangium is a multiple of 2, varying between 16 and 512

Figure 28.27

The coiled "fiddleheads" of fern leaves are rolled up leaf buds. Fiddleheads are formed by a pattern of growth called circinate vernation, as shown in this *Blechnum* fern.

Figure 28.28

The tips of leaves of the walking fern (*Asplenium* sp.) can become meristematic and form new plants. In this photograph, the tip of the long leaf on the plant to the left has formed the new plant (actually a clone) to the right.

A.

B.

C.

Figure 28.29

Fern sporangia. Most ferns have sporangia aggregated into clusters, called sori, on the undersides of the leaves. (a) In some ferns, such as the marginal woodfern (*Dryopteris marginalis*), each sorus is covered by a flap of leaf tissue called an indusium. (b) Other ferns bear uncovered sori, as shown here in *Alsophila sinuata*. (c) In still other ferns, as in the giant maidenhair fern (*Adiantum trapeziforme*), sori are enfolded by the edge of the leaf itself.

(often 16 or 32) in homosporous species. Nevertheless, each plant can produce millions of spores because of the large number of sporangia per sorus and the enormous number of sori per leaf. For example, one mature plant of *Thelypteris dentata* can produce more than 50 million spores each season.

Spores are catapulted from their sporangia by an ingenious method that has attracted much attention from botanists. The flinging action comes from the behavior of an incomplete ring of cells, called the **annulus,** that encircles the sporangium. The thin outer cell walls of the annulus slowly contract as they dry, creating a pulling force that ruptures the thin-walled **lip cells** and the outer walls of the sporangium. As drying continues, the water tension increases in the annulus, sometimes exceeding 300 atmospheres of pressure. Ultimately, the water evaporates completely, and the tension is broken. The sporangium then snaps back into its original position, ejecting the spores forcefully to a distance of about one centimeter.

In most homosporous ferns, spores germinate at first into protonemata. These protonemata eventually differentiate and form green, heart-shaped gametophytes

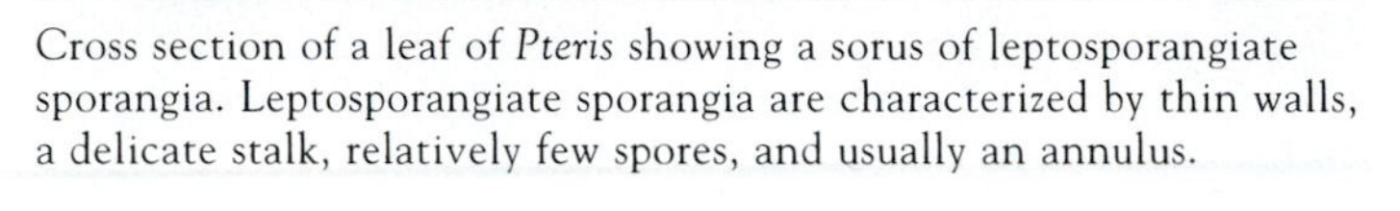

FIGURE 28.30

Cross section of a leaf of *Pteris* showing a sorus of leptosporangiate sporangia. Leptosporangiate sporangia are characterized by thin walls, a delicate stalk, relatively few spores, and usually an annulus.

that are anchored to the soil by rhizoids (fig. 28.31). The gametophytes are usually monoecious, with the sex organs on the lower surface. Archegonia are sunken in the gametophyte tissue near the notch, their necks sticking out slightly; antheridia protrude from the surface near the tip and are intermingled with rhizoids. The multiflagellate sperm released by the antheridia swim into the neck of the archegonium to reach the egg. Once fertilization occurs, the zygote germinates into a young sporophyte that quickly becomes independent of the host gametophyte (fig. 28.32). Shortly thereafter, the short-lived gametophyte dies and begins to decompose.

Heterosporous ferns are decidedly unfernlike in their overall appearance. *Azolla,* for example, is a floating aquatic fern having tiny leaves that are crowded onto slender stems (see fig. 28.26b). Other genera, such as *Marsilea,* grow from roots buried in the muddy bottoms of ponds or ditches (fig. 28.33); only their leaves reach the water surface and float upon it. The heterosporous ferns are leptosporangiate, but their sporangia form in specialized, hardened structures called **sporocarps.** Each sporocarp bears several sori consisting of many sporangia. When the spores are mature, they are either released into the surrounding water or pushed out of the sporocarp by the swelling of a gelatinous, hygroscopic filament.

CONCEPT

More than 12,000 of the approximately 13,000 species of seedless vascular plants are ferns, Division Pteridophyta. The remainder are classified in three divisions: the Psilotophyta (whisk ferns), Lycopodophyta (club mosses), and Equisetophyta (horsetails). Some of the ferns and club mosses are heterosporous; that is, they have one type of spore that grows exclusively into a male gametophyte and one type that becomes exclusively female. All seedless vascular plants produce flagellated sperm and therefore require free water to reproduce sexually.

A.

B.

FIGURE 28.31

Fern gametophyte. (a) A heart-shaped fern gametophyte with rhizoids and reproductive organs (b), which are the dark spots near the notch of the heart.

The Physiology of Seedless Vascular Plants

The physiology of seedless vascular plants is most conveniently studied in the gametophytic phase. Spores can be germinated and gametophytes can be easily grown in laboratory cultures. Because of their independence from the sporophyte phase, these haploid forms are more easily studied in seedless vascular plants than in seed plants. Most of our knowledge of the physiology of seedless vascular plants comes from the largest group, the ferns.

The Influence of Light

Light is the most influential environmental factor controlling the development of ferns, especially of the gametophyte phase. For example, spore germination in ferns is controlled by light: red light, through the action of phytochrome (see Chapter 19),

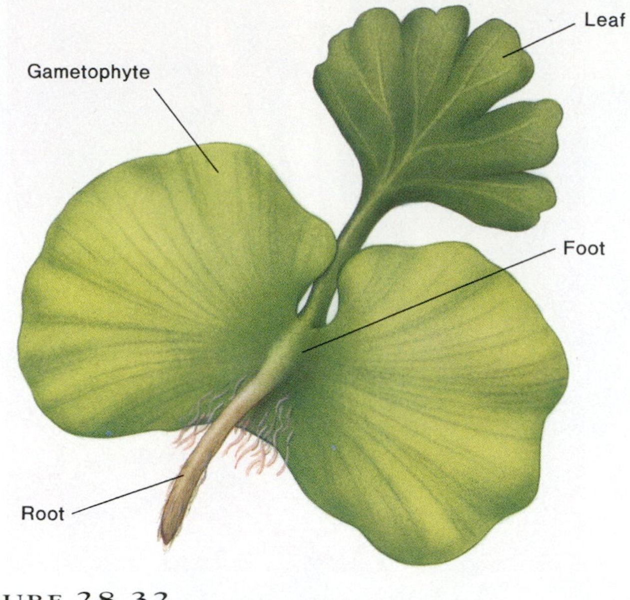

Figure 28.32

Young fern sporphyte growing out of its gametophyte parent.

Figure 28.33

The clover fern (*Marsilea* sp.) is an example of a heterosporous fern. Like many heterosporous ferns, *Marsilea* is aquatic; the roots of these plants are anchored in the muddy bottoms of ponds or ditches.

induces spore germination, and blue light prevents spore germination. Red light also affects protonemata by inducing apical growth and positive phototropism, by increasing the gap phase (G-2) in mitosis, and by delaying cell-plate formation in cytokinesis. All of these phenomena are inhibited by blue light, but the pigment that absorbs in that region is unknown. This unknown blue-absorbing pigment is associated with the nucleus.

The Environment and Gametophyte Sexuality

In *Equisetum*, sex ratios among gametophytes in a population can be affected by variations in temperature or light regimes.

Figure 28.34

(a) The gametophytes of this fern, *Anemia rotundifolia*, secrete gibberellin-like substances, called antheridiogens, that induce other, younger gametophytes nearby to develop antheridia. (b) The chemical structure of an antheridiogen.

Either lower temperatures or higher light intensities cause the ratio of bisexual to male gametophytes to increase. Lower ratios can be induced by red light. In other species, the gametophytes are strictly dioecious.

Reproductive Hormones

When the spores of many ferns germinate, the first heart-shaped gametophytes secrete a substance that induces younger gametophytes to develop antheridia. These substances, called **antheridiogens,** can also break dormancy in darkness, thereby eliminating the usual light requirement for spore germination. Some antheridiogens act as gibberellins in gametophytes, and an antheridiogen from at least one fern (*Anemia*) has a gibberellin-type chemical structure (fig. 28.34). In the genus *Cerotopteris*, however, abscisic acid blocks the formation of antheridia that are induced by antheridiogens.

Sporophyte Physiology

The water relations of sporophytes have attracted the attention of plant physiologists, who have found that, in holly fern (*Cyrtomium falcatum*), the primary factors that affect water conduction are the number and diameter of tracheids. This is essentially the same result obtained for vessels in the vascular tissues of flowering plants.

Physiological studies of seedless vascular plants focus mainly on the gametophyte. Its development is controlled primarily by interactions with light. Sexuality, at least in monoecious fern gametophytes, is controlled by hormones, some of which are chemically and functionally similar to gibberellins.

The Ecology of Seedless Vascular Plants

One of the most interesting ecological aspects of any group of plants involves their interactions with other organisms. For example, the occurrence of mycorrhizal fungi in the gametophytes of *Psilotum* and *Lycopodium* and in the rhizomes and roots of many plants suggests a general symbiotic relationship between fungi and plants. Fungal filaments extract certain minerals from the soil better than do rhizoids or root hairs. Plants obtain such minerals from the fungi associated with them, while the fungi in return receive organic nutrients from their hosts. Fungi found in fossil plants had the same relationship with their hosts more than 400 million years ago. Indeed, this association with fungi hastened the invasion of land by plants.

Several tropical species of ferns harbor ant nests in their rhizomes (fig. 28.35). Roots grow into the ant-carved tunnels and absorb nutrients from the decaying organic matter brought in by the ants, and ants use the sporangia as food. Spores are often shaken loose and dispersed from the parent plant as the ants handle the sporangia. Botanists speculate that the loss of spores to ants is sufficiently offset by the advantage of having at least some of the spores dispersed.

The gametophytes of some homosporous vascular plants secrete chemicals that inhibit the growth and reproduction of other plants (see box 28.1, "Killer Plants"). In many plants, high levels of polyploidy are associated with greater production of various chemicals that can protect plants from insects and fungi. Approximately 95 percent of seedless vascular plants are polyploid, which may explain why ferns, for example, are remarkably resistant to herbivores and fungal parasites.

The ecology of many ferns is well understood because these plants can be such noxious weeds. For example, when people burn native forests to establish pastures, bracken ferns can quickly invade the newly available habitats (see fig. 28.1). Populations of this plant spread rapidly from an extensive network of fast-growing rhizomes. The problem of bracken infestation is worsened by the toxicity of this plant to the cattle raised in such pastures. Herbicides can alleviate the problem somewhat, but this damages other plants and leaves a toxic residue in the soil. Weed scientists are now studying how to control bracken fern by introducing pathogenic fungi into invasive populations.

A.

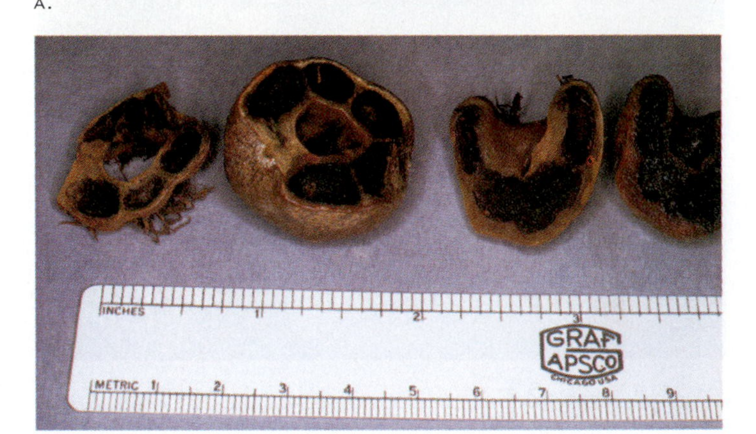

B.

Figure 28.35

(a) *Solanopteris brunei* is a myrmecophytic fern that is native to Costa Rica; (b) ants live in pod-like chambers that are formed by the plant.

Boxed Reading 28.1

Killer Plants

Despite the enormous number of spores produced by each seedless vascular plant, relatively few develop into mature gametophytes in nature. Some of this loss can be attributed to bad luck; most spores are carried by air currents to places that are too dry, too nutrient-poor, or in some other way unfavorable for germination. Spore germination can also be inhibited by substances secreted by other plants, either sporophytes or other gametophytes. In *Thelypteris normalis,* for example, the growth of gametophytes is inhibited by secretions diffusing from the roots of the sporophyte. The two active chemicals, named **thelypterin A** and **thelypterin B,** are indole derivatives that inhibit cell division in the gametophytes of *Thelypteris* and other fern genera. They have no effect on the growth of young sporophytes.

Thelypterins are similar to auxin, both in structure and in function (see Chapter 18). The apparent function of these chemicals is to prevent the growth of gametophytes. The immediate advantage is that competition for potentially limited resources (minerals, water, etc.) is stopped before it even begins.

The secretion by one organism of chemicals that inhibit the growth of another organism is called **allelopathy,** (allelopathy also occurs in seed plants and in fungi). Penicillin, for example, is a classic example of an allelopathic substance. In most instances, however, one species inhibits another species; yet sporophyte-derived thelypterins inhibit gametophytes of the same species. This is a somewhat troubling interaction to decipher, because the sporophyte kills its own offspring. How such an interaction could evolve and be maintained has not been clearly explained. Our understanding of the long-term advantages, if any, of this intraspecific allelopathy remains incomplete.

Box Figure 28.1

The upper gametophyte of *Thelypteris normalis* was grown for 14 days without a nearby sporophyte, whereas the lower gametophyte was grown for 131 days in the presence of a sporophyte (note the absence of a meristem in the lower gametophyte).

Kariba weed (*Salvinia molesta*), an aquatic fern, can be a more serious problem than bracken fern, even though it is not poisonous (fig. 28.36). Trouble with this plant began after it became established in Papua, New Guinea, north of Australia. The invasion of kariba weed overtook waterways and eliminated native species of plants, eventually threatening the availability of water to 80,000 people. Kariba weed is sterile but reproduces asexually very rapidly, doubling its mass approximately every two days. From a genetic viewpoint, millions of tons of this weed may represent a single individual, making it possibly the largest organism on earth. In the 1980s, botanists made a significant breakthrough in understanding the ecology of this organism: its native range was discovered in southeastern Brazil, where a new beetle species was found that feeds exclusively on kariba weed. This beetle is now being used successfully to control invasions of kariba weed. In Papua, New Guinea, the beetles have eaten about two million metric tons (2 billion kilograms) of the plant, reducing the water surface it covered by more than 90 percent.

Figure 28.36

The kariba weed (a) reproduces so fast that it can overtake waterways and crowd out other kinds of plants (b).

C O N C E P T

Seedless vascular plants, especially ferns, are often weedy because they can reproduce rapidly by asexual means. Some species cause problems when they invade pastures (bracken fern) or waterways (kariba weed). Effective biological control of these plants entails introducing their natural enemies, such as pathogenic fungi or herbivorous insects, into areas of rampant growth.

The Economic Importance of Seedless Vascular Plants

Seedless vascular plants are of little economic importance today. Perhaps their greatest economic impact comes from their aid in discovering fossil fuel deposits (see box 28.2, "Spores: 'Instant' Fossils"). Many seedless vascular plants, especially ferns, are often found in greenhouses or are grown as houseplants and groundcovers. Many more of these plants were once useful in ways that are either no longer in vogue or are only important in economically deprived parts of the world. One example, mentioned previously, was the western pioneers' use of *Equisetum* to scrub their dishes. Also, before the invention of flashbulbs, photographers used flash powder that consisted almost entirely of dried *Lycopodium* spores; a pound of spores can still be purchased for only a few dollars from scientific supply companies. In China, where petroleum-based fertilizers are not affordable, *Azolla* is substituted as a rotated crop in rice paddies. This aquatic fern hosts a cyanobacterium, *Anabaena azollae*, that fixes nitrogen from the air, thereby acting as a fertilizer to replenish the nitrates removed from the soil by other crop plants (fig. 28.37).

The evolution of biochemical complexity in land plants coincided with their morphological advancements. Besides structural compounds like lignin, chemicals also evolved to provide protection from ultraviolet light, parasitic fungi, and protozoa and other predators. The continued biochemical diversification of land plants has resulted in a wealth of chemicals that have been useful to humans. For example, Native Americans treated wounds and nosebleeds with spores from one species of club moss, *Lycopodium clavatum*, which have antibiotic and blood-coagulant properties. Resin from the rhizomes of the marginal fern *Dryopteris marginalis* was once taken internally to get rid of intestinal tapeworms. As is true for most medicinal plants, the exact identity of the active ingredient from these plants has not been determined. We do know, however, that many *Lycopodium* species synthesize complex, nitrogen-containing chemicals called **alkaloids** that are potent animal poisons. The dried and powdered leaves containing these chemicals are used directly as pesticides in parts of eastern Europe. More recently, extracts from the fiddleheads of bracken fern, which is considered to be a delicacy by some people, were discovered to cause intestinal cancer.

What are Botanists Doing?

Go to the library and read a recent article about the use of *Azolla* in rice production in China. Is *Azolla* used in rice production in the United States? Why or why not?

Boxed Reading 28.2

Spores: "Instant" Fossils

During the early stages of spore formation, a tough polymer called **sporopollenin** is built into spore walls. Because of the stability and resistance to decay of this substance, spores are essentially "instant" fossils. Indeed, the recovery of spores from sediments entails boiling samples in concentrated acids to dissolve the rocks that contain them. The spores themselves, regardless of whether they are fresh or millions of years old, are unharmed by this process. Their remarkable toughness makes spores, as well as pollen from seed plants, ideal records of plant communities, both past and present.

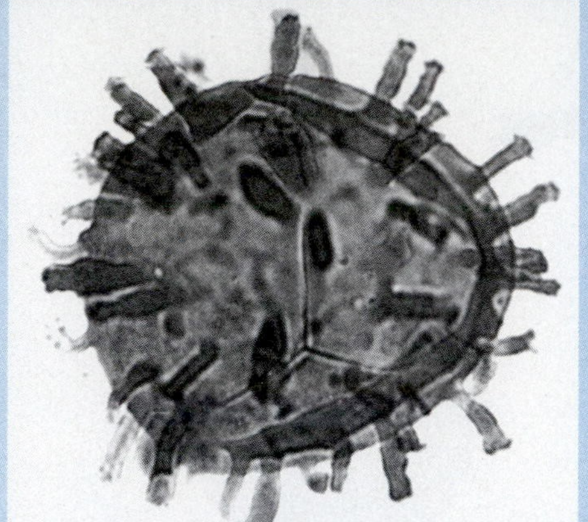
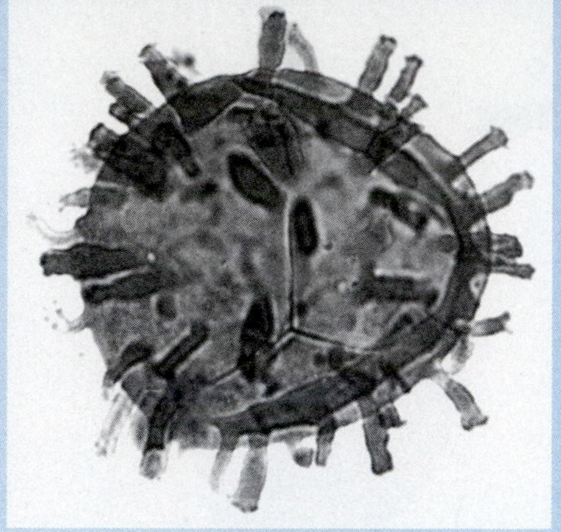

Box Figure 28.2

Raistrickia crocea.

Exploration for fossil fuels, particularly oil, has been made much easier by the finding that plants are abundantly represented by spore and pollen "communities," the remnants of extinct floras that thrived along ancient shorelines. Specific kinds of spore communities are associated with oil deposits. Cores are made by special drills that often penetrate more than several thousand meters below the earth's surface. Oil is usually discovered when the oil-indicator spores and pollen are found in the cores. Much of the analytical success of such efforts depends on **palynologists** (specialists who study pollen and spores); accurate identification of the microfossils in the core samples provides a reliable age for each layer of rocks relatively quickly and inexpensively. These methods have been effective, for example, in explorations of Alaskan oil fields.

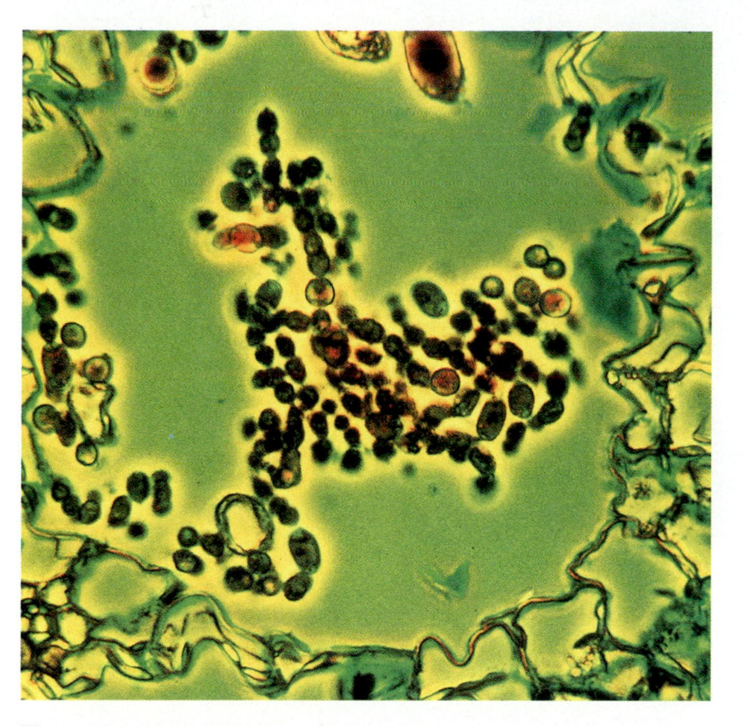

Figure 28.37

Leaf lobes of *Azolla* contain cavities, as shown here, inhabited by the cyanobacterium, *Anabaena azollae*.

The Origin and History of Seedless Vascular Plants

Fossils of vascular plants are much more abundant than those of bryophytes. Vascular tissues, cuticles, and spores are particularly well preserved because of their resistance to decomposition. Although the fossil record is still far from complete, several thousand extinct species have been discovered. Most of these are *form* species or genera; that is, they are known only from spores or from fragments of the plant body, and cannot be matched with an entire plant (see box 28.3, "How Does a Plant Become a Fossil?"). The oldest form genera, consisting of cuticles and spore tetrads, were discovered in late Ordovician rocks in Libya (fig. 28.38). Many botanists are not convinced that these plants had vascular tissue, however, since tracheids or other evidence of vascularization have not been found with them.

The oldest fossils that are unquestionably vascular plants consist of well-preserved vegetative and reproductive structures. These plants, named *Cooksonia*, were rootless and leafless, with slender, dichotomously branched stems (fig. 28.39). They produced homosporous spores in sporangia that developed from the expanded tips of their branches. *Cooksonia* first appeared in the late Silurian period, about 420 million years ago. These first known invaders of land were abundant in the tidal mud flats of New York state, South Wales, the Czech Republic, and Podolia (Ukraine). Their distribution probably included many other areas that experienced periodic flooding, but fossils in these areas were either not preserved or have not yet been discovered.

Cooksonia represents the opening of the floodgates of diversification for vascular plants. There was little competition for colonizing immense areas of land. Although *Cooksonia* was the most successful of the land invaders, it gave way to larger and more complex kinds of plants during the Devonian period. Plants grew taller and thicker, and they began to develop leaves that had a larger surface area for photosynthesis. The development of roots and stem cuticles allowed plants to move ever further from their swampy origins into increasingly drier habitats.

HOW DOES A PLANT BECOME A FOSSIL?

Archaeologists trying to piece together evidence of past civilizations search for tell-tale shards of pottery and other artifacts. Paleobotanists use similar direct evidence, in the form of fossils.

The public perception that fossils are just old plants or animals in rocks doesn't do justice to the wide variety of things that are called *fossils.* Different types of fossils form, depending on where the organisms grew and how fast they were buried in sediment. The preservation of most organisms has occurred in water as sedimentary particles were deposited on plant parts that fell into the water. Heavy sediment flattened leaves or other plant parts, squeezing out water and leaving only a thin film of tissue, called a **compression** fossil. Cellular structure rarely survived this process, but well-preserved cuticles have occasionally been found in deposits of compression fossils. Intact cells (and even DNA) have also been found in compressions, as in the Miocene deposit (ca. 20 million years old) at a fossil site named Clarkia Lake in Idaho.

Other types of fossils usually lack plant tissue. An **impression,** for example, is an imprint of an organism that is left behind when the organic remains have been completely destroyed. Only the contour of the plant remains.

In a third type of fossilization, tissues became surrounded by hardened sediment and then decayed. The hollow negative of the original tissue is called a **mold.** In conditions that allowed the mold to become filled with other sediment that conformed to the contours of the mold, the resultant fossil is called a **cast.** Fossil molds and casts, although formed on a geological time scale, are analogous to those that modern artists make of their subjects.

An interesting but poorly understood process of fossilization involves the replacement of cell contents with minerals. The compaction of such mineralized tissues essentially transforms the organic material into rock. Such fossils, called **petrifactions,** make areas like the Petrified Forest of Arizona famous among botanists and tourists alike.

Plant fossils are rarely found as whole plants. Different plant parts in the same fossil bed are given separate species and genus names, referred to as *form* species or genera. Sometimes two or more parts are connected to each other, and the plant they came from can be drawn as a reconstruction of the original. One such example, the genus *Lepidodendron,* is shown in the accompanying figure. *Lepidodendron* is actually the name given to stem fragments; the rest of the plant is derived from several unconnected organs and other fragments, each of which was given a separate name when it was discovered.

A.

B.

C.

Trunk pattern
Transition zone in bark pattern
Stigmanian root system

D.

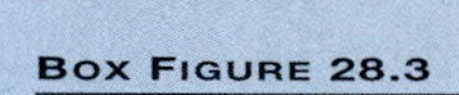

BOX FIGURE 28.3

Types of fossils include (a) compressions, which are made when heavy sediment squeezes out water and leaves only a thin film of tissue, (b) impressions, which are the imprints that are left in rocks after the organic remains of an organism have been completely destroyed, and (c) petrifactions, which form when cell contents are replaced with minerals, thereby transforming the organic material into rock. Petrifactions are so common in some areas that they form "petrified" forests, such as the Petrified Forest of Arizona.

Figure 28.38

Spore tetrad from Ordovician period of Libya.

A.

B.

Figure 28.39

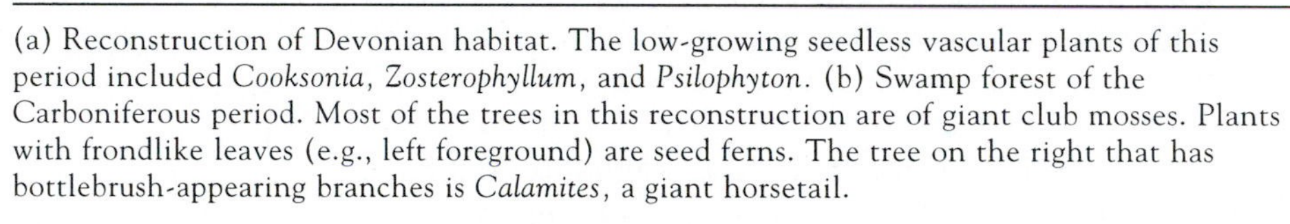

(a) Reconstruction of Devonian habitat. The low-growing seedless vascular plants of this period included *Cooksonia*, *Zosterophyllum*, and *Psilophyton*. (b) Swamp forest of the Carboniferous period. Most of the trees in this reconstruction are of giant club mosses. Plants with frondlike leaves (e.g., left foreground) are seed ferns. The tree on the right that has bottlebrush-appearing branches is *Calamites*, a giant horsetail.

Spores became more resistant to desiccation, which enabled them to tolerate dispersal by wind without drying out.

The First Vascular Plants

Depending on their quality and the extent of preservation, fragments of fossil plants can occasionally be matched so that a significant portion of the original plant can be reconstructed. Perhaps two hundred or so fossil species fit this category. Of these, no more than two dozen are known in superb detail from a large number of species. Our picture of the earliest vascular plants relies heavily on this small number of well-known plants.

Devonian plants reveal two distinct evolutionary lines. Plants in one line, referred to as the **zosterophyllophytes** (Division **Zosterophyllophyta**), superficially resemble the living genus *Zostera*, the marine eelgrasses. Zosterophyllophytes grew in clusters of upright, leafless branches that arose from horizontal stems, much as *Cooksonia* did (see fig. 28.39a). However, their sporangia were attached laterally near the ends of branches, which gave them the appearance of loosely aggregated strobili.

The xylem in zosterophyllophytes was **exarch,** meaning that it developed centripetally; that is, the youngest xylem cells were closer to the periphery of the stem than were the older xylem cells (fig. 28.40a). Even though *Zosterophyllum* and its relatives were extinct by the late Devonian period, the zosterophyllophytes probably gave rise to the ancestors of the Lycopodophyta, including now-extinct trees and present-day club mosses. Botanists believe that the club mosses and their extinct relatives inherited the features of lateral sporangial attachment and exarch protosteles from zosterophyllophyte progenitors.

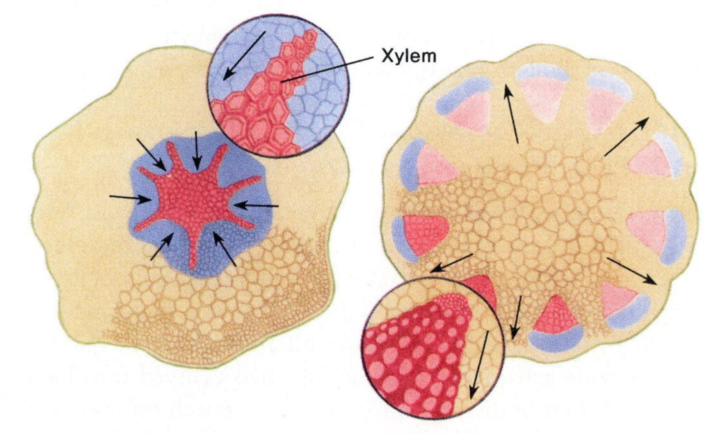

Figure 28.40

Examples of exarch (a) and endarch (b) xylem development. Exarch xylem matures toward the center of the stem (i.e., centripetally), whereas endarch xylem matures toward the perimeter of the stem (i.e., centrifugally).

The second evolutionary line includes *Cooksonia* and its relatives in Division **Rhyniophyta.** This division is named for the remarkably preserved assemblage of Devonian fossils near the village of Rhynie, Scotland. **Rhyniophytes** looked like zosterophyllophytes, and members of the two divisions often occur in the same fossil beds. The main difference between them is that the rhyniophytes had terminal sporangia, whereas the zosterophyllophytes had lateral sporangia. Also, the xylem of rhyniophytes developed centrifugally **(endarch)** from the interior of the protostele (fig. 28.40b), not from the periphery, as in zosterophyllophytes. Although *Cooksonia,* a rhyniophyte, is the oldest known vascular plant, it is unlikely that the rhyniophytes gave rise to the zosterophyllophytes. These two divisions must have shared a common ancestor at some point in their history, but their exact relationship to one another is unknown.

Rhyniophytes occupy an important position in the history of modern plants because they were the ancestral stock for the division **Trimerophytophyta.** Futhermore, because of their large size and structural complexity, the **trimerophytes** are believed to be the most likely precursors of ferns, horsetails, and seed plants. Before these modern groups arose, however, plants had to evolve leaves and roots.

The most widely accepted suggestion for the origin of leaves is that branching patterns gradually changed by the twisting of smaller branches into one plane **(planation),** by the faster growth of some branches over others **(overtopping),** and by the formation of parenchyma tissue between the lateral branches **(webbing)** (fig. 28.41).

The possible origin of roots is less clear than the origin of leaves, but botanists assume that roots came from primitive rhizome branches that developed protective tissue (i.e., root caps) to cover their growing tips. Indirect evidence for such an origin is that roots are generally protostelic, as were primitive rhizomes. In addition, the roots of seedless vascular plants are almost exclusively adventitious, which may also be a holdover from their origins as rhizome branches. However, the xylem of roots is exarch, whereas the rhyniophytes and trimerophytes, the oldest ancestors of modern plants, had endarch protosteles.

Giant Club Mosses, Giant Horsetails, and Tree Ferns: Coal-Age Trees

The first land plants were relatively small and herbaceous. With the advent of leaves, thicker cuticles, roots, and perhaps most important, lignin, plants could grow larger and could more effectively exploit the year-round growth conditions during that period. By the middle of the Carboniferous period (about 300 million years ago), some plant groups had evolved into large trees, and their herbaceous progenitors were well on their way to becoming extinct. These first trees included the earliest seed plants (discussed in the next chapter) as well as several groups of seedless vascular plants. The fossils of these huge plants formed the extensive coal deposits that characterize the Coal Age.

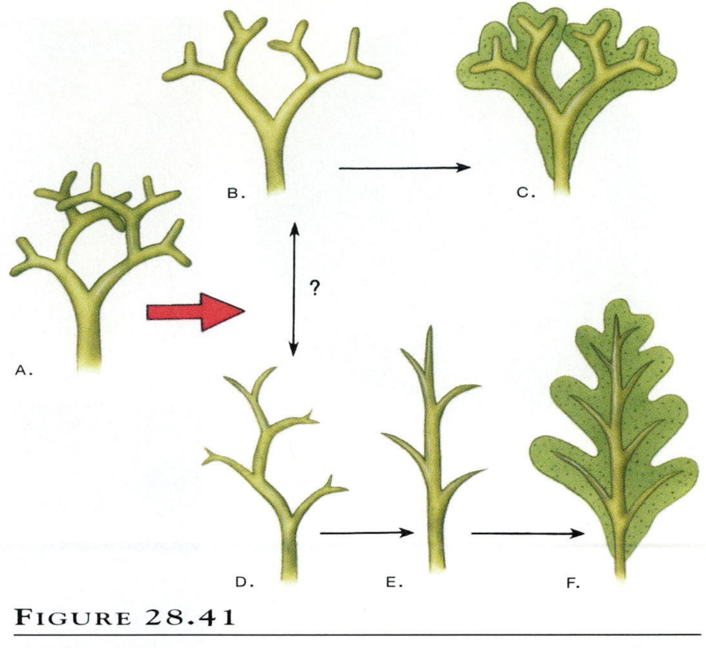

FIGURE 28.41

Possible pathways for the evolution of leaves from branches. (a) The earliest branching patterns consisted of equal, dichotomously-branched (i.e., branches in pairs) systems that were three-dimensional. This pattern changed by planation to flattened branching systems that either remained equally dichotomous (b) or that became unequally branched (d) and then uniaxial with side branches (e) by overtopping. Finally, parenchyma tissue filled in the spaces between branches by webbing (c, f). (The two-headed arrow and question mark between (b) and (d) means that the direction of evolution may have been in either or both directions.)

The swamp-dwelling lycophyte trees, or giant club mosses, grew nearly 40 m tall and dominated the Carboniferous period. Like their zosterophyllophyte progenitors, *Lepidodendron* and related trees were protostelic, although *Lepidodendron* was protostelic only at the base of the stem and the tips of branches; in between, the stem and branches were siphonostelic. The massive stems were supported by extensive periderm tissue and a small amount of secondary xylem. Underground they had a modified branch system with rootlike appendages whose internal structure was like the roots of *Isoetes*. These giant club mosses also had microphyllous leaves, and their sporophylls were packed together into strobili. Trees in the Lycopodophyta were mostly heterosporous. A single strobilus from one of these plants, up to 0.5 m long, may have produced as many as 8 billion microspores, or as many as several hundred megaspores. Although the wastage rate of spores must have been enormous, their dispersal from tall trees enabled them to colonize large areas of land successfully.

Giant horsetails such as *Calamites* lived in swampy forests alongside the giant club mosses. Secondary xylem enabled *Calamites* to grow to almost 20 m tall. With few exceptions, the plants were homosporous, and all the sporangia were borne on sporangiophores in terminal strobili. Unlike the giant club mosses,

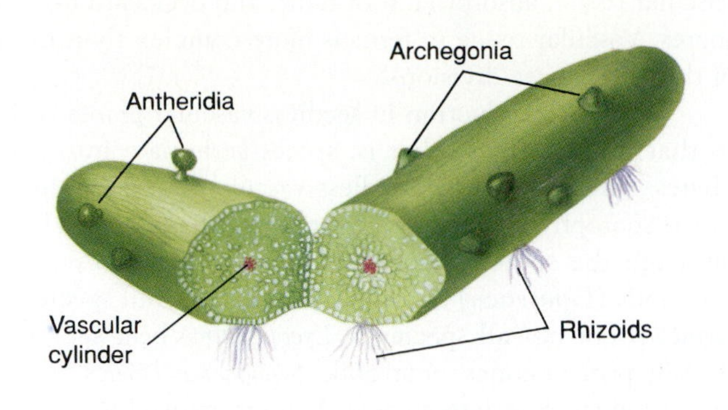

Figure 28.42

Schematic drawing of the "gametophyte" of *Rhynia gwynne-vaughanii*.

the giant horsetails had extensive networks of underground stems that allowed them to spread rapidly by vegetative reproduction. Because of such rapid reproduction by rhizomes, aboveground stems often occurred in dense, bamboolike thickets.

The Carboniferous period also saw the origin and diversification of tree ferns. Although all of the genera of tree ferns from the Carboniferous are extinct, as a group these primitive plants are still well represented in tropical regions of the world today.

At the end of the Carboniferous period, the giant club mosses began to disappear rapidly and were extinct by the end of the Permian period, approximately 250 million years ago. In contrast, the giant horsetails continued to dominate the landscape for another 100 million years or so, until the mid-Cretaceous period, when they also died out. Nevertheless, many of the primitive features of these extinct trees are maintained in the smaller, modern members of these groups.

Ancient Gametophytes

Our knowledge about the structure of extinct plants is based primarily upon features of the sporophytes. Preservation of gametophyte tissue from vascular plants was probably even poorer than for the gametophytes of bryophytes. One intriguing exception to this general pattern may be a controversial Devonian plant named *Rhynia gwynne-vaughanii* (fig. 28.42). This species is known only from vascularized, rhizomelike fragments found in the Rhynie fossil beds. Some botanists interpret the small bumps on the surface of these fragments to be archegonia and antheridia, which means that the fossil fragments are gametophytic. If these structures are indeed sex organs, the gametophytes bearing them resemble those of *Psilotum*. However, the microscopic features of *Rhynia gwynne-vaughanii* are not interpreted as sex organs by all botanists; some botanists believe they are unusual secretory structures on an otherwise ordinary sporophyte rhizome. The argument about whether this plant is a gametophyte or a sporophyte remains unsettled.

C O N C E P T

The first vascular land plants are known from fossil sporophytes of the Silurian period, approximately 420 million years ago. Early vascular plants were small and herbaceous, and they grew in swampy, tropical areas. Like present-day *Psilotum*, they had no leaves or roots but instead had photosynthetic stems and underground rhizomes with absorptive rhizoids. Descendants of these plants evolved leaves, roots, and strong structural support tissues. All groups developed tree forms that dominated the earth for millions of years, and then most died out. The closest living descendants of these early trees are the club mosses, horsetails, and tree ferns. Other evolutionary lines produced the seed plants.

The Puzzle of the Psilotophyta

Progress toward a phylogenetic classification of seedless vascular plants has been enlivened considerably by arguments about how to classify the psilotophytes. The traditional view is that *Psilotum* and *Tmesipteris* comprise a separate division that, with the exception of mosses, may be the most primitive of the vascular plants. This view is based on such assumed primitive features as lack of roots, prophylls instead of leaves, protosteles, an inconspicuous bisexual gametophyte, and dichotomous branching. Starting in 1969, however, David Bierhorst did a series of studies that led him to propose that *Psilotum* and *Tmesipteris* should be classified as ferns. He noted that many of the features of psilotophytes are shared with ferns, especially the primitive genera *Stromatopteris* and *Actinostachys*. Furthermore, by comparison with these ferns, the shoots of the psilotophytes can be interpreted as fronds. At that time, some botanists saw Bierhorst's proposal as an unacceptable departure from traditional classification. Some botanists still do.

Three new kinds of information are now available for further evaluation of the traditional hypothesis versus Bierhorst's alternative hypothesis about the relationships of the psilotophytes to ferns. One kind involves flavonoids: the psilotophytes synthesize a different class of flavonoids than is found in the ferns. Some botanists interpret this as a significant difference that does not support a close relationship between psilotophytes and ferns. The second kind of information comes from a cladistic analysis of all plants, which places the psilotophytes near the ancestral base of seedless vascular plants, just above the mosses (fig. 28.43). Although the cladogram was derived from the same kinds of characters that Bierhorst used for his hypothesis, the conflicting results of the two analyses are probably caused by different assumptions about the importance and evolutionary direction of the characters used.

This conflict may be resolved by yet a third kind of new information, which involves the structure of the chloroplast genome in plants. In a certain region of about 30,000 base-pairs of the genome, all genes point in one direction in some plants

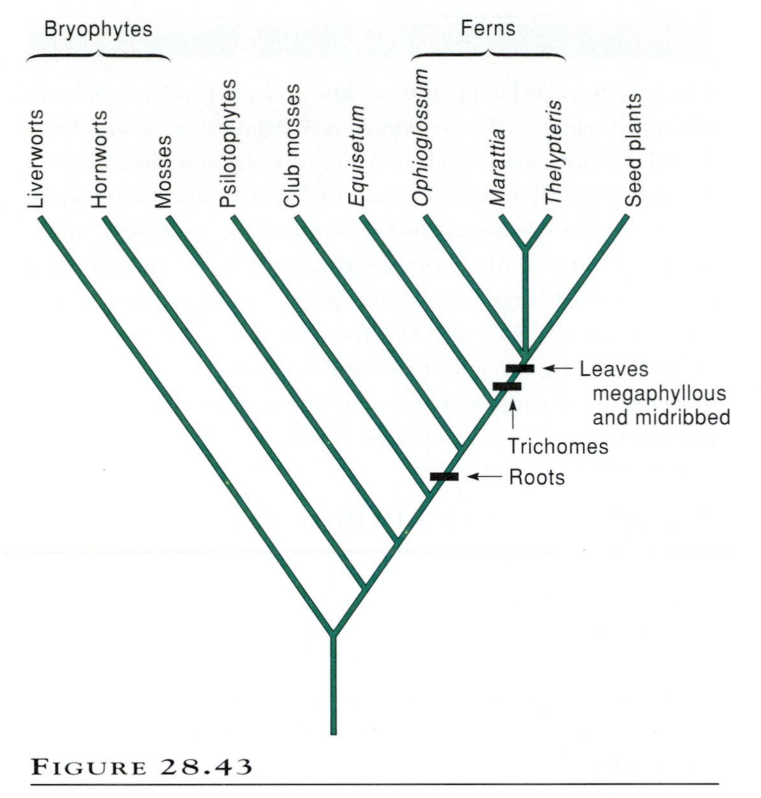

FIGURE 28.43

Cladogram of land plants.

and in the opposite direction in others. Botanists believe that this structural difference in the chloroplast genome is highly significant and that it represents a mutation that probably occurred just once in the evolutionary history of plants. A survey of plants in 1992 found that one orientation occurs in bryophytes and club mosses, while the other orientation occurs in psilotophytes, *Equisetum*, ferns, and seed plants. Assuming that the primitive orientation occurs in the bryophytes and club mosses, then all other plants share a common ancestor that mutated to the other orientation. Although this single feature does not show how close the relationship might be between psilotophytes and ferns, it does provide evidence for rejecting the traditional hypothesis about the primitive position of the Psilotophyta. Furthermore, it also supports the widely held hypothesis among paleobotanists that the club mosses represent the oldest vascular plants.

Chapter Summary

Seedless vascular plants are primarily ferns, but they also include whisk ferns (which are not true ferns), club mosses, and horsetails. Features of seedless vascular plants that enable them to thrive on land include a resistant cuticle, complex stomata, vascular tissue, absorptive root hairs, and desiccation-resistant spores. Vascular tissue in ferns is more complex than in plants of the other three divisions.

Sexual reproduction in seedless vascular plants is similar to that of bryophytes; that is, spores germinate into gametophytes. The life cycle of seedless vascular plants is dominated by the sporophyte phase. Spores are generally dispersed by wind, although the spores of some ferns are dispersed by ants. All horsetails (*Equisetum*) and some club mosses (all species of *Selaginella*, but not all species of *Lycopodium*) bear sporangia in densely packed cones, or strobili. *Selaginella*, *Isoetes*, and some ferns produce dioecious gametophytes from two kinds of spores, a phenomenon called heterospory. The gametophytes of seedless vascular plants are nutritionally independent of their parent sporophytes, but not all are photosynthetic; the gametophytes of whisk ferns, some club mosses, spike mosses, and several ferns are saprophytic. Many seedless vascular plants can reproduce vegetatively, either by the sporophyte or the gametophyte or both.

The seedless vascular plants are of little direct economic value. Some species were used as folk medicines, pesticides, or photographic flash powder. Fossil plants have some indirect economic value because geologists use the location of spores in the earth's strata to judge where oil deposits might occur. Some weedy plants cause problems because they diminish the usefulness of pastureland, clog waterways, and eliminate native plant species.

The fossil history of vascular plants begins with *Cooksonia*, a leafless and rootless progenitor of most modern plant groups. The first plants with leaves and roots appeared more than 150 million years after plants first invaded land. Giant horsetails, giant club mosses, and tree ferns dominated the vegetation of the Carboniferous period. Although these seedless trees ultimately gave way to seed plants, their descendants live on as herbaceous lycopods and as modern tree ferns.

There are conflicting ideas about the evolutionary relationships of the division Psilotophyta. One view is that it is a division of primitive plants, and the other is that psilotophytes are modified ferns.

Questions for Further Thought and Study

1. How are the gametophytes of seedless vascular plants similar to the gametophytes of bryophytes? How are they different?
2. Why is *Cooksonia*, which is the oldest confirmed plant fossil, *not* considered to be the oldest ancestor of all modern plant groups?
3. What are the major distinguishing features of the divisions of extant seedless vascular plants?

4. Most seedless vascular plants are polyploid. What features of their reproductive biology or ecology might explain this phenomenon?
5. *Lycopodium* means "foot of the wolf." What feature of the plant is probably responsible for this name?
6. A tracheid-based vascular system is supposedly less efficient than one based on vessels. Why? How does the geographical distribution of seedless vascular plants support this hypothesis? How does it conflict with this hypothesis?
7. Why are the fertile appendages of *Equisetum* called sporangiophores instead of sporophylls?
8. What are the distinguishing features of leptosporangiate and eusporangiate development?

Suggested Readings

Articles

Banks, H. P. 1975. Early vascular plants: Proof and conjecture. *BioScience* 25:730–737.

Barrett, S. C. H. 1989. Waterweed invasions. *Scientific American* 261(4):90–97.

Graham, L. E. 1985. The origin of the life cycle of land plants. *American Scientist* 73:178–186.

Lumpkin, T. A., and D. L. Plucknett. 1980. *Azolla:* Botany, physiology, and use as a green manure. *Economic Botany* 34:111–153.

Niklas, K. J. 1981. The chemistry of fossil plants. *BioScience* 31:820–825.

Voeller, B. 1971. Developmental physiology of fern gametophytes: Relevance for biology. *BioScience* 21:266–270.

Books

Foster, F. G. 1992. *Ferns to Know and Grow.* Portland: Timber Press.

Gifford, E. M., and A. S. Foster. 1989. *Morphology and Evolution of Vascular Plants.* New York: W. H. Freeman.

Jones, D. L. 1987. *Encyclopedia of Ferns.* Portland: Timber Press.

Stewart, W. N. 1983. *Paleobotany and the Evolution of Plants.* Cambridge: Cambridge University Press.

Thomas, B. 1981. *The Evolution of Plants and Flowers.* New York: St. Martin's Press.

Tryon, R. M., and A. F. Tryon. 1982. *Ferns and Allied Plants.* New York: Springer-Verlag.

West, R. G. 1977. *Studying the Past by Pollen Analysis.* Oxford Biology Reader No. 10. Oxford: Oxford University Press.

Pollen being released into the wind from a cluster of male pine cones. Gymnosperms such as pine depend on the wind to carry pollen from the male cones to the female cones.

CHAPTER 29

Gymnosperms

Chapter Outline

Chapter Overview

A major change occurred in plant evolution when seeds appeared about 360 million years ago. This change signified the increasing protection of the female gametophyte, which was later accompanied by the evolution of pollen. Along with these reproductive changes, vegetative features such as sunken stomata, thicker cuticles, and tougher tissues enabled land plants to thrive as the warm swampy conditions of the Devonian and Carboniferous periods gave way to cooler and drier environments of the Permian period. Changing environments fostered the diversification and extinction of several groups of seed plants; some groups were more successful than others in surviving to modern times. The most dominant of the temporary successes may have been the seed ferns, which lasted about 70 million years, until the environment got too cool and dry during the early Permian period. The cycads, conifers, and cycadeoids thrived after the seed ferns disappeared. Of these, only a few descendants of the cycads and conifers remain. The cycadeoids and cycads were so abundant during the Jurassic that this period is known as the Age of Cycads, at least among botanists (other biologists usually refer to it as the Age of Dinosaurs). The cycadeoids went extinct about the same time as the dinosaurs, about 65 million years ago. The fossil record of gymnosperms provides much information on the earliest of the seed plants—information that helps us understand how the modern gymnosperms evolved.

Introduction

One of the most significant events in the evolution of vascular plants was the origin of the seed. Ideas about how seeds arose are based on interpretations of the earliest fossils that look like seeds. The oldest such fossil is from the late Devonian period (about 360 million years ago), which was more than 40 million years after the first vascular plants appeared. This fossil has been interpreted by some botanists as a loose cluster of seeds, each consisting of a megasporangium surrounded by a protective, integumentlike layer with several fingerlike projections (fig. 29.1). Other fossils, from a little later in the Devonian and from the early Carboniferous period, are more seedlike; they show the megasporangium becoming progressively more enclosed by the integumentary layer (fig. 29.2). Paleobotanists believe that this progression represents the evolutionary development of seeds.

The counterpart of the seed is the pollen grain, which is an immature male gametophyte combined with and contained within a microspore wall (see Chapter 17). Pollen grains evolved by a continued reduction of the male gametophyte and its more secure retention inside the protective spore wall. This description of a pollen grain would also seem to apply to the microspore of *Selaginella* and other seedless vascular plants whose male gametophytes are mostly enclosed in the spore wall, but botanists distinguish true pollen from microspores. Morphologically, the first pollen differed from microspores in the location of the gametophyte's germination from the spore wall. Microspores germinated from the **tetrad scar,** which is the place on a spore where it was attached to three other spores in a meiotic group of four. In contrast, the first true pollen grains germinated from the spore wall opposite the scar. The development of pollen was also accompanied by the disappearance of the tetrad scar entirely (fig. 29.3). The first true pollen did not appear in the fossil record until the Mesozoic era, more than 150 million years after the origin of seeds.

Plant Gender: A Caution about Terminology

The first seedlike fossils contained *mega*sporangia, and the development of pollen was based on reduction of the *male* gametophyte; that is, these plants were heterosporous. In contrast to the heterosporous seedless vascular plants, however, the first seed plants kept their megaspores on the parent sporophyte until after the female gametophytes were fertilized. Unfortunately, for botany students, this evolutionary development also seems to mark the origin of some confusing terminology about the gender of plants. In seedless plants and algae, only the gametophytes are referred to as male, female, or bisexual; their sporophytes have no gender. In seed plants, though, the entire seed is referred to as a female structure, not just the gametophyte within. Furthermore, sporophytes that bear seeds but not pollen are called *female*, and those that bear pollen but not seeds are called *male*. Sporophytes that produce both are bisexual.

The application of such gametophytic terminology to sporophytes is technically incorrect, especially since plants rarely have sex chromosomes. Nevertheless, botanists and nonscientists alike still call a pine a female tree if it only has seed cones, and a mulberry a male tree if it only has pollen. We sometimes try to avoid this technical flaw by using terms like *dioecious* and *monoecious*, but then we may slip up and refer to a dioecious species as having male and female sporophytes or refer to a monoecious species as having bisexual sporophytes. Nevertheless, only gametophytes truly have gender, and in seed plants they are always unisexual; that is, they are either male or female.

FIGURE 29.1

Reconstruction of ovules (with fingerlike projections) of a plant that grew in the late Devonian period (about 360 million years ago). Some botanists conclude that this ovule consists of a loose cluster of seeds.

After G.W. Rothwell.

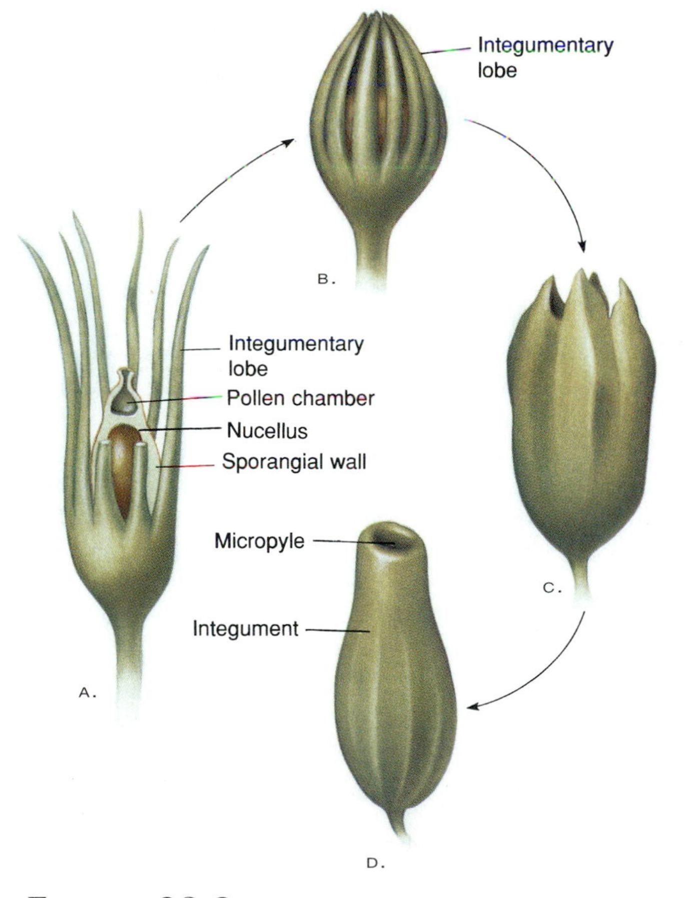

FIGURE 29.2

The evolution of seeds. Reconstructed ovules from the early Carboniferous period are more seedlike; they are progressively more enclosed by the integuments.

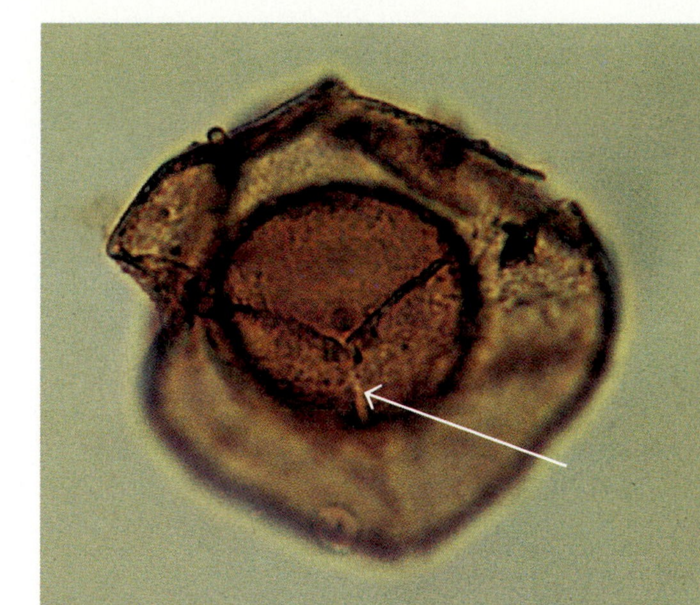

A.

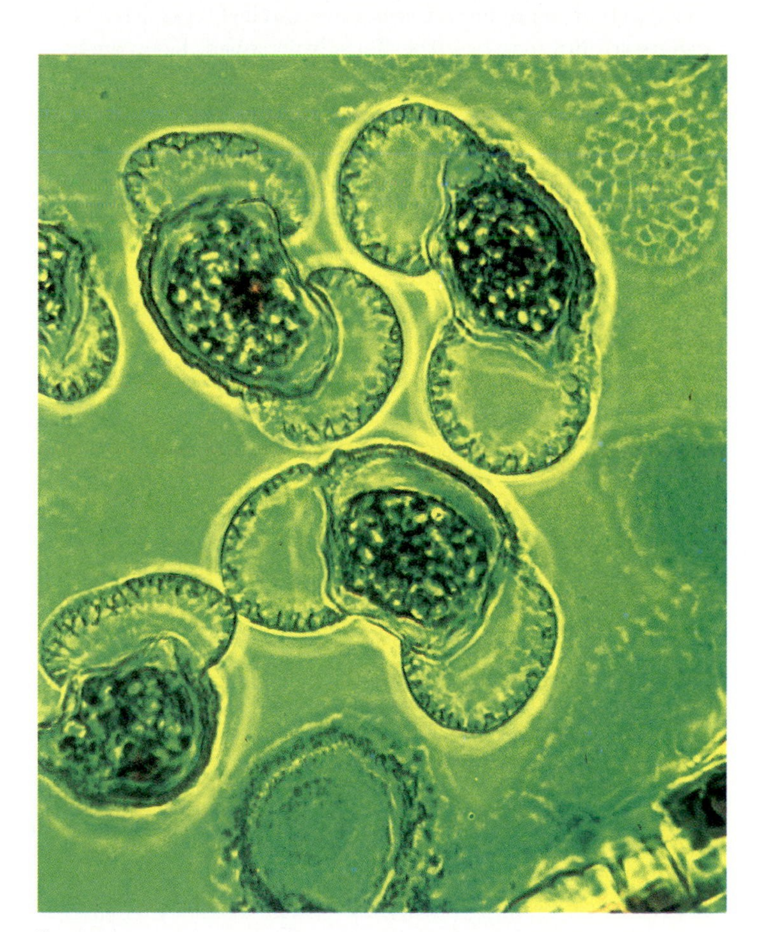

B.

FIGURE 29.3

(a) *Endosporites*, a lycopodophyte spore from the Carboniferous period, has a three-pronged tetrad scar (arrow) that characterizes microspores. (b) In contrast to spores, pollen grains of *Pinus* and other seed plants have no scar.

Features of Gymnosperms

The term *gymnosperm* derives from the Greek word roots *gymnos*, meaning "naked," and *sperma*, meaning "seed."[1] Gymnosperms are plants whose pollen goes directly to ovules (unfertilized seeds) instead of to a stigma (as in the flowering plants), and whose seeds are naked (i.e., are not enclosed in fruits). Thus, by definition, gymnosperms are all fruitless plants. Examples of gymnosperms include maidenhair tree (*Ginkgo*), cycads, conifers, and members of the Gnetophyta (e.g., *Ephedra*, *Gnetum*) (fig. 29.4).

Gymnosperms are characterized by secondary growth that usually forms woody trees or shrubs, but some species are more vinelike. Most gymnosperms lack vessels in their xylem, with the exception of the gnetophytes. In this regard, the gymnosperms are like most of the seedless vascular plants and some primitive angiosperms.

Considering the relatively small number of living gymnosperms (about 720 species in 65 genera), they are remarkably diverse in their reproductive structures and leaf types. Microsporangiate strobili may be either loosely arranged, laden with up to a thousand microsporangia, packed into multiple clusters, or flowerlike (fig. 29.4a–d). Leaf types range from simple, flat blades to needles, compound frondlike leaves, and highly reduced *Equisetum*-like leaves.

Gymnosperms, like angiosperms, differ from the seedless plants in not requiring water for sperm to swim in to reach the egg. Only the cycads and the maidenhair tree have flagellated sperm, but these and other seed plants must be pollinated by wind, animals, or water. This means that the movement of male gametes to female gametes in seed plants relies on airborne transport, not on water transport. Consequently, most gymnosperms produce huge amounts of pollen. For example, each male pine cone annually releases an estimated 1–2 million pollen grains. Similarly, the immense coniferous forests of Sweden have been estimated to release 75,000 tons of pollen every spring.

The most dramatic differences between gymnosperms and other plants involve pollen and seeds and the organs that bear them. These features often differ significantly from those of comparable organs of flowering plants.

The Gymnosperm Pollen Grain

Although pollen evolved by the reduction of the male gametophyte, the pollen grains of gymnosperms still retain vegetative remnants of the ancestral gametophyte thallus. These remnants usually consist of one or two cells, called **prothallial cells,** which often disintegrate before fertilization. The exceptions include *Gnetum* and a few of the Pinophyta, whose pollen grains, like those of angiosperms, have no prothallial cells.

The pollen grains of gymnosperms also apparently contain vestiges of ancestral antheridia. The generative cell divides to form a **stalk cell,** which seems to be all that remains of an antheridial stalk, and a **body cell,** which is the only sperm-producing cell in the male gametophyte. The body cell is essentially a single-celled antheridium. Immediately before fertilization, the body cell divides to form two sperm cells. Among gymnosperms, exceptions to this type of male gametophyte development occur in *Gnetum* and *Welwitschia*, whose generative cells divide directly into sperm cells. The formation of sperm cells directly from generative cells is also a feature of the angiosperms.

The Gymnosperm Seed

The most consistent features shared by all living gymnosperms—features that distinguish this group from seedless plants or other seed plants—involve the seed. Most of the mass of a mature gymnosperm seed consists of an integumentary layer, a multicellular female gametophyte, and one or more embryos; other cells in the seed come from the megasporangium and the megaspore wall. Ovule development begins when a single cell in the megasporangium undergoes meiosis, forming a linear tetrad of haploid megaspores. Three of the spores usually disintegrate, and the remaining **functional megaspore** undergoes repeated mitotic divisions that are not followed immediately by cytokinesis. In *Gnetum*, however, all four megaspore nuclei divide repeatedly inside a single spore wall, as in the tetrasporic embryo-sac development of *Lilium* (see Chapter 17). Either way, the result is a coenocytic stage that is called the **free-nuclear female gametophyte.** The number of free nuclei can be as low as 256 in some species of *Ephedra* to as high as about 8,000 in the maidenhair tree, which is much more than the 8–16 free nuclei in the female gametophytes of angiosperms. Cell walls later form around each nucleus, after which 2 or more (up to 200) archegonia develop at the micropylar end of the ovule. When mature, each archegonium contains a single egg, and all eggs may be fertilized. Archegonia are absent in some of the gnetophytes.

Multiple embryos in the same seed, which occasionally occur in some of the divisions of gymnosperms, form in one of two ways. The less common way, called **simple polyembryony,** occurs when two or more zygotes develop into embryos. Most of the time, all but one of the zygotes fail to develop. The more common origin of multiple embryos occurs when certain cells of a single embryo differentiate into more than one embryo. This mechanism is called **cleavage polyembryony** and produces clonal embryos.

The early stages of embryo development in gymnosperms are characterized by free-nuclear divisions of the zygote nucleus (except perhaps in some gnetophytes). This **free-nuclear embryo** may consist of as few as 4 nuclei in pines to as many as 256 nuclei in cycads. Once the gametophyte becomes cellular, cells near the micropylar end of the embryo elongate into **suspensor cells** in all divisions except the Ginkgophyta.

1. The ancient Greeks went to a *gymnasium* for physical exercise, for which they removed their clothing. We still have gymnasia, but we don't use them in quite the same way as the ancient Greeks did.

A.

C.

D.

FIGURE 29.4

Pollen-bearing cones of gymnosperms. (a) *Ginkgo*; (b) *Zamia*, a cycad; (c) *Pinus*; (d) *Ephedra*.

B.

C O N C E P T

The most significant development leading to the evolution of gymnosperms was the origin of the seed. Seeds probably arose from a megasporangium that became surrounded by an integumentlike layer of tissue with fingerlike projections. Seed-plant pollen arose about 150 million years after the first seeds appeared. Pollen grains evolved from a reduction of the male gametophyte and its more secure retention within the microspore wall.

Seed-Bearing and Microsporangium-Bearing Structures

The simplest seed-bearing structures among gymnosperms are those of the maidenhair tree (*Ginkgo biloba*) and the yew family (*Taxus* and *Torreya*). The seeds of these groups are borne singly at the ends of stalks; *Ginkgo* ovules start out in pairs, but one aborts early in development (fig. 29.5a). The mature ovules of *Taxus* (yews) and *Torreya* (e.g., California nutmeg, *T. californica*) are surrounded by a fleshy, cuplike **aril** that is often bright red (fig. 29.5b). In contrast, most other gymnosperms produce seeds in complex strobili, structures similar to those discussed in Chapter 28. The smallest strobili include those of the junipers (*Juniperus*), which have fleshy scales that are fused into a berry-like structure (fig. 29.5c). The largest seed cones, which may be up to one meter in length and weigh more than 15 kilograms, are produced by cycads (fig. 29.5d).

A.

C.

B.

D.

Figure 29.5

(a) *Ginkgo* seeds attached to tree; (b) yew (*Taxus cuspidata*) seeds. The fleshy bright red coverings are arils; (c) juniper (*Juniperus osteosperma*) "berries"; (d) Seed-bearing cone of a cycad (*Zamia* sp.).

Boxed Reading 29.1

A Fungus That May Save the Yews

In 1989, researchers at the Johns Hopkins University Oncology Center reported that tumors caused by ovarian cancer, which had not responded to traditional therapies (including, in some cases, surgery), had shrunk or disappeared in several patients after treatment with **taxol.** Taxol is a drug obtained from the bark of the Pacific yew (*Taxus brevifolia*), a gymnosperm that grows only in a few areas of the Pacific Northwest and is included among the trees that give shelter to the rare northern spotted owl. Unlike most other cancer drugs, which keep cancer cells from reproducing by damaging their DNA, taxol "freezes" the cancer cells early in the process of cell division. Unable to divide, the cell eventually dies.

Unfortunately, there is far less taxol available than is required to meet the expected need. Pacific yew trees are small and do not occur in extensive stands, and they grow so slowly that they take more than seventy years to attain their full size. Three whole trees produce only enough taxol for a single treatment; a pound of the drug costs $250,000. Furthermore, the extraction of taxol requires removal of the bark, which kills the tree.

Despite its limited availability, scientists are excited about the potential of taxol in the fight against ovarian cancer. Medical science has launched several studies to find

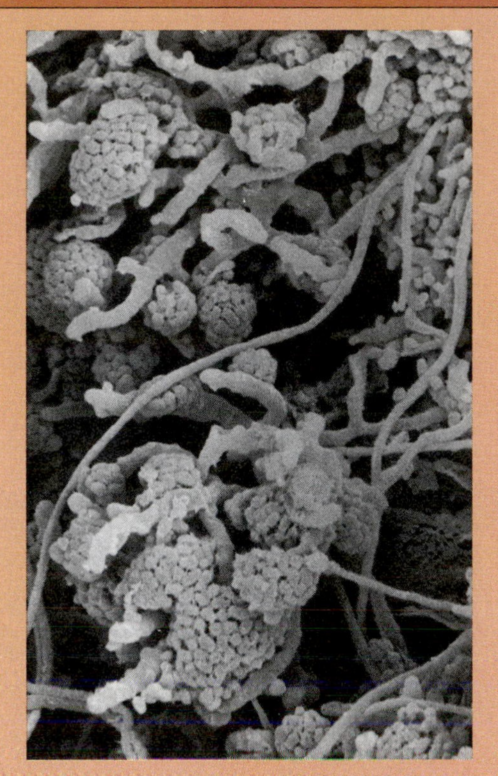

Box Figure 29.1

A scanning electron micrograph of *Taxomyces andreanae*, a recently discovered fungus occurring on bark of the Pacific yew. The fungus produces *taxol*, a cancer-fighting agent, in culture, x2600.

alternative sources of taxol. These include searching for relatives of the yew that may also produce taxol, for economical methods of synthesizing taxol in the laboratory, for ways to grow taxol-producing cell cultures from the plant, and for the genes of taxol biosynthesis, which could be transferred into bacteria for large-scale production of taxol in bacterial culture. Slow but steady progress is being made in all of these efforts, and another, surprising possibility has recently been found. Chemist Andrea Stierle and plant pathologist Gary Strobel of Montana State University have discovered that a fungus that grows in the bark of Pacific yew also contains taxol. More important, the fungus produces taxol in culture, which means that it has its own genetic machinery for making the drug. The fungus, which turned out to be a new genus, was named *Taxomyces andreanae.* In March, 1992, the research team at Montana State University filed a patent application for taxol production by *Taxomyces,* and in January of 1993, the U.S. Food and Drug Administration (FDA) approved taxol—in a record five months—for treating ovarian tumors. Bids for licensing the patent rights for pharmaceutical companies were being solicited in 1993.

The potential for large-scale production of taxol by *Taxomyces* is expected to satisfy additional demand for the drug in treating cancers of the breast, neck, and head. Worldwide, hundreds of thousands of cancer cases may be treatable by taxol. If the Pacific yew were the only source for this drug, the species would be destroyed long before the demand for taxol could be satisfied.

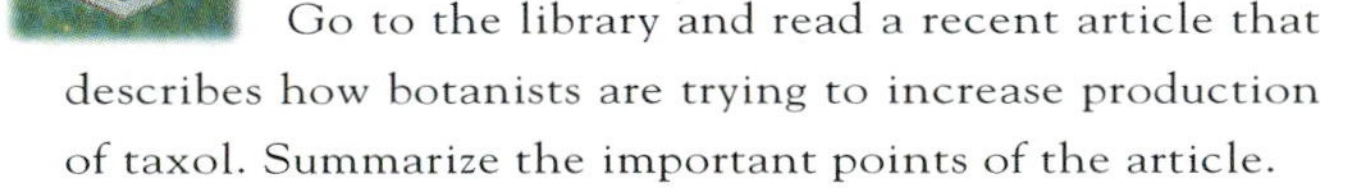

What are Botanists Doing?

Go to the library and read a recent article that describes how botanists are trying to increase production of taxol. Summarize the important points of the article.

Megastrobili show a wide range of complexity, but botanists generally categorize them as either **simple strobili** or **compound strobili.** A simple strobilus consists of an unbranched axis that bears sporophylls, as in *Selaginella* and *Lycopodium* (see Chapter 28). The microstrobili of pines and the megastrobili and microstrobili of cycads are simple (fig. 29.6a). Based on comparative morphology of fossils, current thinking regards the megastrobilus of pines as a reduced branching system. Ovules are borne on reduced shoots that occur in the axils of spirally arranged bracts, which represent branches from the main axis of the cone (fig. 29.6). The entire structure, known as a seed-scale complex, becomes woody at maturity. The compound strobili of *Ephedra* are more reduced (fig. 29.7); the seed cone (megastrobilus) has four to seven pairs of small bracts, but only the uppermost pair bears ovules. Likewise, the compound microstrobilus of *Ephedra* has up to seven pairs of bracts. In addition, microsporangia often occur in clusters on a single stalk (see fig. 29.4d). Botanists view the compound strobili of *Ephedra* as being more like flowers than the strobili of any other gymnosperms. Moreover, the stalked cluster of microsporangia is often referred to as a *stamen.*

Microsporangia form in strobili in all groups of gymnosperms. In *Ginkgo,* microstrobili consist of many paired, antherlike

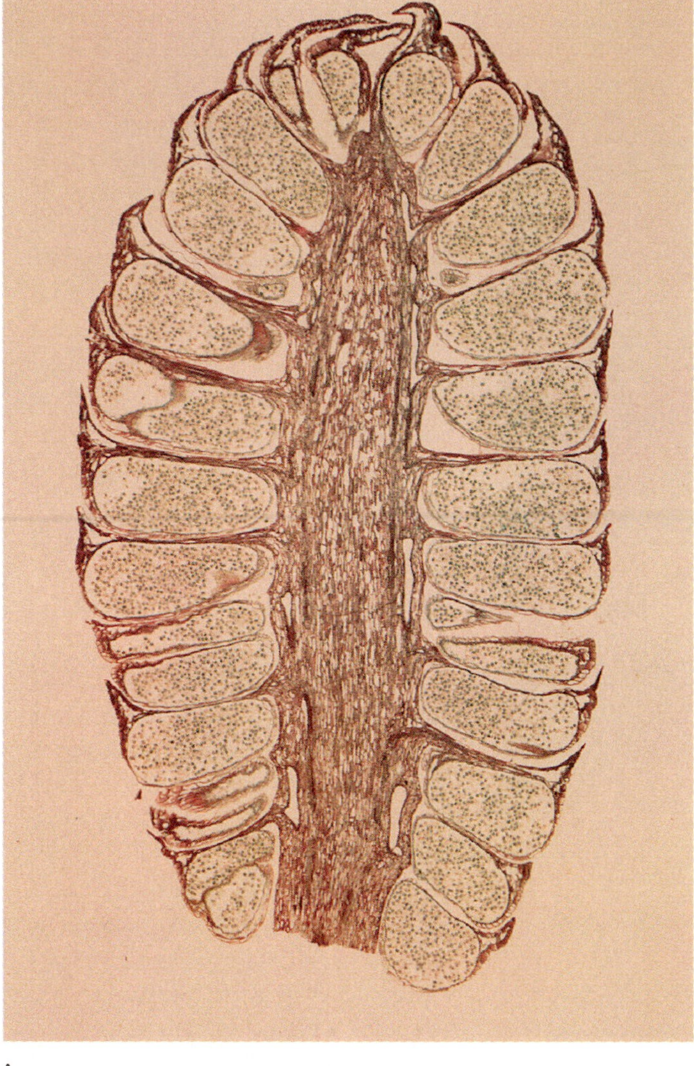

A.

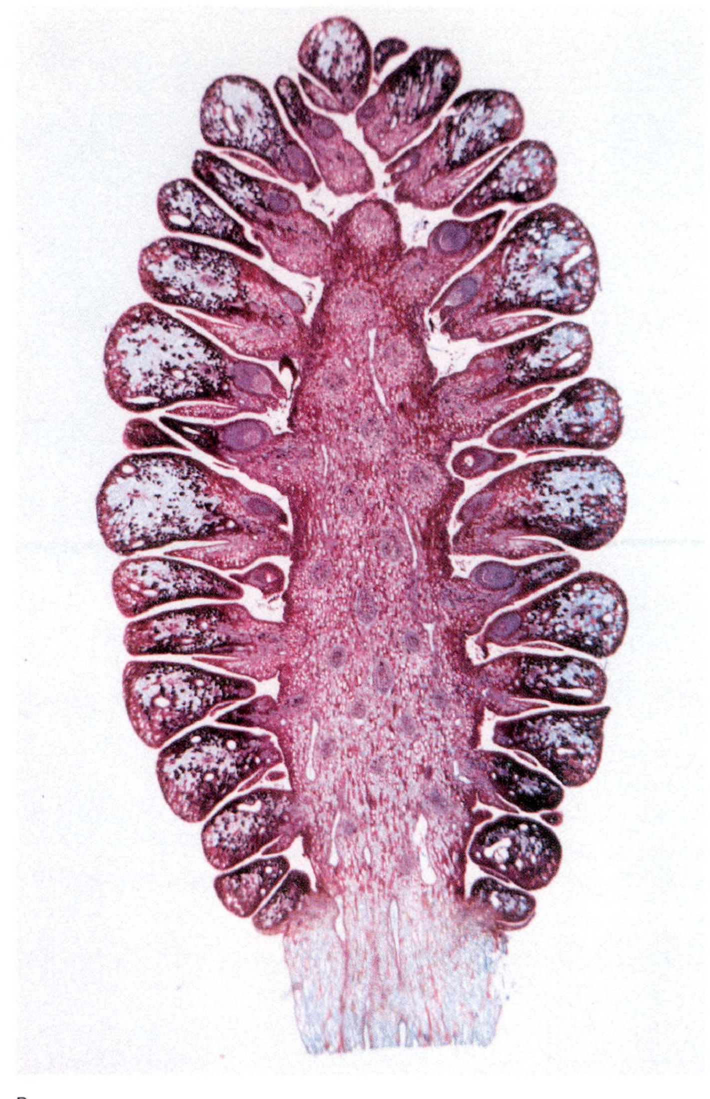

B.

FIGURE 29.6

Pine cones. (a) Longitudinal section through a microstrobilus (pollen-bearing cone). (b) Longitudinal section through a megastrobilus (ovule-bearing cone).

sporangia on short stalks that are arranged in a spiral along a main axis (see fig. 29.4a). Microstrobili are more conelike in other groups, but only the gnetophytes have compound microstrobili. Pine microstrobili are distinguished by having microsporangia borne on the lower **(abaxial)** surface of the sporophylls, instead of on the upper **(adaxial)** surface, as in cycads and in seedless vascular plants (fig. 29.6a). Pollen cones are usually much smaller than seed cones, except in cycads, where the pollen and seed cones may be the same size.

C O N C E P T

All seed plants are heterosporous. Most groups have simple strobili, but some have compound strobili, and others bear seeds singly on individual stalks.

A GENERALIZED LIFE CYCLE OF GYMNOSPERMS

Most of the variation in life cycles among gymnosperms involves the timing of different reproductive events and how long different reproductive stages last. These aspects of gymnosperm life cycles are presented later in this chapter.

The alternation between sporophytic and gametophytic phases in gymnosperms is the same as that in other plants (fig. 29.8). Like the angiosperms, gymnosperms have heterosporous sporophytes, which means that the gametophytes are unisexual. Unlike lycopods and angiosperms, which include species with bisporangiate strobili (e.g., perfect flowers in angiosperms), gymnosperms have only microstrobili and megastrobili. This means that sexual reproduction in gymnosperms

FIGURE 29.7

Seed cones of *Ephedra*.

always requires the transfer of gametophyte material from one strobilus to another.

Pollination in gymnosperms involves a **pollination droplet** that protrudes from the micropyle when pollen grains are being shed (fig. 29.9). This droplet provides a large, sticky surface that catches the normally wind-borne pollen grains of gymnosperms, so that the ovule is more likely to be fertilized. During pollination, dozens of pollen grains may stick to each droplet. After pollination the droplet evaporates and contracts, carrying the pollen grains into the pollen chamber and into contact with the ovule.

THE DIVERSITY OF GYMNOSPERMS

There are considerably fewer gymnosperms than there are angiosperms. Most classifications of gymnosperms include about 65 genera and 720 species. For centuries they were lumped into a single class of seed plants, but botanists now consider them sufficiently diverse to be separated into four extant divisions: Ginkgophyta (maidenhair tree), Cycadophyta (cycads), Pinophyta (conifers), and Gnetophyta (gnetophytes).

Division Ginkgophyta: Maidenhair Tree

Ginkgo, the maidenhair tree, derives its Latin name from two Chinese words meaning "silver apricot." *Ginkgo biloba*, which has remained virtually unchanged for 80 million years, is the only living representative of the division. Its distinctive, fan-shaped leaves with dichotomous venation are produced on two types of shoots: relatively fast-growing long shoots and seedlings produce leaves with a distinct apical notch (hence *biloba*), while slow-growing spur shoots produce leaves without a notch.

Germinating pollen grain
Fertilization
Pollination
Female gametophyte
Archegonium
Egg (n)
Nucellus
Integument
Micropyle
Pollen grain
Tube cell
Spermatogenous cells (n)
Sterile cells
Male gametophyte
2nd sperm nucleus disintegrates
Union of sperm nucle and egg nucleus forr zygote ($2n$)
Pollen grains
Wings
Tube cell
Generative cell
Prothallial cells
Microsporophyll
Meiosis
Tetrad of microspores
Degenerate megaspores
Functional megaspore forms megagametophyte
Ovuliferous scale
Microsporangium
Microsporocytes ($2n$)
Nucellus
Sterile bract
4 megaspores (n)
Integument
Micropyle
Meiosis
Cluster of male cones
Female cone
Megasporangium
Ovuliferous scale
Megasporocyte ($2n$)
Ovule

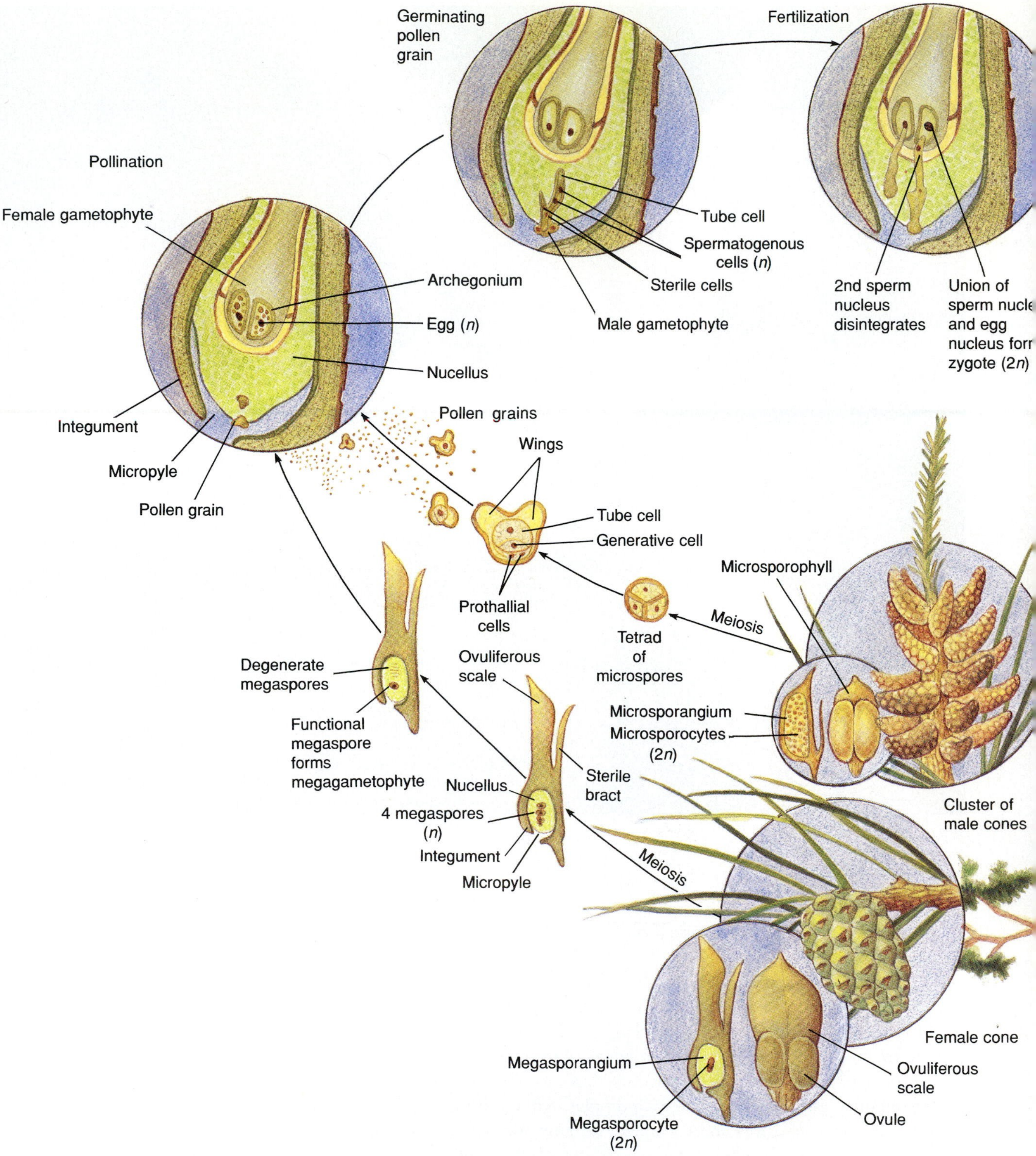

FIGURE 29.8

Life cycle of *Pinus* (pine), a representative gymnosperm.

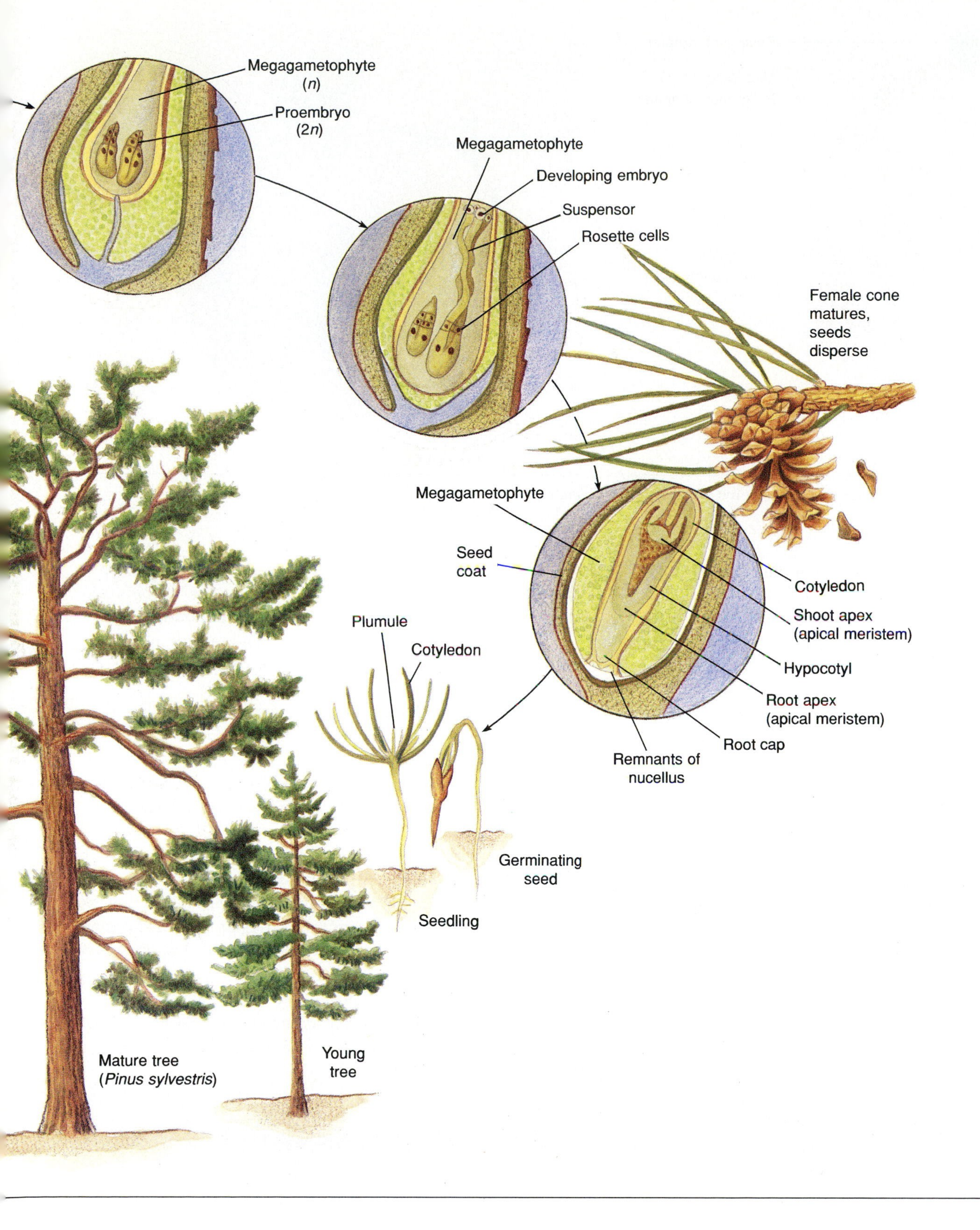
Megagametophyte
(n)
Proembryo
(2n)
Megagametophyte
Developing embryo
Suspensor
Rosette cells
Female cone
matures,
seeds
disperse
Megagametophyte
Seed
coat
Cotyledon
Shoot apex
(apical meristem)
Hypocotyl
Root apex
(apical meristem)
Root cap
Remnants of
nucellus
Plumule
Cotyledon
Germinating
seed
Seedling
Mature tree
(Pinus sylvestris)
Young
tree

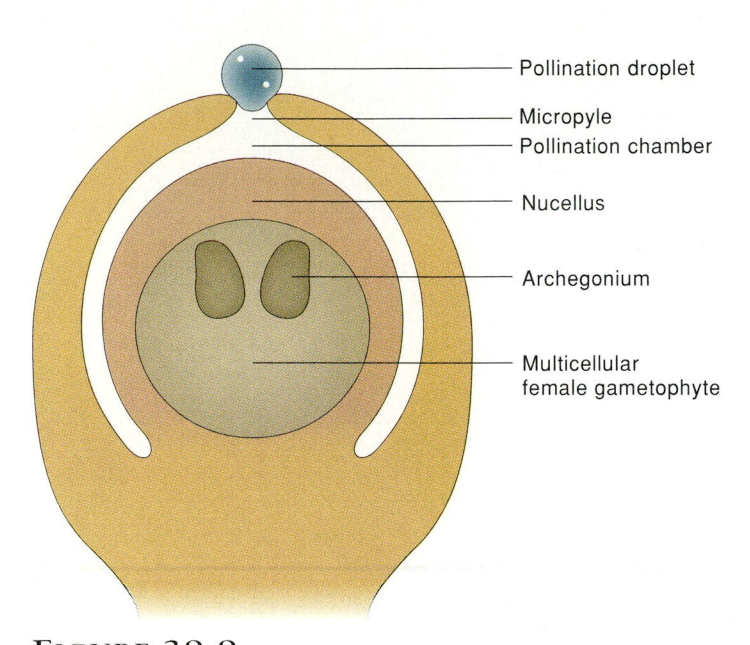

FIGURE 29.9

Diagram of longitudinal section through an ovule of pine (*Pinus*).

FIGURE 29.10

Light micrograph of *Ginkgo* sperm with flagella.

Ginkgo is exclusively dioecious. After the pollination droplet withdraws with its load of pollen, pollen tubes begin to grow by digesting the tissue of the megasporangium. Once the female gametophyte is mature, the egg cells protrude toward the male gametophyte and engulf the sperm cells. Thus, although the sperm cells of *Ginkgo* are flagellated, the flagella are not needed for carrying the sperm to the egg (fig. 29.10). Furthermore, the pollen tubes, which function as digestive filaments, do not convey sperm to egg. This type of pollen tube is referred to as **haustorial,** because it is analogous to the haustorial filaments of parasitic fungi.

The seeds of *Ginkgo* include a massive integument that consists of a fleshy outer layer, a hard and stony middle layer, and an inner layer that is dry and papery. Mature seeds have the size and appearance of small plums, but the fleshy integument has a nauseating odor and irritates the skin of some people. Nevertheless, pickled *Ginkgo* seeds are a delicacy in some parts of Asia.

Ginkgo trees are deciduous, a feature that otherwise occurs only in a few members of the Pinophyta. *Ginkgo* is a popular cultivated tree, but it is apparently extinct in nature. All living *Ginkgo* trees are descendants of plants that were grown in temple gardens of China and Japan, although in the 1950s there were reports of some individuals still living in the wild in eastern China. Much of the genetic diversity of *Ginkgo* has probably been lost in cultivation, because most nurseries propagate only cuttings of microsporangiate trees to avoid the stinky and messy seeds.

Division Cycadophyta: Cycads

There are about 10 genera and 100 species of cycads, distributed primarily in the tropical and subtropical regions of the world. All species of cycads are dioecious. Although they are planted as ornamentals throughout the milder climatic areas of the world, only two species are native to the United States. Both are in the genus *Zamia*, and neither lives in the wild outside of Florida. Many species of cycads are threatened with extinction, and some may soon remain only in cultivation.

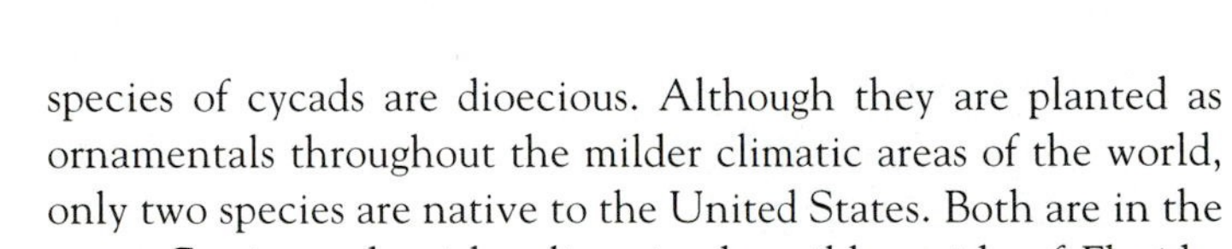

Cycads have palmlike leaves that bear no resemblance to the leaves of other living gymnosperms. Under favorable conditions, cycads usually produce one crown of leaves each year. In some cycads, the roots grow at the surface of the soil and develop nodules containing nitrogen-fixing cyanobacteria.

The strobili of cycads are simple, often with shield-shaped or bractlike sporophylls that may be covered with thick hairs (see fig. 29.5d). The seeds of cycads are more like those of *Ginkgo* than those of any other gymnosperm; they have a three-layered integument, but the inner layer is soft instead of papery.

Also like *Ginkgo*, cycads have flagellated sperm. The sperm cells of cycads are the largest among plants (up to 400 μm in diameter), and can have about 10,000–70,000 spirally arranged flagella. The pollen tubes in cycads are chiefly haustorial, however; the sperm cells are released from the pollen grain and not through the pollen tubes.

Cycads and *Ginkgo* are dioecious and have flagellated sperm, haustorial pollen tubes, and fleshy integuments. Cycads, however, have persistent, palmlike leaves and large strobili, whereas *Ginkgo* has deciduous simple leaves, loose microstrobili, and single-seeded stalks.

BOXED READING 29.2

BIG TREES

Throughout this book, you've read about redwoods (*Sequoia sempervirens*) and giant sequoias (*Sequoiadendron giganteum*). Before the 1840s, the chief admirer of these trees was the chickaree, or Douglas squirrel—a clever-pawed animal that eats the fleshy green scales of sequoia cones. However, the enormous size of these trees soon made them wildly popular after their discovery in 1833. How big are these trees?

- The General Sherman Tree weighs an estimated 12 million pounds.
- The stump of one felled tree, used for a Fourth-of-July cotillion in 1854, accommodated 32 waltzers (half of them in hooped skirts), plus musicians.
- William Waldorf Astor—on a bet—had a dinner table made from a cross section of a redwood tree capable of seating 40 dinner guests.
- The Columbian Exposition of 1893 (in Chicago) displayed "the biggest plank ever sawed"—a piece of coastal redwood 16′5″ wide.

The lumber industry—lured by returns of $1,350 for each $1.25 invested—cleared more than one-third of California's original two million acres of redwood forest between 1850 and 1925. To help preserve the trees, John Muir—the founder of the Sierra Club—ensured federal protection for Sequoia National Park in 1890.* As he said, "Any fool can destroy trees."

* Muir also helped secure federal protection for the Grand Canyon, the Petrified Forest of Arizona, and Yosemite. The 503-acre Muir Woods National Monument (just north of San Francisco) was established in 1907 in his honor.

Division Pinophyta: The Conifers

The informal name of this group, *conifers*, signifies plants that bear cones, even though other divisions of gymnosperms also include cone-bearing species. Members of the genus *Pinus*, considered typical for the Pinophyta, are the most abundant trees in the northern hemisphere; many of the species in coniferous forests are pines. They have also been widely planted in parts of the southern hemisphere, but only the Merkus pine (*Pinus merkusii*), whose distribution barely extends south of the equator in Sumatra, is native south of the equator. Also included in this genus is the bristlecone pine (*P. longaeva*) of the western United States, some of which are the oldest known plants that are not clones (fig. 29.11). One specimen—a tree named Methuselah—is about 4,725 years old; this means that its first sprouts poked out of the stony soil when the Egyptians were building the Great Sphinx at Gizeh. It is 1,000 years older than Stonehenge, 2,000 years older than the Acropolis, and 3,700 years older than Westminster Abbey.

Like *Ginkgo*, pines have short shoots, long shoots, and two kinds of leaves. The more obvious type of leaf is the pine needle, which occurs in groups, called **fascicles,** of generally two to five needles. A few species have as many as eight needles per fascicle, and others have only one. Regardless of the number of needles, a fascicle always forms a cylinder when the leaves are held together. Fascicles are actually short shoots that are surrounded at their base by small, nonphotosynthetic, scale-like leaves that usually fall off after one year of growth. The needle-bearing fascicles are also shed a few at a time, usually every two to five years, so that any pine tree, while appearing evergreen, has a complete change of leaves every five years or less. Bristlecone pines are an exception; their needles last an average of 25–30 years—about as long as the life span of a horse.

Other members of the Pinophyta in the northern hemisphere also have narrow leaves that often have a small point at the tip, but they do not occur in fascicles. These include yews,

FIGURE 29.11

Bristlecone pines (*Pinus longaeva*) are the oldest known plants that are not clones.

firs (*Abies*), larches (*Larix*), spruces (*Picea*), and the coastal redwood (*Sequoia sempervirens*; fig. 29.12a). In contrast, the leaves of cypresses (*Cupressus*) and juniper (*Juniperus*) are scale-like at maturity (fig. 29.12b). In addition, the leaves of some podocarps (*Podocarpus*), araucarias (*Araucaria*), and other

A.

B.

Figure 29.12

Leaves of gymnosperms. (a) Nonfascicled leaves of coastal redwood (*Sequoia sempervirens*). (b) Scalelike leaves of juniper (*Juniperus*). Each branchlet is surrounded by the tiny green leaves.

A.

B.

Figure 29.13

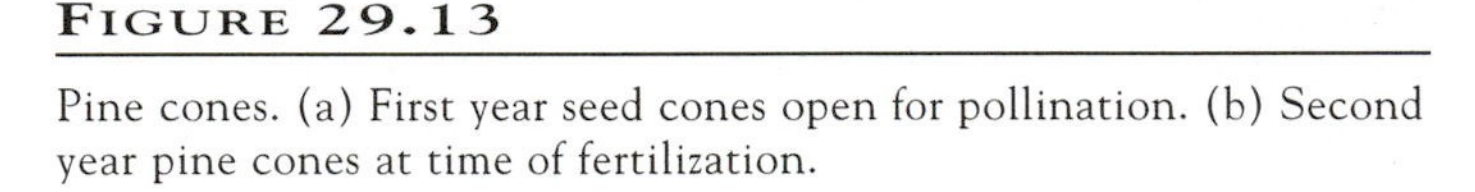

Pine cones. (a) First year seed cones open for pollination. (b) Second year pine cones at time of fertilization.

Pinophyta of the southern hemisphere have flat blades. Like pines, most of these other members of the Pinophyta are evergreen; however, larches, the bald cypress (*Taxodium distichum*), and the dawn redwood (*Metasequoia glyptostroboides*) are deciduous.

Diversity within the Pinophyta is reflected in a wide variety of reproductive structures and in variations of the reproductive cycle. As already noted, seed cones may be absent, as in yews, or have a berrylike appearance, as in junipers. One of the most distinctive variations, however, involves the timing of different reproductive stages in pines. Unlike other gymnosperms, in which pollination, fertilization, and seed maturation occur within the same year, the pines have an extended reproductive cycle. The exact timing varies among species and in different localities.

At the time of pollination, the seed cones are small, and the ovule-bearing scales are slightly separated from one another, enabling pollen grains to reach the pollination droplet (fig. 29.13). After the pollen grains are drawn into the ovule, the micropyle closes, and the seed cones seal up.[2] Meiosis begins in the megasporangium about a month later. The development of the female gametophyte from the functional megaspore is extremely slow, going through an extensive free-nuclear stage and finally becoming cellular about thirteen months after pollination. During this time, pollen tubes digest most of the megasporangial tissue (i.e., as in the cycads and the maidenhair tree). About a month before the female gametophyte becomes fully cellular, the generative cell in each male gametophyte divides and forms a stalk cell and a body cell. Several days before the pollen tube reaches the female gametophyte, the body cell divides into two sperm cells, and each archegonium forms an egg.

2. Romans considered these unopened cones to be symbols of virginity. Virgins, or those who wanted to appear as such, wore garlands of closed evergreen cones at ceremonies.

A.

B.

FIGURE 29.14

Gnetophytes. (a) Leaves and immature seeds of *Gnetum*. (b) *Welwitschia* plants in the Namib desert of southwest Africa.

Although the pollen tube is haustorial in pines, it also transports sperm cells to the female gametophyte. The seed cone is larger and woodier than it was the previous year, but it remains closed while the embryos develop. Following fertilization, the embryo undergoes free-nuclear division, and then becomes cellular and matures before the following winter. By this time, the seed cones have grown into the mature, woody structures that we recognize as pine cones. The time between pollination and seed maturation is usually about eighteen months, but it can range from fourteen to twenty months.

When mature, the cones of most pines become dry, the scales are woody, and they open to release the winged seeds. The cones of some pines do not open so gracefully; these cones often require a fire to cause them to open and release their seeds. The cones of some pines explode like popcorn when heated.

C O N C E P T

Pines are the most abundant species of northern coniferous forests. Many conifers have needlelike leaves that occur in fascicles, while others have broad, flat leaves. Reproduction in pines can require eighteen or more months between pollination and the formation of mature seeds.

Writing to Learn Botany

What might be the adaptive advantage of pine cones that remain closed until heated by fire?

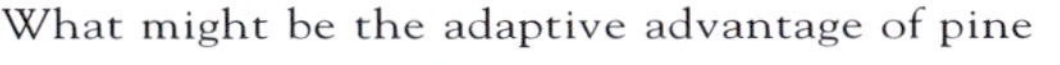

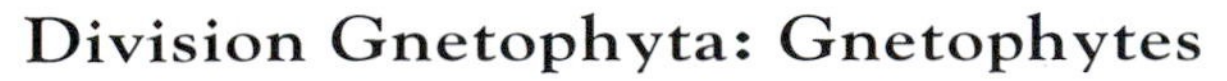

Division Gnetophyta: Gnetophytes

The gnetophytes include some of the most distinctive (if not bizarre) of all seed plants. There are three clearly defined genera and seventy-one species—all woody—that differ from other gymnosperms in that they have vessels in their xylem. These genera are *Ephedra* (forty species), *Gnetum* (thirty species), and *Welwitschia* (one species).

Ephedra, whose species are either monoecious or dioecious, is known as *Mormon tea*, *ma huang*, or *joint fir*. The first two names come from its use as a stimulating or medicinal tea. Ma huang is an Asian species, *E. sinica*, which contains chemicals that are similar to those of human neurotransmitters. The name *joint fir* refers to the leafless appearance of *Ephedra* stems, which resemble the jointed stems of *Equisetum*. *Ephedra* is not leafless, though; its leaves are small and lose their photosynthetic capability as they mature. Photosynthesis in *Ephedra* occurs in green stems.

Members of the genus *Gnetum* inhabit tropical forests. These plants, which are dioecious, are either climbing vines or trees, all with broad, simple leaves similar to those of woody dicots (fig. 29.14a).

Welwitschia mirabilis is the sole living representative of its genus and looks more like something out of science fiction than a real plant (fig. 29.14b). This slow-growing species is confined to the Namib and Mossamedes deserts of southwestern Africa, where most of its moisture is derived from fog that rolls in from the ocean at night. The woody stem, which is concave and bark-encrusted, may be as much as 1.5 meters in diameter and is connected to a large taproot. Mature plants have a pair of large, strap-shaped leaves, which persist throughout the life of the plant. Each leaf has a meristem at its base, which constantly replaces tissue that is lost at its drier, aging tip.

Like many other gnetophytes, *Welwitschia* is dioecious, producing male and female strobili on different plants. Fertilization in *Welwitschia* is unique, in that tubular growths from the eggs grow toward and unite with the pollen tubes; fertilization occurs within these united structures. Reproduction is otherwise similar to that of other gnetophytes.

Boxed Reading 29.3

How Much Paper Do We Need?

Before the 1860s, most newspaper was made from linen or cotton rags (states such as Massachusetts appointed officials to help get rags to mills). Although many people suggested unusual sources for paper (including cabbage, dandelions, and asbestos), the most unusual suggestion was to make paper from rags used to wrap Egyptian mummies. Indeed, because each mummy was wrapped in about 30 pounds of rags, investors calculated that they'd need only 13,500,000 mummies to provide rags for American paper mills. In the 1860s, Augustus Stanwood of Gardiner, Maine—to the dismay of archaeologists—started importing mummies. People in Maine were soon taking their lambchops home in paper made from mummy rags.

When mummies became hard to obtain, investors began looking elsewhere for sources of paper. They quickly settled on wood, especially spruce. Wood pulp was first used to make newspapers in the 1860s, and the first all-wood issue of the New York Times was published on August 23, 1873. Today, . . . Americans use almost 200,000 tons of paper each day, enough to cover 1,350 square miles (an area the size of Long Island). Our uses of paper are endless: food packaging, newspapers, cardboard, toilet paper, and paperback novels are just some of the throw away products that demand a constant supply of wood pulp. Millions of trees are harvested every year to meet this demand. It is only a matter of time before the increasing manufacture of these products outstrips the supply of wood pulp. How can we help? The management of a large American publishing company, in an attempt to find ways of reducing paper consumption in the United States, tried trimming 2.5 centimeters from the width of all rolls of toilet paper in their building. They found that the employees still used the same number of rolls per month as they had previously. From this they calculated that if all rolls of toilet paper in the United States were similarly trimmed, 1 million trees would be saved each year. This is just one example of a simple and painless way to reduce our huge demand for paper goods made from woody trees.

Gnetophytes are unique among gymnosperms because after one of the sperm cells from a male gametophyte fertilizes an egg, the second sperm cell fuses with another cell in the same female gametophyte. Thus, gnetophytes undergo double fertilization, a process otherwise known only in the angiosperms. Unlike double fertilization in angiosperms, however, double fertilization in gnetophytes is not followed by the formation of endosperm. Instead, the diploid cell from fertilization by the second sperm disintegrates.

C O N C E P T

Gnetophytes are the most distinctive group of gymnosperms because of their many similarities with the angiosperms. In various gnetophytes, these include flowerlike compound strobili, vessels in the secondary xylem, loss of archegonia, loss of prothallial cells in the pollen, and double fertilization.

The Physiology and Ecology of Gymnosperms

The physiology and ecology of gymnosperms include the same features described in this text regarding flowering plants. As a smaller group, though, the gymnosperms have a narrower range of features and habitats. Division Pinophyta is dominated by trees and shrubs that are well adapted for temperate or cold climates, especially where free water is scarce for part of the year. Accordingly, northern forests in some areas consist mostly of pines and their relatives, and southern forests consist mostly of araucarias and their relatives. These gymnosperms are adapted to such regions by their sunken stomata, thick cuticle, and tough hypodermis. Some gymnosperms, such as *Ephedra*, *Welwitschia*, and piñon pines, live in deserts. Only a few gymnosperms are tropical; the cycads and *Gnetum* are the main examples.

Doing Botany Yourself

Conifers depend on wind for pollination. How could you estimate the distance that a pollen grain travels?

The Economic Importance of Gymnosperms

The gymnosperms are second only to the angiosperms in their daily impact on human activities and welfare. A detailed account of all they contribute would occupy many volumes, but a few of their main uses are described here.

The greatest economic impact of gymnosperms comes from the use of their wood for making paper and lumber. Indeed, conifers produce about 75% of the world's timber and much of the pulp used to make paper. The chief source of pulpwood for newsprint and other paper in North America is the white spruce (*Picea glauca*). A single midweek issue of a large metropolitan newspaper may use an entire year's growth of 50 hectares of these trees, and that amount may double for weekend editions (see box 29.3, "How Much Paper Do We Need?").

Boxed Reading 29.4

Christmas Trees

Although the origin of the Christmas tree is uncertain (suggestions range from Germans to ancient Druids), today there is no question that Americans consider decorated coniferous Christmas trees to be an important part of the holiday season. Our use of Christmas trees has an interesting history:

- Although the oldest mention of an American Christmas tree dates to 1747, the first Christmas tree didn't appear in the White House until 1856. That tree was decorated by the family of Franklin Pierce, a politician otherwise noted for putting stickum on postage stamps.
- The first commercial ornaments for Christmas trees (i.e., glass icicles and balls) came from Germany. Fake icicles were introduced in 1878.
- The first electric ornaments for Christmas trees were hand-blown light bulbs made in 1882 by Thomas Edison's lab assistants.

Today, people in different parts of the United States have differing preferences for their Christmas trees. Scotch pine is the favorite in the North-central United States; the West prefers Douglas fir; most Californians prefer Monterey pine; Southerners like white pine; and New Englanders prefer balsam fir. There's also a big market for fake trees, the first of which were sold by Sears, Roebuck, and Company in the 1880s. According to their catalog, you could buy a "tree" having 33 "limbs" for only $.50.

Douglas fir (*Pseudotsuga menziesii*), found in the mountains of the west, is not a true fir. In the Pacific Northwest, it grows into giant trees, which are second in size only to the redwoods (see box 29.2 "Big Trees"). It is probably the most desired timber tree in the world today. The wood is strong and relatively free of knots as a result of rapid growth with less branching than in most other conifers; it is heavily used in plywoods and is a major source of large beams. Exploitation for lumber has nearly eliminated old-growth stands, but large numbers of new trees are being grown in managed forests.

Redwoods (*Sequoia* sp.; see photos on pages 354 & 488) are also prized for their wood, which contains substances that inhibit the growth of fungi and bacteria.[3] The wood is light, strong, and soft, but it splits easily. It is used for some types of construction, furniture, posts, greenhouse benches, and for many other purposes.

Spruce wood is especially important to the music industry. The tracheids of spruces have spiral thickenings on the inner walls, which apparently give the wood a resonance that makes it ideal for use as soundboards in violins and related musical instruments.

Important wood products besides lumber and pulp include **resin,** which is the sticky, aromatic substance in the resin canals of conifers. It is a combination of a liquid solvent called **turpentine** and a waxy substance called **rosin.** Both turpentine and rosin are useful products, and a large industry centered in the southern United States and in the south of France is devoted to their extraction and refinement. Turpentine and rosin are often referred to as *naval stores,* a term that originated when the British Royal Navy used large amounts of resin for caulking and sealing their sailing ships and for waterproofing wood, rope, and canvas (most naval stores and a third or more of the lumber used in the United States today come from a group of southern yellow pines, particularly slash pine *P. caribea; P. elliottii*). Pine resin was used by sailors in ancient Greece, Egypt, and Rome. Egyptians sealed their mummy wrappings with pine resin, and the Greeks lined their clay wine vessels with pine resin to prevent leakage. Pine flavoring is still added to Greek wines, giving retsina its distinctive flavor. The unappreciative liken the taste to turpentine.

Turpentine is the premier paint and varnish solvent, and is also used to make deodorants, shaving lotions, drugs, and limonene—the lemon flavoring in lemonade, lemon pudding, and lemon meringue pie. Ballerinas dip their shoes in resin to improve their grip on the stage, and violinists drag their bows across blocks of it to increase friction with the strings. Baseball pitchers use resin to improve their grip on the ball, and batters apply pine tar to the handles of bats to improve their grip. (See box 16.1, "The Bats of Summer: Botany and Our National Pastime.")

The huge kauri pines (*Agathis australis* and *A. robusta;* fig. 29.15a) of New Zealand, which are genetically different from true pines, are the source of a mixture of resins called *dammar.* Dammar is used in high-quality, colorless varnishes, and was also the resin originally used to make linoleum. Dammar, also called *amber,* is the only jewel of plant origin. It comes primarily in fossil form from former or present kauri pine forests, and occurs as lumps of translucent material with a deep orange-yellow tint. These lumps, which weigh up to 45 kilograms, were believed in ancient times to protect the wearer from asthma, rheumatism, and witchcraft. The best amber comes from Russia; however, that supply, which was at its peak at the turn of the century, is now nearing exhaustion. Remarkably lifelike preservations of prehistoric insects in amber still have intact DNA (fig. 29.15b). Other than amber, the most unusual resin product is the nest of the *Dianthidium* bee, which consists of pebbles and pine resin.

CONCEPT

Gymnosperms account for most of the wood that is needed worldwide for lumber and paper. Conifers are the main source of resin and amber.

3. The word "Sequoia" was proposed by Austrian botanist Stephen Erdlicher to commemorate the 18th century Cherokee leader Sequoyah, famed for devising an 83-letter alphabet for the Cherokee language.

A.

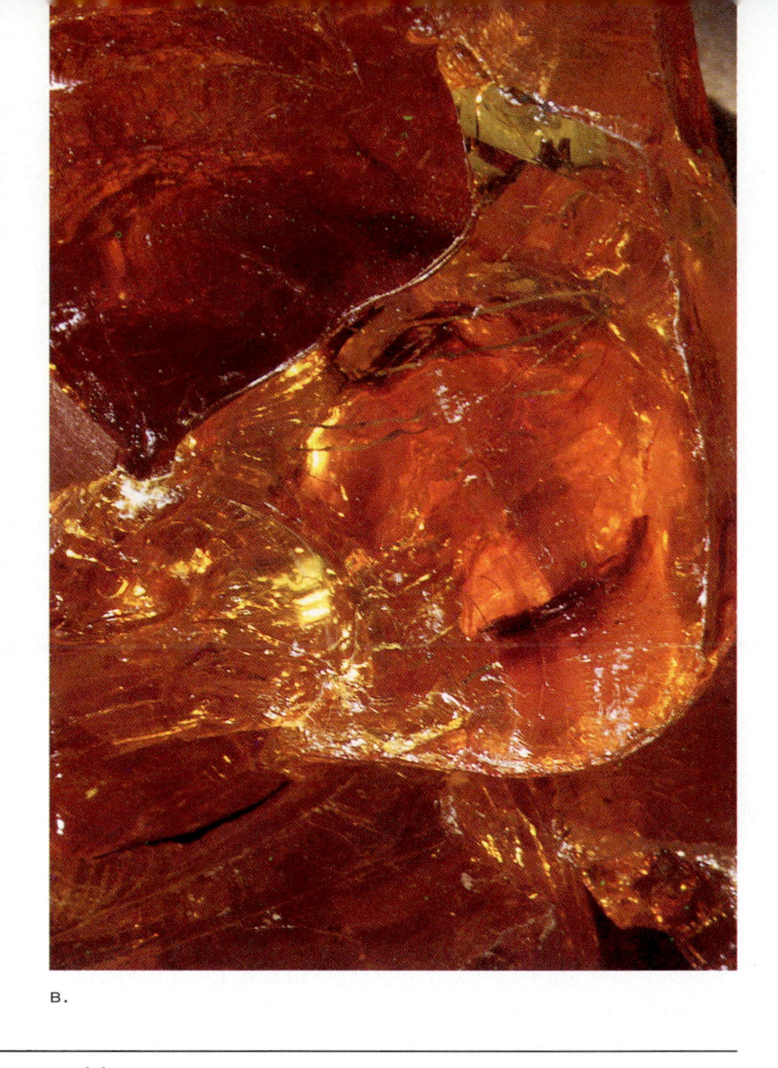

B.

FIGURE 29.15

(a) Kauri pine (*Agathis australis*) of New Zealand. (b) Kauri pine is the source of dammar.

The Lore of Plants

"What the apple is among fruits . . . ," wrote horticulturist Liberty Hyde Bailey, "the pines represent among conifers." This may explain why ten states have chosen the pine as their state tree. Some states (Arkansas and Minnesota, for example) boast a generic pine. The state tree of Alabama is the longleaf pine, Michigan chose the Eastern white pine, Nevada and New Mexico claim the piñon pine, Idaho the Western pine, and Montana the ponderosa pine. The state tree of Maine, the Pine Tree State, is also the Eastern white pine—and its state "flower" is the pine cone.

THE ORIGIN AND HISTORY OF GYMNOSPERMS

Ideas about the ancestry of modern gymnosperms are suggested by the fossil record. The most likely ancestors are the **progymnosperms,** which are classified in Division Progymnospermophyta. The prefix *pro* (first) implies that these plants were not yet gymnosperms, but this depends on how the fossil structure shown in figure 29.1 is interpreted. If it is seen as a pre-seed, or is merely said to be seedlike, then the progymnosperms were not seed plants; however, if it is seen as a primitive seed, then the Progymnospermophyta were the first gymnosperms. Either way, they are the most likely ancestors of the first plants that had seeds.

Some fossils are sufficiently distinctive to be classified in their own divisions, whereas others resemble modern forms closely enough to be classified in one of the extant divisions. All divisions of living gymnosperms have extinct relatives, and two divisions of gymnosperms are entirely extinct: the Pteridospermophyta (seed ferns) and the Cycadeoidophyta (cycadeoids).

Several phylogenetic relationships between extinct and extant gymnosperms have been proposed, based on traditional and modern phylogenetic methods. The following discussion presents an overview of the main features of the most important groups of extinct plants, which is followed by explanations of how past and present gymnosperms might be related.

Division Progymnospermophyta

The progymnosperms may be the oldest seed plants, depending on whether their reproductive structures were seeds (some botanists think they were and others think they were not).

Figure 29.16

Archaeopteris was a woody progymnosperm that grew to more than 8 meters tall; the leaflike blades of this tree were actually flattened branches of simple leaves that were partially or completely webbed together.

Figure 29.17

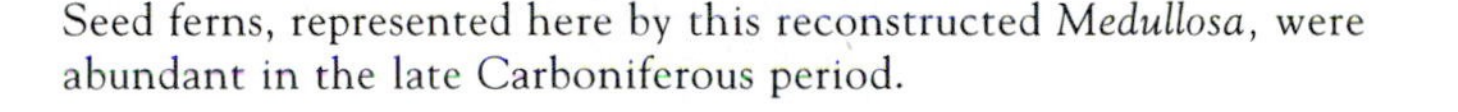

Seed ferns, represented here by this reconstructed *Medullosa*, were abundant in the late Carboniferous period.

Progymnosperms lived from about the middle Devonian to the early Carboniferous periods (ca. 360–310 million years ago) (fig. 29.16). These treelike or shrublike plants had a combination of features that resembled those of the seed plants as well as the seedless vascular plants. Some were heterosporous; they had simple leaves derived from flattened branch systems, and they formed seeds or seedlike reproductive structures. They also had a vascular cambium that could undergo radial divisions. The cambium produced large amounts of secondary xylem and phloem similar to that of modern gymnosperms, and the wood contained long tracheids and bordered pits.

Division Pteridospermophyta

The seed ferns appeared in the late Devonian period. Their leaves were so fernlike that they were originally grouped with the ferns; indeed, it is still not possible to distinguish pteridosperm leaves from fern fronds in the absence of reproductive structures (fig. 29.17). Seed ferns became so abundant during the late Carboniferous that this period became known as the Age of Ferns. In the early part of the twentieth century, however, botanists discovered that the fronds of these fernlike plants bore seeds, which means the Age of Ferns was really the Age of Seed Ferns (fig. 29.18; see also fig. 29.1). Although they are called seed ferns, these plants were not just ferns with seeds. Moreover, they did not evolve from the ferns. Rather, the seed ferns probably evolved from the progymnosperms along an evolutionary line that has no modern descendants.

Like most fossils, the seed ferns are known only from fragments, not from intact plants. In many cases, reproductive structures and vegetative bodies occur in the same strata but not attached to each other. Being appropriately cautious, paleobotanists cannot say which microsporangia or which seeds belonged to which plants. Each organ maintains a separate genus name until it is found attached to a plant in the fossil record. Nevertheless, frondlike fossils are associated with a variety of seed types and reproductive organs that are from plants that would probably be classified into several divisions if they were better known (e.g., fig. 29.19). It is likely, therefore, that the relationships of the seed ferns are more complicated than their classification in a single division would indicate.

Figure 29.18

Seeds were borne on the frondlike leaves of seed ferns, as shown in this drawing of *Emplectopteris*.

Figure 29.19

Reproductive structures of representative seed ferns. (a, b) Seed-bearing cupules of *Lyginopteris;* (c) the probable pollen-bearing organ of *Lyginopteris;* (d) an ovule of *Callistophyton;* (e) the probable pollen-bearing organ of *Callistophyton.*

Division Cycadeoidophyta

Cycadeoids are so named because of their superficial resemblance to cycads (fig. 29.20). Cycadeoids and cycads flourished together in the Jurassic period, which botanists call the Age of Cycads (and zoologists call the Age of Dinosaurs). Although the cycads are still living, the cycadeoids went extinct at the end of the Cretaceous period (ca. 65 million years ago), at about the same time the dinosaurs did. Cycadeoids differ from other gymnosperms, living or extinct, by having bisporangiate strobili, that is, ovules and microsporangia on the same strobilus (fig. 29.21). This situation occurs otherwise only in flowers. The sporangial and vegetative features of cycadeoids traditionally have led botanists to believe that cycadeoids arose from an ancestor among the seed ferns, a lineage generally thought to be unrelated to the flowering plants.

Extinct Plants among Divisions of Living Gymnosperms

The fossil record contains many extinct species and genera of Pinophyta, Cycadophyta, Gnetophyta, and Ginkgophyta. Probably

Figure 29.20

Vegetatively, the cycadeoids such as *Cycadeoidea* were similar to cycads. Internally, like cycads, they had a broad pith, a small amount of xylem, a broad cortex, and a tough outer protective layer formed by persistent leaf bases.

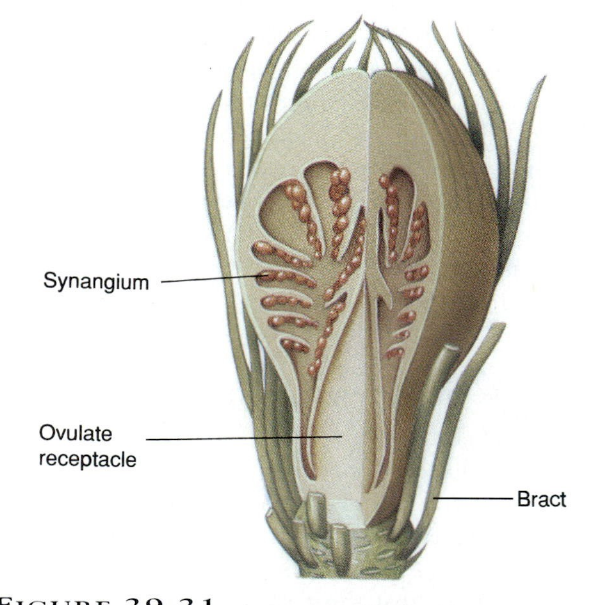

Figure 29.21

Cycadeoid strobilus showing two kinds of sporangia.

the most diverse of these divisions among the extinct gymnosperms is the Pinophyta. This division includes two extinct orders, whereas extinct members of the other divisions all fit into modern orders.

The most abundant plants of the extinct Pinophyta were the cordaites (Order Cordaitales). Cordaites were freely branching trees and shrubs with simple, strap-shaped leaves and compound strobili that bore either seeds or microsporangia. Cordaites were commonly associated with seed ferns in the swamps and lush forests of the Carboniferous period, but they disappeared in the late Permian period (ca. 250 million years ago). At about the time the cordaites were waning, a second order of extinct Pinophyta, the Voltziales, began to flourish. The leaves and branching patterns of Voltziales resembled those of modern araucarias (fig. 29.22). The Voltziales also had compound seed-bearing strobili, but their microstrobili were simple, like those of modern pines.

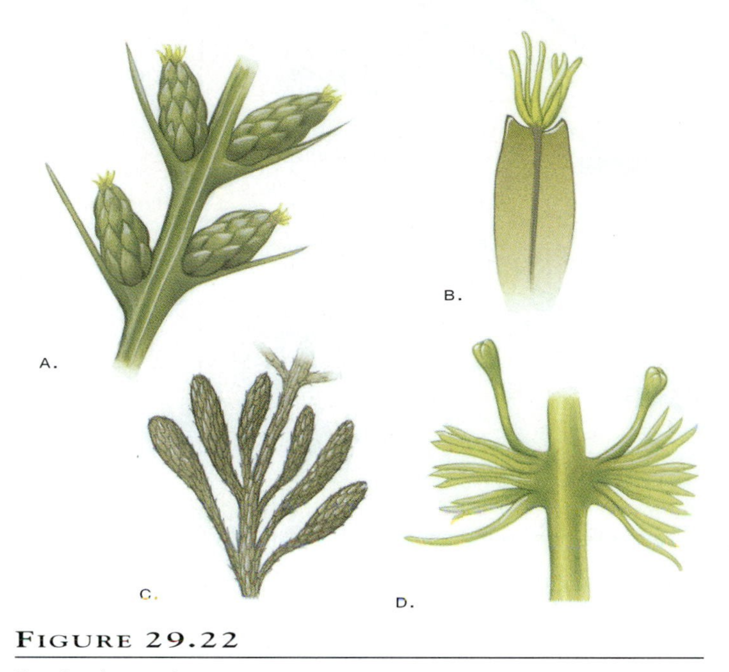

Figure 29.22

Fossilized reproductive structures.

Most groups of gymnosperms are extinct. They included the seed ferns, cycadeoids, at least two orders of the Pinophyta, and probably the progymnosperms. Most ancient cycads and ginkgophytes have also gone extinct.

The Phylogenetic Relationships of Gymnosperms

The phylogeny of gymnosperms interests botanists because of its potential for revealing how reproductive and vegetative structures evolved, as a framework for the interpretation of ambiguous fossil structures, and for answering questions about the origin and evolution of flowering plants. Some of the traditional views about evolutionary relationships of gymnosperms have already been mentioned in this chapter, such as the progymnosperms being ancestral to all seed plants. Many conflicting suggestions about relationships have been made for the gymnosperms, but most have gaps where relationships have not been determined (fig. 29.23). The various proposed phylogenies differ mostly in their interpretations of fossil structures and in the assumed significance of various characters as indicators of ancestor-descendant relationships.

Examples of results from two of the most thorough studies of seed-plant phylogenies are presented as cladograms in figure 29.24. Cladogram A was proposed by Peter Crane (Field Museum of Natural History, Chicago), and cladogram B was proposed by James Doyle (University of California, Davis) and

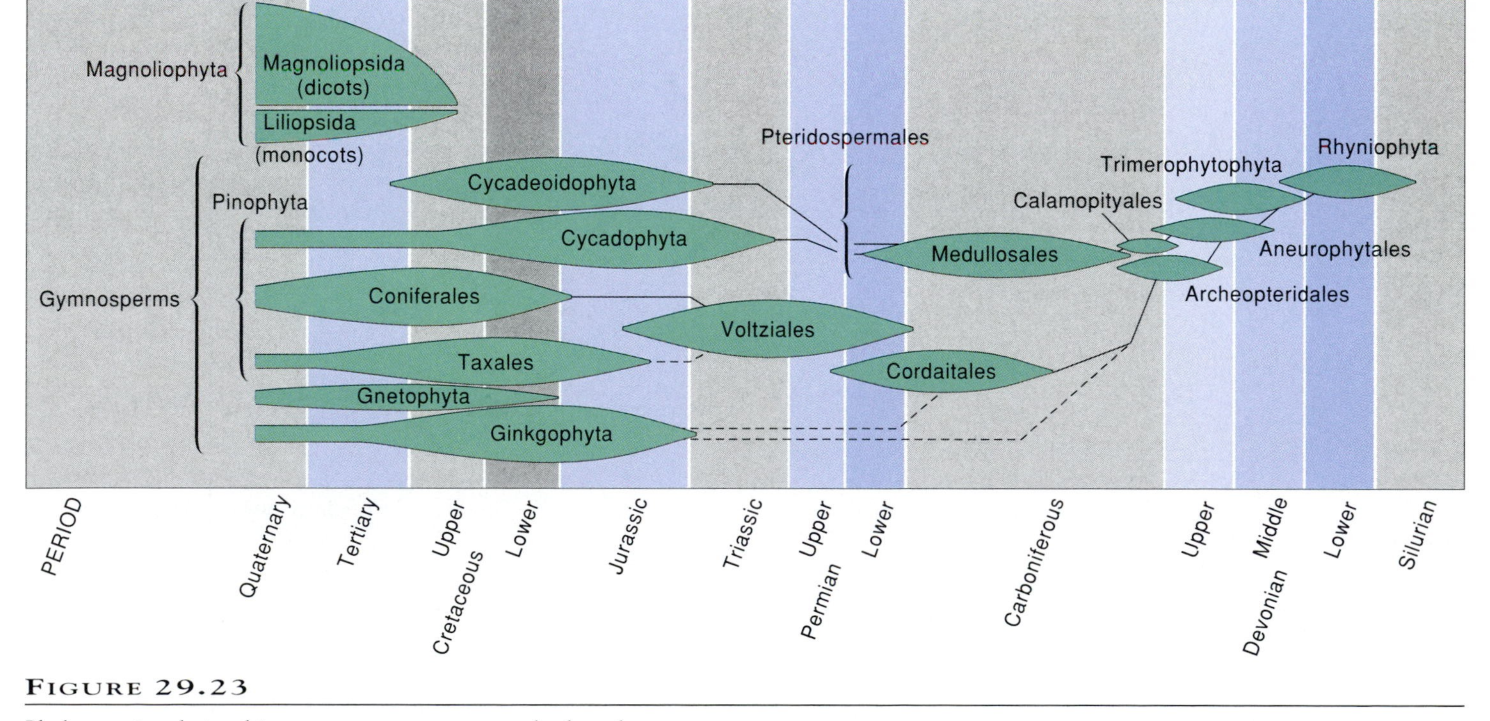

Figure 29.23

Phylogenetic relationships among gymnosperms and other plants.

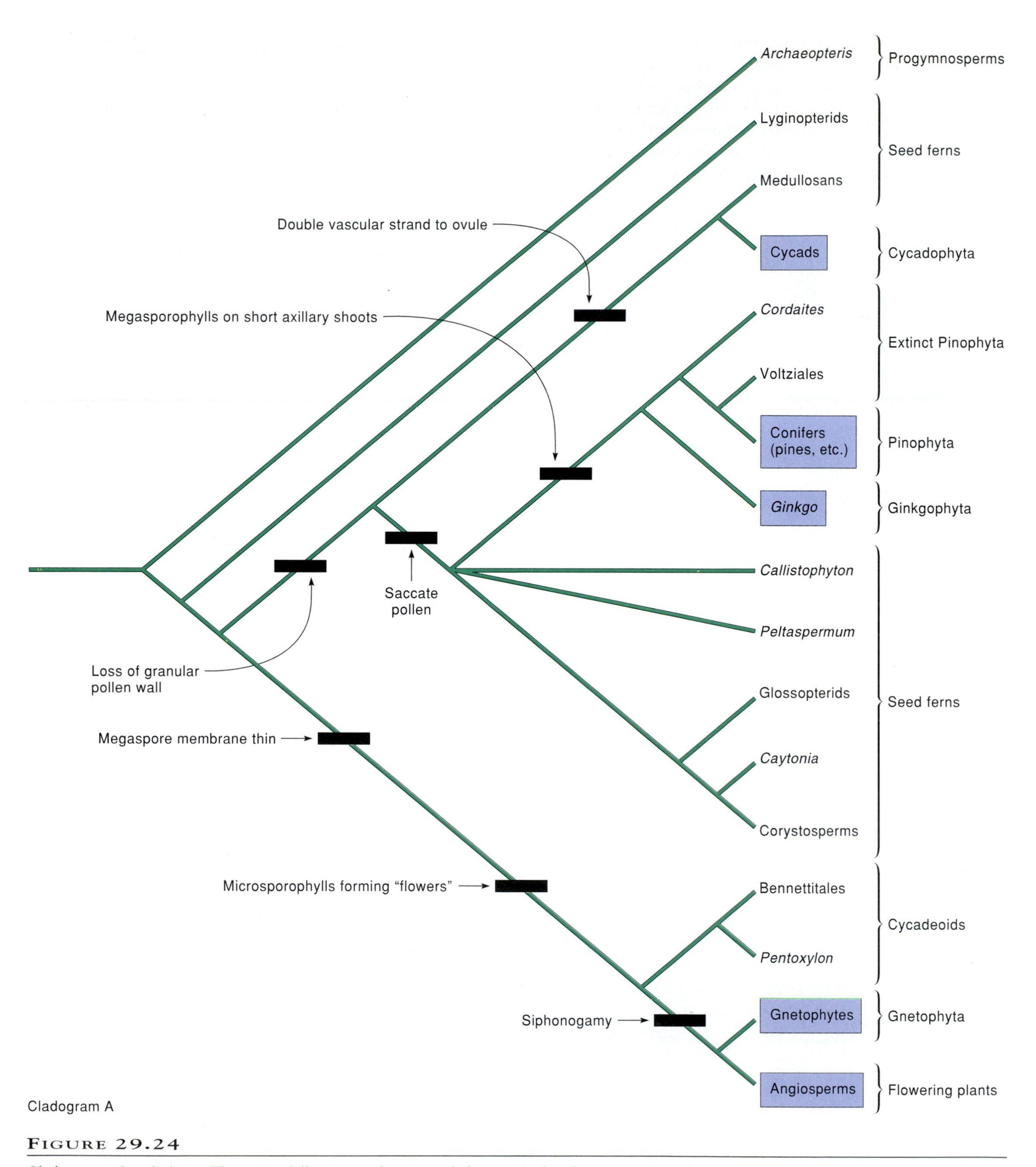

FIGURE 29.24

Cladograms of seed plants. The major difference in these two cladograms is the placement of cycads.

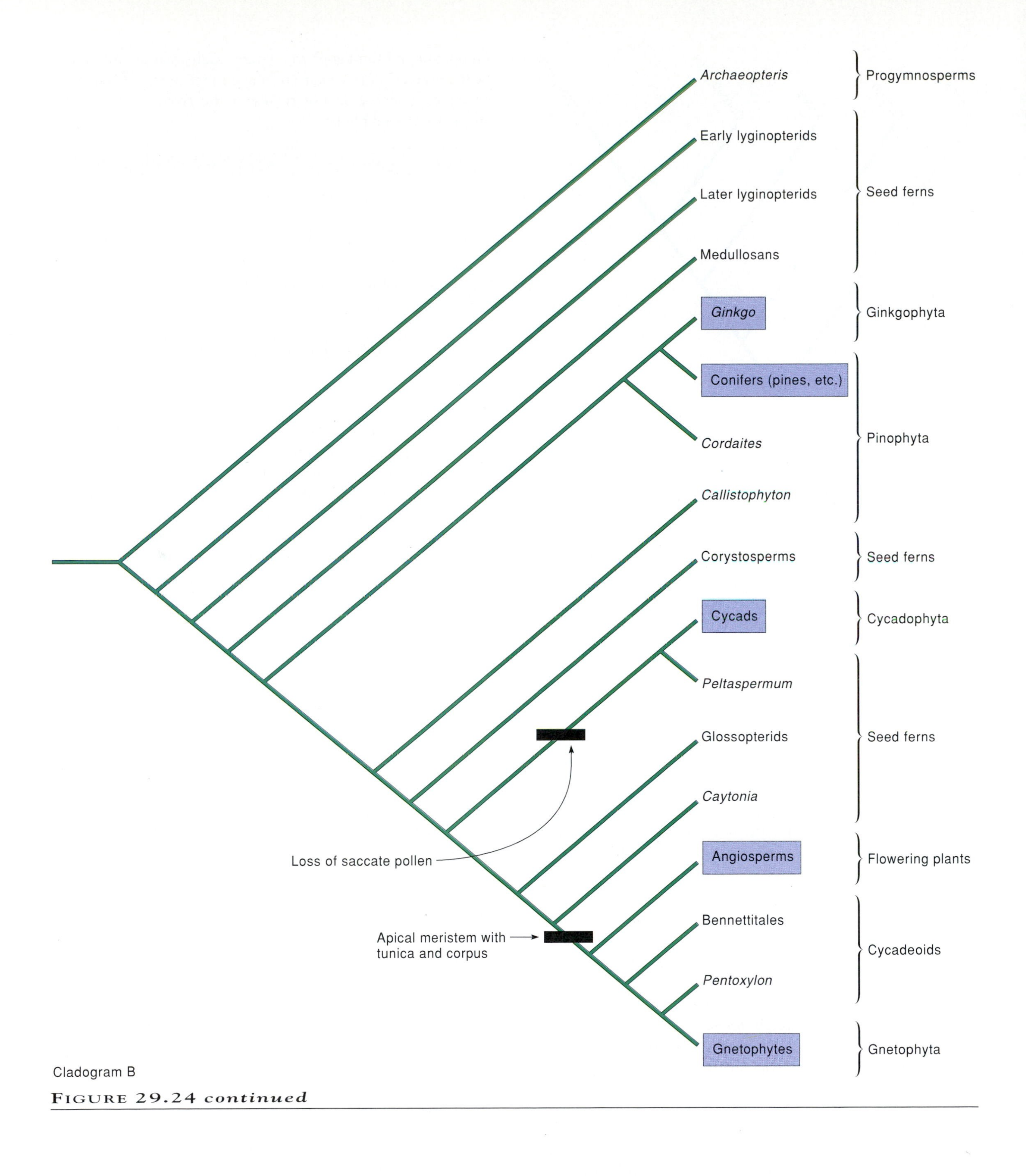
Archaeopteris
Progymnosperms
Early lyginopterids
Later lyginopterids
Medullosans
Seed ferns
Ginkgo
Ginkgophyta
Conifers (pines, etc.)
Cordaites
Callistophyton
Pinophyta
Corystosperms
Seed ferns
Cycads
Cycadophyta
Peltaspermum
Glossopterids
Caytonia
Seed ferns
Angiosperms
Flowering plants
Bennettitales
Pentoxylon
Cycadeoids
Gnetophytes
Gnetophyta
Loss of saccate pollen
Apical meristem with
tunica and corpus

Cladogram B

FIGURE 29.24 *continued*

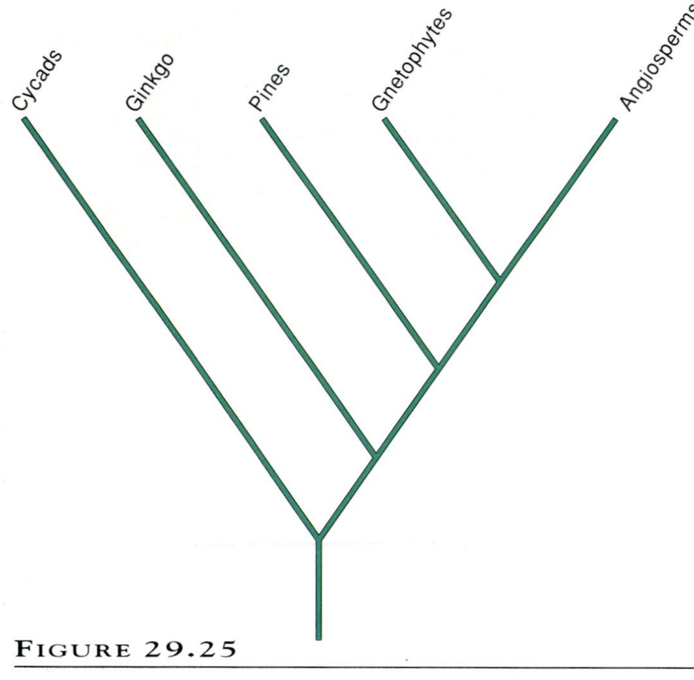

FIGURE 29.25

Cladogram of living seed plants based on *rbc*L/DNA-sequence comparisons.

Michael Donoghue (Harvard University). Both studies used fossils as much as possible, although comparable characters of fossil and living plants are often unknown because of the incompleteness of the fossil record.

The phylogenies shown in figures 29.24a and 29.24b have many similarities, including the ancestral position of the progymnosperms, the wide placement of different seed ferns, the unique common ancestry of Pinophyta and Ginkgophyta, and the close relationships among the cycadeoids, the gnetophytes, and the angiosperms. The main difference between these two cladograms regarding extant divisions is the placement of the cycads. Crane's cladogram shows them to be closer to the medullosan seed ferns, based on a shared derivation of double vascular strands to the ovules. Doyle and Donoghue show cycads as sharing a common ancestry with "higher" seed ferns, based on a loss of saccate pollen (i.e., pollen with air bladders). Other different relationships have also been proposed by these and other botanists. Thus, botanists have not yet settled on what the best phylogeny might be for the seed plants.

At this point it would be nice to have a molecular phylogeny come to the rescue and help resolve the differences among competing cladograms. Unfortunately, DNA has not been preserved in fossils, except for some relatively recent material from about 20 million years ago. Nevertheless, a molecular phylogeny of extant seed plants has been proposed from a comparison of DNA sequences of the chloroplast gene *rbc*L (fig. 29.25). This phylogeny supports a close relationship between gnetophytes and angiosperms. However, it shows the cycads to be near the ancestral root of seed plants, and it fails to support the previously suggested relationship between the Pinophyta and Ginkgophyta. These results may be influenced by the absence of information from extinct species. Cycads, for example, probably do not represent the group closest to the ancestors of seed plants.

C O N C E P T

Various relationships among gymnosperms have been suggested, based on comparisons between living and extinct seed plants. In spite of their differences, they consistently show a close relationship between the gnetophytes and the angiosperms, and a variety of relationships among the seed ferns.

Chapter Summary

Seedlike structures first appeared about 360 million years ago in progymnosperms. These structures consisted of megasporangia surrounded, but not enclosed, by an integumentary layer. Pollen arose more than 150 million years after the origin of seeds. Seed plants diverged and flourished throughout the Carboniferous period as seed ferns; diversification of ancestral Pinophyta and Cycadophyta followed. The Jurassic period was dominated by cycads, cycadeoids, and early members of the Pinophyta. This period is called the Age of Cycads.

Most of the features used for classifying living gymnosperms involve reproductive structures. Most gymnosperms bear seeds and microsporangia on strobili that are either simple or compound. Pollination requires that pollen be transferred from microsporangia to ovules. Ovules are generally exposed (i.e., naked) or temporarily enclosed by sporophylls or other branches from the main axis strobilus.

The Pinophyta, Cycadophyta, Ginkgophyta, and Gnetophyta are four divisions of gymnosperms with living representatives. Pines are abundant in northern coniferous forests and are planted throughout the southern hemisphere. Bristlecone pines are long-lived. Other members of the Pinophyta dominate similar habitats in the southern hemisphere. Pine leaves form on short shoots called fascicles, which usually consist of two to five needles with deciduous, nonphotosynthetic, scalelike leaves at the base. The fascicles themselves are intermittently deciduous, usually lasting for less than five years.

Cycads are slow-growing gymnosperms of warmer climates. Their reproduction is similar to that of the pines, except that the sperm are motile by means of many flagella and their female strobili can be massive. All cycads are dioecious.

There is only one living species of *Ginkgo*, a tree with fan-shaped, dichotomously veined leaves. *Ginkgo* is dioecious and produces seeds that have a nauseating odor.

There are three distinct genera of gnetophytes. *Ephedra* species are monoecious or dioecious, and native to northern drier areas. They are mostly shrubby, and otherwise resemble horsetails in having whorled branches and small, essentially functionless leaves. Their strobili consist of paired bracts, some of which subtend ovules or contain microsporangia. Their life cycle is similar to that of conifers. *Gnetum* species are tropical vines or trees with broad leaves; their reproduction resembles that of *Ephedra*. *Welwitschia* is a bizarre, dioecious plant of southwest African deserts. It produces two large, strap-shaped leaves with basal meristems. The leaves arise from a concave, bark-encrusted, trunkless stem from which a taproot extends into the ground. Strobili form in the leaf axils.

Progymnosperms of the late Paleozoic era appear to have been the progenitors of gymnosperms. Progymnosperms reproduced by means of spores and seedlike organs. Progymnosperms gave rise to seed ferns, which had frondlike leaves but bore seeds. Cycadeoids, which had palmlike leaves like the cycads, were the only gymnosperms to have bisporangiate strobili.

The cordaitales, the earliest conifers, and pteridosperms were abundant during the Carboniferous period. Cordaitalean stems and roots resembled those of modern gymnosperms, and their strobili were simple. They had strap-shaped leaves adapted to the drier conditions that developed in the Permian period.

Several hypotheses have been proposed for the phylogenetic relationships of gymnosperms. These hypotheses differ in the selection of characters and the significance placed on them for inferring ancestor-descendant relationships. Nevertheless, they all show the flowering plants to be closest to the gnetophytes among extant gymnosperms.

Gymnosperms, especially pines and their relatives, are important as a source of lumber, wood pulp, and resin. Resin is used for making turpentine and rosin. Wood pulp is used as raw material for making paper.

Questions for Further Thought and Study

1. The forms of gymnosperm leaves include tiny scales, huge strap-shaped leaves with basal meristems, needles, palmlike leaves, fan-shaped leaves, and simple, broad leaves. Can you think of any functional or ecological significance of the various forms?
2. Cycads, yews, and several other gymnosperms are dioecious; pines are monoecious. Can you think of any survival or other value to the plants in having their reproductive structures in separate cones?
3. In the early evolution of land plants, of what advantage was the development of seeds?
4. Discuss the economic importance of the gymnosperms.
5. How are gymnosperms similar to seedless vascular plants?

Suggested Readings

Articles

Owens, J. N., and V. Hardev. 1990. Sex expression in gymnosperms. *Critical Reviews in Plant Sciences* 9(4):281–294.

Poinar, G. O., Jr. 1993. Still life in amber. *The Sciences* 34(2):34–39.

Stone, R. 1993. Surprise! A fungus factory for taxol? *Science* 260:154–155.

Taylor, T. N. 1982. Reproductive biology in early seed plants. *BioScience* 32:23–28.

Books

Beck, C. B. 1988. *Origin and Evolution of Gymnosperms*. New York: Columbia University Press.

Jones, D. 1993. *Cycads of the World*. Washington, DC: Smithsonian.

Simpson, B. B., and M. Conner-Ogorzaly. 1986. *Economic Botany: Plants in Our World*. New York: McGraw-Hill.

van Geldenen, D. M., and J. R. P. van Hoey Smith. 1989. *Conifers*. Portland: Timber Press.

California poppy (*Eschscholzia californica*) covered with morning dew.

C H A P T E R

30

Angiosperms

Chapter Outline

Chapter Overview

Flowering plants have been the dominant group of plants since the late Cretaceous and early Tertiary periods. Their sudden and abundant appearance in the fossil record has been a mystery since the time of Charles Darwin. To unravel this mystery, botanists have used evidence such as fossils, new ways of analyzing evolutionary relationships, and comparisons of modern molecular data from nucleic acid sequences. Although answers to the questions about when, where, and how the flowering plants arose have not yet been found, botanists have learned much about the evolution of angiosperms in the past ten years.

INTRODUCTION

Flowering plants (angiosperms; Division Magnoliophyta) are the most successful of all plant groups in terms of their diversity. The group includes more than 250,000 species and at least 12,000 genera. Angiosperms live in almost all surface habitats on earth, excluding most marine habitats. Except for coniferous forests and moss-lichen tundras, angiosperms dominate all of the major terrestrial zones of vegetation. Moreover, angiosperms include some of the largest and some of the smallest plants (fig. 30.1); lilies, oak trees, lawn grasses, cacti, broccoli, and magnolias are all flowering plants. Such plants surround us and affect virtually all aspects of our daily lives. It is not surprising that this group of plants attracts the most attention from scientists and the public alike.

Most of the discussions about the genetics, physiology, ecology, structure, and economic importance of plants in this text focus on angiosperms, simply because most of our knowledge of plants is based on the angiosperms. Therefore, this chapter does not include information about these aspects of plant biology. Instead, this chapter presents some of the most interesting and puzzling questions about plants that are not explained in other chapters, including where the angiosperms came from and how they got to be what they are today.

More than a century ago, Charles Darwin referred to the sudden and abundant appearance of angiosperms in the fossil record as "an abominable mystery." In spite of the intense study of angiosperms since that time, we still do not have a completely satisfactory explanation for Darwin's abominable mystery. We do not know where they came from or how they evolved so quickly into such a diverse group. Botanists have used several different kinds of evidence to try to solve some of the mysteries associated with the evolution of flowering plants, including evidence from the fossil record, from the comparative morphology of extant plants, and (more recently) from gene sequencing. The contributions of such evidence to studies of the origin and evolution of flowering plants are the subject of this chapter.

The Fossil Record of Angiosperms

Most plants from the past decomposed without leaving a trace of their existence. Indeed, botanists estimate that the fossil record of plants may be only 1% complete and that as much as 90% of the species that ever existed are extinct. Nevertheless, the fossil record of plants does provide a basis for some general ideas about where flowering plants came from and how they might have evolved.

The first fossils of vascular plants are more than 420 million years old (see Chapter 28), and the first seeds appeared as long as 360 million years ago (see Chapter 29). However, fossils of plant fragments that probably came from angiosperms are not known before the early Cretaceous period, about 135 million years ago. Unfortunately, most of the oldest of these fossils are so fragmented and incomplete that paleobotanists are not certain that they are angiosperms at all. Nevertheless, one particular fossil stands out because it consists of all the parts of a flower attached to a reasonably intact plant. This flowering plant is from a 120-million-year-old fossil deposit near Koonwarra, Australia. Paleobotanists believe that this plant represents the ancestral type of flower. If this is true, then the features shared by the Koonwarra angiosperm and certain modern angiosperms may show which living plants are closest to the ancestral origin of the group.

The Koonwarra Angiosperm

The fossil of the world's earliest known flower was discovered in 1986. At first, it was thought to be a fern, but closer examination showed that the "fern" had carpel-bearing inflorescences (fig. 30.2). The entire plant was less than three centimeters long, and overall it resembled a black pepper plant.

FIGURE 30.1

Angiosperms: (a) *Eucalyptus*, one of the tallest plants. (b) *Wolffia*, the smallest flowering plant. (c) Bunchberry (*Cornus canadensis*).

FIGURE 30.2

(a) The oldest known fossil flower, which is about 120 million years old, was found in the Koonwarra fossil beds near Melbourne, Australia. (b) The entire fossil resembles a small black pepper plant, less than three centimeters long.

The Koonwarra angiosperm had several features that are typical of many modern angiosperms. For example, it had small flowers without petals, a spikelike inflorescence, single-carpel ovaries with short stigmas and no styles, and imperfect flowers with several bracts at their bases. These features occur in present-day members of the lizard's tail family (Saururaceae), the pepper family (Piperaceae), and the chloranthus family (Chloranthaceae), all of which are dicots (fig. 30.3a). In addition, the leaf venation in the Koonwarra angiosperm resembles that of these families as well as the leaf venation of the birthwort family (Aristolochiaceae), the greenbrier family (Smilacaceae), and the yam family (Dioscoreaceae), the latter two being families of monocots (fig. 30.3b).

The Koonwarra angiosperm shows how the ancestor of flowering plants may have looked: a small, rhizome-bearing herb that had secondary growth, small reproductive organs, and simple, imperfect flowers with complexes of bracts at their bases. Families of living plants that share several features with the Koonwarra angiosperm are believed to be primitive members of the dicots and monocots. Furthermore, the appearance of this plant near the apparent beginning of the evolution of angiosperms and its similarity to dicots and monocots suggest that the Koonwarra angiosperm evolved before the divergence between monocots and dicots. This implies that the monocots and dicots separated into two evolutionary lineages less than 120 million years ago, probably from an ancestor similar to the Koonwarra angiosperm. As explained later in this chapter, however, other evidence points to a much earlier evolutionary split between monocots and dicots.

Characteristics of Other Fossil Flowers

During the time that the Koonwarra angiosperm lived, flowering plants were apparently evolving rapidly. Evidence of rapid diversification among angiosperms is based on fossils from different places around the world that show a variety of floral and vegetative features among different plants. These fossils represent how flowering plants may have evolved from their earliest ancestors into the vast array of species that live today. Some of the more significant of these evolutionary developments are described below.

Flower fragments of angiosperms from the early Cretaceous period already show a diversity of floral features. Some fossils apparently have floral parts arranged in a spiral on their axis, as in the flowers and fruits of the modern *Magnolia* (fig. 30.4a). Other fossils have some or all of their floral parts arranged in circles, or **whorls,** around the floral axis, which is the most common arrangement of floral parts among extant angiosperms (fig. 30.4b). Some early fossils have petals and sepals, while others lack a distinct calyx and corolla.

All of the first flowers were radially symmetrical, like poppies and buttercups; and their petals, when present, were free, (i.e., unattached to each other; fig. 30.4c). Flowers with distinct bilateral symmetry, such as those of modern violets and peas, and flowers with fused petals, such as those of cape primose, did not appear until the Paleocene period, less than 65 million years ago (fig. 30.4d–e).

A.

B.

Figure 30.3

(a) Lizard's tail (*Saururus* sp., Saururaceae), showing small apetalous flowers arranged in a spicate inflorescence; (b) leaf of *Smilacina racemosa* (American false spikenard), a member of the lily family. Leaf venation of the Koonwarra angiosperm resembles that of the leaves shown here.

FIGURE 30.4

The earliest flowers had floral parts arranged in spirals (a), as in the carpels of modern magnolia (*Magnolia*), or in whorls (b), as in the petals of the present-day lily (*Lilium*). Many of the oldest fossil flowers also had petal-like sepals, as in the lily, and radial symmetry like the (c) buttercup (*Ranunculus*). More recent flower types include those with bilateral symmetry (d) as in the pansy (*Viola*), and those with petals fused into a tube (e), such as the cape primrose (*Streptocarpus*).

The earliest gynoecia had free carpels, and the first fruits derived from them were apparently follicles or nutlets. Fossils of gynoecia with fused carpels are not as old as those with free carpels. The first gynoecia with fused carpels may have developed into capsules. Fused carpels in more recent fossils show the development of nuts, drupes, berries and pods.

The Koonwarra angiosperm is the earliest intact fossil of a flower. It has features that occur in several modern families of flowering plants. The early evolution of certain floral features is known from fragments of angiosperms, beginning at about the time of the Koonwarra angiosperm and continuing through the Cretaceous period and into the Tertiary period.

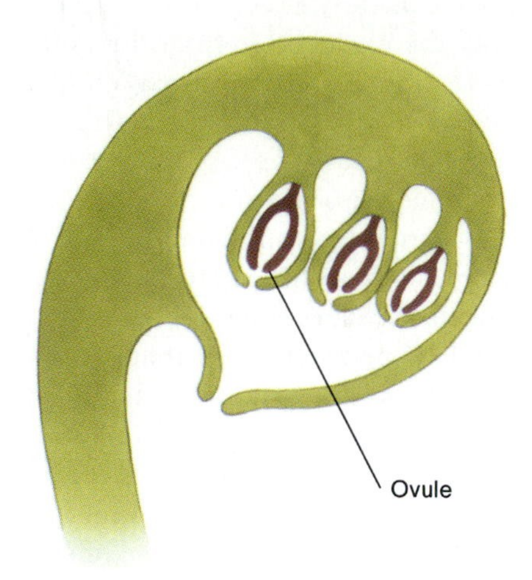

FIGURE 30.5

Diagram of a sectioned *Caytonia* cupule containing three seeds.

The Origin and Radiation of Angiosperms

Although the Koonwarra angiosperm is the oldest known flower, it is probably not the oldest flowering plant. Nobody knows just how long ago the first angiosperm lived, but fossil pollen from the early Cretaceous period, perhaps 10 million years older than the Koonwarra angiosperm, may have come from angiosperms. Pollen has been found that is even older, but it cannot be distinguished from gymnosperm pollen. Thus, we lack direct evidence for the earliest angiosperms. Nevertheless, several suggestions that intertwine the "when, where, and how" of angiosperm evolution have been made from interpretations of the available fossils.

When Did Angiosperms Evolve?

The sudden appearance of a diversity of angiosperms in the Cretaceous period suggests that the evolution of flowering plants began much earlier, perhaps as much as 100 million years before the oldest known angiosperm fossil. If so, then the beginnings of the angiosperms may be found among plants that are believed to be more like the ancestors of angiosperms. Therefore, paleobotanists have often focused on comparing the reproductive features of angiosperms with those of different kinds of nonflowering plants.

Of special importance in explaining when angiosperms evolved is determining when the carpel arose. One hypothesis is that the carpel developed from the cupule of a seed fern like *Caytonia* (fig. 30.5). According to this hypothesis, cupule tissue surrounding the seeds fused to form a closed carpel. Seed ferns were prominent in the Carboniferous period, but few persisted into the Mesozoic era. This means that the carpel, or a precarpel, may have originated as early as 200 million years ago. Another hypothesis is that the carpel arose from the longitudinal folding of an ovule-bearing leaf, that is, from a folded megasporophyll (see Chapter 17). This hypothesis lacks support from the fossil record, however, because there are no good examples of such ancient reproductive structures.

Cycadeoids were once considered to be the ancestors of angiosperms, partly because the microsporangia and ovules of cycadeoids occur in the same cone (see Chapter 29). Such an arrangement simulates perfect flowers, that is, flowers with both stamens and carpels on the same receptacle, which are the most common type of flower in angiosperms.[1] (Recall, however, that the Koonwarra angiosperm had imperfect flowers.) Further evidence for a cycadeoid-angiosperm relationship is that the ovules of cycadeoids are believed to have contained linear tetrads in the megasporangia, which is a characteristic of angiosperms.

More recently, botanists have used the methods of cladistics to show how angiosperms may have descended rather directly from seed-fern ancestors in a line parallel to cycadeoids. If this were the case, then the origin of angiosperms could have occurred in the Triassic period at about the same time the cycadeoids first appeared in the fossil record (see box 30.1, "Angiosperms and Dinosaurs"). This hypothesis is discussed in more detail later in this chapter.

Where Did Angiosperms Evolve?

Botanists believe that pre-Cretaceous angiosperms were well adapted to cool, dry climates. These plants were also probably small, with tough leaves and seed coats, and with vessels in their secondary xylem. Most of them were probably deciduous and thus avoided seasonal drying. These hypotheses represent guesswork based on the fossils of more recent, Cretaceous angiosperms, and if correct, they suggest that the most likely places for angiosperms to have evolved were in the semiarid central regions of Western Gondwanaland (fig. 30.6). Unfortunately, the drier conditions in these upland regions, unlike the wet conditions along shorelines and in lowland basins, did not allow much chance for plants to be preserved in the fossil record.

Plants of lowland basins from the Jurassic and Triassic periods were mostly gymnosperms, but angiosperms apparently began to invade these areas by the early Tertiary period, less than 65 million years ago. The more recent invasion of angiosperms into these lowland areas can be explained by climatic and geologic changes at the end of the Mesozoic era. At that time, the average temperature over the earth dropped as much as 20°C. Also, mountains were forming rapidly, and continents were breaking apart and drifting northward. Although these drastic changes drove some kinds of organisms to extinction, they also provided new opportunities for the migration and evolutionary radiation of angiosperms into lowland habitats. Thus, although the first angiosperms probably evolved in cool, dry climates of uplands sometime before the Cretaceous period, the explosive diversification of flowering plants seems to have occurred much later in Tertiary lowlands.

1. Although the arrangement of microsporangia and ovules on cones simulates a perfect flower, the term *flower* is not used to describe the cones of cycadeoids.

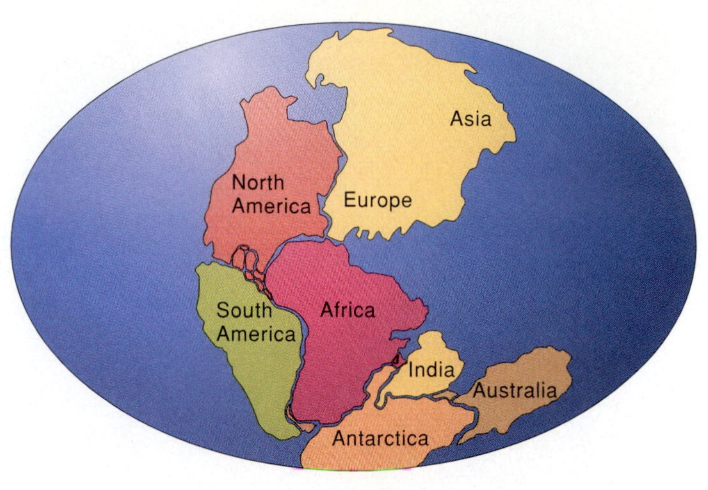

FIGURE 30.6

Positions of the earth's land masses in the early Triassic period (ca. 220 million years ago).

CONCEPT

Although the first fossils of angiosperms are no older than about 135 million years, the angiosperms probably arose much earlier. Indirect evidence from the possible ancestors of angiosperms indicates that they may have originated as much as 200 million years ago. Angiosperm fossils of that age are unknown, probably because they evolved in dry uplands that were not conducive to fossilization. The rapid evolution of angiosperms into the most diverse group of plants on earth occurred much later in the lowlands of the Tertiary period.

How Did Angiosperms Evolve?

Any discussion of how angiosperms evolved must include a topic we have discussed throughout this book: the role of insects. Early in seed-plant evolution, insects became pollen carriers as they searched for food. In turn, plants evolved floral nectar and odors for attracting insects to carry pollen. The earliest, unequivocal angiosperm nectaries are from the late Cretaceous period, but they probably evolved even earlier.

The earliest insects that could have been pollinators were probably beetles. Long before the appearance of angiosperms in the fossil record, cycadeoids were already specialized for pollination by beetles. They had tiny ovules that were protected by scales, which is a feature that is consistent with pollination by insects with chewing mouthparts, such as beetles. By analogy with the cycadeoids, angiosperms that may have lived with the first beetle-pollinated cycadeoids were probably also adapted for pollination by beetles.

Boxed Reading 30.1

Angiosperms and Dinosaurs

Dinosaurs were spectacular animals, some of which reached massive size. They lived throughout the Mesozoic era in the Age of Dinosaurs, or the Age of Cycads (the term preferred by botanists for the same period) and during the rise of angiosperms. Of these three groups of organisms, only the angiosperms remain as a highly diversified and dominant life-form in modern times. The dinosaurs were all extinct by the early Tertiary period, which coincided with the disappearance of the cycadeoids. By this time, the number of different cycads and other gymnosperms had also begun to diminish. Recent studies by paleobotanists and paleontologists have focused on how the gymnosperms, angiosperms, and dinosaurs may have interacted and even influenced their respective evolutionary histories. The main question asked by botanists is, "How did the angiosperms become dominant at the expense of the gymnosperms?" One possible explanation involves adaptation to insect pollination by angiosperms, as already discussed in the text. Another explanation involves dinosaurs.

The largest herbivores ever to walk the earth were the enormous sauropods, such as *Diplodocus, Brachiosaurus,* and *Apatosaurus* (popularly known as Brontosaurus). These animals probably roamed in great herds, browsing on the tops of coniferous trees and other tall gymnosperms. Such browsing would not have killed large trees. Large herbivores did not destroy seedlings by browsing, either, because they could not reach down to them. Thus, gymnosperms could flourish without being destroyed by large herbivores.

The coexistence of large, herbivorous dinosaurs and tall gymnosperms began to disintegrate at the beginning of the Cretaceous period, when the body size of dinosaurs mysteriously began to shrink. By the late Cretaceous period, large herbivores had been replaced by smaller, lower browsers of the ornithischian type (e.g., *Ankylosaurus, Parasaurolophus,* and *Montanoceratops*). These dinosaurs had relatively large, muscular heads with flat, grinding teeth for chewing plant tissues. Although the Jurassic period, when the largest dinosaurs lived, is known as the Age of Dinosaurs, there were actually more kinds of dinosaurs toward the end of the Cretaceous period. The greater diversity of smaller, late Cretaceous dinosaurs may indicate a higher degree of specialization in plant-dinosaur relationships than in earlier periods.

The long-term decrease in body size and increase in diversity of the dinosaurs coincided with the evolutionary radiation of angiosperms and the decline of gymnosperms. Gymnosperms were probably devastated by the low browsers, which would have eaten small seedlings before they were able to reach maturity and produce seeds. Since the first angiosperms were smaller and herbaceous, they probably grew and reproduced more rapidly than woody gymnosperms, and therefore stood a better chance of producing seeds before being eaten. Furthermore, the destruction of gymnosperms by small dinosaurs opened up new habitats for the invasion and evolution of angiosperms.

While the dinosaur-gymnosperm-angiosperm connection seems plausible, the suggestion that angiosperms diversified by adapting to insect pollination is equally so. It may be that each of these explanations is correct and that dinosaurs and insects both influenced the rapid evolution of angiosperms in the Cretaceous period. However, the greatest diversification of angiosperms occurred in the Tertiary period, after the dinosaurs were already extinct.

Insect pollination was not associated with rapid diversification of angiosperms until the appearance of specialized lepidopterans (e.g., butterflies, moths) and hymenopterans (e.g., bees) between the late Cretaceous and early Tertiary periods. The rise to dominance by angiosperms in the Tertiary, therefore, seems to have been greatly influenced by adaptations for pollination by an increasing diversity of flying insects. Thus, insects probably played an important role in the evolution of angiosperms into the largest and most diverse group of plants.

One of the main forces behind the evolutionary radiation of angiosperms was adaptation for pollination by insects. The first insect-pollinated flowers probably arose at the same time beetle-pollinated cycadeoids were alive. Pre-Cretaceous, beetle-pollinated angiosperms gave way to a much greater diversity of angiosperms that evolved by adaptation for pollination by butterflies and bees in the Tertiary period.

Angiosperm Diversity

The angiosperms are such a large group that a full appreciation of their diversity would be a monumental undertaking. Systematists normally specialize in some part of the division, such as a group of related species, a genus, a family, or even a group of families. If estimates are correct that even the brightest botanists can know only about 2,000 kinds of plants in significant detail at one time, then what hope is there of keeping track of more than a quarter of a million species? How can we even know all of the species of just one state, such as California, where more than 6,000 species occur, most of which are angiosperms? Part of the answer is to maintain lists, keys, descriptions, and classifications of plants just as botanists have done since before the time of Linnaeus. Books that contain this kind of information are called **floras.** (Floras also include nonflowering plants, but these are almost always a small minority of the species included, unless the floras are devoted to bryophytes or some other special nonflowering group.) The most recent flora lifornia includes all the relevant information for identifying ... classifying plants native to that state in about 1,400 pages. If this proportion held true for all flowering plants, then at least forty books of that size would be needed to cover the entire division.

A flora has not been completed for all species of flowering plants for at least two centuries, nor is one likely ever to be completed. Most species live in tropical areas, which occur primarily in third-world countries that do not have enough money to finance the work needed to do their floras. Even in the United States, work on an updated flora of North America began only relatively recently. It is expected to be completed by the year 2,000, depending on how fast research budgets allow the work to go. It is a 10–20 year project being done by hundreds of taxonomists.

Another way to know about angiosperm diversity is to study it from the top down, that is, by looking at the highest levels of the taxonomic hierarchy first, then examining lower levels one at a time to get a better picture of the whole division. You have already done this to some extent by learning the differences between monocots and dicots, the only two classes in the division. The next step would be to study the characteristic features of each order and then of each family. However, only a few systematists have proposed comprehensive classifications of the orders and families of flowering plants, and their classifications differ not only in the grouping of genera into families and families into orders but also in their various ideas about the phylogenetic relationships of these groups. Table 30.1 gives you an idea of some of these differences by showing how many taxa are included in each category by different systematists.

The features of most importance in classifying plants in floras, comprehensive treatments of divisions, or in any group of flowering plants are usually those of the reproductive organs. For example, plants that have small flowers with no sepals, five petals fused into either a tubular or strap-shaped corolla, inferior ovaries, five anthers fused into a cylinder around the style, two stigmas, and achenes for fruits are all classified in the sunflower family (Asteraceae) (fig. 30.7). Sometimes, however, vegetative features are also important because of wide variability in reproductive structures or the similarity of such features among families. For instance, the pineapple family (Bromeliaceae) is unique because of its absorptive leaf scales, nearly spherical crystals of silica, twisted stigmas, and base chromosome number ($n = 25$). These features unite such seemingly disparate plants as the pineapple (*Ananas comosus*) and the Spanish moss (*Tillandsia usneoides*) (fig. 30.8).

CONCEPT

There are so many kinds of flowering plants that systematists usually study only a few of them at a time, depending on the number of species in a genus, or the number of genera in a family. Floras may cover thousands of species in a state or region, but there is no such treatment for all of the world's flowering plants, or even for all of the plants in North America. Comprehensive treatments of the flowering plants mostly involve the classification of families and higher taxonomic categories.

TABLE 30.1

Summary of Different Classifications of Angiosperms

Classification	Number of Orders	Number of Families
Cronquist	83	387
Dahlgren	73	403
Takhtajan	92	410
Thorne	69	440

Note: Systematists identify different systems of classification by the names of the principal authors who wrote them. Those summarized in the table are the four most widely accepted classifications today. More information about them appears in Chapter 25.

Doing Botany Yourself

Choose any three species of angiosperms on your campus. How could you find out which of these species are more closely related? That is, which two species evolved from more recent common ancestors that did not give rise to the third species?

Testing Taxonomic Hypotheses: The Case of the Velloziaceae

As discussed in Chapter 24, modern classifications attempt to be phylogenetic, that is, to show natural relationships. This approach to explaining diversity automatically presents hypotheses about relationships, which can be tested by cladistic analysis or by the application of new kinds of data (e.g., gene sequencing). Ideally, systematists look at competing taxonomic hypotheses and find evidence to reject all but the best one. Systematists worldwide use this approach regularly, so there are many examples to choose from to explain how it works. Our example involves a family of angiosperms called the Velloziaceae.

The Velloziaceae, tropical monocots, includes about 250 species in six genera (fig. 30.9). The Dahlgren classification includes the Velloziaceae with families like the pineapple family, the cattail family (Typhaceae), and the ginger family (Zingiberaceae), among others (fig. 30.10a). In contrast, the Cronquist system proposes that the Velloziaceae are more closely related to the families of the order Liliales (e.g., lilies, Liliaceae) (fig. 30.10b). This case was resolved by a 1993 study that was based on a cladistic analysis of DNA sequences from the chloroplast gene *rbc*L. The results were used to reject the relationship proposed in the Dahlgren system. This study also corroborated earlier ideas about the occurrence of fluorescent chemicals in the cell walls of some monocots: fluorescent chemicals, such as ferulic acid, bind to cell-wall polymers in the species of some families but not others (fig. 30.11). They include the pineapple, ginger, and cattail families but not the Velloziaceae. The occurrence of such UV-fluorescent cell walls is now considered to be a shared-derived character-state that shows common ancestry of this group of monocots, excluding the Velloziaceae.

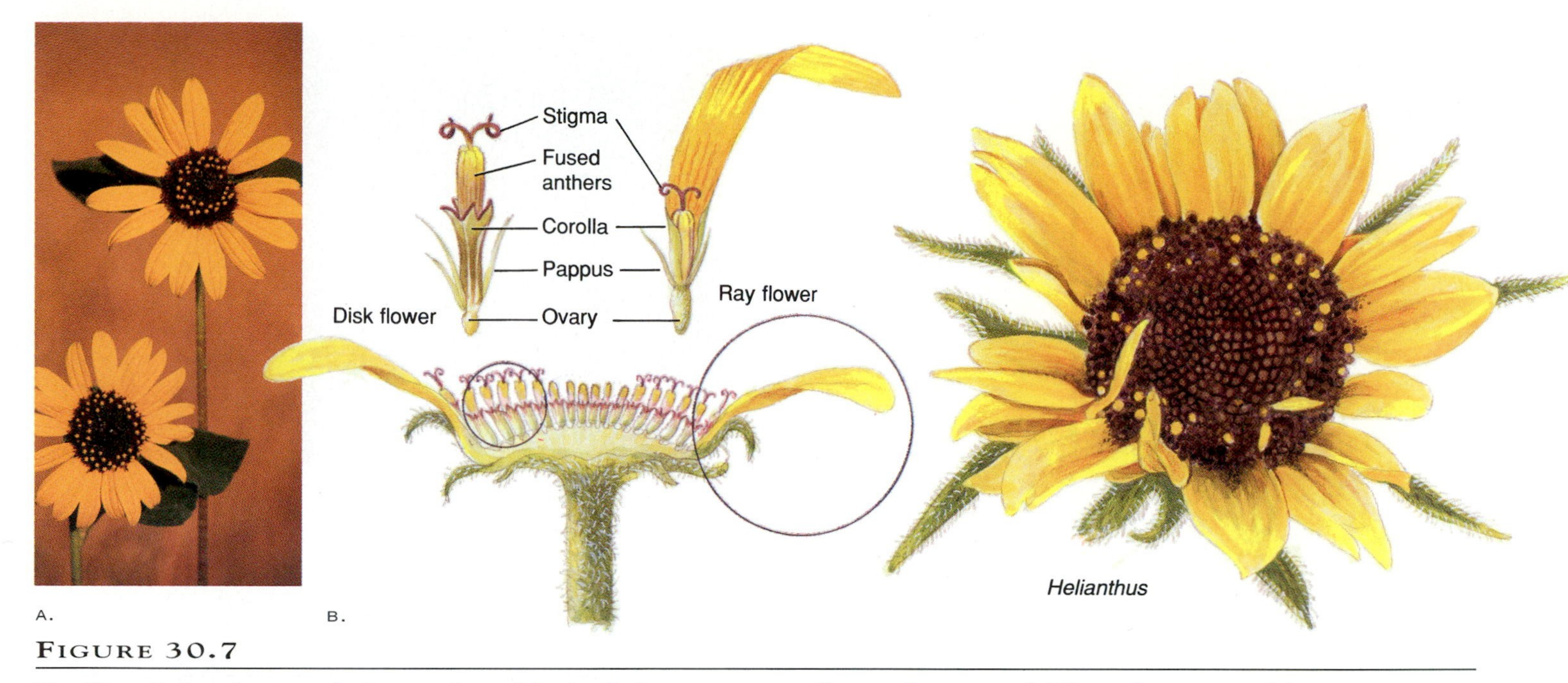

FIGURE 30.7

The "flower" of sunflowers and other members of the family Asteraceae is actually an inflorescence. (a) Two inflorescences of the common sunflower, *Helianthus annuus*. (b) Each sunflower inflorescence is comprised of dozens of each of two kinds of flowers: ray flowers around the perimeter of the inflorescence and disk flowers toward the interior.

FIGURE 30.8

Pineapple plants, showing the stiff, narrow leaves and a pineapple that has developed from inflorescence of 100 or more flowers whose ovaries and other structures have become united into a single, multiple fruit.

FIGURE 30.9

Vellozia is a large genus (ca. 100 species) of tropical Africa and Madagascar.

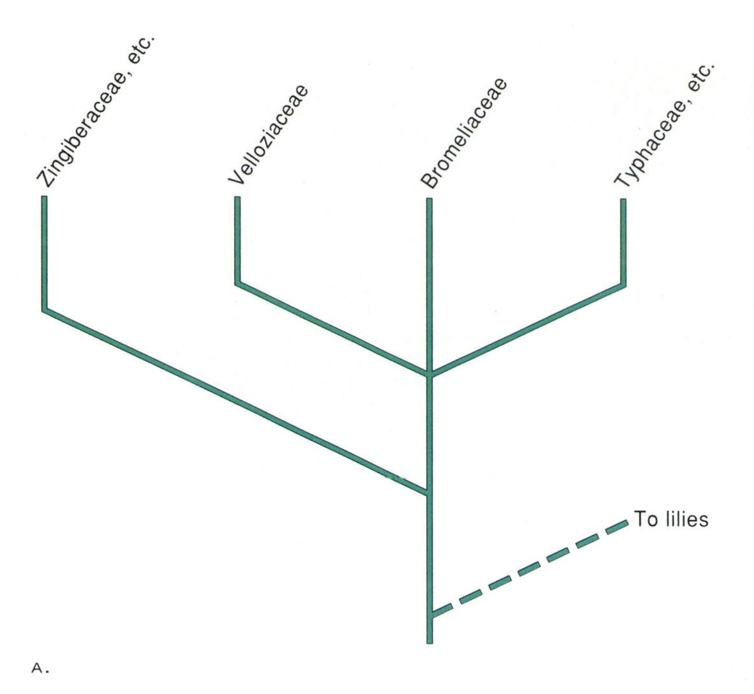

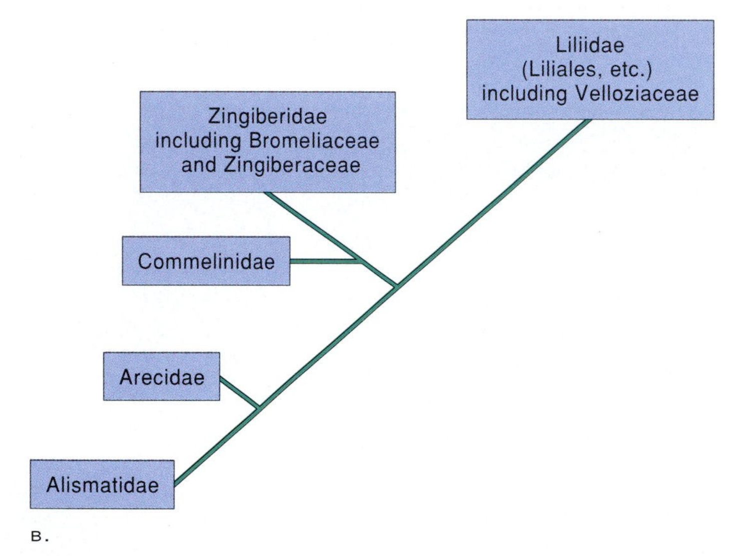

FIGURE 30.10

(a) Proposed relationships of certain monocot families based on the Dahlgren system of classification. In Dahlgren's system, the Velloziaceae are closely related to the Bromeliaceae. (b) According to the classification system of Cronquist, the Velloziaceae are more closely related to lilies than to the Bromiliaceae. (The -idae ending indicates subclasses in the Cronquist system.)

Angiosperm diversity is explained by competing taxonomic hypotheses about phylogenetic relationships in different classifications of the flowering plants. Systematists test these hypotheses by newer methods or by comparisons of DNA sequences or other new kinds of data.

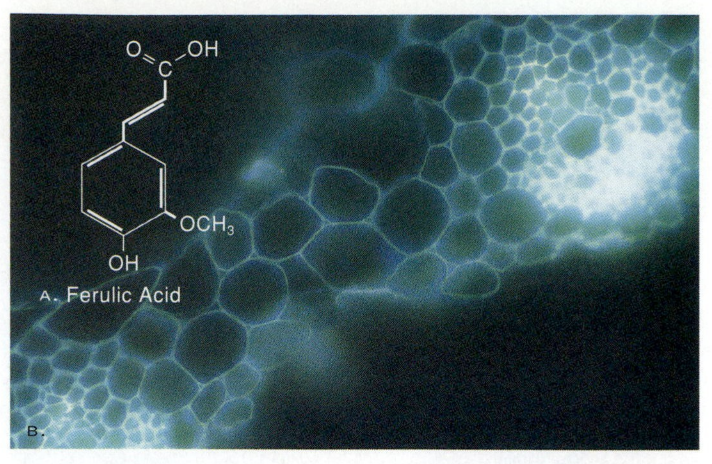

FIGURE 30.11

(a) Chemical structure of ferulic acid. (b) UV-fluorescent cell walls of *Lachnanthes caroliniana*. The linkage of UV-fluorescing compounds to cell wall polymers is now considered to be a shared-derived character state for certain monocots.

THE PHYLOGENETIC RELATIONSHIPS OF ANGIOSPERMS

A phylogenetic analysis of all of the species or genera of flowering plants cannot be done using available computer technology. Furthermore, the relevant taxonomic information from 250,000 species or 12,000 genera for such an analysis could not be obtained in a reasonable period of time, even if hundreds of taxonomists did nothing but gather data. It has recently become possible, however, to construct a complete phylogeny of flowering plants for the four-hundred or so families and their higher categories. More data are needed to complete the job, but as of 1993 a single gene, *rbc*L, had been sequenced for almost five-hundred species of seed plants, including representatives of many families of monocots and dicots and all other divisions of seed plants.

Phylogenies of the Flowering Plants

The gene *rbc*L is not quite 1,500 nucleotides long, which means it can potentially provide 1,500 characters, each with four states (A,T,C,G). We can expect, however, that some of these characters are unchanging due to strong natural selection against their mutation. Conversely, other nucleotide positions, which may not be so important for the activity of the protein, may change often enough to show multiple parallel changes (homoplasies) in the phylogeny of a large set of taxa. The analysis of such data, for example, showed that there are at least 3,900 equally parsimonious cladograms, each with 16,305 character-state changes. The high number of character-state changes indicates that there are many homoplasies.

Cladistic methodology does not provide a way to choose the best cladogram from among thousands of equally good ones, so systematists generally look for common branch-points among

FIGURE 30.12

Simplified molecular phylogeny of flowering plants based on sequences of the chloroplast gene *rbc*L.

FIGURE 30.13

According to a recent analysis based on molecular phylogenies, the hornwort *Ceratophyllum* may be the most primitive living angiosperm.

different phylogenies, from which the strongest suggestions about relationships can be made. The placement of the Velloziaceae, for example, was distant from the Bromeliaceae in all 3,900 cladograms. This provides a strong basis for rejecting the relationships shown by the Dahlgren classification system.

Another consistent and interesting result is the relationship of the monocots to the dicots. Although different versions vary somewhat, all results show the monocots to be embedded among the dicots, which means that the dicots are paraphyletic (fig. 30.12). Monocots seem to have diverged from a common ancestor of the **eudicots,** along an evolutionary line that also gave rise to other kinds of dicots and to the so-called paleoherbs. These other dicots, which include magnolias (Magnoliales) and laurels (Laurales), are therefore more closely related to monocots than to eudicots. The paleoherbs in this analysis include most of the plants that resemble the Koonwarra angiosperm, which have been suggested as the most primitive angiosperms.

While the molecular phylogeny of angiosperms gives some consistent answers about the relationships of monocots, paleoherbs, and dicots, it also poses a new mystery. The most primitive flowering plant in the latest analysis seems to be *Ceratophyllum*, a weedy hornwort that grows in waterways worldwide (fig. 30.13). You may recognize it as a common aquarium plant. Does its position at the base of the flowering plant cladogram indicate that *Ceratophyllum* is more like the first flowering plant than any other extant angiosperm? If so, does this mean that the angiosperms evolved in an aquatic habitat, not in the cool, dry uplands of the late Cretaceous period as previously thought? Answers to these questions are not yet available, but trying to find them will keep botanists busy for quite a while.

C O N C E P T

Phylogenies based on a cladistic analysis of DNA sequences provide a basis for supporting or rejecting competing notions about classification. They also present new puzzles that show the paraphyly of dicots and the potential origin of angiosperms from aquatic ancestors.

Molecular phylogenies of plants are being made more often now from comparisons of DNA sequences. How many such phylogenetic studies can you find in *Biological Abstracts* or other library resources that were published in the past twelve months? What were the top three genes among these studies; that is, which three genes were used the most for inferring phylogenetic relationships?

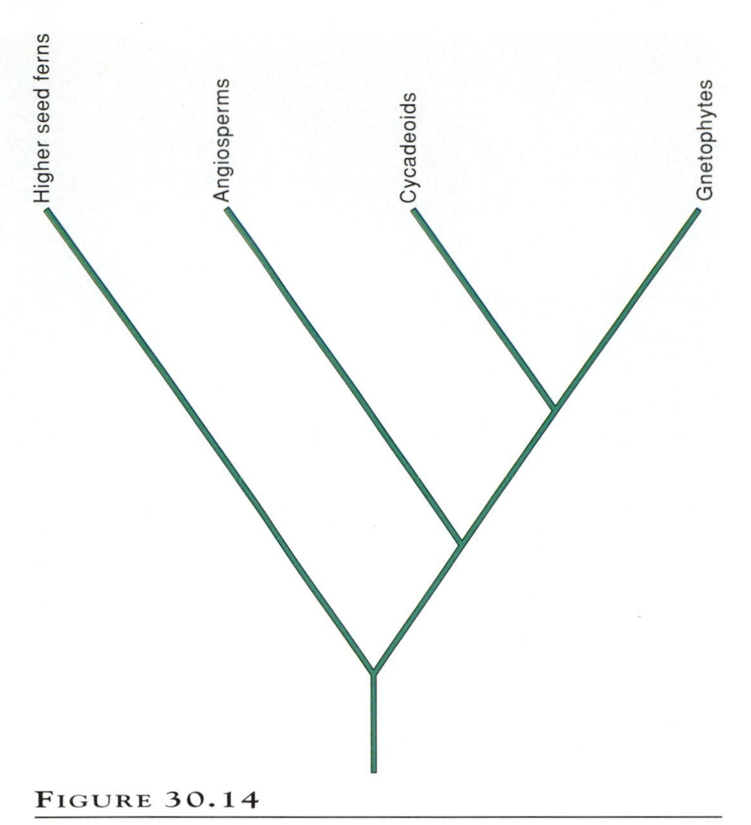

FIGURE 30.14

Simplified phylogeny of these seed plants shows an angiosperm-cycadeoid-gnetophyte evolutionary line arising from the same ancestor that gave rise to the higher seed ferns.

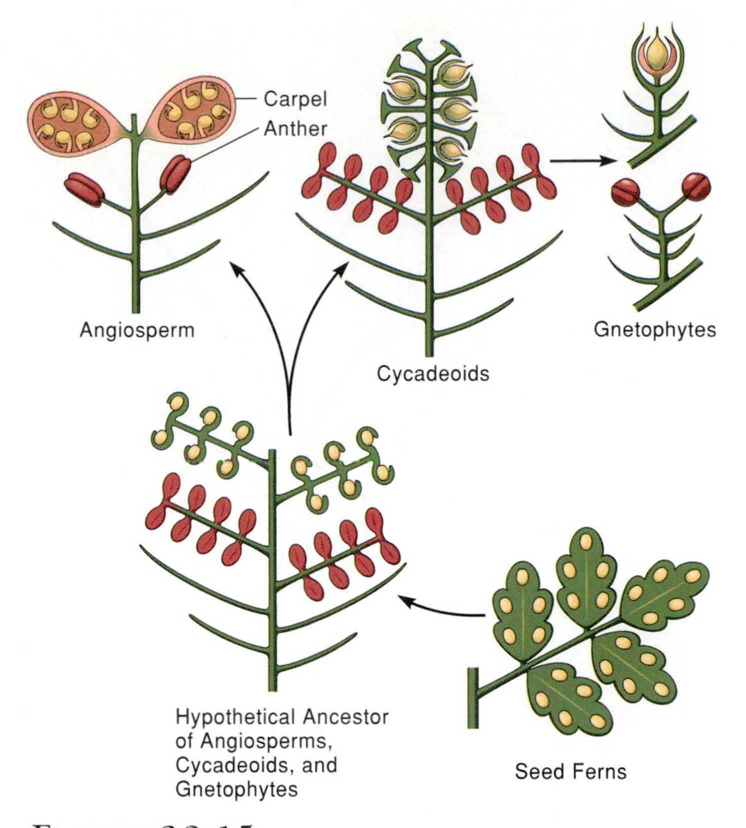

FIGURE 30.15

Proposed origin of flowers.

The Origin of Flowers

If a *Ceratophyllum*-like flower is the most primitive type among living angiosperms, then the ancestral flower may also have had simple, monoecious flowers, numerous stamens, and a simple, single-chambered carpel. How did this type of flower arise? Recent cladistic analyses of all seed plants, past and present, have approached this question by trying to find out which nonangiosperm seed plants evolved from the same ancestor as the angiosperms. Results from one of these studies have been used to generate a hypothesis about the origin of flowers from preflowering ancestors.

The origin of flowers was suggested from the cladistic analysis of seed plants by James Doyle and Michael Donoghue, which was discussed in Chapter 29. A simplified portion of this phylogeny shows the relationships among angiosperms, cycadeoids, and gnetophytes (fig. 30.14). These relationships show that the angiosperm-cycadeoid-gnetophyte line shared ancestor with the higher seed ferns, and that the angiosperms evolved earlier than the cycadeoids or the gnetophytes. Based on this phylogeny, Doyle and Donoghue proposed an evolutionary origin of flowers from a bisporangiate compound strobilus that arose from the fertile leaves of seed ferns (fig. 30.15). Furthermore, this ancestral strobilus also gave rise to the bisporangiate strobili of cycadeoids, and later to the monosporangiate strobili of gnetophytes. Unfortunately, there are no fossils of such a preangiosperm strobilus to provide evidence for this hypothesis. Nevertheless, indirect evidence that angiosperms, cycadeoids, and gnetophytes evolved from a common ancestor with bisporangiate strobili comes from *Welwitschia mirabilis*. The pollen come of this species has an abortive ovule, which suggest that gnetophytes could originally have been monoecious.

In addition to providing ideas about the origin of flowers, the phylogeny shown in figure 30.15 also gives us a clue as to how old the angiosperms are. If the origin of angiosperms occurred before the cycadeoids appeared in the fossil record, as the cladogram indicates, then flowering plants existed in the late Triassic period, at least 200 million years ago. This suggestion matches that of paleobotanists who have long thought that the angiosperms and cycadeoids arose at about the same time.

C O N C E P T

The most parsimonious evolutionary tree showing the origin of angiosperms suggests that they evolved before cycadeoids and gnetophytes. This phylogeny has been interpreted to mean that flowers arose from an extinct, bisporangiate ancestor that also gave rise to cycadeoids and gnetophytes. Considering the age of cycadeoids in the fossil record, this phylogeny also means that the angiosperms arose at least 200 million years ago.

Dating the Monocot-Dicot Divergence

The **molecular clock hypothesis,** first mentioned in Chapter 11, states that the amount of dissimilarity between two nucleic acid sequences should be proportional to the amount of time through which they have evolved from their common ancestral gene. A nucleic acid sequence can therefore be viewed as a clock: the rate of mutational change in the sequence is like the rate at which the clock ticks. Thus, the "age" of *rbc*L sequences, for example, can be calculated by comparing differences among sequences of various plants to see how long the clock has been ticking. Several biologists have attempted to use this principle to determine when the monocots and dicots evolved from their common ancestor. The best estimate, which was based on comparisons of sequences from *rbc*L and eleven other chloroplast genes, suggests that this divergence occurred approximately 200 million years ago. A different estimate, which was based on comparisons of sequences from two nuclear genes, suggests that the monocot-dicot divergence may have occurred as much as 320 million years ago.

That the monocots and dicots could have diverged as long ago as 320 million years is incomprehensible to paleobotanists, but the history of scientific discovery abounds with surprises that eventually became accepted. However, the estimate of 200 million years is more consistent with the time of origin of angiosperms according to the seed-plant phylogeny proposed by Doyle and Donoghue, and it is currently considered to be the better one.

The Lore of Plants

The informal competition among scientists to find the oldest fossil DNA seemed to be won in 1990, when 17-million-year-old DNA was found in fossils of *Magnolia* leaves. In 1992, however, the record was broken by 30-million-year-old DNA from insects that had been preserved in amber. Even this record may be shattered soon, though, if a report of DNA in 100-million-year-old fossil plant from Nebraska can be confirmed.

Writing to Learn Botany

What explanations can you give for the 80-million-year discrepancy between the age of the oldest fossil flower (120 million years) and the age of the monocot-dicot divergence based on the estimated mutational rates of *rbc*L (200 million years)?

Chapter Summary

The angiosperms, which are the most dominant plants on earth, have a flower that includes seeds in a carpel. Fossils of carpels and other parts of flowers are known from Cretaceous deposits that are at least 135 million years old, but the first complete fossil flower is about 120 million years old. This flower resembles several plants in different families of monocots and dicots, on the basis of simple features of the flower and its herbaceous type of body.

The main force behind the rapid evolutionary radiation of angiosperms may have been pollination by insects and the availability of habitats left open by the disappearance of many gymnosperms. The first flowers were probably pollinated by beetles; later angiosperms attracted butterflies and bees.

Another hypothesis that explains the rapid evolution of angiosperms involves the influence of dinosaurs. Large, high-browsing dinosaurs mysteriously gave way to smaller, low-browsing dinosaurs by the end of the Mesozoic era. The low-browsing dinosaurs probably devastated gymnosperms by eating young seedlings, but allowed low-growing, herbaceous angiosperms to survive and diversify.

The diversity of angiosperms is so great that most of the knowledge about this group is divided into floras or taxonomic treatments of smaller subsets of taxa. Traditional comprehensive classifications are created primarily by organizing families and higher levels of the taxonomic hierarchy. Modern methods of cladistic analysis and new types of data, such as gene sequences, are now being used to evaluate competing hypotheses from different traditional classifications.

Estimates of rates of mutational change in gene sequences show how old some genes may be and, by inference, how old different plant groups may be. The molecular clocks of dicots and monocots, for example, show that these two groups probably diverged from one another about 200 million years ago, more than 65 million years before the first unequivocal angiosperm fossil.

Questions for Further Thought and Study

1. What is the fossil evidence that angiosperms originated at least 200 million years ago?
2. What correlation is there between the seed-plant phylogeny estimated by Doyle and Donoghue and the date of the monocot-dicot divergence calculated from the chloroplast gene clock?

3. If angiosperms evolved in the Triassic or Jurassic period, why are there no angiosperm fossils from those periods?
4. Dinosaurs and most other animals went extinct at the end of the Cretaceous period. Some scientists have proposed that such mass extinctions were caused by geological catastrophes. If this is true, how might angiosperms have escaped such catastrophes?
5. In addition to coevolution with insect pollinators, what other kinds of coevolutionary interactions may have occurred between flowering plants and animals?
6. Does the fact that angiosperms do not appear in the fossil record until the Cretaceous period mean that they were not around before then? Explain your answer.
7. The evolution of diversity among angiosperms depended heavily on insects. What other partnerships have angiosperms and other plants established with other organisms to ensure mutual success?
8. Angiosperms usually have a shorter life cycle than do other plants. How might this have affected the evolution of angiosperms?

Suggested Readings

Articles

Anonymous. 1990. World's earliest flower discovered. *Earth Science* 43 (2):9–10.

Clegg, M. T. 1990. Dating the monocot-dicot divergence. *Trends in Ecology and Evolution* 5:1–2.

Taylor, D. W., and L. J. Hickey. 1990. An Aptian plant with attached leaves and flowers: Implications for angiosperm origin. *Science* 247:702–704.

Books

Fernholm, B., K. Bremer, and H. Jörnvall, eds. 1989. *The Hierarchy of Life: Molecules and Morphology in Phylogenetic Analysis.* New York: Excerpta Medica.

Friis, E. M., W. G. Chaloner, and P. R. Crane, eds. 1987. *The Origins of Angiosperms and Their Biological Consequences.* New York: Cambridge University Press.

Greyson, R. I. 1994. *The Development of Flowers.* Oxford: Oxford University Press.

Heywood, V. H. (ed.). 1993. *Flowering Plants of the World.* Oxford: Oxford University Press.

UNIT NINE

Ecology . . .

In the previous unit we discussed the great diversity that has evolved among plants, fungi, algae, and bacteria, and also introduced some of the remarkable adaptations for survival that have developed among these organisms.

In this unit you examine these and other adaptations in more detail, and develop a better appreciation and understanding of the sometimes intricate interrelationships between animals, other organisms, and their environment. The discussions include an examination of ecosystems and the flow of energy within them, the cycling of water and nutrients, and the impact humans have had, and continue to have, on the natural world around them. Also included is an introduction to forest, grassland, desert, and other major natural plant associations.

Tundra in northern Manitoba, Canada.

CHAPTER 31

Population Dynamics and Community Ecology

Chapter Outline

Chapter Overview

This chapter deals with basic ecological concepts, and includes an introduction to populations, communities, and ecosystems. You'll read about the interactions of producers, consumers, and decomposers, as well as the factors involved in succession. The chapter concludes with discussions of the results of human disruption of ecosystems; the topics covered include the greenhouse effect, acid rain, ozone depletion, and water pollution.

INTRODUCTION

Ecology is the broad discipline of biology that deals not only with the relationships of organisms to each other but also with the relationships of organisms to their environment. Ecologists investigate and add to our understanding of a host of processes and events, including the growth or decline in numbers of individuals of a particular species, how the species are affected by climatic and other environmental changes, how organisms interact with other species and with other members of the same species, and how nutrients are recycled within an ecosystem.

The word *ecology* (from the Greek *oikos*, meaning "home") was first proposed by the German biologist Ernst Haeckel in 1869 (fig. 31.1). Although ecology was recognized as a biological discipline at the beginning of this century, its origins date back to early civilizations when humans, using tools and fire, first learned to modify their environment. Since the 1960s, *ecology* has become a household word, and it is now a vast area of study. This chapter will introduce you to some of the major aspects of ecological study and concern, especially as they relate to plants.

In the past two decades, topics such as the effects of pollution on land, water, and people have filled thousands of books. During this time, attempts to stop environmental damage have resulted in the requirement that various construction projects file environmental impact reports before proceeding. These reports provide information that helps various agencies to evaluate the possible effects of a proposed project on the flora, fauna, and physical environment and to determine how a project should be modified before it can be approved. The process has often produced much controversy and emotional debate between industrial developers and those who feel that preservation of the environment should be considered first. Effectively resolving these controversies requires an understanding of ecology.

We begin our discussion of ecology by considering populations, communities, and ecosystems.

POPULATIONS, COMMUNITIES, AND ECOSYSTEMS

Plants, animals, and other organisms are associated in various ways with one another and with their physical environment. For example, the lichens and mosses on a rock constitute a community, as do the seaweeds in a tidepool. These communities also have animals and other organisms associated with them in larger communities. It is preferable, therefore, to refer to the unit composed of all the populations of living organisms in a given area as a **biotic community.** Considered together, the communities and their physical environments constitute **ecosystems,** which are interconnected by physical, chemical, and biological processes. Some populations, communities, and ecosystems may be microscopic, whereas others are much larger—even global.

Populations

A population is a group of individuals of a single species. Populations may vary in numbers, in density, and in the total mass of individuals. Depending on circumstances, a biologist may measure the importance of a population in various ways. If, for example, she is concerned about the preservation of a rare or threatened species, she may simply count the number of individuals, although this may not always be feasible. The biologist may also estimate **population density,** which is the number of individuals in a given area (e.g., five blueberry bushes per square meter). If the individuals in a population have different sizes or distributions, a better estimate of the population's importance to the ecosystem might be its **biomass** (total mass of the individuals present). Population biologists may also study physiology, seed dispersal, germination, survival, and pollination in evaluating the importance of populations to their ecosystem.

FIGURE 31.1

Ernst Haeckel (1834–1919).

FIGURE 31.2

Creosote bushes (*Larrea divaricata*) growing evenly spaced because of inhibition of competing vegetation by toxins produced by these plants.

Populations within a relatively small geographic area (e.g., whose pollinators have ready access to all individual members) are referred to as **local populations.** If, however, there are many individuals in several local populations distributed over a relatively large geographic area, the population may be an **ecological race** within a species. Individuals of a population may be clumped together as a result of uneven distribution of seeds, or they may be distributed more evenly. Some species produce chemicals that inhibit the germination or growth of competitors in their vicinity, resulting in a uniform distribution of plants (fig. 31.2). Uniform distributions are, however, rare in nature. Because dispersal mechanisms often operate more slowly than do changing environmental conditions, most species never completely occupy their potential geographic ranges, which change over time.

Virtually all plants produce seeds or other dispersible reproductive structures (e.g., spores). When the seeds or spores germinate in a new habitat, we might expect population growth to accelerate. The rate of population growth is, however, slowed by several factors. Annuals complete their life cycles within a single growing season, at the end of which the population stops growing. Some perennials may require several years to complete their life cycles. Also, most seeds and spores either do not germinate, or the seedlings do not survive long enough to reproduce themselves.

The number of offspring under ideal conditions that live long enough to reproduce is referred to as the **biotic potential.** Populations of annuals that produce few seeds per plant often have a greater biotic potential than do trees that produce many thousands of seeds because the **generation time** (i.e., the time from seed germination to reproductive maturity) of the annuals is so much shorter. In addition, the maximum number of individuals in any population that can survive and reproduce in an ecosystem is limited, and constitutes the **carrying capacity** of the ecosystem. As the number of individuals in a population increases, competition for nutrients, water, and light increases. When the germination and survival of offspring equal the death rate of mature plants in a population, population growth stops.

FIGURE 31.3

A grassland community in spring.

FIGURE 31.4

A coulter pine (*Pinus coulteri*) growing in California.

Communities

Communities consist of populations of different species living in the same location (fig. 31.3). Similar communities occur under similar environmental conditions, although the actual species composition can vary considerably and can change between the boundaries of an area. A community is difficult to define precisely, however, because members of one community can also occur in other communities. For example, the Coulter pine (*Pinus coulteri*) of California occurs in yellow pine forest communities and in foothill woodland communities (fig. 31.4). Furthermore, members of one community may be genetically adapted to that community. If individuals are transplanted to a different community where the same species occurs, they may not necessarily be able to survive alongside their counterparts, which are adapted to their own community. Individuals adapted to specific communities within their overall distribution are called **ecotypes.** Ecotypes are often so distinct from one another in their appearance that botanists argue about whether to classify them as separate species.

Ecosystems

Living organisms that interact with one another and with the nonliving environment constitute an ecosystem. The nonliving, or **abiotic, factors** of the environment include light, temperature, oxygen level, air circulation, precipitation, and soil type. The distribution of a plant species is controlled mostly by temperature, precipitation, soil type, and the effects of other organisms **(biotic factors).** Species in arid ecosystems, for instance, are adapted to low precipitation and high temperatures. Such plants are called **xerophytes.** Cacti often have small leaves or nonphotosynthetic leaves in the form of spines, which reduce transpiration and thus adapt them to their particular environments (fig. 31.5). Xerophytes may also have special forms of photosynthesis (such as CAM photosynthesis; see Chapter 7). Similarly, plants such as water lilies, which grow in water (**hydrophytes;** fig. 31.6), have modifications that adapt them to their aquatic environment (see Chapter 14). The leaves of hydrophytes, for example, have thin cuticles and more stomata on their upper surface than on their lower surface.

FIGURE 31.5

A barrel cactus (*Ferrocactus covillei*). The spines are modified leaves that protect the plant and reduce the surface area available for transpiration.

The mineral content of soils influences the distribution of plant species (see Chapter 20). The high concentrations of metals such as magnesium, iron, nickel, and chromium and the low amounts of calcium and nitrogen found in serpentine soils characterize a relatively inhospitable environment in which only a few plants can survive. Similarly, the diversity of tree species in rain forests often varies with the level of phosphorus and magnesium in the soil. Biotic factors such as competition for light, mineral nutrients, water, and space also influence the distribution of plant species, as does the direct removal of plants by grazing animals.

C O N C E P T

Ecologists view the interrelationships of organisms as a hierarchy. The lowest level is a population, which is a group of individuals of a single species. Populations of different species make up a community, and communities and their physical environments compose an ecosystem. Ecosystems are the most complex level of organization.

FIGURE 31.6

Water lilies (*Nymphaea*). The floating leaf blades have thin cuticles; nearly all their stomata are on the upper surface.

Doing Botany Yourself

Plant ecologists sometimes employ a simple means of analyzing vegetation in an area. They drive a stake into the ground and tie string to the stake. Then they roll out the string in a straight line for a measured distance (e.g., 100 meters), drive in another stake, and attach the other end of the string to it. This straight line is called a *line transect*. Next, they take careful notes about all the vegetation living a measured distance (e.g., 10 cm) on either side of the string. They may then form another line transect in a second nearby area and compare the vegetation along the two transects.

You can relatively easily construct at least two line transects somewhere near where you live. You may not know all the names of the plants, but you can at least discern features such as small, medium, and large woody plants, and short, medium, and tall herbaceous plants. If you have time, you could perform a little more sophisticated transect study involving meter-wide transects. To do this you would run two parallel transects exactly 1 meter apart. Then divide the transect into 1-meter squares, and count all the kinds of plants in every third square.

Characteristics of Ecosystems

Trophic Levels

Ideally, ecosystems can sustain themselves entirely by photosynthesis or chemosynthesis and the recycling of nutrients. Autotrophic organisms either capture light energy and convert

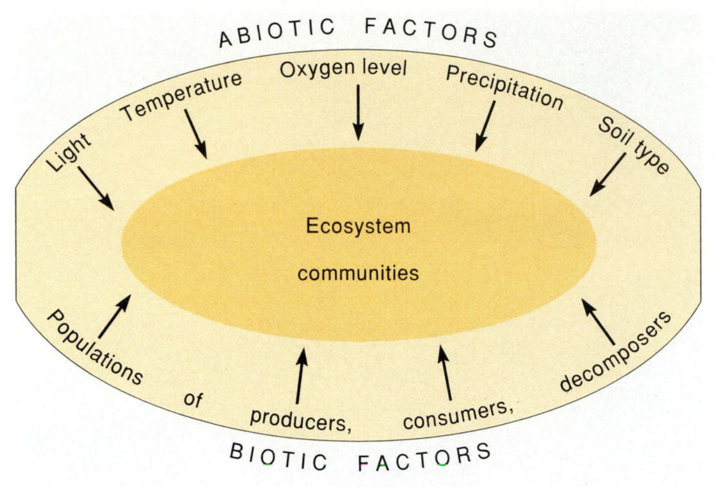

FIGURE 31.7

The basic components of an ecosystem.

it, along with carbon dioxide and water, to energy-rich sugars, or they oxidize chemicals as a source of energy. These autotrophic organisms are called **producers.** Heterotrophic organisms such as herbivores (e.g., deer and sheep), which eat only producers, are called **primary consumers. Secondary consumers,** such as carnivores (e.g., eagles, wolves, and lions), eat primary consumers. Omnivores such as bears may eat plants and animals. **Decomposers** break down organic materials to forms that can be reassimilated as mineral components by the producers. The foremost decomposers in most ecosystems are bacteria and fungi (fig. 31.7).

In any ecosystem, the producers and consumers form *food chains* or interlocking **food webs** that determine the flow of energy through the different levels. Since most organisms have more than one source of food and are themselves often eaten by a variety of consumers, there are considerable differences in the length and complexity of food chains or webs (fig. 31.8).

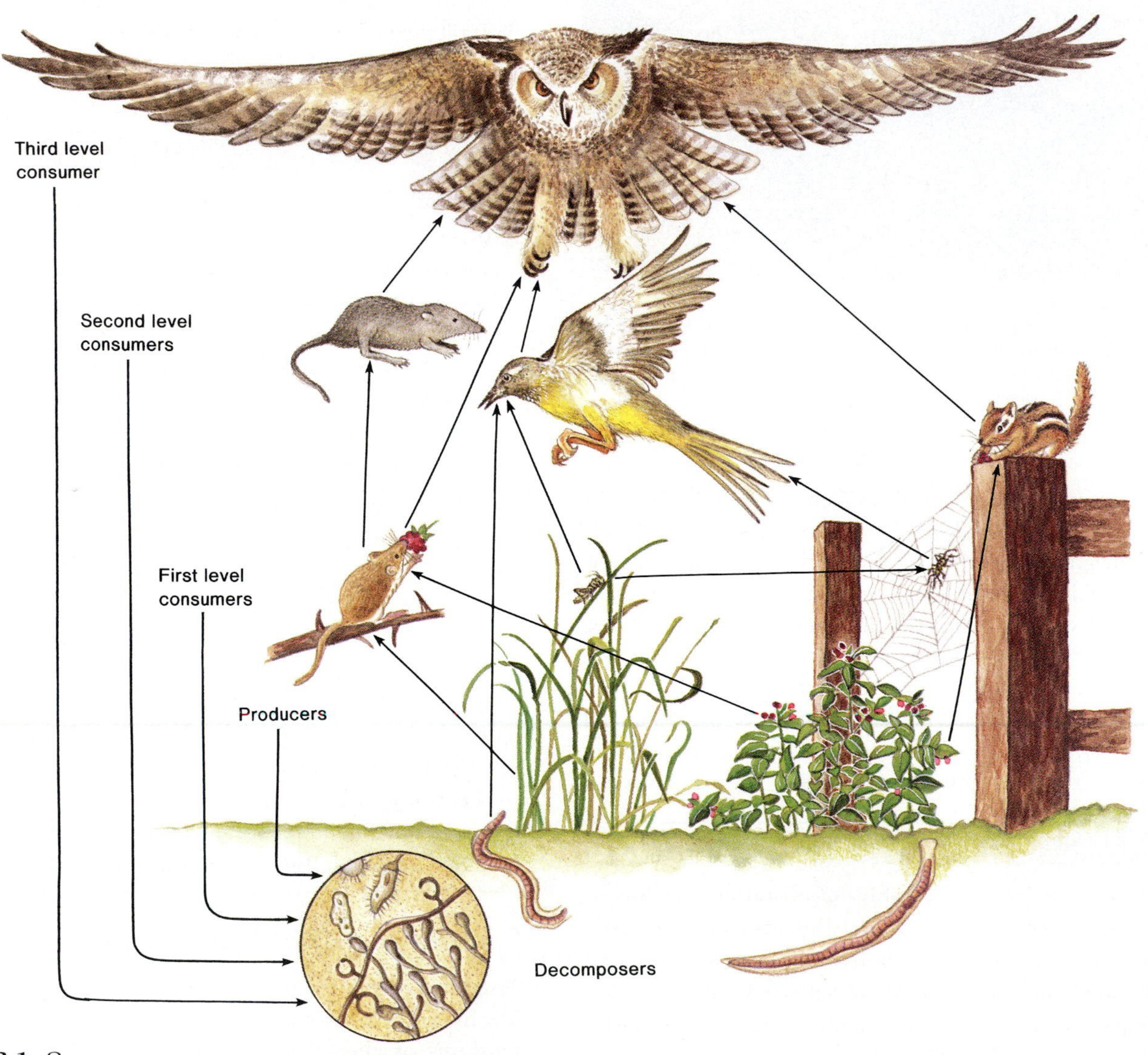

FIGURE 31.8

A hypothetical food web. Although no real food web would be so simple, this diagram shows that plants are the producers and that animals and decomposers are the consumers in an interlocking food web.

The technique of *stable isotope tracing*, which determines the relative amounts of specific isotopes in tissue samples, enables ecologists to unravel food sources and consumption within food webs. Producers each have a typical ratio of ^{12}C to ^{13}C (referred to as the *carbon signature;* see Appendix A). A device called a *mass spectrometer* is used to determine relative amounts of the carbon isotopes from carbon dioxide gas obtained from cleaned bones, fur, or other material that has been burned, revealing the specific types of foods previously consumed by the animals. Carbon signatures from the past can also be determined from fossils by this method, which has recently revealed that kelps, and not plankton (as previously believed), are and were the principal primary producers in some coastal ocean ecosystems.

Energy, which enters a food chain at the producer level, cannot be recycled in an ecosystem (see Chapter 5). Only about 1% of the light-energy striking a temperate-zone community is converted to organic material. As the organisms at each level respire, energy gradually dissipates as heat into the atmosphere. Additional energy is stored in organisms that are not consumed. For example, the energy in leaves that fall from a plant before a herbivore grazes on it is not released until decomposers (bacteria and fungi) degrade the leaves.

Only about 10% of the energy stored in green plants that are eaten by cattle is converted to animal tissue; most of the remaining energy dissipates as heat. When we eat beef, our bodies use about 10% of the beef's stored energy for growth, maintenance, and reproduction. The remaining energy is converted to heat. If 90% of the energy is lost as heat at each level of a food chain, then only about 0.1% of the original energy captured by the producers will be used in a typical food chain with three levels of consumers. Thus, the longer the food chain, the greater the number of producers necessary to provide energy for the final consumer. Conversely, the shorter the food chain, the smaller the number of producers necessary to provide energy for the final consumer. To appreciate this consider the following example.

Assume that for every 100 units of light energy striking corn plants each day, 10 units are converted to plant tissue (including corn). Then suppose that the corn is fed to cattle. Only 10%, or 1 unit, of the original energy is converted to beef. When we eat beef, our bodies, in turn, use only 0.1% of the energy originally converted to plant tissue. If we eat the corn directly, however, we receive the same amount of energy from 90% fewer plants than it took to produce the beef if we ate the beef instead.

The concept just described has important implications. For example, a vegetarian diet uses energy more efficiently than a diet based on meats. Consequently, where food is scarce or humans are abundant (as in India or Ethiopia), many humans become vegetarians. In terms of the numbers of individuals and the total mass, there is a sharp reduction of usable energy at each level of the food chain. In a given part of ocean, for example, there may be billions of microscopic algal producers supporting millions of tiny crustacean consumers, which, in turn, support thousands of small fish, which meet the food needs of scores of medium-sized fish, which are finally eaten by one or two large fish. In other words, one large fish may depend on a billion tiny algae to meet its energy needs every day (fig. 31.9).

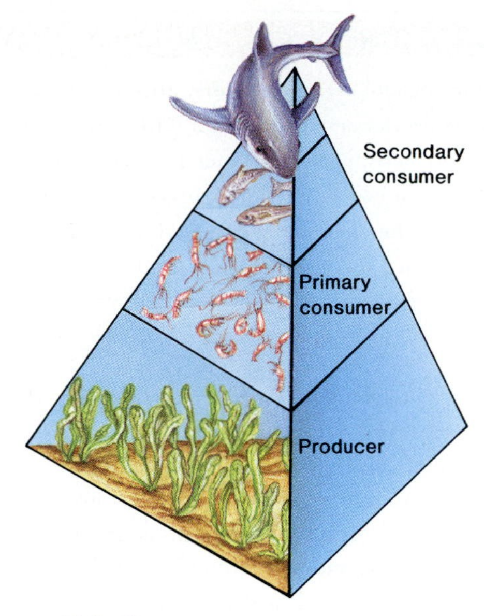

FIGURE 31.9

This pyramid represents how much energy from lower levels of the food web is required to support successively higher levels. In this example, each kilogram of the shark (top carnivore) indirectly requires as much as 1,000 kilograms of algae (producers).

The interrelationships and interactions among the components of an ecosystem can be complex, but over the long term there is a balance between producers and consumers. An increase in food made available by producers can increase the number of consumers. This increased number of consumers reduces the available food, which then inevitably reduces the number of consumers. The result is sustained self-maintenance of the ecosystem.

Ecosystems exhibit considerable variation in **net productivity,** which is defined as the energy produced by photosynthesis minus that lost in respiration. Productivity (in terms of biomass produced) is usually measured as grams per square meter. Forest ecosystems, for example, produce 3–10 grams of dry matter per square meter of land per day, whereas grasslands produce 0.5–3 grams, and deserts less than 0.5 gram. Crops grown throughout the year (e.g., sugar cane) produce up to 25 grams.

Ecosystems are dynamic; they undergo daily changes, seasonal changes, and changes that may take many years. Photosynthesis occurs only during daylight hours, whereas many animals in ecosystems are active during either the day or night but not both. Even in tropical rain forests, some plants lose their leaves for part of the year, and seasonal periods of dormancy and rejuvenation in woody plants are common outside of the tropics.

C O N C E P T

In any ecosystem, producers, consumers, and decomposers form food chains, with producers at the base. The interrelationships often result in food webs, with a balance between producers and consumers. Energy cannot be recycled, and is lost at each level. Huge numbers of producers may be required to meet the needs of a single secondary consumer at the end of a long chain.

Diversity

Each ecosystem includes a variety of organisms that are distributed in specific patterns determined by the physical environment and their relationships to each other. Some herbaceous members of the ecosystem produce rhizomes, tubers, or bulbs, whereas others grow as epiphytes, vines, or parasites. Woody components may be deciduous or evergreen, thin-leaved or thick-leaved, and occur as tall trees, low shrubs, or in many intermediate forms. Ecosystems with similar climates in different parts of the world usually have similar organisms. Ecosystems that include chaparral scrub communities, for example, occur in southern California and in central Chile (see figs. 32.26 and 32.27).

Processes and activities such as expansion of buds, flowering, fruiting, and abscission of leaves are coordinated within an ecosystem. In the Amazon Basin, for example, the fruits of certain trees fall just as the trees' bases become seasonally flooded. Fish that swim under the trees during the flooding often catch the fruits in their mouths as they fall, and the seeds, which pass unharmed through their digestive tracts, are deposited elsewhere.

Species diversity depends on the number of species and the number of individuals per species in an ecosystem. Harsh climates with poor soils, such as those in arctic regions and deserts, limit species diversity, whereas mild or tropical climates with fertile soils permit extensive diversity. Indeed, about half of all living species occur in tropical rain forests, which occur on less than 7% of the earth's land masses. In rain forests there is intense competition for light, minerals, and other resources, but each species is well enough adapted to its habitat to allow for great diversity.

What are Botanists Doing?

Go to the library and read an article or research paper that directly or indirectly addresses this hypothesis: A small number of abundant species account for a large fraction of ecosystem function. What do you conclude from this article?

How Much Does Diversity Matter?

Humans and other organisms depend on ecosystems for a variety of important services (e.g., preservation of soil fertility, control of pest outbreaks). But what does it take to preserve a working ecosystem? Would the loss of biodiversity—currently a hot topic among environmentalists—change an ecosystem's ability to do those services? Despite all of the talk about the importance of biodiversity, virtually no one has tried to answer that question.

At the heart of the problem is how ecosystem ecologists and population biologists approach their work. Most ecosystem ecologists study how energy and materials flow through an ecosystem, and give little thought to the individual organisms involved. Conversely, population biologists study individual species and the food webs they form, and generally ignore these organisms' roles in the larger flow of energy and materials. Consequently, there's little known about how biodiversity affects the functioning of an ecosystem. Although increased diversity typically increases the production of plant biomass (to a point), there's little evidence about the importance of species diversity in nutrient cycling and decomposition. Ecologists know that we can sustain some loss of species diversity without a loss in the productivity of an ecosystem. This leads to an incredibly tough task: formulating policies for deciding which species really are crucial, and which ecosystem functions they're needed for.

Shaping the debate about this issue are two contradictory theoretical predictions about the importance of species diversity. One, presented in 1981 by Stanford ecologists Paul and Anne Ehrlich, is known as the "rivet popper" hypothesis. According to this hypothesis, the diversity of life is something like the rivets on an airplane: each species plays a small but important role in the working of the whole. The loss of each rivet weakens the plane by a small but noticeable amount; the loss of enough rivets causes the plane to crash.

The opposing idea, called the redundancy hypothesis, was presented in the early 1990s by Brian Walker, an Australian. The hypothesis asserts that most species are superfluous—that is, they are more like passengers than rivets—and that only a few key species are needed to keep the system in motion.

Which is more accurate? Several studies now support the "rivet" hypothesis. One experiment in 1993, in controlled-environment chambers, tested the effect of diversity on productivity by setting up 14 artificial ecosystems. The result: As species numbers went up, the mean production of plant biomass also increased. A possible explanation is that more species usually generate a more diverse plant architecture, which allows the system to capture more light and produce more plant biomass. This conclusion is supported by studies of multicrop agriculture: The best way to increase the productivity in a corn field is not to pack in more corn plants, but to add other plants such as melons and nitrogen-fixing beans.

Species diversity also increases the ecosystem's resilience: diverse ecosystems recover from stress (e.g., drought) much faster than do ecosystems with less diversity. This is supported by the work of David Tilman and his colleagues at the University of Minnesota, who studied how as many as 250 kinds of plants could thrive in midwestern grasslands. Tilman created 207 four-meter-square plots distributed among one native prairie and three abandoned fields of different ages. Each season, they clipped a different section in each plot and analyzed its species diversity and biomass. They left some plots alone and added nitrogen fertilizer to others. Nitrogen fertilizer tends to reduce species numbers even as it boosts productivity. During a drought in 1987 to 1988, the productivity in all of the plots fell drastically. However, the drop in species-rich plots was only a fourth that of species-poor plots. Moreover, the species-rich plots recovered in only one season, rather than in the four seasons required for species-poor plots to recover. Tilman concluded that "Biodiversity is a way to hedge bets against uncertainty, even in managed systems."

Interestingly, the enhancing effects of biodiversity on ecosystem resilience and productivity become saturated after a certain level of diversity. For example, one study showed that the largest gains in stability came with the first 10 species; beyond that, adding more species didn't add much stability, perhaps because the essential functional niches had already been filled. Similarly, some evidence suggests that the growth and photosynthetic rates in tropical forests top out with only 10 or so tree species. Thus, when it comes to productivity and resilience, some species in ecosystems may be more like passengers than rivets.

Although biodiversity is valuable to a certain point, most ecosystems contain more diversity than is needed to reach peak productivity. This conclusion holds, even when considering large ecosystems. For example, although the temperate forests of the Northern Hemisphere are vastly different in terms of biodiversity (the forests of Europe include 106 tree and shrub species, those of North America 158, and those of East Asia 876), their productivity is virtually identical.

Apparently, the random loss of species does not impair the productivity of an ecosystem, for such extinctions leave behind a few species in each structural category—for example, vines, canopy trees, and understory ferns in a tropical forest. So should we be concerned about the loss of biodiversity and extinctions that human activities are now causing?

Yes, because these extinctions are not random: Logging, burning, and grazing always affect a specific subset of species. Thus, these activities probably have a much greater impact on an ecosystem than do random extinctions.

Habitats

Specific sets of environmental conditions in which organisms live are referred to as **habitats.** Habitats contain organisms adapted to combinations of environmental conditions; the organisms live, reproduce, and die in the specific environments of their habitats. As indicated earlier, living factors such as soil microbes and fungi, plants, and animals are said to be *biotic*, whereas some of a habitat's nonliving physical components such as light, oxygen, elevation, latitude, climate, fire, and avalanches are *abiotic*. Soil components and moisture, shade, associated organisms, and other features of a habitat that affect a plant directly make up its *operational habitat;* other factors of the habitat may not affect a plant at all. For example, certain species of plants may occur both in an area with a stream running through it and in another area without a stream; the stream apparently has no direct impact on the distribution of specific plants. The stream is, however, a part of the operational habitat of aquatic plants within it.

Roles of Biotic Factors in the Habitat

As a plant respires and photosynthesizes, it produces by-products that modify its habitat (e.g., CO_2 that combines with water to form carbonic acid; see Chapter 15). The plant also absorbs minerals and creates shade, and it may provide incentives for visits by pollinators, produce substances that limit competition by other plants, or promote interaction with other plants in a mutually beneficial way. The plant may also provide shelter and food for other organisms and in other subtle ways create a **niche** to which it is better adapted than are other plants.

Many organisms occupy ranges with a diversity of features such as topography, available moisture, soil types (referred to as *edaphic factors*), and sunlight. In a given part of the range, a species may compete for water, while in another part edaphic factors or light may be more critical to competition. If gene flow throughout the range of the species is not widespread and rapid, localized populations may become genetically adapted to local conditions, resulting in ecotypes. Because ecotypes may be morphologically identical to plants in other parts of the range, an experiment would be necessary to determine if natural selection had changed their genetic composition. A simple experiment to answer this question might involve transplanting the suspected ecotypes to a common growing area within the range. Genetic change is evident if the transplanted individuals die or exhibit a different morphology than plants of the same species occurring at the site. If the transplanted individuals show no differences from the local plants, genetic factors are probably not involved.

The extent to which competition is believed to determine the species components of a habitat is controversial. Some biologists believe that the best-adapted species exclude less well-adapted species, whereas others believe that **sympatric** species (i.e., those occupying the same range) overlap in their tolerances of each other, and that each uses areas of the range not occupied by the other. Where two species overlap, the better-adapted species dominates the overlap area, while the species that is not as well adapted occurs in only part of its potential niche. If either species is removed, the other species may take over the entire range.

Pollinators and other organisms are important factors in a plant's habitat. Many associations between plants and other organisms are **mutualistic** (mutually beneficial). For example, several species of bleeding hearts (*Dicentra* spp.) have oil-bearing appendages on their seeds. Ants carry the seeds to their nests, strip off the appendages for food, and then deposit the seeds outside the nest, where they germinate. Thus, ants help distribute the plants (fig. 31.10).

FIGURE 31.10

Seeds of the Pacific bleeding heart (*Dicentra formosa*). The glistening white appendages are *elaiosomes*, which are stripped from the seeds and eaten by ants.

Roles of Abiotic Factors in the Habitat

Habitat features, such as the mineral content of soils, that are not directly related to life or living organisms are abiotic. As indicated in Chapter 20, digging a deep hole in an undisturbed area where soil has developed often reveals a soil profile of intergrading regions. The upper layer, which typically extends 10 to 20 cm down, is called the *A horizon*, or *topsoil*. Water washes nutrients from the topsoil downward to the next layer, known as the *B horizon*, or *subsoil*. Subsoil is rich in nutrients; it is usually lighter in color than topsoil and often contains more clay. The lowermost layer, or *C horizon*, is also referred to as *parent material*, and extends down to *bedrock*. Plants absorb essential elements from the soil, while other elements are leached away. The essential elements are constantly recycled as the plants die; decay organisms release them, and they are once again used by the plants.

The first plants to become established on new soils are called *pioneers*. Pioneers must often adapt to rapid runoff or leaching of water, lack of protection from wind, high solar radiation, and other conditions that many other plants cannot tolerate. The pioneers gradually change the soil's composition by adding organic matter and breaking up bedrock as they penetrate cracks. As already mentioned, carbon dioxide released by respiring roots may combine with water to form carbonic acid, which further breaks down rock particles. Over time, this process, called *succession*, changes a soil's profile. You'll learn more about succession later in this chapter.

Climate plays a vital role in the distribution of nearly all organisms. Tropical plants cannot tolerate freezing or lengthy periods of drought or low humidity. Many temperate-region plants that become dormant in the fall shed their leaves. They may even require periods of freezing to break their bud dormancy. The amount of precipitation and the average temperatures of a region are not, however, as important as their timing and nature. For example, the precipitation in a Mediterranean climate occurs primarily in the winter and may be similar in amount to that of temperate regions that receive precipitation throughout the year. Some plants adapted to Mediterranean climates may not tolerate sustained summer rainfall, whereas plants adapted to precipitation throughout the year can seldom adapt to the long, dry summers of Mediterranean regions.

The total solar radiation available each year for photosynthesis varies considerably with latitude. Cloud cover may reduce the amount of radiation from the sun at the equator, but the sun is directly overhead there for longer each day than anywhere else; sunlight can reach plants at the equator for twelve hours every day of the year. Between the Tropics and the Arctic and Antarctic Circles, the differences in daylength between summer and winter can vary by as much as ten hours, and the plants adapted to these regions usually have photoperiods that determine flowering times (see Chapter 19). At high latitudes, there is light available continuously in midsummer and no sunlight at all in midwinter. In addition, the sun is overhead only briefly during the year, thereby reducing the amount of radiation reaching the plants. Because of light limitations, cold temperatures, poor soil, and lack of available soil water, the growing season may be less than three months long in arctic regions, and may also be hampered by high winds and waterlogged soil. Similar conditions often prevail at high elevations, regardless of latitude. Landslides, fires, floods, tornadoes, and other disturbances can disrupt an ecosystem by removing large numbers of established organisms.

C O N C E P T

An environment in which an organism lives includes biotic factors such as soil microbes and fungi, and other plants and animals. Abiotic (i.e., nonliving) factors include light, oxygen, climate, and physical disturbances.

Interactions between Plants, Herbivores, and Other Organisms

In a given habitat, the amount of nutrients, light, water, and other resources needed for plant growth are shared by all the plants in that particular environment. When two species grow together in a controlled environment, one of them may eventually be eliminated. In nature, however, many environmental factors affect the plants. When resources are limited, two different species may coexist as long as environmental factors remain constant. The individual plants of both species, or at least one of them, however, are smaller in number or in size, and coexistence depends on their occupying different *microhabitats*.

Competition for light among plants has resulted in the evolution of several mechanisms that allow species to adapt to different light intensities. Depending on the light that is available, different forms of photosynthesis may be involved (see Chapter 7), leaf orientation may change throughout the day,

FIGURE 31.11

Boston ivy (*Parthenocissus tricuspidata*) growing on a wall. Note how each leaf is oriented so that it has maximum exposure to light.

plants may grow taller, and the thickness of the leaf blades and the number of chloroplasts in the mesophyll may change. The shapes, sizes, arrangements, and orientations of leaves ensure maximum exposure to the light (fig. 31.11). The relationship between light, photosynthesis, and biomass is obvious. Although the total mass of consumers is determined largely by the total mass of food made by the producers, the interactions among producers themselves and among the decomposers and the other members of the ecosystem, are usually more subtle. In an ecosystem, the defenses that producers and consumers have against each other have developed through a process of coevolution resulting from natural selection and are maintained in a delicate balance. Flowering plants have evolved a variety of substances that either inhibit or promote the growth of other flowering plants or function as defenses against herbivores. For example, black walnut trees make a substance (juglone) that wilts tomatoes and potatoes and injures apple trees that contact the black walnut roots. The production of natural antibiotics that kill or inhibit the growth of fungi or bacteria by other plants makes them resistant to various diseases (see Chapter 2).

Plant/herbivore interactions are widespread. For example, large herbivores such as deer and moose eat a variety of plants, each having a different nutritional value. Each plant species also produces different combinations, types, and amounts of chemicals in addition to proteins, fats, and carbohydrates. Many of these compounds are toxic to consumers, but the animals are not affected because their digestive systems break down and eliminate the toxic compounds (fig. 31.12). The limitations imposed by such compounds cause the consumers to vary their diet, seek familiar foods, and be wary of new ones. If a plant species lacked some natural defense, such as chemical compounds or structural modifications like spines, primary consumers might soon overexploit that species and threaten it with extinction.

FIGURE 31.12

A Koala eating *Eucalyptus* leaves. Koalas' digestive systems break down substances in the leaves that are toxic to other animals.

Many plants secrete substances that either offend insect predators or function as insecticides. For example, the volatile oils produced by members of the mint family (Lamiaceae) deter insect larvae and several higher animals from eating the leaves. Some members of the sunflower family (Asteraceae; e.g., chrysanthemums) produce the biodegradable insecticide *pyrethrum*, which is sold commercially. Some plant pests, however, have evolved enzymes that break down these chemicals, enabling them to thrive on plants that are toxic to other predators. Members of the carrot family (Apiaceae), which includes carrots, parsley, and celery, produce aromatic substances that serve as insect repellents. However, larvae of the anise swallowtail butterfly feed exclusively on members of that family; the larvae have specific enzymes that degrade the repellent chemicals (fig. 31.13).

Various inhibitors produced by some bacteria and fungi limit the growth of higher plants, and parasitism by other bacteria, fungi, and flowering plants limits population size. The degree of parasitism varies considerably with the organisms involved. About 15% of the 3,000 members of the figwort family, Scrophulariaceae (which includes snapdragons [*Antirrhinum*] and *Penstemon*), exhibit various degrees of parasitism. Species of

Figure 31.13

Anise swallowtail butterfly, whose larvae feed exclusively on members of the parsley family (Apiaceae). The larvae have specific enzymes that break down chemicals that repel other insects.

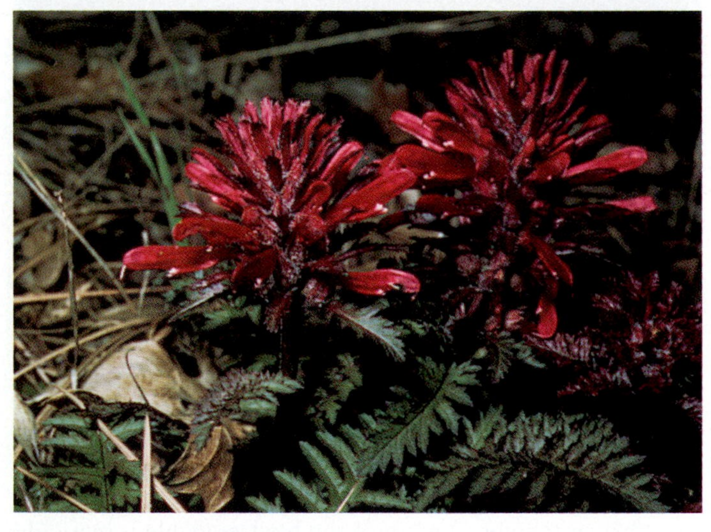

Figure 31.14

An Indian warrior (*Pedicularis densiflora*) plant. Although the leaves are green, there is insufficient chlorophyll for all the plant's energy requirements. The balance of nutrients needed is obtained through parasitism of the roots of trees and shrubs.

Harveya and *Hyobanche* lack chlorophyll and are entirely dependent on their flowering-plant hosts for their energy and other nutritional needs. *Gerardia aphylla* has pale leaves with about 10% of normal chlorophyll content; additional energy and nutrients come from their host plants. Parasitic species of *Pedicularis* and *Odontites rubra* have green leaves that furnish about half of their energy needs (fig. 31.14). Still other figworts (e.g., *Castilleja* sp.) may parasitize the roots of certain plants but can also live independently. For a more thorough discussion of parasitic plants, see Chapter 20.

C O N C E P T

Producers and consumers defend themselves in ecosystems by such means as producing substances that inhibit competitive growth or making tissues unpalatable or toxic to consumers. Some consumers produce enzymes that enable them to break down toxic substances, giving them an advantage over competitors. Over time, a delicate balance of defenses has coevolved through natural selection.

The Lore of Plants

Cobra plants (*Darlingtonia californica;* see fig. 20.24) occur in a few swampy places in the mountains of Oregon and northern California. These interesting plants, whose principal leaves form insect-trapping pitchers, have apparently survived because the swamps to which they are adapted have magnesium salts in concentrations high enough to kill most of their competitors.

A pitcher leaf, which may be nearly 1 meter long, resembles a cobra head with its hood inflated; it even has "fangs" that insects follow right into the mouth of the trap. Once the insect is in the mouth, it encounters stiff, downward-pointing hairs that facilitate its descent and hinder its escape. Escape is made even more difficult by numerous small patches of transparent tissue on the back of the hood. The false windows mislead the victim in its efforts to find an escape route. Eventually the insect drowns in fluid at the base of the pitcher, where its soft parts are digested by bacteria.

Nutrient Cycles

As you learned in Chapters 5 and 7, nutrients cycle in ecosystems. This section describes the cycling of nitrogen, carbon, and water.

The Nitrogen Cycle

As noted in Chapter 2, much of the mass of living cells consists of protein. The most abundant element in our atmosphere, nitrogen, constitutes about 18% of the protein. There are nearly 69,000 metric tons of nitrogen in the air over each hectare of land, but the total amount of nitrogen in the soil seldom exceeds 3.9 metric tons per hectare and is usually considerably less. This discrepancy exists because the nitrogen of the atmosphere is chemically inert; that is, it does not combine readily with other molecules. Thus, atmospheric nitrogen is largely unavailable to plants and animals.

Most of the nitrogen in plants and, indirectly, in animals comes from the soil in the form of inorganic ions absorbed by the roots. These ions are released by bacteria and fungi that

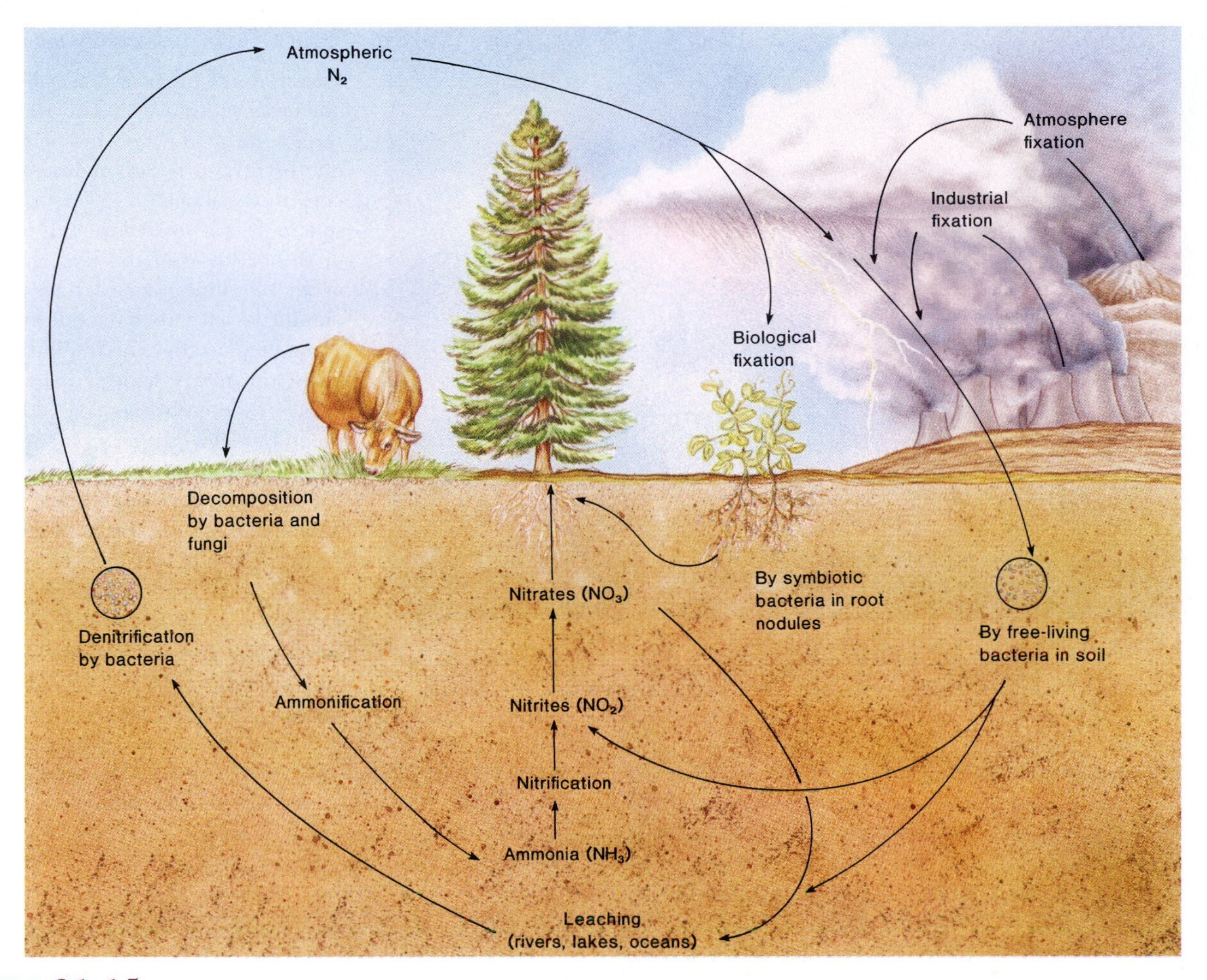

FIGURE 31.15

The nitrogen cycle. Most nitrogen occurs as atmospheric nitrogen (N_2), which is not usable by plants and animals. Nitrogen fixation and decomposition make nitrogen available for plant metabolism as nitrates. Nitrogen from these sources is incorporated into organic compounds in plants and other organisms in the food web.

break down the complex molecules in dead plant and animal tissues. Some nitrogen from the air is also *fixed,* that is, converted to ammonia or other nitrogenous substances by *nitrogen-fixing bacteria.* Some of these organisms live symbiotically in plants, particularly in members of the legume family (Fabaceae; e.g., peas, beans, clover, alfalfa), while others live free in the soil (see Chapter 25).

As shown in figure 31.15, nitrogen flows from dead plant and animal tissues into the soil and from the soil back into the plants. Bacteria and fungi can break down huge amounts of dead leaves and other tissues to tiny fractions of their original volumes within a few days to a few months. If their activities were to stop abruptly, the available nitrogen compounds would be exhausted within a few decades, and the supply of carbon dioxide needed for photosynthesis would be seriously depleted as the respiration of decomposers ceased. Forests, jungles, and prairies would die as accumulations of shed leaves, bodies, and debris would bury the living plants and keep the light from reaching their leaves.

Much nitrogen is leached from soil by water. More is removed with each harvest (the average crop contains about 25 kilograms per hectare per year). This nitrogen can be recycled and used as fertilizer if vegetable and animal wastes are returned to the soil. While bacteria are decomposing tissues, they use nitrogen, and little is available until they die and release their accumulations into the soil. Accordingly, crops should not be planted in soils to which only partially decomposed materials have been added until bacteria have broken down the organic matter (see Chapter 25).

FIGURE 31.16

Burning rice stubble in the San Joaquin Valley of California. Such burns reduce the amount of nitrogen in the soil.

Weeds and stubble are frequently controlled by burning (fig. 31.16), which depletes soil nitrogen. The annual combined loss of nitrogen from the soil in the United States from fire, harvesting, and other causes is estimated to exceed 21 million metric tons, and only 15.5 million metric tons are replaced by natural means. To offset the net loss, some 32 million metric tons of inorganic fertilizers are applied each year (see "Putting Things Back," Chapter 20). If organic matter is not added at the same time, the application of inorganic fertilizers combined with the annual burning of stubble may eventually create a *hardpan* soil. Hardpan develops through the gradual accumulation of salt residues, which dissolve humus and disrupt the structure of the soil. Clay particles then clump and produce colloids that are impervious to moisture. In hardpan soils and others low in oxygen (e.g., soils in flooded areas), *denitrifying bacteria* use nitrates instead of oxygen in their respiration, thus rapidly depleting the nitrogen remaining in the soil.

Precipitation returns some nitrogen to the soil from the atmosphere, where it accumulates as a result of the action of light on industrial pollutants, fixation by flashes of lightning, and the diffusion of ammonia released by decay. The activities of nitrogen-fixing bacteria and volcanoes also replenish the nitrogen supply by converting it to forms that can be used by plants.

C O N C E P T

Microorganisms and fungi are involved in the cycling of nitrogen from dead tissues into the soil and back into plants. Precipitation returns a little nitrogen to the soil from the atmosphere.

The Carbon Cycle

Bacteria and fungi also recycle carbon and other substances. As noted in Chapter 7, one of the two raw materials of photosynthesis is carbon dioxide, which constitutes about 0.035% of our atmosphere. It is estimated that all the plants of the oceans and the land masses use about 14.5 billion metric tons of carbon obtained from carbon dioxide every year. This is replaced by the respiration of all living organisms, with perhaps as much as 90% or more being released by bacteria and fungi as they decompose tissues. Lesser amounts of carbon dioxide are released by the burning of fossil fuels, and a small amount originates from fires and volcanic activity.

All of the carbon dioxide in our atmosphere would be consumed in about twenty-two years if it were not constantly replenished (fig. 31.17). Oceans act as a buffer by absorbing and storing excess carbon dioxide as carbonates, but their capacity is limited. Some biologists believe if this storage capacity is exceeded, the carbon dioxide in the atmosphere will rise dramatically (see box 31.2 "Curbing Methane Emissions to Curtail the Greenhouse Effect" later in this chapter).

C O N C E P T

Decomposers, through respiration, constantly produce most of the carbon dioxide needed by green organisms for photosynthesis. All other living organisms, as well as abiotic processes, are the sources of the remaining carbon dioxide produced. Excess carbon dioxide is stored in oceans.

The Water Cycle

More than two-thirds of the earth's surface is covered by water, which is accumulated in oceans, freshwater lakes, ponds, glaciers, and polar ice caps. Evaporation occurs as surface waters are warmed by the sun. The water vapor rises into the atmosphere, where it cools, condenses, and returns as precipitation. Air currents move the moisture-laden air around the globe.

When precipitation occurs over land, some of the water returns to the oceans, lakes, and streams in the form of runoff, and some percolates through the soil to underground water tables. When plants absorb water from the soil, all except about 1% of it is transpired through the leaves and stems. Any water retained by the plants is stored in cells, where it is used in metabolism and the maintenance of turgor. The transpired water vapor combines with other evaporated water, and the cycle is repeated (fig. 31.18).

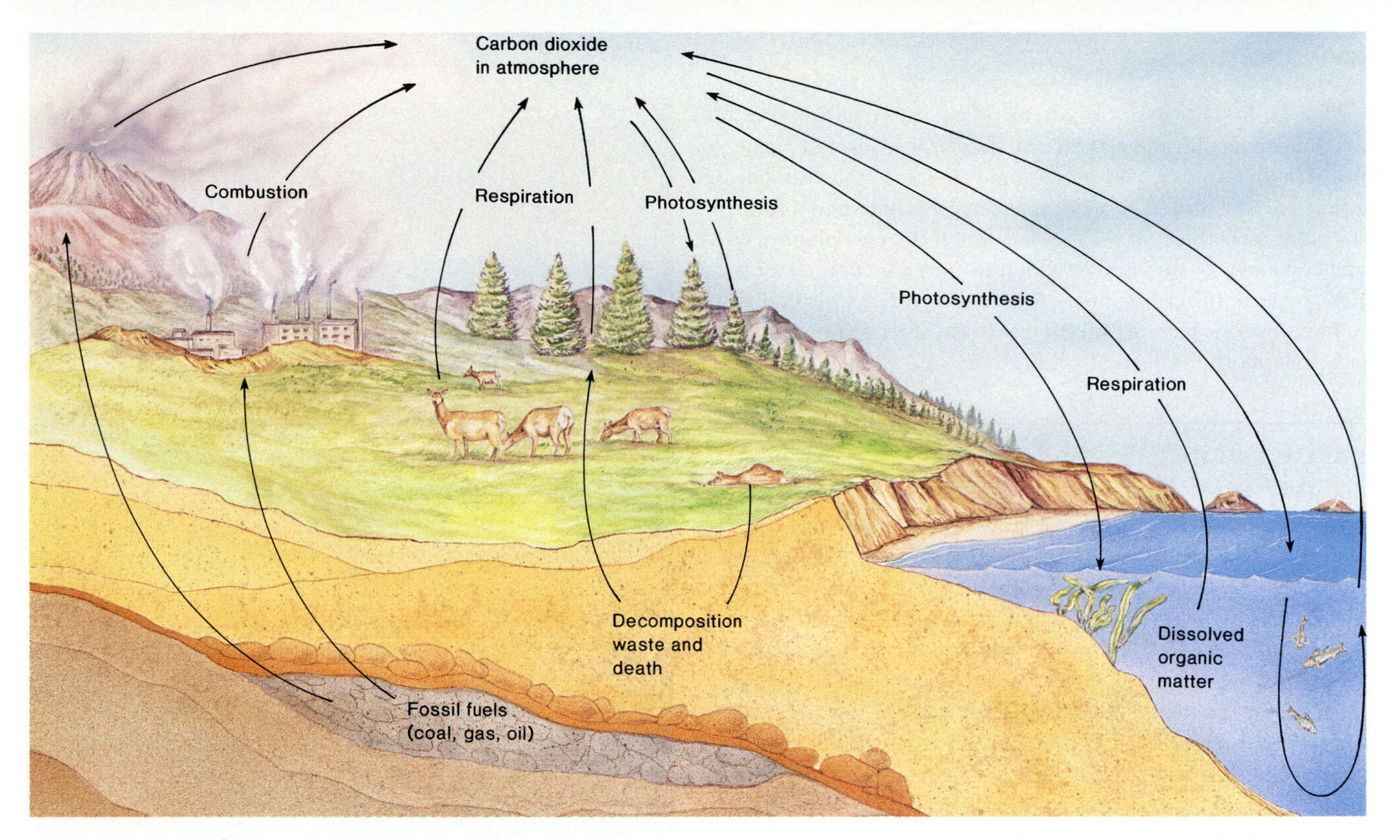

FIGURE 31.17

The carbon cycle. Carbon dioxide is formed from respiration by plants and animals, from decomposition of organic matter by bacteria and fungi, from burning fossil fuels and wood by humans and natural fires, and from volcanic activities. Carbon dioxide is used by plants and other photosynthetic organisms for photosynthesis.

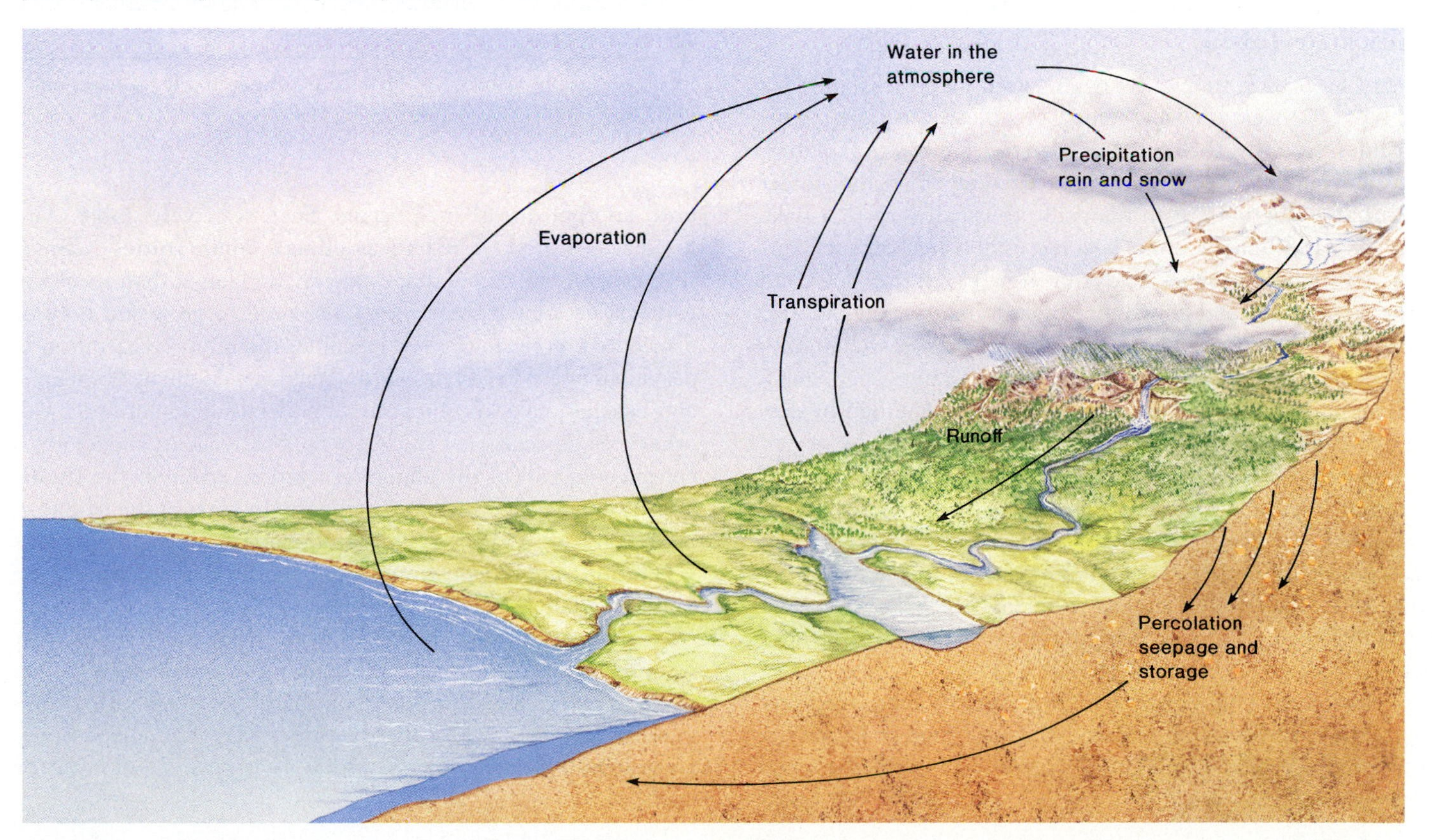

FIGURE 31.18

The water cycle. Atmospheric water falls mostly as rain or snow and either evaporates, runs off, or percolates into the soil. Some of the percolated water reaches underground storage and some of it is absorbed by plant roots. About 99% of all water taken in by roots is re-released into the atmosphere by transpiration.

For million of years, carbon, nitrogen, water, phosphorus, and other molecules have been passing through cycles. Some molecules that were a part of a primeval forest that became compressed and turned to coal may have become part of another plant after the coal was burned. Then the new plant may have been eaten by an animal, which in turn contributed molecules to yet another organism. Molecules in our own bodies may have been a part of some prehistoric tree, a dinosaur, a woolly mammoth, or even all three.

Succession

After a volcano spews lava over a landscape, or after an earthquake or a landslide exposes rocks for the first time, there is initially no life on the lava or rock surfaces. Within a few months, or sometimes within a few years, organisms appear, and a sequence of events known as **primary succession** occurs. During primary succession these initial organisms, including pioneer plants, gradually alter their environment as they grow and reproduce. Over time, accumulated wastes, dead organic material, and inorganic debris promote soil formation. Other changes (e.g., changes in shade and water content) favor different species, which may replace the original ones. These, in turn, modify the environment further, so that still other species become established (fig. 31.19).

Figure 31.19

Plant succession near Mt. St. Helens, Washington. Fir seedlings and other vegetation have reappeared 12 years after complete destruction by volcanic ash.

Primary Succession Initiated on Rocks or Lava

Primary succession frequently begins with bare rocks and lava that have been exposed by glacial or volcanic activity or by landslides. Initially, the rocks are sometimes subjected to alternate thawing and freezing, at least in temperate and colder areas. These fluctuations in temperature usually crack or flake the rocks, and lichens often become established on such surfaces. The lichens produce acids that slowly etch the rocks, and as the lichens die and contribute organic matter, they are replaced by other, larger lichens. Certain rock mosses adapted to long periods of desiccation also may become established, and a small amount of soil begins to accumulate, augmented by dust and debris blown in by the wind.

Eventually, enough of a mat of lichen and moss material is present to permit some ferns or even seed plants to become established, and the pace of soil buildup and rock breakdown accelerates (fig. 31.20). If deep cracks appear in the rocks, the larger seeds may widen them further as they germinate and their roots expand. Indeed, seedlings have been known to split rocks that weigh several tons (fig. 31.21).

As soil buildup continues, larger plants take over, and eventually the vegetation reaches an equilibrium in which the associations of plants and other organisms remain the same until another disturbance occurs. Such relatively stable plant associations are referred to as **climax communities.** Climax species are usually more shade-tolerant and larger than species in earlier successional stages; they also tend to grow and become distributed more slowly. For example, the climax vegetation of deciduous forests in eastern North America is dominated by maples and beeches, oaks and hickories, hemlocks and white pines, or other tree associations. In desert regions, various cacti form a conspicuous part of the climax vegetation, while in the Pacific Northwest, large conifers predominate. In parts of the Midwest, prairie grasses and other herbaceous plants form the climax vegetation, and in wet, tropical regions, a complex association of jungle plants constitutes the climax.

Succession, particularly in its more advanced stages, may not proceed in one direction. Sometimes it may appear to become reversed, depending on the influence of neighboring communities, and it may be difficult to interpret what one observes. Occasionally when a volcano produces ash instead of lava, the

FIGURE 31.20

Crustose lichens living on rocks.

FIGURE 31.21

Expansion in girth of the roots of this yellow birch has split a large rock.

ash buries the landscape and associated vegetation. As a result, secondary succession may occur; some of the successional stages involving lichens and mosses may be bypassed, and larger plants may become the successional pioneers. This occurred after the series of ash eruptions of Mount St. Helens in the state of Washington during the early 1980s.

Primary Succession Initiated in Water

In the northern parts of midwestern states such as Michigan, Wisconsin, and Minnesota, ponds and lakes of various sizes abound. Many were left behind by retreating glaciers and are often not drained by streams. The water that evaporates from them is replaced annually by precipitation runoff. The ponds and lakes also shrink each year as a result of primary succession. This succession often begins with algae either carried in by the wind or transported on the muddy feet of waterfowl and wading birds. The algae multiply in shallow water near the margin, and with each reproductive cycle the dead parts sink to the bottom. Floating plants such as duckweeds may then appear, often encircling the water. Next, water lilies and other rooted aquatic plants with floating leaves become established. Each group of plants contributes to the organic material on the bottom, which slowly turns to muck. Cattails and other flowering plants that produce their inflorescences above the water often take root in the muck around the edges. As they become established, the accumulation of organic material accelerates.

Meanwhile, the algae, duckweeds, and other plants move farther out, and the surface area of exposed water gradually shrinks. Grasslike sedges become established along the damp margins and sometimes form floating mats as their roots interweave with one another. Dead organic material accumulates and fills in the area under the sedge mats, and herbaceous and shrubby plants then move in. As the margins become less marshy, coniferous trees whose roots can tolerate considerable moisture (e.g., tamaracks or eastern white cedars) gain a foothold, eventually growing across the entire area as the pond or lake disappears. Trees continue to help form soil, and finally the climax vegetation takes over. No visible trace of the pond or lake remains, and the only evidence of its having been there lies beneath the surface, where fossil pollen grains, bits of wood, and other materials reveal the area's history. Such succession may take hundreds or even thousands of years and has never been witnessed from start to finish. The evidence that it does occur, however, is overwhelming (fig. 31.22).

Under natural conditions, stream-fed lakes and ponds eventually fill with silt and debris, although this, too, may take thousands of years. The streams that feed these lakes gradually bring in silt, and the nutrient content of the water slowly rises as dissolved organic and inorganic materials are brought in. This gradual enrichment, called **eutrophication,** stimulates the

FIGURE 31.22

Succession in a pond.

FIGURE 31.23

Quaking aspen (*Populus tremuloides*) trees, which often become established in mountainous areas after fires. Climax vegetation later may replace the aspens.

growth of algae and other organisms, which also add their debris to the bottom of the lake. When sewage and other pollutants enter the lake, eutrophication accelerates. Eutrophication may also accelerate when trees are cleared from land around lakes before the construction of homes and resorts. The cleared land erodes more readily, and precipitation runoff carries soil into the water. Regardless of size, all bodies of water, including rivers, are subjected to these processes.

Secondary Succession

Succession occurs whenever and wherever there has been a disturbance of a natural area. It proceeds at varying rates, depending on the climate, the soils, and the organisms in the vicinity. When an area is disturbed by fire, floods, or landslides, some of the original soil, plant, and animal material may remain. Ecologists refer to the pattern of changes that follow such disturbances as **secondary succession.**

Secondary succession, which has fewer phases and usually arrives at climax vegetation faster than primary succession, may occur if soil is present and there are surviving species in the vicinity. Indeed, survivors strongly affect subsequent succession. Many secondary successions follow human disturbances, such as the conversion of land to farmland. Other secondary successions follow fires. Grasses and other herbaceous plants become established on burned (or logged) land. These are usually followed by trees and shrubs that have widely dispersed seeds (e.g., aspen and sumac in the Midwest and East, and chaparral plants such as chamise and gooseberries in the West) or that sprout from root crowns. After going through fewer stages than are typical of primary succession, the climax vegetation takes over, often within less than one hundred years (fig. 31.23). Some biologists contend that in fire-dependent systems such as chaparral, no true climax communities exist.

C O N C E P T

Succession is the progressive change in community composition that occurs as vegetation develops in an area. Soil forms as a result of primary succession, which is initiated on freshly exposed rocks or lava, or in bodies of water. Secondary succession takes place after disturbances occur in areas where soil already exists. The end of a successional series is the climax community.

Boxed Reading 31.1

A Rogue's Gallery of Pests

Despite the efforts of quarantine officers, exotic species continue to invade new territories. Although the Africanized honeybee and zebra mussel grab the headlines, a variety of exotic plants have disrupted ecosystems. You learned about one of these pests—water hyacinth (*Eichhornia crassipes*)—in Chapter 14 (see boxed reading 14.2, "Clogging the Waterways," p. 321). There are more:

- **Kudzu** (*Pueraria lobata*) is a Chinese vine that was intentionally planted throughout the South in the 1930s to control erosion. Today, it covers between 2 and 4 million acres, and grows as fast as 30 cm **per day**. It grows over everything that it encounters. Today, kudzu is responsible for more than $50 million in lost farm and timber production each year. The latest idea for using kudzu is to harvest the starch from its huge roots.
- **Purple loosestrife** (*Lythrum salicaria*) is a pretty European weed that was brought to the United States in the late 1800s. Although the plant is used in some areas to make bouquets, people elsewhere are splashing purple loosestrife with herbicides in hopes of saving native wetlands plants. The USDA hopes to check the spread of the invader with insects now being tested as biological controls.

Other invaders are not so obvious. For example, eucalyptus is not a native species of California, and has disrupted many natural ecosystems. However, it may boost the productivity of California scrubland by tapping into deep water.

Fire Ecology

Natural fires started primarily by lightning and the activities of indigenous peoples have occurred for thousands of years in North America and other continents. Trees such as giant redwood and ponderosa pine, although scarred by certain types of fire, often survive, and the dates of fires can be determined by the proximity of the scars to specific annual rings (see fig. 13.35). Growth-ring studies from western North America show that forests of ponderosa pine burned about every six to seven years. The fires, climate, topography, and soil profoundly affected various biomes. Humans eventually tried to control the fires when they threatened life and property, and this in turn altered the vegetation. We have learned that trying to eliminate fires, at least in certain areas, disrupts natural habitats more in the long run than allowing them to occur. Agencies such as the U.S. National Park Service and the U.S. Forest Service now may allow some fires to burn under prescribed conditions.

Fires benefit grasslands, chaparral, and forests by converting accumulated dead organic material to mineral-rich ash, whose nutrients are recycled within the ecosystem (fig. 31.24). If the soil has been burned, some of its nutrients and organic matter may have been lost, and the composition of microorganisms originally present is likely to have changed. Losses are offset by the fact that nitrogen-fixing soil bacteria, including cyanobacteria, increase after a fire, and there is a decrease in fungi that cause plant diseases.

Figure 31.24

Forest fires convert dead organic matter to mineral-rich ash, whose nutrients are recycled within the ecosystem.

At least some of the North American grasslands originated and were maintained by fire. Since grassland fires have largely been controlled, many of these areas have now been invaded by shrubs. In some areas, such as the prairies of the Midwest, grasses are better adapted to fire than are woody plants. These grasses produce seeds within a year or two after germination. Perennial grass buds at the tips of rhizomes usually survive the most intense heat of fires, producing new growth the first season after a fire. Thus, a fire destroys only one season's growth of grass, often after reproduction has been completed (fig. 31.25).

Shrubs, however, have much of their living tissue above ground, and a fire may destroy several years' growth. Also, woody plants, unlike grasses, often do not produce seeds until several

FIGURE 31.25

Grassland regenerating after a fire.

years after a seed germinates. Many shrubs sprout from burned root crowns, particularly in chaparral areas, but repeated burning keeps them small. Most chaparral species, both woody and herbaceous, are so adapted to fire that their seeds will not germinate until fires remove accumulated litter and inhibitory substances produced by the plants during growth.

Fires also play a role in the composition of forests. In the mountains of east-central California, gooseberry and deerbrush are abundant after a fire, but their numbers stabilize within 15–30 years when larger trees return to the area. Ponderosa, jack and southern longleaf pines, and Douglas firs (which do not tolerate shade) are among the species that repeatedly replace themselves after fires, and the seeds of some species rarely germinate until they have been exposed to fire. In view of the long-range beneficial effects of fires in some ecosystems, wise land-management practices of the future will include guiding succession in many plant communities for the greatest utility and safety.

HUMANS IN THE ECOSYSTEM

The total human population of the world was less than 20 million in 6,000 B.C. During the next 7,750 years it rose to 500 million; by 1850 it had doubled to 1 billion, and 70 years later it had doubled again to 2 billion. The 4.48 billion mark was reached in 1980, and within five years the population had grown to 4.89 billion. The estimate for the year 2,000 exceeds 6.25 billion. The earth remains constant in size, but humans have occupied more of it over the past few centuries, and their population density has also greatly increased. In feeding, clothing, and housing themselves, humans have greatly affected their environment. We have cleared natural vegetation from vast areas of land and drained wetlands. We have polluted rivers, oceans, lakes, and the atmosphere. We have killed pests and plant-disease organisms with poisons, which have also killed natural predators and other useful organisms. In general we have disrupted the delicately balanced ecosystems that existed before we began our depredations.

If we are to survive on this planet, we must control the size of our population. Also, many foolish agricultural and industrial practices that have accompanied population growth must be replaced with practices that will restore some ecological balance. Agricultural practices of the future must include returning organic material to the soil after each harvest, instead of adding only inorganic fertilizers. Timber and other crops must be harvested in ways that prevent erosion. The practice of clearing brush with chemicals must be changed or abolished. Industrial pollutants will have to be rendered harmless and recycled whenever possible. Energy conservation will need to become universally practiced, and wasteful packaging and consumption practices must be stopped or sharply curtailed.

Many substances that now are discarded (e.g., organic waste, paper products, glass, metal cans) must be recycled on a larger scale. Biological controls will have to replace the use of toxic chemical controls whenever possible. Water and energy must be conserved. Rare plant species, with their largely unknown genetic potential for medicine and agriculture, will need to be saved from extinction by the preservation of their habitats. The general public must understand the urgent need for wise land management and conservation to resist the influential forces that promote unwise measures in the name of progress—before additional large segments of our natural resources are irreparably damaged or forever lost. The alternative appears to be nothing less than death from starvation, respiratory diseases, poisoning of our food and drink, and other catastrophic events endangering the survival of humanity.

Writing to Learn Botany

If you could introduce congressional legislation to restore disrupted ecosystems, what bills might you propose that could do this without increasing taxes?

The Greenhouse Effect

The **greenhouse effect** refers to the maintenance of an equable temperature over the planet, and is linked to *global warming*. Global warming is a global rise in temperature due to the accumulation in the atmosphere of gases that permit radiation from the sun to reach the earth's surface, but prevent the heat from escaping back into space. Although there is much controversy about the extent and impact of the greenhouse effect, many biologists believe that it could significantly alter our environment in the future.

Gases such as chlorofluorocarbons are involved in producing the greenhouse effect. These gases are the relatively recent by-products of the manufacture of refrigerants, plastics,

Boxed Reading 31.2

Curbing Methane Emissions to Curtail the Greenhouse Effect

In an effort to limit global warming, the United States government proposed in early 1991 that a comprehensive framework for limiting greenhouse gases would be preferable to concentrating on a single gas. Hogan, Hoffman, and Thompson (Nature 354 [1991]:181–82) observed that methane is often ignored in debates about the greenhouse effect. They argued that reducing methane emissions would be relatively easy, and could significantly reduce global warming. They point out that reducing methane emissions would be 20–60 times more effective in reducing the potential warming of the earth's atmosphere over the next century than reducing carbon dioxide emissions. They also noted that methane released by human activities is generally a wasted resource, opening the possibility that reductions might even be profitable.

Much methane is generated in landfills by the anaerobic decomposition of wastes. By collecting this gas, existing recovery systems can reduce emissions by 30%–60%. Coal mining emits similar amounts of methane, although relatively few mines emit most of the gas that is released. If vertical wells are used before and during the mining operations, methane emissions from underground mine areas could be reduced by more than 50%, and the cost of mine ventilation would also be reduced.

The production of oil and natural gas also produces much methane. Improved technologies such as leak-proof pipelines can reduce methane emissions in a cost-effective manner. The second largest source of methane related to human activities involves the cattle industry; the cattle themselves produce large amounts of the gas. Some feeds that may reduce methane emissions while increasing productivity have been identified. One hormone (bovine somatotropine), if approved for use, could reduce methane emissions by 10% while increasing milk production. Recovery systems can profitably recover 50%–90% of the methane generated from animal wastes. Such recovery systems are already in use in India, where they generate enough electricity to power lights for a few hours each evening. However, animal wastes might be more appropriately used as soil fertilizer.

Rice cultivation is considered to be the single largest source of methane emitted by human activities. An integrated approach to irrigation, fertilizer application, and cultivar selection could reduce methane emissions by as much as 30%. Alternative approaches to agricultural practices involving the clearing of land and crop stubble by fire could also reduce methane emissions, but trade-offs clearly would be involved.

and aerosol propellants. Other gases that promote the greenhouse effect have been part of our atmosphere for millions of years—carbon dioxide and methane.

Carbon Dioxide

D. L. Lindstrom of the University of Illinois at Chicago and D. R. MacAyeal of the University of Chicago recently examined records of cores of ancient ice from Siberia, Scandinavia, and the Arctic Ocean. Using computers to simulate the status of ice and atmosphere going back 30,000 years, they found that levels of carbon dioxide had increased enough at the end of the most recent ice age to melt the ice and raise the earth's temperature. Their findings suggest that cycles of ice ages followed by shorter warm periods may have been caused solely by rising and falling levels of carbon dioxide in the atmosphere.

In 1986, worldwide carbon dioxide emissions from transportation and industrial sources totaled somewhat less than 5 billion tons. By 1987 the total was more than 5.5 billion tons and has continued to rise. The burning of fossil fuels and deforestation (which destroys the major recyclers of carbon dioxide) have caused a 25% increase of carbon dioxide in the atmosphere since 1850. In the last twenty-five years alone, the increased insulation of carbon dioxide has resulted in the earth's atmosphere becoming 0.4° C warmer, and between 1983 and 1990 the surface temperature of the ocean rose about 0.8° C.

These increases in temperature may seem insignificant, but during the last ice age in North America, when ice covered the northern United States and Canada, the average temperature of the earth at sea level was only 4° C colder than it is now. In 1989, Mostafa Tolba, head of the United Nations Environment Program, estimated that if the current levels of gas release into the atmosphere continue, the earth's temperature will probably rise between 1.4 and 4.3° C in the next fifty years. These higher temperatures will melt the polar ice caps; the released water will raise sea levels and inundate low-lying, often densely populated, coastal areas. The Environmental Protection Agency estimates that for each 30 cm the ocean rises, it moves 30 m inland. During the past century, worldwide ocean levels have risen 12.7 cm (5 inches), and it is estimated that more than 1,800,000 hectares of land in the United States alone will be flooded if the temperatures rise as predicted. Higher temperatures can also affect winds, currents, and weather patterns, causing droughts and creating deserts in some areas, while causing heavy rainfall in others. The greatest grain-production areas of the world lie in the interiors of continents. If they become warmer and drier, world food supplies may be affected.

FIGURE 31.26

Termites in wood.

Methane

Swamps and wetlands have long been known to be sources of methane produced by anaerobic bacteria. Many animals produce methane during digestion, and large amounts of this gas are released by wood-digesting organisms in the guts of termites (fig. 31.26). The total annual production of methane in the atmosphere has also increased slowly in recent years. A small part of this increase may stem from the increased numbers of termites in cleared tropical rain forest areas which, in 1990, were being destroyed at the rate of more than 58,000 hectares per day.[1] This means that we clear an area of tropical rain forest equivalent to that of more than 1,100 basketball courts *every second,* twenty-four hours per day.

Acid Rain

Acid rain occurs after the burning of fossil fuels releases sulphur and nitrogenous compounds into the atmosphere. There, sunlight converts these compounds to nitrogen and sulphur oxides, and they combine with water to become acid rain (mostly nitric acid and sulfuric acid). Acid rain changes the pH of lakes and streams and kills many organisms in them. It also injures plants upon which it falls (fig. 31.27). About half of the Black Forest in Germany has succumbed to its effects. Acid rain has also stunted or killed trees growing downwind from industrial sites. Acid rain also affects nonliving materials. For example, the natural weathering of ancient Mayan ruins in southern Mexico, the Parthenon in Greece, and monuments in Washington, D.C. has been accelerated by acid rain during the past decades.

Acid rain is not responsible for all dead or dying trees in the world's forests. Some trees have perished as a result of insufficient rainfall during successive dry years. Others have succumbed to insect infestations or salt scattered to melt ice and snow on roads, and still others have been weakened by disease.

1. 1 hectare = 100 ares = 11,960 yd^2 = 10,000 m^2 = 2.47 acres

FIGURE 31.27

Damage caused by acid rain to spruce and fir trees.

Ozone

Ozone (O_3) is a form of oxygen in the stratosphere that is more effective than ordinary oxygen (O_2) in shielding living organisms from intense ultraviolet radiation. Ozone is produced from oxygen with the aid of ultraviolet light. Chlorofluorocarbons, which are inert chemicals used for refrigeration and other industrial purposes, are broken down by sunlight at high altitudes into active compounds that destroy ozone. Destruction of ozone in the stratosphere results in increased exposure to ultraviolet radiation, which increases the incidence of skin cancers, genetic mutations, and damage to vegetation—especially crops.

The accelerating destruction of the ozone shield has been recognized as a serious global problem by both the United States and the European Economic Community. In 1987 the United States proposed a 50% reduction in the production and uses of chlorofluorocarbons by the year 2,000, and in 1989 the European Economic Community proposed a total ban on chlorofluorocarbons, also by the year 2,000. In 1990, however, developing nations such as India, China, and Brazil had plans to expand the production of chlorofluorocarbons and contended that a ban would place them at an economic disadvantage. Because global cooperation on this matter is urgently needed, the major

FIGURE 31.28

Toxic waste pouring into a river.

industrial nations are seeking ways to allay the economic concerns of developing countries by, for example, producing viable alternatives to chlorofluorocarbons. One such alternative was introduced in 1994, when nonCFC refrigerants became standard in the air-conditioning systems of most new automobiles sold in the United States and elsewhere.

Chlorofluorocarbons are not the only danger to the protective ozone layer. Bromine-based compounds called *halons*, which are commonly found in electronic equipment and in portable fire extinguishers, can be up to ten times more destructive of ozone than are chlorofluorocarbons. Halon concentrations in the atmosphere increased about 20% per year between 1980 and 1986, according to the Environmental Protection Agency. Some scientists believe the concentrations are as much as 50% higher than that, and are recommending that powders and other inert gases be substituted for halons in fire extinguishers.

Contamination of Water

Much of the pollution in our lakes and streams comes from the dumping of toxic industrial wastes and runoff from polluted land (fig. 31.28). Other sources include the spraying of pesticides, the exhaust and other emissions from aircraft and ships, and airborne pollutants originating from the combustion of fossil fuels. Groundwater supplies become polluted from the infiltration of the soil by pesticides and wastes, septic tank effluents, and garden and farm fertilizers. Even deep wells, which seldom became polluted in the past, are becoming contaminated with substances that damage the health of humans and animals that depend on the water. In response to the problem, many communities are improving their water treatment plants, and many individuals have installed water filtration systems in their homes.

Genetic engineering and bacteria will probably play a major role in reducing water pollution. For example, a bacterium has been bred that can remove more than 99% of 2,4,5-T (a major component of Agent Orange) from a contaminated environment, and several other bacteria are being genetically engineered to break down other toxic wastes.

C O N C E P T

The great increase in numbers of humans has been accompanied by serious disruption or destruction of ecosystems. To ensure human survival, currently widespread individual, industrial, and agricultural practices must be modified or changed. Emissions of gases or other substances that contribute to global warming or destroy stratospheric ozone must be curbed.

Chapter Summary

Ecology deals with the relationships of organisms to one another and to their environment. Ecologists view these interrelationships as a hierarchy. At the lowest level of the hierarchy are populations, which are groups of individuals of the same species. Populations of different species are associated into a community, and communities and their physical environment form an ecosystem.

Populations vary in numbers, in density, and in the total mass of individuals. The variations depend on both biotic and abiotic factors.

Ecosystems ideally sustain themselves entirely through photosynthetic activity and the interactions of producers, consumers, and decomposers. In any ecosystem, the producers and consumers form food chains of various lengths; these determine the flow of energy through the different levels.

Energy is not recycled in an ecosystem. At each level of a food chain, only about 1% of the light-energy available is converted to organic material, and the energy gradually escapes in the form of heat as it passes from one level to another. Variations in food production in an ecosystem are balanced by corresponding variations in consumption, so that the ecosystem is self-maintaining. When two species are grown together in a controlled environment, one of the species often disappears. In nature, if two species coexist, at least one is reduced in size and/or numbers, and survival depends on their occupying different microhabitats.

Living components of an ecosystem may also influence its composition by parasitism and by secreting growth-inhibiting substances, growth-promoting substances, and predator repellents. The chemical compounds or structural modifications coevolved by both producers and consumers are maintained in a delicate balance in an ecosystem.

Some nitrogen from the air is fixed by bacteria found in legumes and other plants. Nitrogen flows from dead plant and animal tissues into the soil and from the soil back to the plants. Water leaches nitrogen from the soil and carries it away when erosion occurs. Other nitrogen is lost from harvesting crops, but the loss can be reduced if wastes are decomposed and annually returned to the soil. Replacing lost nitrogen with chemical fertilizers can eventually create hardpan. Carbon, water, and other substances also undergo cycling.

Succession, which is a directional change in the species composition of a given area over a period of time, occurs whenever there has been a disturbance of natural areas on land or in water. Primary succession involves the formation of soil, beginning with either a dry or wet environment. Secondary succession occurs in areas previously covered with vegetation. A stable vegetation (climax vegetation) becomes established at the conclusion of succession and remains until a disturbance disrupts it. Fires alter the nutrient and organic composition of the soil and open up areas in which trees such as aspens may become established until slower growing climax species return.

Human populations have grown rapidly. The disruption of ecosystems by human activities directly or indirectly associated with feeding, clothing, and housing billions of people threatens the survival not only of humans but other organisms as well. Survival will ultimately depend on improved agricultural practices, reduction of wasteful activities, conservation, and recycling.

The greenhouse effect refers to the maintenance of an equable temperature over the planet. A global rise in temperature due to the insulating effect of gases has been noted. Acid rain, which damages or kills organisms, occurs when sulphur and nitrogen compounds released by the burning of fossil fuels are converted to nitric and sulphur oxides by sunlight and then mix with water to become acids. In the stratosphere, sunlight converts chlorofluorocarbons into compounds that destroy the ozone shield that protects us from intense ultraviolet radiation. Halons, used in fire extinguishers, also destroy ozone.

Water contamination occurs when toxic wastes, pesticides, effluents from septic tanks, and fertilizers wash or leach into surface and groundwater. Bacteria may soon be used to break down various contaminants.

Questions for Further Thought and Study

1. Distinguish among populations, communities, and ecosystems.
2. What is meant by the phrase, *balance of nature?*
3. Discuss the nitrogen cycle.
4. What is the greenhouse effect? Why should we be concerned about it?
5. What concept or idea discussed in this chapter do you think is most important? Why?
6. Energy and nutrients move differently in ecosystems. What are the consequences of this difference?
7. Today it is fashionable for people to refer to themselves as *environmentalists*. What does this term mean to you?
8. In industrialized countries, humans have disrupted ecosystems everywhere they have settled. Could humans also improve an ecosystem? If so, how?
9. What vital services do ecosystems provide?
10. If a plant species is unnecessary for the functioning of an ecosystem, should we be concerned if it is eradicated? What might be some economic, moral, or aesthetic reasons for protecting an "unnecessary" plant?
11. How have human activities affected the distribution of plants and animals where you live?
12. Discuss the meaning of the following quotations:

    "What makes it so hard to organize the environment sensibly is that everything we touch is hooked up to everything else."

    —Isaac Asimov, writer and biochemist

    "Eventually, we'll realize that if we destroy the ecosystem, we destroy ourselves."

    —Jonas Salk, medical researcher

    "We prefer economic growth to clean air."

    —Charles Barden, environmentalist

    "Destroying species is like tearing pages out of an unread book, written in a language humans hardly know how to read, about the place where they live."

    —Rolston Holmes III, philosopher

13. Recent evidence suggests that an ecosystem's productivity and stability—long treated as independent of each other—are, in fact, inextricably linked. How would you test this hypothesis?

Suggested Readings

Articles

Appenzeller, T. 1993. Filling a hole in the ozone argument. *Science* 262:990–991.

Baskin, Y. 1994. Ecologists dare to ask: How much does diversity matter? *Science* 264:202–203.

Bormann, F. H. 1985. Air pollution and forests: An ecosystem perspective. *BioScience* 35:434–441.

Bushbacher, R. J. 1986. Tropical deforestation and pasture development. *BioScience* 26:22–28.

Finegan, B. 1984. Forest succession. *Nature* 312:109–114.

Holloway, M. 1993. Sustaining the Amazon. *Scientific American* 269:90–99.

la Rivière, J. W. M. 1989. Threats to the world's water. *Scientific American* 261:80–94.

Swetnam, T. W. 1993. Fire history and climate change in giant sequoia groves. *Science* 262:885–889.

Books

Barbour, M., et al. 1987. *Terrestrial Plant Ecology*. 2d ed. Menlo Park, CA: Benjamin/Cummings.

Begon, M. 1990. *Ecology: Individuals, Populations and Communities*. 2d ed. Cambridge, MA: Blackwell Scientific.

Caldwell, M. M., and R. W. Pearcy, eds. 1994. *Exploitation of Environmental Heterogeneity by Plants. Ecophysiological Processes Above and Below Ground*. San Diego, CA: Academic Press.

Dafni, A. 1993. *Pollination Ecology*. Oxford: Oxford University Press.

Glenn-Lewin, D. C., et al., eds. 1992. *Plant Succession: Theory and Prediction*. New York, NY: Chapman and Hall.

Kozlowski, T. T., and C. E. Ahlgren, eds. 1974. *Fire and Ecosystems*. San Diego, CA: Academic Press.

Lovett-Doust, J., and L. Lovett-Doust. 1988. *Plant Reproductive Ecology*. Oxford: Oxford University Press.

Rice, E. L. 1984. *Allelopathy*. 2d ed. San Diego, CA: Academic Press.

Silvertown, J., and J. L. Doust. 1993. *Introduction to Plant Population Ecology*. 3d ed. Cambridge, MA: Blackwell Scientific.

Southwick, C. H., ed. 1985. *Global Ecology*. Sunderland, MA: Sinauer Associates.

Trona pinnacles and desert plants on a dry lake bed in south central California.

CHAPTER

32

Biomes

Chapter Outline

Chapter Overview

Ecosystems in different areas of the world are so similar that ecologists classify them into groups called *biomes*. The same biome may therefore span several continents. Although the distribution of biomes is determined mostly by climate, each biome is characterized primarily by its vegetation. Classification of the major vegetation types of the world entails the recognition of anywhere from seven to seventeen biomes, depending on the classifier. In this text, the major ecosystems of the world are grouped into eight terrestrial biomes: tundra, taiga, temperate deciduous forests, grassland and savannas, deserts and semideserts, mediterranean scrub, mountain forests, and tropical rain forests. This classification includes the major types of vegetation that occur on land, but it excludes aquatic and marine ecosystems.

INTRODUCTION

Most biomes cover large areas, and most also occur on more than one continent. A biome usually includes several ecosystems, each with its unique combination of plants and animals, that share similar climates and soils. Most of the important biomes in North America also occur on other continents (fig. 32.1).

If you have a keen eye, a little knowledge about plants, and a good imagination, you may already know how the types of vegetation in different parts of the world are similar. For example, desert vegetation in the southwestern United States looks like desert vegetation in northern Africa,[1] and the grasslands of western China look like the grasslands of the central United States. Ecologists recognize such similar types of vegetation as convenient indicators of comparable ecosystems.

A group of similar ecosystems is called a **biome.** Succulent plants such as cacti and low-growing trees and shrubs with spines and small leaves are examples of typical desert vegetation; an area where these plants grow is called a *desert biome*. Grasslands of central North America, western and central Asia, southeastern South America, and northeastern Australia are, collectively, a grassland biome.

Viewing the earth as a collection of biomes is somewhat artificial, because biomes overlap and ecosystems usually include local areas with plants that do not fit the vegetation types of the biome according to our definition. Furthermore, biomes usually exclude aquatic and marine ecosystems, because we do not know enough about them to organize them into groups. Nevertheless, biomes remain a convenient way to organize the terrestrial ecosystems of the world for discussion. The characteristics of individual biomes are influenced mostly by abiotic factors such as precipitation, temperature, and soil type. Furthermore, seasons and climate are influenced by the amount of solar energy that reaches a given area, which changes throughout the year with the angle of the sun's rays.

The wide bands of air currents that occur over the earth (fig. 32.2) and the location of mountains affect vegetation patterns. At the equator, warm air that is more or less saturated with water vapor rises and condenses at higher altitudes, resulting in rain. Moisture-laden air may also be carried by winds toward mountains, where much of the moisture falls on the windward side as the air is forced upward. The leeward side of the mountains usually lies in the "rain shadow," where little precipitation occurs. For example, on the west coast of Hawaii, the largest of the Hawaiian islands, there are areas that receive less than 22 cm of rain annually, while precipitation only 10 km away on the windward side of the mountains exceeds 250 cm per year.

The organization of abiotic factors and the types of vegetation associated with them have led ecologists to recognize anywhere from seven to seventeen biomes. In this text, we have chosen to be conservative. We emphasize the similarities among eight groups of ecosystems; that is, we group the earth's vegetation types into eight biomes. We exclude aquatic and marine ecosystems because there is no adequate classification of underwater ecosystems on a worldwide scale. Biologists have no idea whether the lakes and oceans of the world can be organized into one biome or a hundred biomes.

1. Botanists get extra enjoyment out of movies that are set in certain areas of the world but filmed where the vegetation is not what it is made out to be. For example, there is a scene in *The Young Lions* where a German army officer, played by Marlon Brando, peeks over a hill at an enemy encampment in North Africa during World War II. You don't have to look very closely to see the branches of an ocotillo (*Fouquieria splendens*) nearby; this species grows in a southern California desert, where that scene was filmed, but not in the Sahara Desert of Africa, where that scene was supposed to occur.

TUNDRA—ARCTIC AND ALPINE

The word **tundra** is derived from a Russian word meaning "treeless, marshy plain." Tundra (figs. 32.3 and 32.4) is a vast, mostly flat biome, whose terrain is marshy in the summer and frozen for much of the remainder of the year. It occupies about 25% of

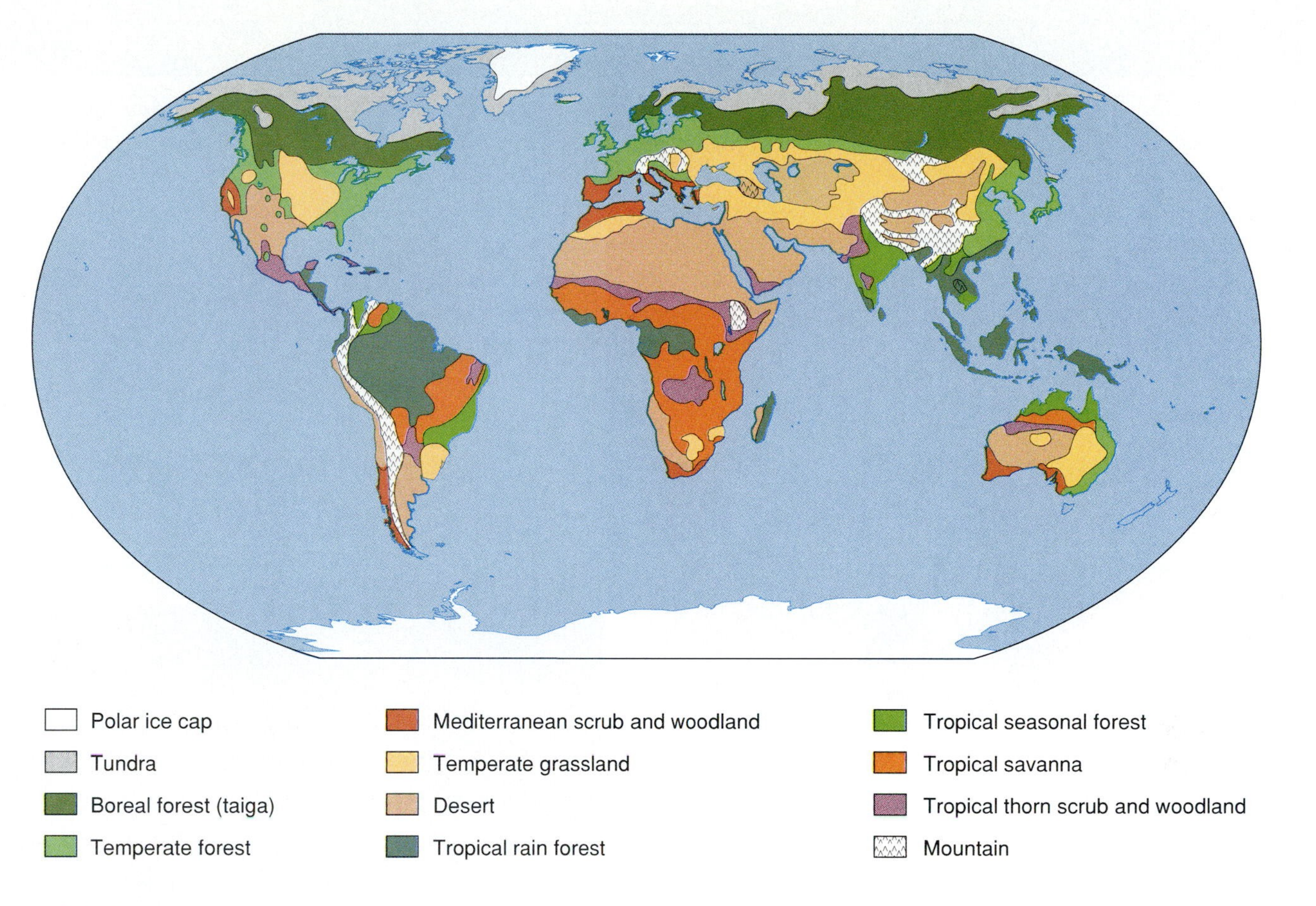

FIGURE 32.1

Major biomes of the world.

the earth's land surfaces, primarily above the Arctic Circle, with some, however, extending further south. Alpine tundra, which occurs in patches above timberline on mountains below the Arctic Circle, is seldom flat, and also differs from arctic tundra in having less annual variation in daylength, less humidity, and more direct solar radiation. The climate and soil of tundra are usually not suitable for agricultural activities.

Fierce, drying winds and freezing temperatures can occur in tundra on any day of the year, but temperatures can also reach 27° C or higher during a midsummer day. In addition to low temperatures, tundra is characterized by shallow (5.0–7.5 cm deep), nutrient-poor clay soils that are waterlogged during the growing season, which lasts only two to three months. The cold, anaerobic conditions of the soil prevent any significant recycling of the extensive organic components, which are produced primarily by peat mosses. Although annual precipitation averages less than 25 cm per year, waterlogging occurs primarily because of the flat terrain, and also because **permafrost** (permanently frozen soil) beneath the surface prevents water from draining into the soil. Permafrost also causes water to accumulate throughout the biome in the form of numerous shallow lakes.

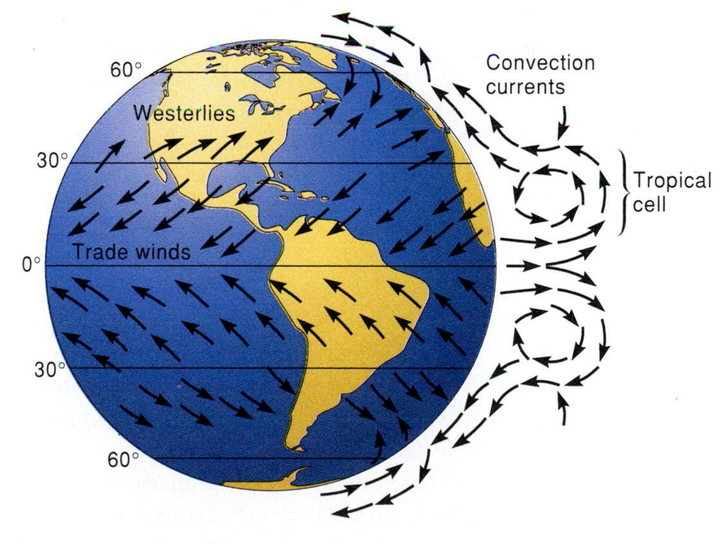

FIGURE 32.2

Global wind patterns.

BOXED READING 32.1

BIOMES: ELEVATION VS. LATITUDE

High mountain regions have climates and vegetation types that are similar to those of the arctic tundra; mountain areas a little lower in elevation are similar to the high latitude coniferous forests known as taiga. These observations may seem obvious now, but the similarities between ecosystems in northern latitudes and ecosystems in high mountains were not correlated until the early nineteenth century. The correlation between elevation and latitude was first noted by the German explorer and naturalist Alexander von Humboldt, who observed in 1805 that climbing a high mountain was analogous ecologically to traveling further north or south from the equator.* Von Humboldt based his observations on his exploration of Mount Chimborazo in Ecuador, although he could have discovered the same relationships much closer to home—between the high Alps and far northern Europe.

* This observation was first published in Paris in his *Essai sur la geographie des plantes* (*Essay on the Geography of Plants*).

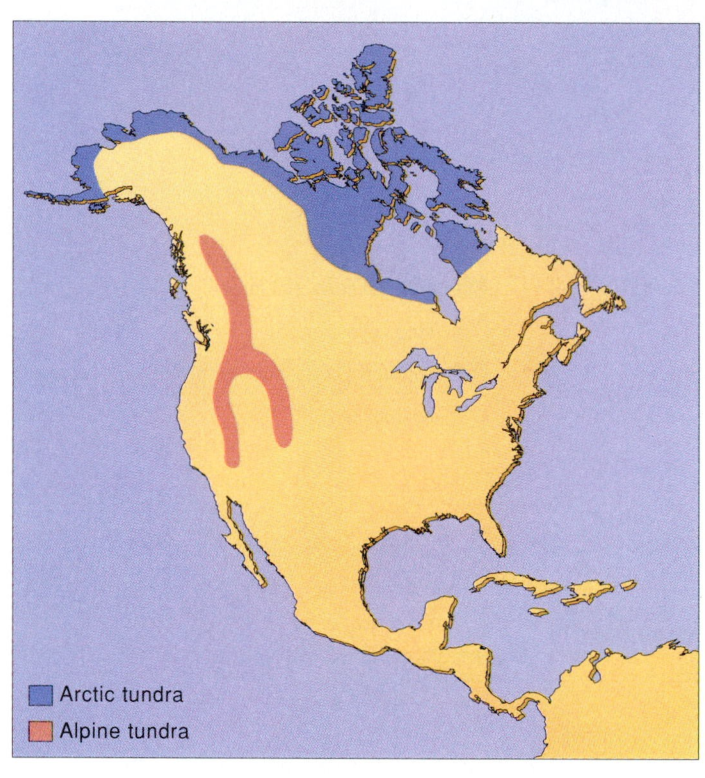

FIGURE 32.3

The distribution of tundra in North America.

Permafrost, which usually occurs at a depth of a few cm to about 1 m below the surface throughout arctic tundra, may be up to 500 m deep in northern Alaska and Canada; at Nordvik in northern Siberia it is more than 600 m deep. Permafrost is often absent from alpine tundra.

The vegetation of arctic tundra consists primarily of grasses, sedges, mosses, and lichens, and it is mostly evergreen. Grasses and sedges tend to dominate in alpine tundra, whose environment differs in several ways from those of arctic and antarctic areas. Solar radiation is more intense at high altitudes because humidity is lower, and the sun's rays are more direct than in arctic areas. Daylengths are not as variable, especially in mountains toward the equator.

A.

B.

FIGURE 32.4

Tundra. (a) Alaskan tundra in early fall. (b) Alpine tundra in Glacier National Park, Montana. Tundra plants have a short but spectacular growing season near and above the timberline.

Tundra is treeless, but low shrubs survive and even dominate in many parts. Such plants include willow (*Salix*), birch (*Betula*), and blueberry (*Vaccinium*), which are mostly less than 25 cm tall and rarely more than 1 m tall at maturity. Leaves, which tend to be small, are protected by thick cuticles and dense hairs. The stems are often covered with lichens (which can photosynthesize vigorously at −10° C) that protect the plants from drying winds and frigid temperatures. When temperatures drop below freezing, ice crystals form in intercellular spaces. The cells are exceptionally tolerant of the dehydration that occurs as the freezing draws water from the cells. This form of protection from cold is so effective that the dormant stems of arctic woody plants can survive immersion in liquid nitrogen, which has a temperature of −196° C.

The flora of tundra also includes low perennials that produce brightly colored flowers during the brief growing season and form brilliant mats over the topsoil. Some plants growing in tundra reproduce vegetatively much of the time, producing seeds only during exceptionally mild and long growing seasons, (which occur once or twice every century). Their seeds remain viable for long periods—sometimes for centuries. Few of the seeds that germinate encounter the exceptionally long, relatively warm growing seasons necessary for development to maturity; however, just enough do survive to perpetuate the species. Many of these perennials are adapted to the harsh growing conditions by having as little as 10% of their mass above ground, the remainder being in the form of rhizomes, tubers, bulbs, or fibrous roots below the surface. The level of the permafrost determines the depth to which the roots can grow.

Flower buds usually form by the end of the previous growing season and remain dormant, developing rapidly and opening during the first thaws of the following summer. Some of the bowl-shaped flowers exhibit heliotropic movements, so that the sun's energy is concentrated on the stamens and pistils throughout the day (see Chapter 19). The temperature inside such flowers may be as much as 25° C higher than the air temperatures outside, and the warmth attracts insect pollinators.

During midsummer, photosynthesis can occur throughout all of the twenty-four-hour day, and virtually all of the carbohydrate resulting from photosynthetic activity is produced within a few weeks each year. Most of the carbohydrate is stored beneath the surface, with stem and leaf growth being minimal. Many of the grasses and sedges use C_4 photosynthesis (see Chapter 7).

Arctic-adapted plants of alpine sorrel (*Oxyria digyna*) flower only after daylengths have exceeded twenty hours, and other arctic species probably have similar photoperiods. The arctic sorrels also reach peak photosynthetic rates at lower temperatures than do plants of the same species adapted to warmer climates. Fruits and seeds of arctic plants mature in as little as three weeks. Annuals are rare in tundra because the cold environment usually inhibits the germination of tiny seeds and the subsequent completion of the life cycle in such plants, even during summers that are milder than usual.

FIGURE 32.5

A barren area similar to a blowout in Alaskan tundra.

Tundra is exceptionally fragile. A truck or car driven across it compresses the soil enough to kill roots, and the tracks remain for many years. Occasionally, sheep grazing on tundra pull up patches of the matted vegetation, leaving exposed edges. High winds catch the exposed edges and rip away larger segments of mat, leaving barren patches called *blowouts* (fig. 32.5).

CONCEPT

Terrestrial ecosystems are classified into biomes on the basis of the similarities of their vegetation and physical environments. Tundra is characterized by freezing temperatures, permafrost, and the virtual absence of trees and annuals. The tundra biome includes the northernmost ecosystems, which are dominated by low-growing plants. It occupies about one-fourth of the earth's land surfaces.

What are Botanists Doing?

Find out what kind of research is being done on biomes by looking through journals such as the *Journal of Environmental Management*, the *Journal of Vegetation Science*, and *Conservation Biology*. Is there any pattern to the kind of research being published about ecosystems, areas of the world, or kingdoms (plant or animal) as the main subjects, and so on?

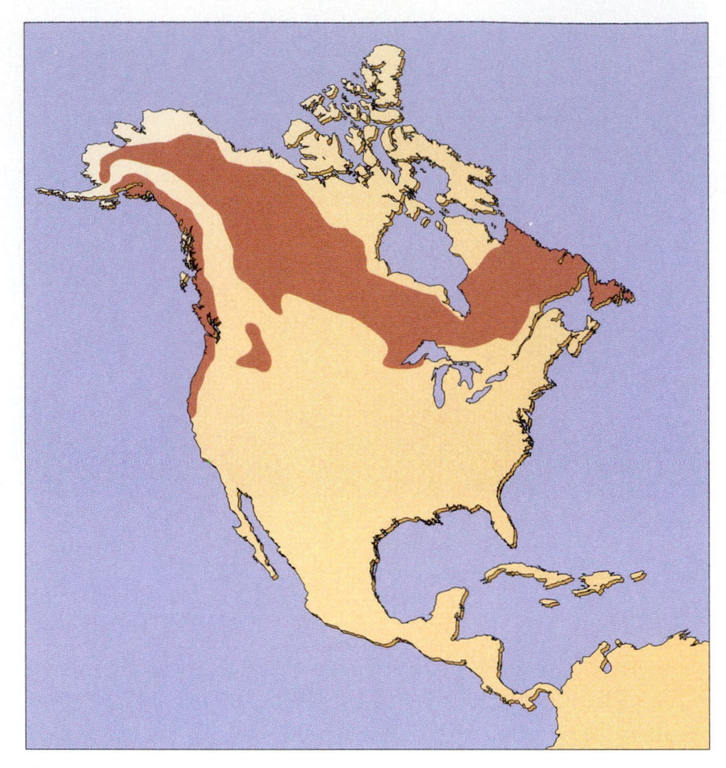

FIGURE 32.6

Distribution of taiga in North America.

TAIGA

Taiga (figs. 32.6 and 32.7), also referred to as *northern coniferous* or *boreal forest,* occurs mostly adjacent to and south of the arctic tundra across large areas of North America and Eurasia. (Similar vegetation occurs in high mountains of many parts of the world, but these ecosystems are classified as the mountain forest biome.) Permafrost occurs within more than 65% of the taiga, usually less than a meter from the surface. Lightning often starts fires, which melt the permafrost and stimulate plant growth in the burned area for several years.

The soils of taiga are usually acidic and nutrient-poor, making them unsuitable for most agricultural activities other than timber farming. Snow accumulates in the taiga during the winters, which are long and cold. In midwinter there may be only six hours of light per day, and temperatures drop to −50° C or lower during the coldest months. In summer, the temperatures may reach 27° C, and daylight lasts for up to eighteen hours. Most precipitation occurs in the summer, ranging from about 25 cm to more than 100 cm in parts of western North America.

The vegetation of taiga is relatively uniform and dominated by a few genera of coniferous trees, including spruce (*Picea*), fir (*Abies*), and pine (*Pinus*). Deciduous trees such as birch, poplar, aspen, willow, alder (*Alnus*), and tamarack (*Larix*) often occur in some of the wetter areas, such as the margins of the

FIGURE 32.7

A taiga scene. Plants that grow in the taiga are adapted for long, cold winters and heavy snowfall.

many lakes, ponds, and marshes. Nitrogen-fixing actinomycetes (*Frankia*) associated with the roots of the alders enable these trees to survive in otherwise comparatively barren soils. A thick-walled epidermis and hypodermis, sunken stomata, and a thick cuticle all adapt the leaves of taiga conifers to the rigors of the harsh winter climate. Many (mostly bulbous) perennials and a few cold-hardy shrubs occur, but annuals, which usually have small seeds and delicate seedlings, are generally prevented from becoming established by the severe climatic conditions.

Taiga consists of coniferous forest ecosystems that are mostly adjacent to the tundra biome.

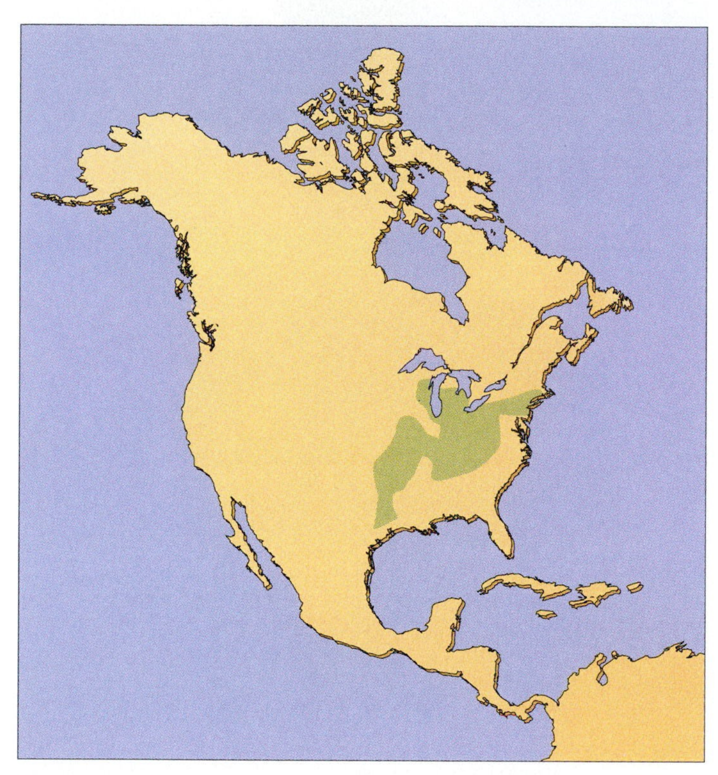

FIGURE 32.8

The distribution of temperate deciduous forests in North America.

FIGURE 32.9

An opening within the eastern deciduous forest.

TEMPERATE DECIDUOUS FORESTS

Deciduous trees are broad-leaved species that shed their leaves annually during the fall and remain dormant during the winter. These trees tend to be most abundant in areas where the summers are warm and the winters are relatively cold. Like the taiga, nearly all **temperate deciduous forests** (figs. 32.8 and 32.9) occur on large continental masses in the Northern Hemisphere. In North America, this type of forest occurs from the Great Lakes region south to the Gulf of Mexico and extends from the general vicinity of the Mississippi River to the eastern seaboard. Temperate deciduous forests also occur in western Europe and Asia. Temperatures within the area vary greatly, but normally fall below 4° C in midwinter and rise to above 20° C in the summer. The trees, which usually have thick bark and become dormant before the onset of cold weather, are well adapted to subfreezing temperatures, particularly if the cold is accompanied by snow cover that prevents the ground from freezing down to the root zone. Precipitation averages between 90 and 225 cm per year, and occurs mostly during the summer.

Some of the most beautiful of all the broad-leaved trees occur in temperate deciduous forests. In the upper midwest, sugar maple (*Acer saccharum*, fig. 32.10) and American basswood (*Tilia americana*) predominate. Sugar maple also occurs to the northeast, where it is often associated with the stately American beech (*Fagus grandifolia*). In the west and west-central part of the forest, oak (*Quercus* species) and hickory (*Carya*

FIGURE 32.10

Sugar maples (*Acer saccharum*) in the fall. Sugar maples are tapped to make maple syrup (see box 21.3, "Tapping Wood for Maple Syrup").

Figure 32.11

New growth sprouting from the base of an American chestnut (*Castanea americana*) killed by chestnut blight.

Figure 32.12

An eastern white pine (*Pinus strobus*).

species) are dominant. Oaks are also abundant along the eastern slopes of the Appalachian Mountains, where American chestnut (*Castanea americana*) was once a conspicuous part of the forest. The chestnuts have now virtually disappeared, having been killed by chestnut blight disease (*Cryptonectria parasitica*, an ascomycete accidentally introduced from China to North America in 1904 (fig. 32.11). Oak trees extend into the southeastern United States, where they are associated with pine trees and other species such as the bald cypress (*Taxodium distichum*).

Before the arrival of European immigrants, a mixture of large deciduous trees that included maple (*Acer* species), ash (*Fraxinus* species), basswood, beech, buckeye (*Aesculus* species), hickory, oak, tulip tree (*Liriodendron tulipifera*), and magnolia (*Magnolia* species) occurred on the eastern slopes and valleys of the Appalachian Mountains. Some of these trees were over 30 m tall and had trunks up to 3 m in diameter. Except for a few protected pockets in the Great Smoky Mountains National Park, the largest trees of this rich forest have been all but eliminated by logging.

An extension of the temperate deciduous forest, in which the trees are smaller than those of the same species elsewhere in the biome, occurs in an area northeast of Baton Rouge, Louisiana, in western Tennessee and Kentucky, and in southern Illinois. American elm (*Ulmus americana*), also once a part of the forest, is rapidly disappearing as Dutch elm disease kills both trees in the wild and those planted along city streets and on college campuses. Several midwestern towns, which had hundreds of elms planted along their streets, were left with few live trees within a year or two after Dutch elm disease was introduced in the vicinity. Dutch elm disease is caused by *Ophiostoma ulmi*, an ascomycete (see Chapter 25) that was introduced to North America from Europe in 1930. The spores are spread by the elm bark beetle, which infects the phloem while boring into the inner bark.

A mixture of deciduous trees and evergreens occurs on the northern and southeastern borders of the temperate deciduous forest. Hemlock (*Tsuga canadensis*) and eastern white pine (*Pinus strobus;* fig. 32.12) occur from New England west to Minnesota and south to Alabama along the Appalachians. The once-vast stands of eastern white pine are now almost gone,

FIGURE 32.13

Trilliums (*Trillium* sp.) carpeting the floor of part of the eastern deciduous forest before the leaves of the trees have formed a dense canopy that blocks sunlight from reaching the ground.

their valuable lumber having been used for construction and other purposes. Some have been lost to still another tree disease, white pine blister rust (*Cronartium ribicola,* a basidiomycete with two hosts; see Chapter 25), but scattered trees remain. Various pines dominate the Atlantic and Gulf coastal plains from New Jersey to Florida, and west to east Texas.

During the summer, the trees of deciduous forests form a relatively closed canopy that prevents most direct sunlight from reaching the floor. Many of the showiest spring flowers of the region, such as bloodroot (*Sanguinaria canadensis*), hepatica (*Hepatica* species), Dutchman's breeches (*Dicentra cucullaria*), buttercups (*Ranunculus* species), trilliums (*Trillium* species), and violets (*Viola* species), flower before the trees have leafed out fully, and complete most of their growth within a few weeks (fig. 32.13). Other plants that can tolerate more shade, such as waterleaf (*Hydrophyllum* species), flower after the canopy has formed. Several members of the sunflower family, such as asters (*Aster* species) and goldenrods (*Solidago* species), flower in succession in forest openings from midsummer through fall.

The temperate deciduous forest biome, which occurs where summers are warm and winters are relatively cold, is characterized by broad-leaved trees that lose their leaves annually in the fall.

GRASSLANDS AND SAVANNAS

Natural **grasslands** (figs. 32.14 and 32.15) occur toward the interiors of continental masses and along arid coastlines in areas having temperate climates. Grasslands tend to intergrade with forests, woodlands, or deserts at their margins, depending on the amounts and patterns of precipitation. Grasslands may receive as little as 25 cm of rainfall, or as much as 100 cm annually in some areas. Air temperatures can range from 50° C in midsummer to −45° C in midwinter.

Savannas (fig. 32.16), which, like grasslands, are dominated by grasses, occur in areas having subtropical to tropical

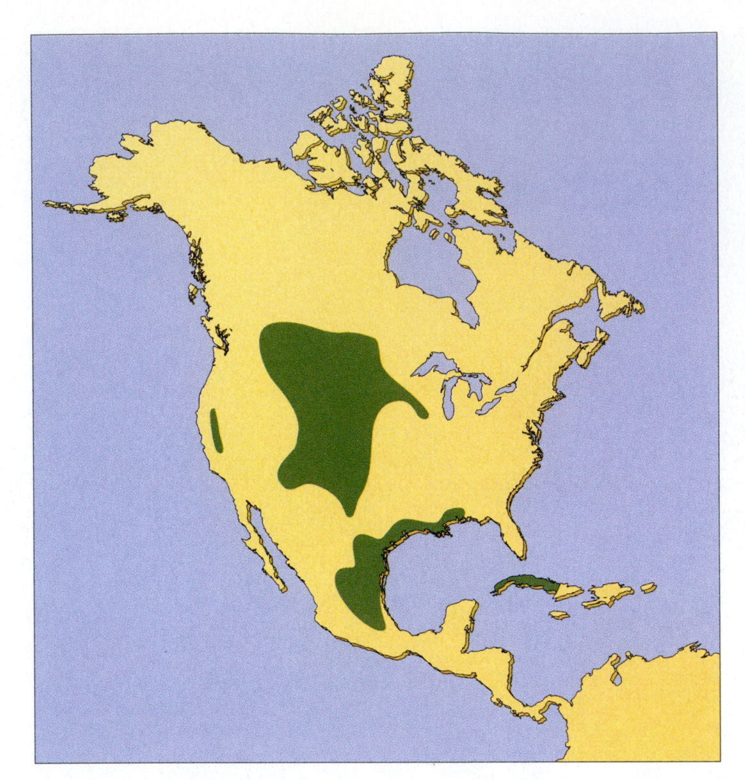

FIGURE 32.14

Distribution of grasslands and savannas in North America.

A.

B.

C.

FIGURE 32.15

Grasslands. (a) Bunch grasses of midwestern plains. (b) Short grass prairie in northern Nebraska. (c) Tall grass prairie in Illinois.

climates, where long, dry periods are interspersed with seasonal precipitation that may sometimes exceed 150 cm per year. In South America, savannas extend from south of the Amazon Forest to near the Uruguayan border; in Africa, they occur both north and south of the tropical forests; in Australia, they occur on both sides of the large interior desert; and in North America, they occur from the drier parts of Texas south to the interior areas of eastern Mexico.

Savannas characteristically include widely scattered, thick-barked deciduous trees that lose their leaves during the dry periods. Baobabs (*Adansonia* species; fig. 32.17) are common trees of African and Australian savannas. The larger trees usually give way in the drier areas to smaller scrubby trees (mostly *Acacia* species) that typically produce many conspicuous thorns and seldom exceed 15 m in height. Herbaceous perennials with bulbs, rhizomes, and other storage organs are common in grasslands and savannas.

In North America, the natural grasslands, or prairies, were once grazed by huge herds of bison. The bison disappeared as the settlers cultivated more and more of the land and hunters slaughtered more and more of the large animals. By 1889, there were only 551 bison left. Large areas of prairie are now used for growing cereal crops (particularly corn and wheat) and for grazing cattle.

Before it was destroyed, the American prairie was a remarkable sight. In Illinois and Iowa, the grasses grew over 2 meters tall during an average season and another meter taller during a wet one. A dazzling display of wildflowers began before the

FIGURE 32.16

Savanna in Emas National Park, Brazil.

FIGURE 32.17

A baobab tree (*Adansonia digitata*) in an east central African savanna.

young perennial grasses emerged in spring and continued throughout the growing season. Even today, more than fifty species of flowering plants can be observed flowering simultaneously in the middle of spring on as little as one hectare of undisturbed natural grassland (fig. 32.18).

Grasslands in areas with a Mediterranean climate (e.g., the Great Central Valley of California), where most of the precipitation occurs during the winters, may include *vernal pools*. These are temporary pools of rainwater that often accumulate in areas with clay soil or hardpan beneath them. The water evaporates and disappears after the rains stop. Their spring floras include an orderly succession of flowering plants unique to the habitat, some appearing initially at the pool margins, with each species forming a distinct zone or band until the water is gone (fig. 32.19). Some species flower only in the damp soil and drying mud that remains. The seeds of several species germinate under water.

African savannas are susceptible to how people manage big game animals such as elephant, giraffe, lion, cape buffalo, zebra, and a large variety of antelopes and deer (fig. 32.20). Elephants, for example, feed almost continuously for much of their lives, with each adult daily consuming between 150 and 175 kg of vegetation. The destruction of vegetation by elephants

FIGURE 32.18

A mixed-grass prairie in South Dakota.

FIGURE 32.19

Meadowfoam (*Limnanthes douglasii*) in flower at the edge of a vernal pool in northern California.

can become a serious problem when the animals are confined to parks. As the animals multiply, park herds are systematically thinned to prevent devastation of the habitat. If the large animals are not artificially confined, however, they travel large distances in search of seasonal fruits; significant destruction of a restricted habitat is thus avoided.

C O N C E P T

The most common plants in the grassland biome and the savanna biome are grasses. The grassland biome occurs in temperate climates, and the savanna biome occurs in subtropical to tropical climates where there may be long, dry periods each year.

Writing to Learn Botany

Ecosystems in the grassland biome occur in central and western Asia, eastern Europe, Argentina, New Zealand, central and eastern Mexico, and the central United States. If this biome is so widespread, does it matter that the ecosystems in eastern Europe and in the central United States have been almost totally destroyed by human activities? After all, we still have plenty of the grassland biome in the other ecosystems.

FIGURE 32.20

Elephants in a savanna.

DESERTS AND SEMIDESERTS

Sand, heat, mirages, oasis, and *camels* are features almost universally associated with **deserts** (fig. 32.21). These features are typical of the Sahara and other large deserts that occur both north and south of the equator in the interior of Africa and Eurasia, primarily in the vicinity of 20° to 30° latitude. Camels, however, are not typical of the deserts or semideserts that cover roughly 5% of North America, nor are they found in the Australian desert. They are also absent from other smaller deserts such as those of Namibia, Chile, and Peru, which occur in coastal areas where there are cold offshore ocean currents.

Deserts may occur whenever precipitation is consistently low or where water passes quickly through the soil (fig. 32.22). Many deserts get less than 10 cm of precipitation per year, but some desert areas with porous soil, such as those of the Sonoran and other deserts in the southwestern United States and northwestern Mexico, may receive 25 cm or more annually (see box 32.2, "Desertification: What Is It and What Causes It?").

The low humidity of deserts causes large fluctuations in daily temperatures. During the summer, for example, daytime temperatures can exceed 35° C and often fall below 15° C the same night. Solar radiation is greater in the dry air than it is in areas where atmospheric water vapor filters out some of the sun's rays. Many desert plants have adapted to these conditions through the evolution of crassulacean acid metabolism (CAM) photosynthesis; C_4 photosynthesis is also more common in desert plants than in those of other areas (see Chapter 7).

Other adaptations of desert plants include thick cuticles, fewer stomata, water-storage tissues in stems and leaves, leaves with a leathery texture and/or reduced size, and even the total absence of leaves. The roots of the bizarre *Welwitschia* plants of southwest African deserts (see fig. 29.14b) get all the water they need from fog drip; these plants thrive in the absence of

FIGURE 32.21

Sand dunes and an oasis in the Sahara Desert, Algeria.

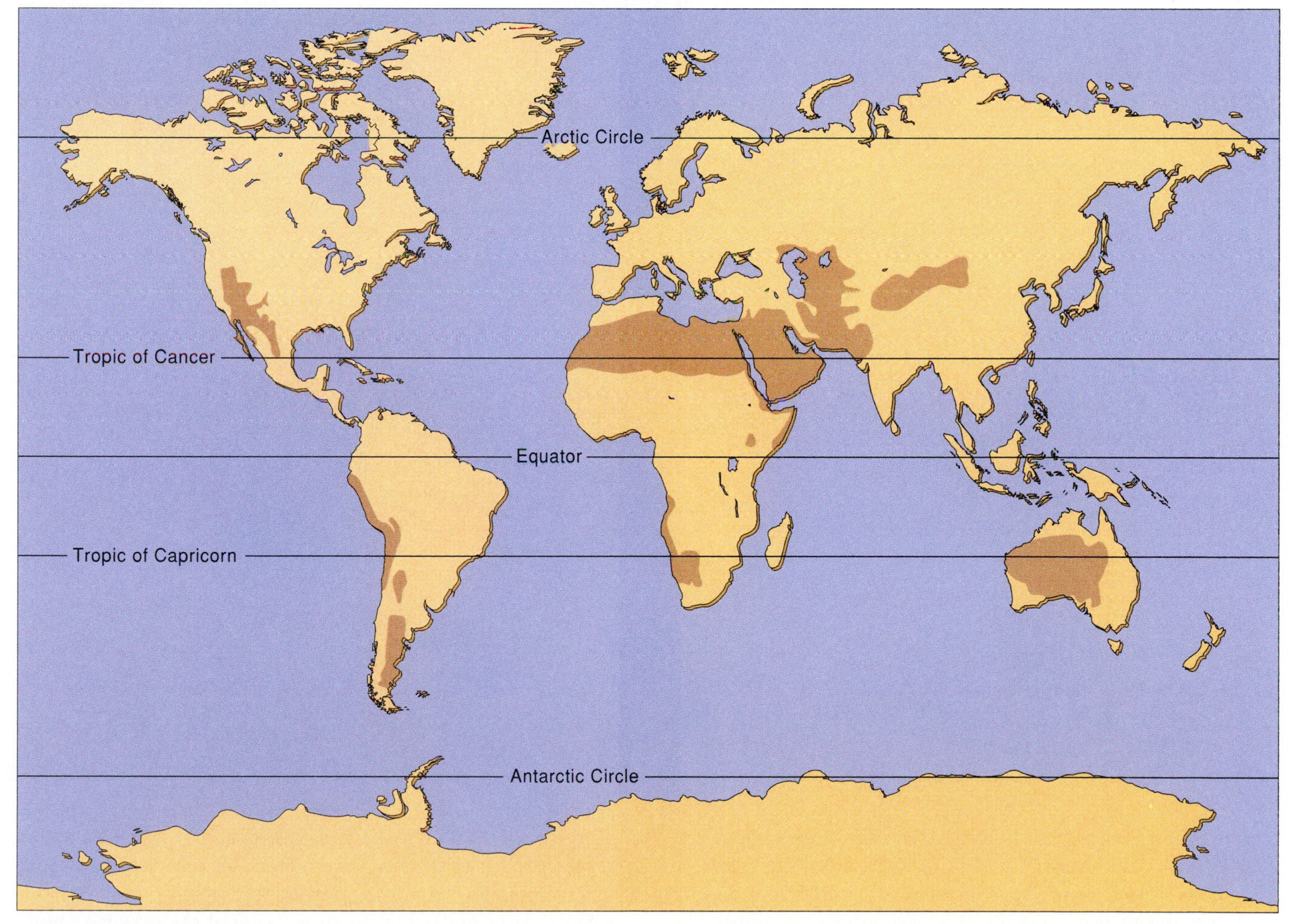

FIGURE 32.22

Distribution of world's deserts.

Boxed Reading 32.2

Desertification: What Is It and What Causes It?

Desertification is the conversion of grasslands or other biomes into deserts. Deserts have been expanding for the past 15,000 years due to worldwide drying since the most recent period of glaciation; thus desertification is not new. Nevertheless, it has been accelerated in certain parts of the world because of human activities. One of the causes of desertification is the irrigation of arid habitats for farming. Irrigation water always contains salt, in some cases more than a ton (ca. 900 kg) per acre-foot.* As crop plants absorb the water, the salt is left behind. Salt has accumulated in some areas over the years to the point that these areas are about to undergo desertification. For instance, many farmlands in the southwestern United States that are irrigated with water from the Colorado River are becoming polluted with salt and are expected to be useless for farming within a few years.

Desertification in the Sahel region of the southern Sahara Desert is another story. Desertification in this semiarid region, which is correlated with the apparent drying out of adjacent savanna ecosystems, has caused tremendous human suffering, most notably the widespread famines of 1972 and 1984. The most widely accepted hypothesis to explain desertification in the Sahel is the *drought hypothesis:* —the Sahel is drying out because of the lack of rain. "Drought" and "famine" are used in the same breath so often that cause and effect are almost fixed in our minds. But let's look a little closer. Widespread famine is a relatively recent phenomenon, yet the low rainfall of the 1970s and 1980s has occurred several times earlier in this century without causing desertification. Furthermore, a ranch of more than 1,000 km^2 in the heart of the Sahel has remained green since it was started in 1968, despite receiving the same low rainfall as adjacent areas undergoing desertification. These observations contradict the drought hypothesis.

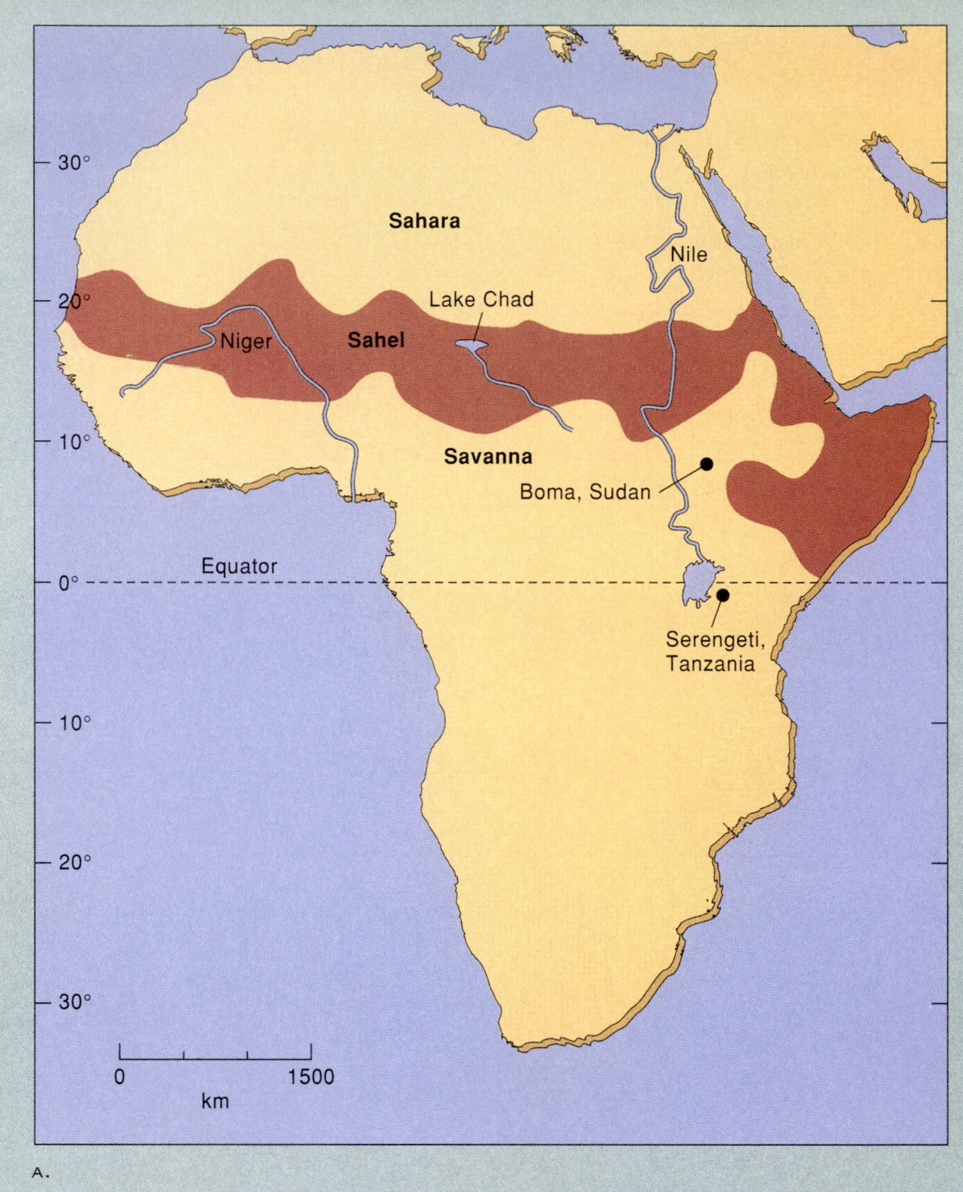

A.

Box Figure 32.2

(a) The Sahel is a semiarid belt that spans Africa between the Sahara Desert and the savanna; (b) Rainfall records from the Sahel show that, since 1900, periods of below average rainfall have been relatively frequent, although widespread drought and famine in the area have only happened since 1973.

subterranean water or precipitation. In cacti and other succulents without functional leaves, the stems are photosynthetic (fig. 32.23). Such plants have widespread, shallow root systems that can absorb water rapidly after the infrequent rains. The water is then stored for long periods inside the stems. Other perennials grow from bulbs that are dormant for much of the year. Some desert trees, such as mesquite (*Prosopis*), have long taproots that can reach several meters down to the water table.

Annuals provide a spectacular display of color and variety, particularly during an occasional season with above-average precipitation (fig. 32.24). The seeds of annuals often germinate after a fall or winter rain and then grow slowly for

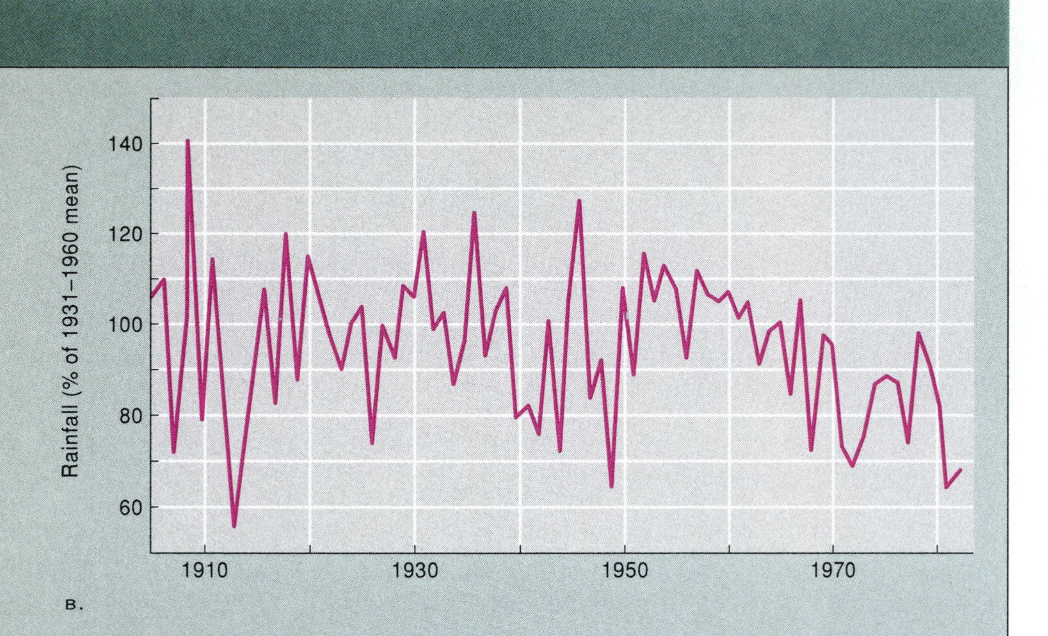

B.

What is the difference between the green ranch area and the desertifying areas of the Sahel that might explain why desertification is occurring? Several interrelated factors, all pointing to human activities, are cited as the cause of desertification. This explanation is referred to as the *settlement-overgrazing hypothesis.* Native people of the Sahel used to migrate seasonally, driving small herds of cattle to the north when rains came and the annual grasses grew, then back to the wetter southern areas as the rainy season passed. The cattle gained weight and reproduced based on a diet of the high-protein annual grasses that grew in the fertile soils of the north. When water became scarce, they returned to the low-protein perennial grasses that grew in the nutrient-poor soil of the south; this diet is not nutritious enough for growth and reproduction.

The pattern of seasonal migration of people and their cattle has been replaced by settlements and agriculture over the years since the Second World War, when Western countries increased their aid toward agricultural development for developing African countries. Unfortunately, crops in the southern areas of the Sahel were unsuccessful because of the poor quality of the soils there. With the exception of the green ranch, cattle were allowed to overgraze the northern annual grasses to compensate for the destruction of these areas by failed farms until the grasses could no longer survive. Both the north and the south are now so severely denuded that they look like they have suffered through an extensive drought. However, the effects of the change in life-style of the native people from a migratory one to a sedentary one are evidence that overgrazing is the real culprit (box fig. 32.2).

*An acre-foot is the amount of water that covers an acre to a depth of 1 foot. It is the preferred unit of measure for water usage in agriculture. There is no metric equivalent to acre-foot, unless you convert acre to hectare and foot to centimeter, but there is no use for such a unit.

Source: Sinclair, A. R. E. and J. M. Fryxell. The Sahel of Africa: Ecology of a Disaster. In *Canadian Journal of Zoology* 63:987–94, 1985. Copyright © 1985 National Research Council of Canada, Ottawa, Canada.

several months before producing flowers in the spring. Hundreds of different species of desert annuals can occur within a few square kilometers of desert in the southwestern United States.

In the colder parts of North American deserts, sagebrush (*Artemisia* species) is common, while in the warmer desert areas creosote bush (*Larrea tridentata*) and many species of cacti predominate. In areas with more precipitation, palo verde (*Cercidium* species), mesquite, ocotillo (*Fouquieria splendens*), and junipers (*Juniperus* species) are common.

Desert ecosystems, like those of the tundra, are fragile, and recovery from disturbances often takes many years. Large numbers of off-road vehicles and wild donkeys have devastated several desert areas in the southwestern United States; current efforts to curb the destruction thus far have met with very limited success.

CONCEPT

Deserts occur wherever precipitation is consistently low or the soil is too porous to retain water. Ecosystems of the desert biome are hotter and drier than those of other ecosystems. Plants of the desert biome include succulents and plants with small, tough leaves and deep taproots or extensive, shallow root systems.

Mediterranean Scrub

Areas around the Mediterranean Sea have dry, hot summers and cool, wet winters. Similar areas occur along parts of the west coasts of North and South America, the southwestern tip of Africa, and in parts of southern and southwestern Australia. Unique scrubby vegetation that is either evergreen or deciduous in summer has evolved in these areas with Mediterranean climates (figs. 32.25 and 32.26). Most growth occurs in the relatively short, wet winters and early spring; plants are dormant for the remainder of the year.

Many of the plants in **Mediterranean scrub** biomes are well adapted to fires, which are frequent. Such plants have thick roots that resprout after the above-ground parts have been burned (fig. 32.27); others have rhizomes that lie close enough to the ground to survive when a fire races through the erect vegetation above them. Still others, such as ear drops (*Dicentra chrysantha; D. ochroleuca*), have seeds that usually do not germinate unless seared by fire.

The dense Mediterranean scrub of lower elevations of the Pacific Coast is known as *chaparral,* a word derived from *chabarro,* the Basque name for scrub oak. Dominant chaparral

FIGURE 32.23

A saguaro cactus in southern Arizona.

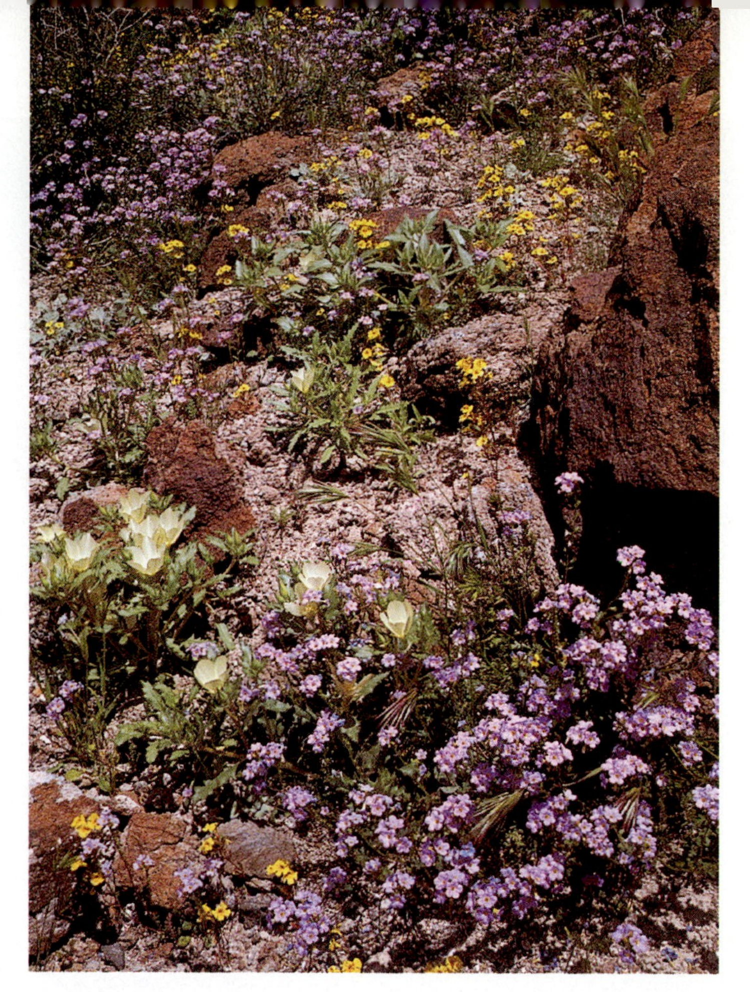

FIGURE 32.24

Spring wildflowers in the Mojave Desert.

FIGURE 32.25

Distribution of chaparral in North America.

FIGURE 32.26

A chaparral scene. California lilac (*Ceanothus integerrimus*) is flowering.

FIGURE 32.27

Chaparral shrubs resprouting from the base a year after a fire swept through the area.

species include buckbrush (*Ceanothus cuneatus*), scrub oak (*Quercus dumosa*), silk-tassel bush (*Garrya fremontii*), chamise (*Adenostoma fasciculatum*), manzanita (*Arctostaphylos* species), California bay (*Umbellularia californica*), and poison oak (*Toxicodendron diversilobum*).

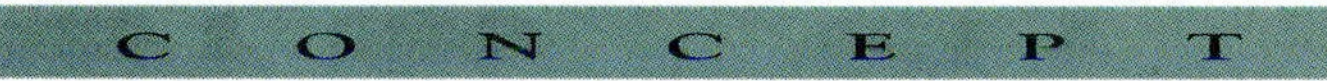

Ecosystems that have hot, dry summers and cool, wet winters are dominated by scrubby vegetation types, which comprise the Mediterranean scrub biome.

MOUNTAIN FORESTS

Mountain forests occur in widely scattered areas of the world, including Norway, the Himalayan region, the Andes of South America, the Pyrenees and other central European mountains, the Caucasus region, Mongolia and eastern Siberia, the Russian Sikhote Alin' and Ural mountain ranges, the Atlas mountains of Morocco, central African mountain ranges, southwest Saudi Arabia, central New Guinea, and western North America. The **mountain forest** biome is similar to the taiga biome in being characterized by coniferous forests. However, permafrost is absent in mountain forests and, unlike taiga, which is restricted to the Northern Hemisphere, this biome occurs in both the Northern and Southern Hemispheres. A brief discussion of North American mountain forests follows.

In the geologic past, deciduous forests extended to western North America. As the climate changed and summer rainfall was reduced, conifers largely replaced the deciduous trees, although some (e.g., maple, birch, aspen, oak) still remain, particularly at the lower elevations. Today, coniferous forests occupy vast areas of the Pacific Northwest and extend south along the Rocky Mountains and the Sierra Nevada and California coast ranges (figs. 32.28 and 32.29). Isolated pockets of this biome also occur in other parts of the West, particularly toward the southern limits of the mountains. The trees tend to be large, particularly in and to the west of the Cascade Mountains of Oregon and Washington and on the western slopes of the Sierra Nevada. Part of the reason for the huge size of trees such as Douglas fir (*Pseudotsuga menziesii*) is the high annual rainfall, which exceeds 250 cm in some areas. The world's tallest trees, the coastal redwoods of California (*Sequoia sempervirens*; fig. 32.30), however, apparently depend more on moisture from fog for their size and longevity than on large amounts of rain. The fog, which condenses on the foliage and drips into the soil, also reduces transpiration rates.

As moisture-laden air is forced upward along mountain ranges, annual precipitation changes between sea level and the higher slopes. Temperatures also drop about 1° C for each 100 m of elevation. Thus, the western and southern sides of mountain ranges in the Northern Hemisphere are often warmer and drier than the eastern and northern slopes because the eastern slopes receive direct sunlight only in the mornings when it is cooler and there is less solar radiation; at the same time the northern

FIGURE 32.28

Distribution of mountain forests in North America.

FIGURE 32.29

A mountain forest scene in California's Sierra Nevada. The grayish green trees are gray pines (*Pinus sabiniana*).

slopes receive no direct sunlight. It is not surprising, therefore, that even at the same latitudes, different associations of plants are common at different elevations (see box 32.1 "Biomes: Elevation and Latitude").

At lower elevations in both the Rocky Mountains and the Sierra Nevada, the predominant conifer is ponderosa pine (*Pinus ponderosa*). At lower elevations in the northern part of the Cascades, Douglas fir, western red cedar (*Thuja plicata*), and western hemlock (*Tsuga heterophylla*) are more common. At intermediate elevations in the Sierra Nevada, the established conifers include sugar pine (*Pinus lambertiana*), white fir (*Abies concolor*), and Jeffrey pine (*Pinus jeffreyi*), while at higher elevations different conifers, such as red fir (*Abies magnifica*), predominate.

Most of the North American mountain forest biome has comparatively dry summers. Lightning often starts fires, and did so long before human carelessness became a leading cause of forest fires. Several tree species are well adapted to survival after being partially burned. Douglas fir, for example, has a thick protective bark that can be charred without transmitting enough heat to the interior to kill the cambium. Moreover, Douglas fir seedlings thrive in open areas after a fire (fig. 32.31). When the bark of the giant redwoods (*Sequoiadendron giganteum*) of the Sierra Nevada is burned, the trees are rarely killed. This has undoubtedly contributed to the great age and size of many of the trees.

The cones of some pine trees, such as knobcone pine (*Pinus attenuata;* fig. 32.32), remain closed and do not release their seeds until a fire causes them to open. Similarly, the seeds

FIGURE 32.30

Coastal redwoods (*Sequoia sempervirens*) in northern California.

FIGURE 32.31

A large Douglas fir (*Pseudotsuga menziesii*). Note the fire scars on the trunk.

FIGURE 32.32

The closed seed cones of knobcone pine (*Pinus attenuata*) do not open until seared by fire.

of several other species germinate best after they have been burned. Because of these attributes of members of the mountain forest biome, rangers in some of our national parks occasionally allow fires at higher elevations to run their natural course. The above-average incidence of fires occurring since humans have come in large numbers to the forest has made it necessary, however, for most fires to be controlled, even though doing so interferes with natural cycles that would otherwise occur in the biome.

CONCEPT

The mountain forest biome is similar to the taiga biome in vegetation type; however, the mountain forest biome occurs both north and south of the equator and has no permafrost. North American mountain forests have a fairly conspicuous zonation of species.

Doing Botany Yourself

Identify the biome you live in or closest to. How could you estimate the biomass of a square meter or a hectare in your biome? How might you be able to use this information to distinguish your biome from other biomes?

TROPICAL RAIN FORESTS

Tropical areas of the world include many biomes, from seasonally dry areas to grasslands to high mountains. The greatest biological diversity on earth, however, occurs in **tropical rain forests.** The principal tropical rain forests of the world occur in the Amazon Basin of Brazil, the Congo Basin of central Africa, much of central America, and in or near the equatorial regions of southeast Asia and Indonesia. A narrow band also occurs along the northeast coast of Australia. About 7% of the earth's surface, representing nearly half of the forested areas of the earth and 25% of the earth's species, are included in this biome (figs. 32.33 and 32.34). Tropical rain forests have existed for about 200 million years and, unlike other biomes, have escaped the effects of past ice ages.

Rain forests occur throughout areas of the tropics where annual rainfall normally ranges between 200 and 400 cm and temperatures range between 25° C and 32° C, with night temperatures seldom dropping more than 5° C lower than those at noon. Although monthly rainfall varies, there is no dry season, and some precipitation occurs every month of the year, frequently in the form of afternoon cloudbursts. The humidity seldom drops below 80%. Such climatic conditions favor and support a diversity of plants and animals so great that the number of species in tropical rain forests exceeds that of all the other biomes combined.

FIGURE 32.33

Distribution of tropical rain forests in North America.

FIGURE 32.34

A scene in a tropical rain forest in Costa Rica. These biomes contain a vast array of diverse plants and animals.

Rain forests are dominated by broadleaf evergreen trees, whose trunks are often unbranched for as much as 40 meters or more, with luxuriant crowns that form a beautiful, dark green and multilayered canopy (fig. 32.35). Small parts of the forest, particularly along rivers, are jungles where the canopy is so dense that little light penetrates to the floor. The few herbaceous plants that survive are generally confined to openings in the forest. There are hundreds of species of trees, each usually represented by widely scattered individuals. Root systems are shallow, and the trees are often buttressed, the broader bases compensating for lack of root depth (fig. 32.36). Organic matter is sparse in tropical soils, which tend to be acidic. The soils are often deficient in important nutrients such as potassium, magnesium, calcium, and phosphorus, or the minerals (particularly phosphorus) are in forms that the plants cannot use. Despite the lush growth, there is little accumulation of litter or humus, because decomposers rapidly degrade leaves and other organic material on the forest floor; the nutrients released by decomposition are quickly recycled or leached by the heavy rains.

Most of the plants in rain forests are woody flowering plants; no conifers occur naturally in these ecosystems. Although evergreen plants predominate, some are deciduous species that can shed their leaves from some branches, retain leaves on others, and flower on yet other branches all at the same time, while branches of adjacent trees of the same species are losing their leaves or flowering at different times. Many *lianas* (woody vines that are rooted in the ground) hang from tree branches, and even more numerous *epiphytes* (especially orchids and bromeliads) are attached to limbs and trunks (fig. 32.37). The epiphyte roots, which are not parasitic, do not touch the ground; the plants are sustained entirely by rainwater that accumulates in their leaf bases and by their own photosynthetic activities. Traces of minerals, also necessary to the growth of the epiphytes, accumulate in the rainwater as it trickles over decaying bark, bird-droppings, and dust.

C O N C E P T

Tropical rain forests include broad-leaved, mostly evergreen plants in a hot, humid environment. More species occur in tropical rain forests than in any other kind of ecosystem.

The Lore of Plants

The southernmost recorded flowering plant is the Antarctic hair grass (*Deschampsia antarctica*), which was found at Latitude 68 ° 21′ S on Refuge Island, Antarctica. The northermost plants are the yellow poppy (*Papaver radicatum*) and the Arctic willow (*Salix arctica*), which grow on the northernmost land at Latitude 83 ° N. The highest altitude at which any flowering plants have been found is 6,400 m (21,000 ft) on Kamet; the plants were *Ermania himalayensis* and *Ranunculus lobatus*.

FIGURE 32.35

Levels of plant life in the rain forest. A few tall trees comprise the emergent layer, which overgrows the shorter trees of the dense canopy. Shrubs and low trees occupy the shady understory, and fallen leaves and seedlings are scattered on the forest floor.

FIGURE 32.36

Buttress roots on a tropical tree in Ecuador.

FIGURE 32.37

Orchids and other epiphytes on the branches of a tree in a South American rain forest.

The Future of the Tropical Rain Forest Biome

In the 1960s, major plans were developed to convert the Amazon rain forest into large farms, hydroelectric plants, and mines. At the beginning of the 1990s, gold mining activities were filling the rivers with silt, and the tropical rain forests were being destroyed or damaged for commercial purposes at the rate of more than 35 hectares per minute (fig. 32.38). Stated another way, human activities destroy or damage an area equivalent to that of about 45 football fields *per second.* This damage and destruction has had a devastating effect on wildlife. For example, a 1989 study found that populations of low-flying bird species in a 10-hectare section of the Amazon rain forest fell by 75% just six weeks after adjacent land had been cleared. Ten of the forty-eight bird species studied disappeared completely. This situation has been repeated on many occasions. Indeed, dividing large areas of rain forest into smaller pieces separated by as little as 10 m of cleared land can have disastrous effects on the ecology of the entire forest. Most of the bird and other animal species originally present disappear permanently.

FIGURE 32.38

Destruction of tropical rain forest in Belize, Central America. These biomes are rapidly disappearing.

Only a small part of the rain forest biome is now protected from commercial development. The rapid loss of biodiversity results in the permanent loss of gene pools with potential for medicine and agriculture. The cleared land, with its poor and eroded soils, often becomes unfit for agriculture after only two or three years. Even if the land were now to be left fallow, the plant and animal life that has already become extinct will never return. The loss of massive amounts of vegetation directly affects photosynthesis, respiration, and transpiration. If losses continue on a large enough scale, global climate will also be affected. Many more organisms are doomed to extinction before they have even been seen or described for the first time. The rain forest biome will vanish within twenty years if governments and individuals do not stop or slow the large-scale destruction.

Writing to Learn Botany

As you have read in this chapter, the tropical rain forest biome is being destroyed at a rate that could render it extinct shortly after we enter the twenty-first century. How could you educate the public about this serious problem, and what could you do to slow or halt the destruction?

Chapter Summary

Biomes are land-based groupings of ecosystems considered on a global or a continental scale. Most biomes are extensive; they include several ecosystems and occur on more than one continent.

Tundra, which occurs primarily above the Arctic Circle, includes shrubs, many lichens and grasses, and tufted flowering perennials. It is also characterized by the presence of permafrost (permanently frozen soil) below the surface. Tundra is fragile and easily destroyed.

Taiga is dominated by coniferous trees such as spruces, firs, and pines, with birches, aspens, tamaracks, alders (which have nitrogen-fixing actinomycetes in their roots), and willows in the wetter areas. Many perennials but few annuals occur in taiga. Most precipitation occurs in the summer. Lightning frequently starts fires in this biome.

Temperate deciduous forests are dominated by deciduous trees. In North America such trees include sugar maple, American basswood, beech, oak, and hickory; evergreens such as hemlock and pine occur toward the northern and southeastern borders. In early spring a profusion of wildflowers carpets the forest floor before tree leaves expand.

Grasslands occur primarily in temperate areas toward the interiors of continents. Savannas tend to be in similar locations in areas with subtropical to tropical climates. Savannas also have widely scattered trees. Grasslands located in Mediterranean climatic zones usually include vernal pools that support unique annual floras. Many grasslands have been converted to agricultural use.

Deserts have low annual precipitation and widely fluctuating daily temperatures. Desert plants are adapted both structurally and metabolically to the environment, which has more solar radiation than do other biomes.

Mediterranean scrub has unique, mostly evergreen vegetation adapted to cool, wet winters and hot, dry summers. The vegetation is also adapted to frequent fires.

Mountain forest occupies many mountain areas on all of the continents. In North America it occurs in the Pacific Northwest and extends south along the Rocky Mountains and the Sierra Nevada and Cascades of California and Oregon. Mountain forests have mostly dry summers, and some of the trees (e.g., Douglas fir, giant redwood) have thick bark that protects them from frequent fires. Other trees (e.g., knobcone pine) depend on fires for the release and germination of seeds.

The tropical rain forests constitute nearly half of all forest land and contain more species of plants and animals than all the other biomes combined. Many woody plants and vines form multilayered canopies, which keep most light from reaching the forest floor. Soils are poor, and nutrients released during decomposition are rapidly recycled. Tropical rain forests are being destroyed so rapidly that they will disappear within twenty years if the destruction is not slowed or stopped.

Questions for Further Thought and Study

1. Off-road recreational vehicles have devastated parts of the desert biome in recent years; the tracks of any vehicle driven across tundra are visible for many years after. Should we be concerned about such activities? Explain.
2. It has been estimated that at least 25% of all plant species are likely to become extinct in the next thirty years. So far, we seem to be getting along without a number of plants that have become extinct in the past ten years. Considering what you learned in this and the previous chapter about biodiversity, why should we try to slow the rate of extinction?
3. The word "deforestation" usually brings to mind chainsaws, bulldozers, and fires in the Amazon rain forest. However, other forests—including those in Central America and Southeast Asia, both of which are biodiversity "hot spots"—are being cleared two to three times faster than are Amazon rain forests. How should this problem be addressed?
4. How do mountain forests differ from temperate deciduous forests?
5. When Mediterranean scrub plants are carefully transplanted to mountain forest habitats with somewhat similar climates, they often die within a year or two. Can you suggest reasons for this?
6. Much of the tropical rain forest biome is in developing countries, which have weak economies. Bearing this in mind, what could be done to halt the destruction of this biome?

Suggested Readings

Articles

Monastersky, R. 1990. The fall of the forest. *Science News* 138:40–41.

Perry, D. R. 1984. The canopy of the tropical rain forest. *Scientific American* 250:138–147.

Repetto, R. 1990. Deforestation in the tropics. *Scientific American* 262:36–42.

Books

Allan, T., and A. Warren. 1993. *Deserts*. Oxford: Oxford University Press.

Barbour, M. G., and W. D. Billings, eds. 1988. *North American Terrestrial Vegetation*. New York: Cambridge University Press.

Forsyth, A., and K. Miyata. 1984. *Tropical Nature: Life and Death in the Rain Forests of Central and South America*. New York: Charles Scribner's Sons.

Furley, P., et al., ed. 1992. *The Nature and Dynamics of Forest-Savanna Borders*. New York: Chapman and Hall.

Koopowitz, H., and H. Kaye. 1983. *Plant Extinction: A Global Crisis*. Washington, DC: Stone Wall Press.

Moore, R., and D. S. Vodopich. 1991. *The Living Desert*. Piscataway, NJ: Enslow.

Myers, N. 1992. *The Primary Source: Tropical Forests and Our Future*. Rev. ed. New York: W. W. Norton.

Odom, E. P. 1989. *Ecology and Our Endangered Life-Support Systems*. Sunderland, MA: Sinauer.

Packham, J. R., et al. 1992. *Functional Ecology of Woodlands*. New York: Chapman and Hall.

Riely, J. O., and S. Page. 1990. *Ecology of Plant Communities*. New York: Halsted Press.

Weaver, J. E. 1991. *Prairie Plants and their Environment: A Fifty-Year Study in the Midwest*. Lincoln, NE: University of Nebraska Press.

Medicinal plants for sale in a market in Brazil. Plants are used for medicinal and superstitious purposes in many parts of the world.

Plants and Society

Epilogue Outline

Epilogue Overview

It is difficult to name an area of our existence that is not affected by plants. Plants feed us, house us, clothe us, entertain us, generate oxygen that we breathe, and cure our illnesses. However, plants are not entirely beneficial to humans: plants also damage crops, cause severe allergic reactions, and produce some of the most potent poisons known. Not surprisingly, humans have learned to recognize plants that help us and those that harm us. Plants have been integral parts of all civilizations.

A.

B.

FIGURE E.1

Plants used for fragrances and spices. (a) Romans used lavender (*Lavandula*) to scent and disinfect their baths. The genus name of lavender, *Lavandula*, is from the Latin word *lavare*, meaning "to wash." (b) A woman in Madagascar cleaning cloves. Cloves are the dried flower buds of *Syzygium aromaticum*.

A.

B.

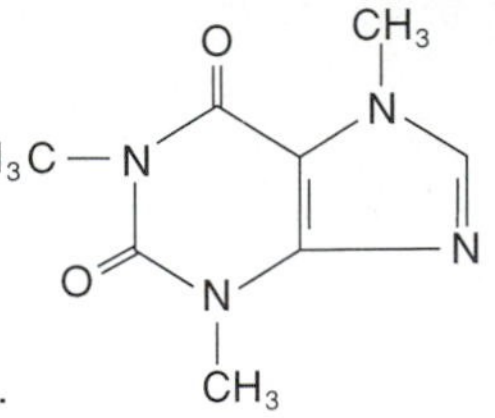

C.

FIGURE E.2

Tea. (a) Tea comes from *Camellia sinensis*, a plant that originated in subtropical Asia. (b) Tea leaves being processed. (c) Tea and coffee contain caffeine, a compound that stimulates the central nervous system.

Plants and Society

Throughout this book you've seen many examples of how plants affect our lives. We eat grains, fruits, and vegetables, and are clothed by fibers from stems and leaves. Plants generate the oxygen we breathe, and trees provide us with lumber, paper, and welcome shade on hot days. Many plants make medicines, brilliant dyes, industrial chemicals, and useful oils. The colors and fragrances of flowers and foliage satisfy our aesthetic senses, while plant-derived spices such as black pepper, nutmeg, ginger, cloves, and cinnamon have enhanced our enjoyment of food since before the time of the Roman Empire (fig. E.1). Spices can also preserve foods, which made possible the colonization of the New World. Tea and coffee are the world's most popular beverages (fig. E.2). From the body paints of Amazon Indians to modern cosmetics, and from early Egyptian papyrus (fig. E.3) of more than 5,000 years ago to today's pulp mills that produce more than 200 million tons of paper each year, plants affect all aspects of our lives. Here are some of the plant families that we use for other, more unusual, purposes (fig. E.4):

- **Laurel family** (Lauraceae: camphor, sassafras, sweet bay, avocados) We use laurel leaves to crown winners of athletic contests and to bestow academic honors. Sassafras is used to make toothpaste, gum, mouthwash, tea, beer, and gumbo, while sweet bay is used as a spice. Avocados, which contain up to 60% oil, contain more energy per unit weight than does red meat.

Figure E.3

Papyrus (*Cyperus papyrus*) growing at the edges of a lake. Papyrus was once used to make paper.

A. B. C. D. E.

Figure E.4

Economically important plants. (a) Sassafras (*Sassafras albidum*) is used to flavor gum and mouthwash, and to make gumbo. (b) Spearmint (*Mentha spicata*) is the source of spearmint oil. (c) Peyote (*Lophophora williamsii*) is the source of various hallucinogens. (d) Monkshood (*Aconitum columbianum*) is the source of aconite, a poisonous drug once used as a sedative. (e) Peppers (*Capsicum* species) are favorites of people who like spicy food.

- **Mint family** (Lamiaceae: spearmint, peppermint, *Coleus*) These plants are important sources of oils, such as peppermint oil. Menthol from peppermint is used to make candy, gum, and cigarettes.
- **Carrot family** (Apiaceae: dill, celery, carrot, parsnip, parsley) Poison hemlock contains coniine, the Athenian State Poison used to kill Socrates in 399 B.C. Ever the scientist, Socrates insisted on having the stages of his poisoning accurately observed and recorded. Elixir of Celery advertised in the 1897 Sears and Roebuck catalog was used to cure nervous ailments, and crushed celery seeds mixed with soda are still marketed as Cel-Ray by Canada Dry. We use some members of the carrot family to make liquor.
- **Cactus family** (Cactaceae: cacti) The fruits of many cacti taste like pears, which explains their common name, prickly pear. Peyote (*Lophophora williamsii*), a small, spineless cactus with carrotlike roots, is the source of mescaline. The Aztecs used dried slices (i.e., *buttons*) of this hallucinogenic plant for religious purposes, as do Indians in Mexico and the southwestern United States.
- **Pumpkin family** (Cucurbitaceae: pumpkin, squash, cantaloupe, watermelon) These plants have been used to make everything from jack-o-lanterns and dishes to laxatives.
- **Lily family** (Liliaceae: asparagus, sarsaparilla, *Aloe*, chives, garlic, meadow saffron) Sarsaparilla is used to make soft drinks, and *Aloe*, which was once used as a source of phonograph needles, is used to treat burns; *Aloe* is also a common ingredient of many cosmetics. Meadow saffron (fall crocus; *Colchicum autumnale*) is the source of colchicine, an important drug once used to treat gout and now used in biological research (see page 218). Meadow saffron is different from the true saffron, a member of the Iris family and the source of the world's most expensive spice.
- **Buttercup family** (Ranunculaceae: columbine, monkshood, larkspur) Columbine, the state flower of Colorado, has been used as an aphrodisiac. Monkshood and wolfsbane contain aconitine, a powerful poison that humans have used to kill a variety of animals, including wolves. Monkshood, an attractive plant grown in many gardens, was used to kill Roman Emperor Claudius and Pope Adrian VI.
- **Spurge family** (Euphorbiaceae: cassava, Pará rubber tree, candelilla, poinsettias) The Pará rubber tree is a source of latex for rubber, and candelilla is the source of several waxes, including candle wax. Cassava, whose roots are used to make tapioca, alcohol, and acetone, is a staple in the tropics.

  Despite its widespread use in December, the poinsettia is one of the newest Christmas plants. Joel Poinsett became interested in the plant while serving as American ambassador to Mexico from 1825–29. He brought the plant home, where it was named in his honor.
- **Poppy family** (Papaveraceae: bloodroot, opium poppy) All members of this family produce drugs, the most popular of which are morphine and codeine. To learn more about opium poppies, see box E.1, "Poppies, Opium, and Heroin."
- **Nightshade family** (Solanaceae: tomato, potato, tobacco, petunia, chili pepper) These plants produce drugs such as atropine (once used to dilate pupils and as an antidote to nerve-gas poisoning), hyoscaymine (a sedative), and scopolamine (a tranquilizer). Hot peppers dried and crushed yield cayenne pepper, and sweet peppers are used to make paprika.[1] Fiery chili peppers get their heat from capsaicin, a chemical that's perceived not by our taste buds, but rather by pain receptors in our mouth. We can taste it at concentrations as low as 1 molecule per million. That's why only a trace of capsaicin can make an unwary diner grab frantically for water.

  Many of the pilgrims considered tomatoes to be evil—on a par with dancing and card playing. Interestingly, tomato juice is said to neutralize butyl mercaptan, the nose-shriveling ingredient of skunk spray. Joseph Campbell brought out his famous Tomato Soup in 1897, soon after chemist John Dorrance, at a weekly salary of $7.59, figured out how to condense it. In 1962, Andy Warhol's painting of a can of that soup sold for $50,000.

Although modern civilization would collapse without plants, not all plants benefit people. For example, plants such as castor bean, mistletoe, dumb cane, caladium, elephant's ear, philodendron, English ivy, rhubarb, poinsettia, oleander, and yew (the plant used by Robin Hood to make his bows) contain potent toxins. Stinging nettle (*Urtica dioica*), poison ivy (*Toxicodendron radicans*), and the less common poison oak (*T. diversilobum* and *T. toxicarium*) also elicit strong reactions in some people, while plants such as ragweed, grasses, and many fungi cause hay fever and other allergic reactions. Overgrowth of some plants clogs streams and water pipes, and weeds damage crops.

1. Vitamin C was purified in 1928 by Hungarian biochemist Albert Szent-Györgyi from peppers in a rejected supper dish of sweet paprika peppers (before then, Szent-Györgyi had tried to purify the vitamin from bovine adrenal glands). The purified vitamin C from those peppers earned Szent-Györgyi a Nobel prize nine years later.

Boxed Reading E.1

Poppies, Opium, and Heroin

What shall we do about poor little Tigger
If he never eats nothing he'll never get bigger;
He doesn't like honey and haycorns and thistles
Because of the taste and because of the bristles
And all the good things that an animal likes
Have the wrong sort of swallow or too many spikes.

—A. A. Milne, *The House at Pooh Corner,* 1928

Throughout this book you've seen examples of what Pooh Bear perceptively noted: that mechanical defenses such as thorns are not plants' only defenses. Many species have developed a chemical arsenal that helps deter herbivores. Among them is poppy, *Papaver somniferum,* an annual herb that grows about a meter tall. One of the deterrents in poppy leaves is opium. The euphoric effects of opium were known as early as 4,000 B.C., and by the 1800s opium was a common ingredient in medications. For example, Godfrey's Cordial, an English concoction of opium, molasses, and sassafras, was used as a cough remedy and a cure for diarrhea.

Box Figure E.1

Poppy (*Papaver somniferum*), the source of opium. Opium, the dried latex that oozes from cut fruits, is the source of alkaloids such as morphine and codeine.

Today, poppies are often grown innocently in gardens for their pretty purple and white blossoms. Their ripe seeds can be toasted and used as ornaments on crackers or cakes, or crushed and used as a cooking oil. Because the seeds do not contain many alkaloids, their oils produce no euphoric effects. Most of the alkaloids are in latex in the urn-shaped fruits. When the fruits are cut, the latex oozes from the fruits and dries to a brown residue called *opium.* Opium contains more than twenty-five alkaloids, the most abundant of which is morphine (another abundant ingredient is codeine). The illicit sale of heroin, which is made from morphine, is a 6-billion-dollar industry.

Worldwide production of opium exceeds 10,000 tons, of which only 400 are used for medicine. India is the largest and only legal producer of opium.

Nature's Botanical Medicine Cabinet

Just as people learned to exploit plants for food, so they learned to use plants as medicine. Plants represent a huge storehouse of drugs: they produce more than 10,000 different compounds to protect themselves from hungry animals. People that search for these compounds are natural-products chemists, such as Isao Kubo:

> *Several years ago, I was on a plant collecting trip near Serengeti National Park in East Africa. I was searching for the rare medicinal plant,* Kigeria africana. *After 3 days of combing the Serengeti plain, I came upon an outcrop of* K. africana *trees. Feeling very lucky indeed, I climbed up onto one of the tree's branches to collect the sausage-shaped fruit.*
>
> *But as I busily went about my collecting, I realized I was being watched. Two menacing eyes from an adjacent tree stared at me through the branches. Those eyes belonged to a rather large leopard. I tried with all my might to keep from falling from my perch as my knees knocked and sweat poured from my body. This brief encounter ended abruptly, however, when the leopard decided I was not a suitable meal and disappeared into the bushes. Well, I thought, another day in the life of a natural products chemist.*
>
> —Isao Kubo, Professor of Natural Products Chemistry, University of California at Berkeley, *From Medicine Men to Natural Products Chemists* (1985)

Boxed Reading E.2

Chewing on Resin

Wounded stems of red or black spruce (*Picea,* Latin for "pitch") release a resin that flows into the wound and hardens upon contact with air, thereby sealing the wound. This resin is called *spruce gum,* and was the first commercial chewing gum. The novelty of spruce gum made it popular: rural Europeans have been chewing it for centuries, and in early Sweden spruce gum was considered a present for a potential sweetheart—the Scandinavian equivalent of a box of chocolates. Among the devoted chewers was Mark Twain's Tom Sawyer, who shared some with Becky Thatcher one day after school.

Although the spruce gum business originally thrived—the first factory employed more than 200 people and produced almost 2,000 boxes of gum per day—its popularity soon waned; indeed, only one company in the United States still makes the gum. One reason for spruce gum's decline may have been its unusual taste, which one expert likened to "sinking your teeth into frozen gasoline." Moreover, it turns hard and crumbly when stuck to the bedpost overnight. However, the death of the industry was also due partly to the increased demand for newspapers, which prompted foresters to clear vast forests of spruce to make newsprint (see Chapter 16). Faced with dwindling supplies of spruce, manufacturers started looking for alternatives. Their first choice was paraffin, a by-product of petroleum distillation. However, paraffin never really caught on, despite being the first gum to be sold in packets including picture cards. Today, paraffin is still available in semiedible novelties such as wax fangs, wax lips, and wax buck teeth. A more successful substitute for spruce gum was chicle, the chewable latex of the Mexican sapodilla tree.

Box Figure E.2

Resin oozing from a wounded spruce tree.

Boxed Reading E.3

David Douglas

Although Isao Kubo's comments earlier in this chapter describe "another day in the life" of a natural-products chemist, they don't come close to describing the life and times of indomitable Scottish botanist David Douglas. Douglas is famous for describing hundreds of plants, including the Monterey pine, digger pine, sugar pine, and ponderosa pine (interestingly, he did not describe Douglas fir, the tree named after him). Under the aegis of the Royal Horticultural Society of London, Douglas led many collecting trips to North America. These trips were fraught with disasters: he was attacked by grizzlies, Indians, and red ants; he braved blizzards and avalanches; he suffered all sorts of diseases; he twice had to eat his horse to avoid starvation; and he was shipwrecked and washed out to sea by a hurricane. Douglas accepted it all, entering only this understated complaint in his diary: "Traveled thirty-five miles, drenched and bleached with rain and sleet, chilled with a piercing north wind; and then to finish the day experienced the cooling, comfortless consolation of lying down wet without supper or fire. On such occasions I am very liable to become fretful." At age 35, Douglas died in Hawaii in 1834, as spectacularly as he lived. While crossing mountains en route to Hilo, he fell into a pit dug as a cattle trap. There, he was gored to death by a wild bull.

Box Figure E.3

David Douglas

A.

B.

C.

D.

Figure E.5

Drugs derived from plants. (a) Many members of the snakeroot family contain alkaloids that help lower blood pressure. (b) Alkaloids from *Ephedra* are used to treat asthma and other allergies. (c) Yams contain a variety of steroidal drugs. (d) Woman gathering leaves of *Erythroxylon coca,* the source of cocaine.

Natural-products chemists such as Isao Kubo search through the bounty of chemicals made by organisms for substances that people can use. Here are some of the more common drugs derived from plants and a fungal parasite of a plant (fig. E.5):

- Reserpine is extracted from snakeroot (*Rauwolfia serpentina,* a low evergreen shrub) and is used as a sedative and to decrease blood pressure. In India, snakeroot is used to make an antidote for snakebites.[2] Many schizophrenics and others with mental disorders can lead nearly normal lives after being treated with reserpine. In the 1990s, annual prescriptions for reserpine in the United States alone exceeded $100 million.
- Ephedrine, which is extracted from *Ephedra,* is used to ease bronchitis.
- Steroids such as estrogens and testosterone are extracted from yams (*Dioscorea*). The progesterone used to make the first birth control pills was extracted from 10 tons of yams harvested from the Mexican jungle by Russell Marker and his coworkers.
- Digitoxin, extracted from *Digitalis purpurea* (foxglove, a garden ornamental), is taken as a heart stimulant by more than 3 million Americans each day.
- Cocaine comes from the leaves of *Erythroxylon coca.* This plant, the "divine plant" of the Inca civilization, is native to the eastern slopes of the Andes. Cocaine, an ingredient of Coca-Cola until 1904, is grown mostly in Peru and Bolivia. Cocaine is a stimulant and hunger depressant, which explains why poor laborers can work for two or three days without food while chewing coca leaves. The laborers even measure distances in *cocadas,* the distance they can walk on one chew. Today, cocaine is marketed illegally in highly addictive forms, with enormous costs to society.
- Tetrahydrocannabinol (THC) is derived from hemp (marijuana; *Cannabis sativa*) and is used to make hashish. Today, the former Soviet Union is the world's leading producer of hemp.
- Lysergic acid (the chemical used to make lysergic acid diethylamide, or LSD) comes from the spore-producing fruiting body of *Claviceps purpurea,* a fungus that infects wheat and rye.
- Nicotine, which comprises 1%–3% of the weight of tobacco, is an ingredient of many insecticides.

2. Most drugs are poisons when consumed in large amounts. For example, white snakeroot killed thousands of people in the midwest during the 1800s; in some areas, such as in Dubois County, Indiana, the death toll may have been one out of every two people.

TABLE E.1

Some Important Medicinal Plants and Fungi

Plant	Plant Parts Used	Active Compounds	Uses in Medicine
Atropa belladonna (belladonna)	Leaves, roots	Atropine, hyoscyamine	Cardiac stimulant, pupil dilator, antidote for organophosphate poisoning
		Scopolamine	Motion sickness, antiemetic
Cannabis sativa (marijuana)	Leaves, inflorescence	Tetrahydrocannabinol (THC)	Treatment of glaucoma, relief of nausea from chemotherapy
Catharanthus roseus (Madagascar periwinkle)	Leaves	Vinblastine, vincristine	Treatment of leukemia, Hodgkin's disease, and other cancers
Cinchona sp. (fever bark tree)	Bark	Quinine	Treatment of malaria
Colchicum autumnale (autumn crocus)	Corm	Colchicine	Treatment of gout
Digitalis purpurea (foxglove)	Leaves	Digitoxin, digoxin	Cardiac stimulant, diuretic
Dioscorea sp. (yam)	Tubers	Steroids	Production of cortisone, sex hormones, and oral contraceptives
Ephedra sp. (Mormon tea)	Stems	Ephedrine	Decongestant, treatment of low blood pressure and asthma
Erythroxylon coca (coca)	Leaves	Cocaine	Local anesthetic
Hydnocarpus kurzii (chaulmoogra tree)	Seeds, fruits	Ethyl esters of chaulmoogra oil	Treatment of leprosy and related skin diseases
Hydrastis canadensis (goldenseal)	Roots, rhizomes	Hydrastine	Treatment of inflamed mucous membranes
Nicotiana tabacum (tobacco)	Leaves	Nicotine	Stimulant
Papaver somniferum (opium poppy)	Latex from capsule	Morphine, codeine	Narcotic, analgesic, cough suppressant
Penicillium notatum (penicillin)*	Hyphae	Penicillin	Antibiotic
Podophyllum peltatum (May apple)	Roots, rhizomes	Podophyllotoxin	Treatment of venereal warts
Rauwolfia serpentina (rauwolfia)	Roots	Reserpine	Treatment of high blood pressure and psychosis
Salix sp. (willow)	Bark	Salicin	Analgesic, antiinflammatory, treatment of rheumatoid arthritis and headaches

* Fleming's discovery of penicillin (see Chapter 1) involved *Penicillium notatum*, but most commercially produced penicillin today is derived from x-ray induced mutants of *Penicillium chrysogenum*, which produce more than 1,000 times the penicillin originally derived from *P. notatum*.

TABLE E.2

Some Important Hallucinogenic Plants and Fungi

Plant	Plant Parts Consumed	Psychoactive Compounds
Acorus calamus (sweet flag)	Rhizomes	α-asarone, β-asarone
Amanita muscaria (fly agaric mushroom)	Basidiocarp	Muscimole, muscazone
Atropa belladonna (belladonna)	Leaves, roots	Scopolamine, hyoscyamine, atropine
Claviceps purpurea (ergot)	Sclerotia	Ergine, isoergine
Datura sp. (jimson weed)	Stems, leaves, seeds	Hyoscyamine, scopolamine
Hyoscyamus niger (henbane)	Leaves, roots	Hyoscyamine, scopolamine
Ipomoea violacea (morning glory)	Seeds	Ergine, isoergine
Lophophora williamsii (peyote or mescal)	Shoots (crowns)	Mescaline
Mandragora officinarum (mandrake)	Leaves, stems	Hyoscyamine, scopolamine
Psilocybe sp. (psilocybe mushroom)	Basidiocarp	Psilocybine
Psychotria viridis (psychotria)	Leaves	Dimethyltryptamine (DMT)
Rivea corymbosa (morning glory)	Seeds	Ergine, isoergine
Trichocereus pachanoi (San Pedro cactus)	Stems	Mescaline

All chemicals in plants are potential drugs. A chemical that oozes from a tree's bark to discourage hungry caterpillars from eating it may also be an effective drug. For example, a chemical in the bark of the Indian neem tree keeps desert locusts off the tree. The people of Serengeti National Park in east Africa chew the twigs of these trees to prevent tooth decay. Other more familiar plants have healing properties as well (tables E.1, E.2).

Today's natural-products chemists are a modern version of the medicine man (or woman) who traditionally explored the healing power of plants (fig. E.6). Herbal medical practices may have begun in prehistoric times, when some individuals became botanical experts by sampling plants themselves. Clay tablets carved 4,000 years ago in Sumeria list several plant-based medicines, as do records from ancient Egypt and China. Roman philosopher Pliny the Elder wrote in the first century A.D., "If remedies were sought in the kitchen garden, none of the arts would become cheaper than the art of medicine." Modern-day medicine men called the *Bwana mgana* practice herbal medicine in the region of East Africa explored by Isao Kubo.

FIGURE E.6

Mark Plotkin, a scientist from Conservation International, collecting medicinal plants with the advice of a Wayana medicine man in southeastern Surinam. Despite the importance of such plants (e.g., as sources of medicine), the opportunities for gathering these plants is rapidly disappearing as tribal cultures are lost. Because written records about plants do not exist in many traditional societies, we are losing much valuable knowledge.

A good example of medicines derived from plants used today are the chemicals called *alkaloids*, which come from the periwinkle plant and other species. Alkaloid-derived drugs such as leucocristine and vincaleukoblastine (both from periwinkle; *Vinca roseus*) have helped revolutionize the treatment of some leukemias (blood cancers), and alkaloid narcotics derived from the opium poppy, including morphine, are excellent (but addictive) painkillers. Today, nearly half of all prescription drugs contain chemicals manufactured by plants, fungi, or bacteria, and many other drugs contain compounds that were synthesized in a laboratory but modeled after plant-derived substances. The many medicines that we use from the plant kingdom are a compelling reason why we must stop destroying the world's tropical rain forests, where plant life is so abundant and diverse that all of the species have not even been identified, much less studied.

Case Study: Plants and Malaria

One disease whose spread has been greatly influenced by plants is malaria, which kills more people worldwide each year than any other disease. Malaria starts with chills and violent trembling and progresses to an extremely high fever accompanied by delirium. Finally, the person sweats profusely, is completely exhausted, and develops a dangerously enlarged spleen. The disease strikes in a relentless cycle, with symptoms returning every 2–4 days. Within weeks, the sufferer either dies or manages to marshall the body's immune defenses against the invading parasite that causes the disease.

Malaria is an interesting example of the relationship between plants and human disease, because plants contribute to both its spread and to its cure. Like many parasitic diseases, the malaria parasite (*Plasmodium falciparum*, *vivax*, *malariae*, or *ovale*) must spend part of its life cycle within an intermediate host organism—mosquitoes of the genus *Anopheles*. This insect must bite humans for the disease to spread. The type of vegetation in an area determines whether the mosquito, and the parasite it carries, will thrive. Unfortunately, agriculture often ushers in malaria by replacing dense forests with damp rice fields that are a haven for the mosquitoes. Indeed, in heavily infested areas of Africa, residents are bitten by disease-carrying mosquitoes *every night*.

Some plant products can kill the malaria parasite. In the sixteenth century, natives of Peru gave Jesuit missionaries who were on their way to Europe their secret malaria remedy—bark of the cinchona tree (fig. E.7). However, it was not until 1820 that two French chemists extracted the active ingredient from

A.

B.

FIGURE E.7

Malaria and quinine. (a) Bark of cinchona contains quinine, a drug used to combat malaria. (b) Quinine being extracted from bark of cinchona.

cinchona bark. This substance, which they called quinine, was the standard treatment for malaria until the 1930s. As malaria parasites developed resistance to quinine, malaria spread through the developing nations as more and more land was cleared for farming. New natural antimalarial drugs were sought, and a promising one has come from ancient Chinese medicine.

In 1967, researchers in the People's Republic of China began a systematic study of all plants known to have medicinal properties in search of a new drug to fight malaria. In a document entitled "Recipes for 52 Kinds of Diseases" unearthed from a Mawangdui Han dynasty tomb from 168 B.C., a plant called *qinghao* was described as a treatment for hemorrhoids. A reference from A.D. 340 cited the same plants (also known as *Artemisia annua*, or sweet wormwood, a relative of tarragon and sagebrush; fig. E.8) as a treatment for fever. In 1596, a Chinese herbalist prescribed qinghao to combat malaria.

In the 1970s, chemists isolated the active ingredient from sweet wormwood and called it *artemisinin*. By 1979 it had been tested on more than 2,000 malaria patients, in whom it cured the fever in 72 hours and eliminated parasites from blood within 120 hours. The drug is more than 90% effective in treating cerebral malaria, the most severe form of the disease. In animal tests, artemisinin is proving effective against other parasitic diseases as well. Not surprisingly, other animals use plants as drugstores (see box E.4, "Nature's Medicine Cabinet").

FIGURE E.8

Dried leaves and flowers of *Artemisia annua* contain artemisinin, an antimalarial drug used in China for centuries.

BOXED READING E.4

NATURE'S MEDICINE CABINET

To learn more about animal behavior, ecologist Holly Dublin spent most of 1975 doing something rather unusual: tracking a pregnant elephant in Kenya. Dublin noticed that the sixty-year-old expectant mother seldom changed her routine of walking about 5 kilometers per day. However, one day the elephant changed her routine: she walked 28 kilometers to a riverbank and began eating leaves from a species of tree that Dublin had never seen an elephant eat before. Before leaving, the elephant ate the entire tree. Four days later, the elephant had her baby.

Dublin was puzzled by the elephant's abrupt change in behavior and unusual meal. Did the tree eaten by the elephant have anything to do with inducing birth? To her surprise, Dublin later learned that pregnant Kenyan women induced labor by drinking tea made from the tree's bark and leaves.

Dublin's observations are among a growing number of studies suggesting that animals use plants as drugstores:

- Chimps often eat leaves of the shrub *Vernonia amygdalin* when they're tired and sick. The plant is used by African tribes to cure the same symptoms. Similarly, chimps eat leaves of *Aspilia,* a member of the sunflower family. These leaves contain thiarurbine-A, a red, sulfur-containing oil that kills pathogenic bacteria and parasitic worms. Humans use extracts of the oil as anticancer drugs.
- Wild rhesus monkeys often eat dirt with their food. That dirt contains much kaolin, a clay that detoxifies many poisons and is the active ingredient in Kaopectate, an antidiarrheal medicine.

Most biologists don't think that these and other examples are coincidence. Rather, they suspect that animals doctor themselves by using plants as preventive medicine.

AGRICULTURE: PLANTS AS FOOD

Obtaining food today is as simple as a trip to the supermarket or a visit to the garden or farm stand, but for our ancestors living 12,000 years ago, finding a meal was quite a challenge. Tribes of these people moved constantly in search of food (fig. E.9). In what is now the United States, some of our forebearers hunted the large plant-eating mammals that roamed the great plains, while others collected seeds, nuts, roots, fruits, and grasses. These hunter-gatherers were at the mercy of the environment. If there was no food where they were, they had no choice but to move on or starve.

From Hunter-Gatherers to Farmers: The Dawn of Agriculture

People first began to domesticate plants in the Fertile Crescent (today's Lebanon, Jordan, Syria, Israel, Turkey, and Iraq) of the eastern Mediterranean about 12,000 years ago. The first plants to be grown agriculturally in these regions were probably barley and wheat, followed soon thereafter by lentils and peas. Several factors contributed to this first spark of civilization. For example, the last ice age was ending. As the glaciers receded, the land that was revealed became inhabited by a vast assemblage of animals and flowering plants. Humans no longer had to follow herds of animals or gather berries to eat, because food could be found in more areas. People could stay in one place, eating readily available small game, fish, and wild plants.

By about 10,000 years ago, humans learned to take care of certain animals, such as wild sheep and goats, so that they

FIGURE E.9

Hunter-gatherers in central Africa. The woman carries the family dog; the man has caught a turtle. Hunter-gatherers such as these people cannot build cities.

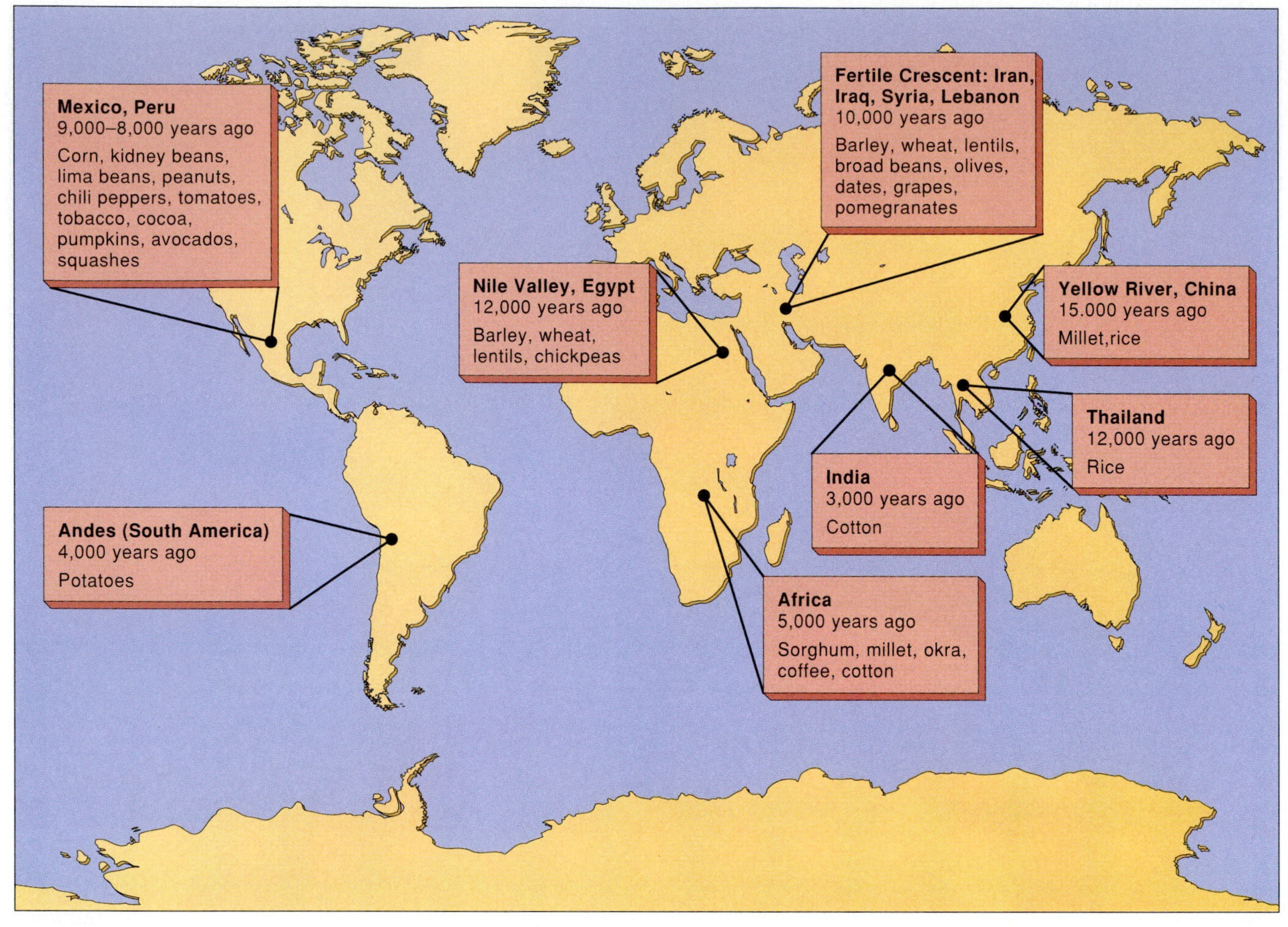

FIGURE E.10

The domestication of agriculturally important plants. Farming arose independently in many places throughout the world.

could have a continual supply of milk and meat. People also learned to manipulate the wild weeds that sustained many tribes. For example, they learned to leave seeds in the ground to ensure a future supply of food. Gradually, these early farmers discovered more about plant reproduction and used their knowledge to improve their food supply. They deduced by trial and error when and how to plant the correct seeds in the right soil and to see that the seeds got enough water to sprout. They learned to recognize when plants were ready to be harvested. People saved the seeds from one season's most useful plants to sow the next year's crop, thereby encouraging certain combinations of traits (a practice called *artificial selection*) that might not have predominated in the wild. Later on, people learned to preserve their harvested food by drying it, thereby reducing even further their dependence on the unpredictable weather and climate. This domestication of animals and the intentional planting and cultivation of crops marked the birth of agriculture.

Farming probably arose independently in many places around the world (fig. E.10). Archeological evidence indicates that domestication of sheep and goats and cultivation of wheat and barley began in the Fertile Crescent about 12,000 years ago. Gradually, agriculture spread to eastern Europe about 8,000 years ago, and to the western Mediterranean and central Europe about 7,000 years ago. Egyptian and Assyrian tombs, mummy wrappings, paintings, and hieroglyphics from 4,000 years ago depict a rich agricultural society that cultivated pomegranates, olives, grapes, figs, dates, and cereals. A similar spread of agriculture occurred in the Americas, based upon native corn. Yams and cassava were early crops in Africa (fig. E.11). Tomatoes and coffee are the only two crops domesticated in the last 2,000 years.

FIGURE E.11

Cassava (*Manihot esculenta*) is an important food crop in the tropics, particularly in Africa. The large tuberous roots are peeled, boiled, and mashed. Tapioca is made from cassava.

With the growth of towns and then cities, human influence on cultivated plants spread. The distribution of crops across the planet today indicates that when humans colonized new lands, they often brought their native plants with them and likewise carried new plant varieties from the new land back to the old (for example, corn was introduced to Europe from the Americas by Christopher Columbus in the late fifteenth century).[3]

3. Columbus brought back the kernels and the native word for the grain: *mahiz*, which survives today as the common name for American corn, *maize*. Only in the United States is the plant called *corn*.

Tracing the origins of crop plants is difficult because archeological evidence is often destroyed over time. The available evidence, however, suggests that different crop plants were domesticated in the Old and New Worlds. Today, thanks to efficient transportation, many different kinds of plants are grown far from where they originated. Moreover, our relatively efficient farming methods free most people to pursue other activities. For example, only about 3% of the U.S. workforce are farmers; these people provide almost all of our food as well as more than 25% of our exports.

The Lore of Plants

Agriculture among Native Americans differs significantly from modern agriculture. Native Americans' agriculture, like the rest of their culture, is based on a harmony with nature. Native American agriculture is effective and sophisticated:

- About 100 of the 120 known food crops were domesticated by Native Americans, including potatoes, avocados, beans, squash, pumpkin, bell pepper, tomato, peanuts, pecans, cashews, many berries, and grapes. They made chocolate, flavored their food and drinks with vanilla, grew pineapples (which later were sent to Hawaii), and even coated their popcorn with maple sugar, thus making a treat similar to Cracker Jacks.
- Native Americans used selective breeding to develop a wide variety of crops. For example, they developed more than three hundred varieties of corn from teosinte, a wild grass.
- Native Americans set up huge farms. Columbus reported corn fields more than 18 miles long that included irrigation canals, crop rotation, and fertilizers.

Most aspects of Native American agriculture were motivated by necessity. Today we continue to learn from Native Americans, the first true environmentalists.

Plants as Food

Our most important use of plants is as food. Today, as much as 90% of the total calories consumed by humans comes from crop plants:

Grains: wheat, rice, corn, sorghum, millet, barley, oats, rye
Tuber and root crops: potato, yam, sweet potato, cassava
Sugar crops: sugar cane, sugar beets
Protein seeds: beans, soybeans, peas, lentils
Oil seeds: olive, soybean, peanut, coconut, sunflower, corn
Fruits and berries: citrus, mango, banana, apple
Vegetables: cabbage, lettuce, onion

Most of the calories in our diet come from wheat, rice, corn, potatoes, yams, and cassava (the source of tapioca). Cereals provide much dietary carbohydrate, and the seeds of legumes (beans, lentils, peas, peanuts, and soybeans) are rich in protein. Cereals have different types of amino acids than legumes, so eating these two foods together provides a good balance of proteins. The best plant-derived nutrition combines a cereal (a rich source of carbohydrate) with a legume (a source of protein), a green, leafy vegetable (rich in vitamins and minerals), and perhaps small amounts of sunflower oil, avocado, or olives (which provide fats).

FIGURE E.12

(a) Bananas growing in southern Mexico. Bananas are an important crop in the tropics. (b) The 50 to 100 coconuts that grow on a palm tree each year provide carbohydrates, oils, and proteins.

In the tropics, bananas and coconuts are staples of peoples' diets (fig. E.12). Bananas have been cultivated in tropical Africa for the past 2,000 years; there, starchy bananas called *plantains* are a major part of the diet (sweeter varieties are popular in the United States). The 50 to 100 coconuts that grow on a palm tree each year provide proteins, oils, and carbohydrates, while the leaves and trunks are excellent building materials.

Boxed Reading E.5

Jojoba

Jojoba (*Simmondsia chinensis*) is one of several recently developed crops that has great potential. Its large seeds contain up to 60% of a light yellow, odorless, liquid wax. This wax is used as a lubricant, an ingredient of cosmetics, and even a non-cholesterol cooking oil. It is also an excellent substitute for the sperm-whale oil needed in heavy machinery and to make ballistic missles. Jojoba grows in hot deserts that are unsuitable for other crops.

Although jojoba is popular today, before 1980 it was an obscure, low-spreading, grayish green bush growing in the southwestern United States and Mexico. In that year, seeds from wild-growing plants were collected and used to start the first cultivated plantings. The next season, researchers took cuttings from the healthiest bushes and used them to start the next crop. During recent years, the jojoba crop in the United States has grown by at least 40% per year.

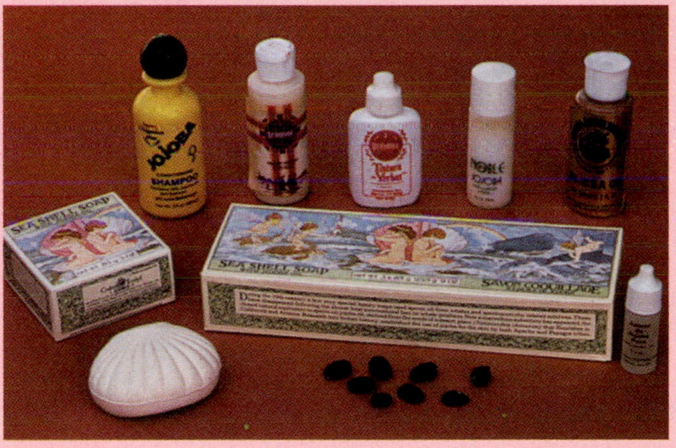

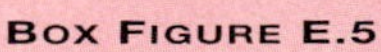

Box Figure E.5

Jojoba (*Simmondsia chinensis*) and products made from jojoba. Jojoba can grow in hot deserts that are unsuitable for other crops.

Although we eat many kinds of plants, cereals such as corn and wheat influence our lives more than do other plants. The importance of cereals cannot be understated. All of the world's great civilizations have been based on the cultivation of cereals: maize was the basis of the Inca, Aztec, and Mayan empires; rice was the staple food in ancient China, Japan, and India; and the civilizations of Egypt, Rome, Greece, and Mesopotamia were based on wheat. Without an abundant and reliable source of these cereals, villages could not have grown into cities, nor cities into empires. Our lives would be very different without cereals.

Cereals: Staples of the Human Diet

Cereals such as wheat, rice, and corn include nine of the ten most economically important groups of plants and provide about half of all the protein in our diets. These plants are all members of the grass family (Poaceae). Rice is the primary source of food in the Orient, while corn is a major part of the human diet in South America. In the United States, many of our foods are based on wheat.

Wheat (Triticum aestivum)

Wheat (fig. E.13) grows in a range of climates, from the Arctic to the equator, from below sea level to 10,000 feet (3,048 meters)

A.

B.

C.

FIGURE E.13

Wheat. (a) Wheat (*Triticum*) fields in Colorado. Grasses such as wheat provide a staple food for humans. (b,c) Harvesting and winnowing of wheat in Tunisia, North Africa. Wheat has been cultivated for at least 12,000 years.

above it, and from areas with less than 2 inches (5 centimeters) of annual rainfall to places with more than 70 inches (178 centimeters) of rain per year. Most wheat is used for food. The grain of the "hard" bread wheats contains 11%–15% protein, mostly of a type called *gluten*. When ground and mixed with water, it forms an elastic dough that is excellent for making bread. The grain of the "soft" bread wheats contains 8%–10% protein and is best for making cakes, cookies, crackers, and pastries.

Wheat was probably the earliest domesticated plant; we've cultivated it for at least 12,000 years. Today there are three kinds of wheat, based upon their number of chromosomes: 14 (diploid), 28 (tetraploid), or 42 (hexaploid) (fig. E.14). Bread wheats are hexaploid, and the durum wheat used to make macaroni is tetraploid. Geneticists study the number of chromosomes in modern wheats to reconstruct the evolution of these important crop plants. It is a fascinating and complex story. Wheat as we know it apparently evolved from an accidental merger between a distant wheat ancestor and a weedy grass. It may have happened as follows. About 10,000 years ago, wandering peoples in what is now Jericho, in Israel, came upon a rich oasis, a spring in the desert ringed by hills covered with wild grass. While looking for food, the people discovered that the grain held in the grass, when ground, made a fine flour. For many years, they did not know how to encourage the grass's growth intentionally, so each season they would simply forage for whatever nature provided.

Then about 8,000 years ago, something happened that was to have profound effects on agriculture and human civilization. The grass that the people had grown fond of, really an ancient wheat called *Einkorn*, crossbred with another type of grass that was not good to eat (fig. E.14). The hybrid grass then underwent a genetic accident (nondisjunction; see Chapter 10) that prevented the separation of chromosomes in some of the developing germ cells. The result was a new type of plant that

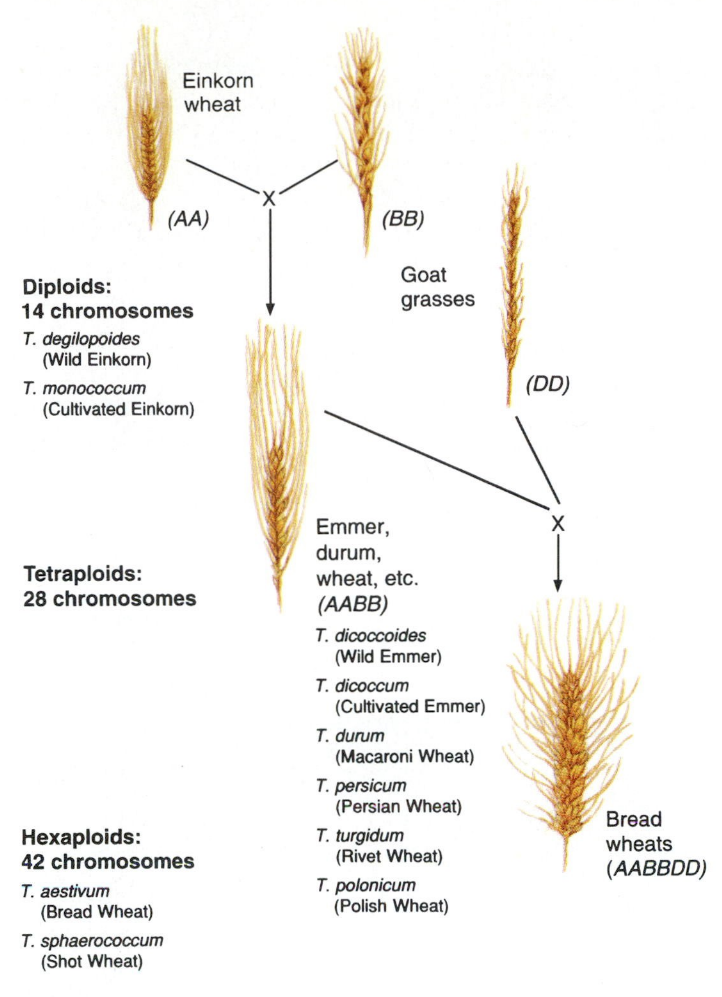

FIGURE E.14

Evolution of wheat.

had twice the number of chromosomes as either parent plant (i.e., twenty-eight chromosomes total). This plant was called *Emmer wheat*, and it was a better source of food than the parent wheat whose grains had become a staple. The doubled number of chromosomes in each cell produced a plant with larger grains. Also, the grains were attached to the plant in such a way that they could be easily loosened and spread by the wind, so that the new wheat was soon plentiful. It was probably about this time that early farmers learned to select seed from the most robust plants to start the crops of the next season.

Then about 6,000 years ago, another "mistake" of nature further improved the quality of wheat. Emmer wheat crossed with another weed, goat grass, and after another fortuitous "accident" of chromosome doubling, led to bread wheat, which has forty-two chromosomes (fig. E.14; also see box 10.1, "Polyploidy in Plants" on p. 219). Bread wheat has even larger grains than the Emmer wheat that gave rise to it, but at a cost. Its ears are so compact that, on its own, the grain cannot be released. However, farmers collected the rich seed each season for food, and kept a certain percentage to be planted the next season. The interdependence of humans and crops that is the basis of modern agriculture was thus born. Today, interesting hybrids

FIGURE E.15

Triticale, a popular grain that is a hybrid of wheat (genus *Triticum*) and rye (genus *Secale*). Cells of triticale are polyploid: they each contain a complete set of chromosomes from wheat as well as a set from rye. Triticale can tolerate the harsh environment of rye while producing the high yields of wheat.

add variety to our diets, such as triticale, a combination of wheat (*Triticum*) and rye (*Secale*) (fig. E.15). The cells of triticale are polyploid, containing a complete set of chromosomes from wheat and one from rye. Triticale has the high yield of wheat and can cope with the harsh environments of rye.

Doing Botany Yourself

Devise an experiment to test whether the scenario for the evolution of bread wheat from tetraploid and diploid relatives by breeding with wild grains is possible.

Corn (*Zea mays*)

The tasty, sweet, or starchy kernels of the corn plant, *Zea mays*, sustained the Incas, Aztecs, and Mayan Indians of South America and the pilgrims of colonial Massachusetts. Today, corn continues to be a dietary staple from Chile to Canada. The corn crop in the United States presently exceeds 9 billion bushels a year and is used to make food products, drugs, and industrial chemicals, as well as to feed humans and animals (fig. E.16). Corn is also being used to help solve a troubling environmental problem: plastics. After years of being dumped into our oceans, plastics that were made to last forever are coming back to haunt us. For example, merchant ships dump about 500,000 plastic containers into international waters every day. These and other plastics maim and kill marine life (last year, about 40,000 seals died after becoming entangled in plastics). Other plastics are filling our landfills and polluting our water. In one of the many examples of how plants are used to combat such problems, one company has recently patented a technique for making biodegradable plastics containing corn starch. This technique involves inserting tiny pellets of starch into the plastic, which helps the plastic to decompose to dust in only a few months. If this technology becomes popular, biodegradable containers

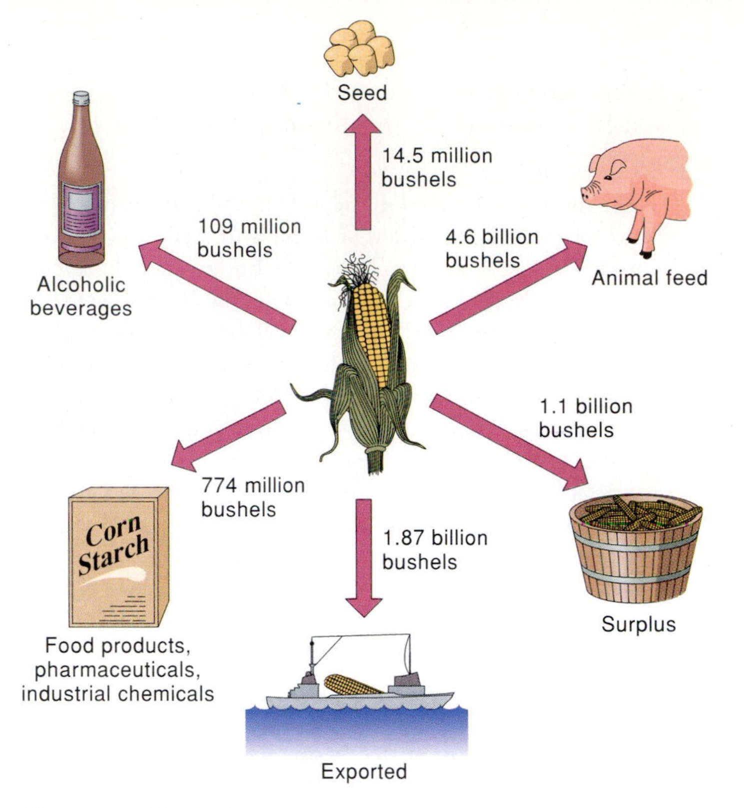

FIGURE E.16

The fate of corn produced in the United States.

and cups will soon be common. In the meantime, most products advertised as "biodegradable" fail to live up to their promise.

Like wheat, modern corn probably arose from a naturally occurring wild grass that initially produced a grasslike type of corn with small ears. Fossil evidence exists of such a primitive corn. A clue to the origin of modern corn comes from "teosinte" a group of wild grasses in the genus *Zea* that grow in Mexico today and probably have for thousands of years. The Indians call teosinte *madre de maize*, which means "mother of corn." Unlike other cereals, corn does not look like its wild ancestors (fig. E.17a). However, there are a few similarities. Both species have twenty chromosomes and produce fertile hybrids when they are crossed. Wild teosinte sometimes grows on the outskirts of cultivated corn fields.

Evolutionary biologists have suggested hypotheses for corn's origins, based on the interfertility and morphological similarities between modern-day corn and teosinte. The first of these two hypotheses is that the ear of corn evolved directly from the ear of teosinte (fig. E.17). The cause of this change occurred about 7,500 years ago, when a population of wild teosinte faced an environmental stress that was endured by only a few plants. Farmers noticed that the unusually hardy plants had larger and better-tasting kernels, and they chose those plants to cultivate. Part of the teosinte tassel might have enlarged and became the corn tassel. Over the centuries, modern corn, with

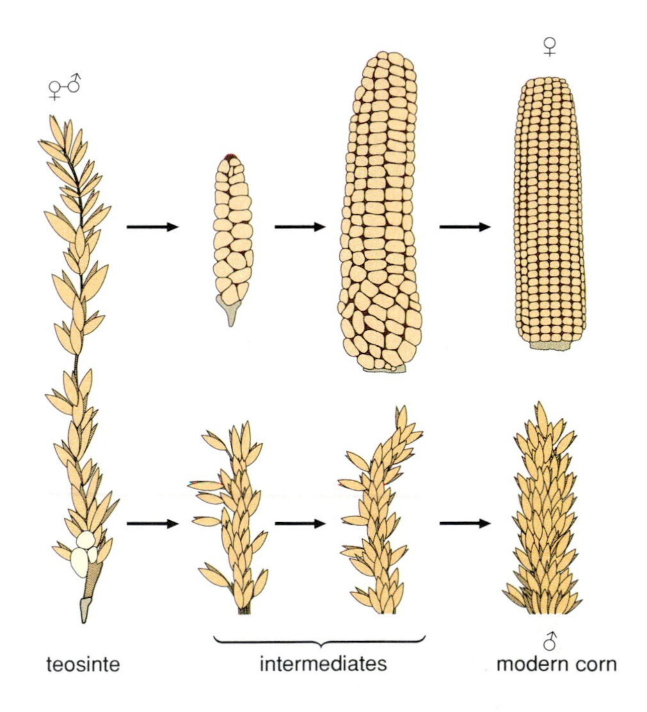

A.

B.

C.

FIGURE E.17

Corn and teosinte. (a) Teosinte *(Zea mexicana)*, a wild grass, has structures that may have given rise to modern corn *(Zea mays)*. The seeds of teosinte evolved into corn kernels, and the tassels eventually became the familiar tassels of corn. (b) Although no one knows what the earliest corn looked like, this modern ear represents a reasonable guess. (c) Modern corn probably evolved from teosinte. Note the double row of seeds in single husk.

female ears and male tassels, was artificially selected by farmers seeking plants with plump and tasty kernels. Indirect evidence for this "ear conversion" hypothesis is that a single mutation will change the hard case around a teosinte kernel into the tunicate form of corn (fig. E.17b).

The second hypothesis for the origin of corn from teosinte is that the male tassel of teosinte underwent a catastrophic sexual change into a kernel-bearing inflorescence (fig. E.17b). Indirect support for this "sexual transmutation" hypothesis comes from the common appearance of sexual abnormalities in modern corn. These abnormalities include the formation of ears by male inflorescences and the occurrence of tassels in female ears. Moreover, environmental stress is known to induce such sexual reversals in teosinte.

Although modern-day maize may have arisen by the domestication of teosinte, the ears of teosinte and maize differ markedly: teosinte ears produce many fewer kernels that are enclosed by harder fruit cases that make harvest difficult (fig. E.17). The *tga1* gene, which regulates the structure of the fruit case, alone accounts for much of the difference between teosinte and maize. Thus, the maize ear was derived from the teosinte ear by a series of modifications, each principally controlled by one or two genes.

Charles Darwin, the father of evolution, was also interested in the origin of corn. Darwin used breeding experiments in his greenhouses to show that when corn plants continually self-fertilized, they produced increasingly weaker offspring—that is, inbred plants were more prone to disease and produced ears with fewer rows of kernels. Darwin noted, however, that the offspring of cross-pollinated plants were strong and healthy, with many rows of plump kernels. Although Darwin knew little about the genetic discoveries of his contemporary, Gregor Mendel, what he had demonstrated was the genetic phenomenon of hybrid vigor. Plants with unrelated parents were more vigorous than self-fertilized plants because they had inherited new combinations of genes.

In the first decade of the twentieth century, George Shull, at the Station for Experimental Evolution in Cold Spring Harbor, New York, extended Darwin's greenhouse experiments, thanks to the additional insight provided by the rediscovery of Mendel's laws of inheritance. Shull continually bred plants from the same ears of corn, artificially selecting highly inbred lines. The inbred corn plants were sickly looking, with small ears. But when Shull crossed two different inbred lines to one another, the offspring—hybrids—were exceptionally vigorous. Shull had developed hybrid corn, which revolutionized corn output worldwide. In the United States, corn production has jumped from 21.9 bushels per acre in 1930 to 95.1 bushels per acre in 1979 and well beyond that today, thanks largely to hybrid corn. Indeed, hybrid corn is the largest seed business in the world today. Many of us start our days with flakes made from this tasty grain (see box 1.2 "Breakfast at the Sanitarium: The Story of Breakfast Cereals," p. 13).

Today's harvest of corn is about 9 billion bushels per year. We feed about half of that to animals. Two centuries ago, however, much was used to make corn whiskey, which was consumed as a cure for colds, coughs, toothaches, and arthritis. These ailments must have been rampant: between 1790 and 1840, people drank an average of five gallons of the whiskey per person per year.

Rice (Oryza sativa)

Rice is the most popular food in the world. It's been cultivated in Thailand for more than 12,000 years, and in Asia it comprises 80% of the human diet. In Taiwan, a hungry person is not asked, "Are you hungry?" or "What would you like to eat?" but "Would you like some rice?" In much of the world, as in Taiwan, rice is so much a dietary staple that it is synonymous with food. The 400 million metric tons of rice produced each year are grown in many types of environments, ranging from 53° north latitude to 40° south latitude. Today's rices include twenty wild species plus many cultivars, which are domesticated species produced by agriculture rather than by the evolutionary forces of natural selection (i.e., cultivars do not grow in the wild). The oldest species of rice for which we have fossil evidence, *Oryza sativa,* originated about 130 million years ago in parts of South America, Africa, India, and Australia, which were then joined into one land mass. Unlike wheat and corn, rice is an ancient plant that evolved in a tropical, semiaquatic environment.

Human cultivation of rice began about 15,000 years ago in the area bordering China, Burma, and India. By 7,000 years ago, efforts to control rice growth had spread to China and India; by 2,300 years ago, the crop was growing in the high altitudes of Japan. By 300 B.C., rice was a staple throughout Asia. It has only been over the past 700 years that rice has been eaten in West Africa, Australia, and North America.

As migrating peoples took their native rices with them to new lands, the plants adapted to a wide range of environments, from deep salty water to the driest of drylands. From the 1930s to the 1950s many nations collected hundreds of native rice varieties, growing small amounts of each type every season, just to keep the collections going.

Other cereals used extensively by humans include barley, millet, oats, and rye. **Barley** was first grown in the eastern Mediterranean, and today is grown mostly in Europe, North America, and Australia. Most barley is used as cattle fodder and to make malt used in distilling and brewing. **Millet** was first cultivated in China in about 2,700 B.C. Although it is now grown mostly for cattle fodder in the United States, its resistance to drought is also being exploited to make it a food crop in tropical Africa. **Oats** probably originated as a weed growing with other cereals, such as barley or wheat. It was domesticated about 2,500 years ago. Today, it is used primarily as cattle fodder and in breakfast cereals. **Rye** originated as a weed growing among other cereals. It was first cultivated in southwest Asia in about 1,000 B.C. and now is used as flour in rye bread and as food for cattle.

Feeding the World

The Struggle to Produce Enough Food

Throughout history, famines have killed millions of people. For example, Indian famines killed 10 million in 1769–1770, 1 million in 1866, 1.5 million in 1869, 5 million in 1876–1878 (as another famine in China killed 10 million), and another

Boxed Reading E.6

Our Fascination with Plants: A Final Look

Throughout this book you've read about unusual plants and the equally unusual things we do with those plants. Here's a final look at some of our favorites:

- The world's oldest potted plant is the cycad *Encephalartos altensteinil* brought from South Africa in 1775 and now housed at the Royal Botanic Gardens, Kew, Great Britain.
- The largest hanging basket had a volume of 120 m^3 (4,167 ft^3), weighed about 4,000 kg (4.4 tons), and contained 600 plants.
- The longest single unbroken apple peel on record is 52.2 m (172′ 4″) long, peeled by Kathy Wafler in 11.5 h from an apple weighing 570 g (20 oz).
- The longest distance at which a grape thrown from level ground has been caught in someone's mouth is 99.8 m (327′ 6″).
- The largest jack-o-lantern in the world was carved from a 375 kg (827 lb) pumpkin.
- The smallest seeds are those of epiphytic orchids; 1.2 million seeds weigh only 1 gram.
- The duration record for staying in a tree is more than 24 years by Bungkas, who went up a palm tree in 1970 and has been there ever since. He lives in a nest that he made from branches and leaves.
- The largest rutabaga ever grown weighed 22.1 kg (48 lb 12 oz), the largest cabbage 56.3 kg (124 lb), and the largest squash 372 kg (821 lb). No word about who ate those whoppers.
- If you've ever wondered how many pickled peppers were in Peter Piper's peck, the answer—in jalapeños—is about 22.5. It's anyone's guess, however, how many seashells he sold down by the seashore.
- The average American eats about 52 kg of potatoes each year, and more than eleven billion kilograms of potatoes go to make French fries. Luckily for potato lovers, German philosopher Friedrich Nietzsche was wrong when he suggested that "A diet which consists predominantly of potatoes leads to the use of liquor."

million people in 1900. Similarly, the Irish potato famine of 1846–1847 killed 1.5 million people and caused a mass exodus from Ireland. If you're Irish, chances are this famine drove your ancestors to North America.[4]

There are many factors involved in feeding the world's population, the most important of which is the size of the population. Our population is growing extremely fast. For example, at the beginning of agriculture about 12,000 years ago, only about 5 million people lived on earth. By the time Christ was alive, the population had grown to 250 million. Thereafter, it doubled to 500 million in 1650, doubled again to 1 billion in 1850, doubled again to 2 billion in 1930, and doubled again to 4 billion in 1976. Today's ever-increasing population of more than 5.6 billion demands huge amounts of food.

Botanists have struggled to produce enough food for the population. One of the pioneers in this struggle was Norman Borlaug, a plant geneticist. Borlaug began creating new varieties of plants in 1944 and soon achieved remarkable results. For example, in 1944 Mexico imported wheat to feed its citizens. By 1964, however, Mexico was exporting wheat. Since 1950, production has quadrupled. Similar advances have been made in Pakistan and other countries. This dramatic increase in production was called the Green Revolution. For his work, Borlaug received a Nobel prize in 1970—not so much for the technology that produced the high-yielding crops, but more for his humanitarianism as he tried to help feed the world. However, even his Green Revolution hasn't been able to keep up with the ever-increasing demand for food. Moreover, critics claim that the Green Revolution actually worsened the problem because it created social and environmental havoc.

The Green Revolution and similar programs have not eliminated world hunger because they have not addressed the problem that drives world hunger: overpopulation. Consider these facts:

- Every year, 100 million people—more than three people per second—are added to the earth's population. That's the equivalent of adding 27 cities the size of Los Angeles.
- Earth's population is now at 5.6 billion people. By the year 2,000, it will exceed 6.3 billion. More than one billion of those people—that's equivalent to the entire population at the beginning of the Industrial Revolution—will be added in the 1990s.
- More than 95% of the earth's population lives in developing countries.

4. Before the famine, a typical Irish family (two adults and four children) ate about 250 pounds of potatoes per week. Today, the average American eats about 115 pounds of potatoes per year.

A.

B.

C.

D.

FIGURE E.18

Deforestation and reforestation of a rain forest. (a) LANDSAT photo showing Rondonia fires. (b) Fire in a Guatemalan rain forest. (c) Destruction of a rain forest by fire. (d) Tree seedlings grown for reforestation of rain forests in Brazil.

- Each day, the average urban resident of the United States uses about 150 gallons of water and 15 pounds of fossil fuel while generating nearly 120 gallons of sewage, 3.4 pounds of garbage, and 1.3 pounds of pollutants.
- Earth's population has doubled in less than thirty-five years and now grows at an annual rate of 1.8%. This means that each day there are 260,000 more people to feed.

The consequences of these facts are devastating. For example, in 1990, more than 1.1 billion people—that's one in every five—lived in abject poverty. More than 500 million people were getting less than 80% of the recommended intake of calories; during the hour or so that it takes for us to eat our Thanksgiving Day meal, more than 1,600 people (mostly children) die of hunger. Moreover, our efforts to feed those people are damaging the environment: we're quickly losing topsoil (see box 20.2, "Losing the Soil," p. 466) and are polluting the soil and water with herbicides and insecticides. As we bring more land into cultivation, we destroy habitats and threaten many native species of plants and animals (fig. E.18). Earth cannot continue to support such an increasing population.

Prospects for the Future: Beyond the Green Revolution

Although many of the wonders you have learned about in this text have greatly increased crop yields and our ability to produce food, more people are starving than ever before. Our rapidly

growing population, combined with ineffective governmental policies and food-distribution methods, has overwhelmed our agricultural system. Solving this problem is a tremendous—perhaps impossible—challenge. However, all hope is not lost.

Throughout this book you've read about one of the most promising tools for helping to feed the world: biotechnology.[5] Just as producing hybrid corn in the 1930s doubled the corn harvest, biotechnology is being hailed as the "second green revolution" that will produce plants that protect and nourish themselves and, in the process, help feed the world. For example, in 1994 a group of researchers used genetic engineering to identify and manipulate a self-incompatibility gene (also called the S gene). A plant having an active S gene recognizes and rejects its own pollen, and therefore cannot fertilize itself. The ability for growers to prevent plants from fertilizing themselves could double the yield (and cut labor costs by half) of many vegetables and flowers.

Other botanists are now using genetic recombination to create high-yielding crops that resist disease, drought, and pests. Still others are improving the caloric and nutritional value of crops. In years ahead, genetically engineered plants will become a leading source for increasing food production. However, biotechnology is not the only tool in the war on hunger.

FIGURE E.19

Amaranth is grown for its cereal-like grain and edible leaves. The plants shown here are growing in Pennsylvania.

Another strategy to increase our supply of plant foods is to look for new crops among the many naturally occurring plants. Fewer than 30 of the 240,000 species of flowering plants provide more than 90% of plant-based foods eaten by people. One of the most promising plants that botanists are studying as a new source of food is the majestic-looking amaranth, a member of the pigweed family (fig. E.19). These plants stand about 8 feet (2.4 meters) tall and have broad purplish green leaves and massive seed heads. Each plant produces about a half million mild, nutty-tasting, protein-rich seeds, each the size of a grain of sand. The flowers are a vivid purple, orange, red, or gold. Interestingly, amaranth was cultivated extensively in Mexico and Central America until the arrival of the Spanish conquistadors in the early 1500s, who banned the plant from use as a crop because of its importance in the Aztec religion.

Amaranth is being grown experimentally at the Rodale Research Center in Kutztown, Pennsylvania, and is available from commercial seed catalogs for home garden use. It grows rapidly, can tolerate a wide range of environmental conditions (high salt, high acid, high alkalinity), yields many seeds, and comes in many varieties. Problems with cultivating amaranth include controlling weeds and pests and harvesting the tiny seeds.

Nutritionally, amaranth is superb. Its seeds contain 18% protein, compared to 14% or less for wheat, corn, and rice. Amaranth is rich in amino acids that are poorly represented in the major cereals. The seeds can be used as a cereal, a popcornlike snack, and a flour to make graham crackers, pasta, cookies, and bread. The germ and bran together are about 50% protein, making them ideal to add to prepared foods and animal feeds. The broad leaves of amaranth are rich in vitamins A and C as well as the B vitamins riboflavin and folic acid. They can be cooked like spinach or eaten raw in salad.

---

5. Investors are also banking on the promise of biotechnology. For example, Calgene, Inc. of Davis, CA spent $25 million to produce their "Flavr Savr" tomatoes. These tomatoes contain an antisense ("reverse") copy of the gene that codes for the expression of polygalacturonase, an enzyme that softens the cell walls during fruit development. This delays ripening, thereby improving the flavor and doubling the shelf-life of the tomatoes to about two weeks. The U.S. Food and Drug Administration approved the tomatoes on 18 May 1994, thereby making it the first food to be genetically altered via molecular biology. (Although foods such as milk and cheese already benefit from biotechnology to enhance production or processing, Calgene's tomato is the first food to have new genes that could not be produced by conventional plant breeding.) The tomatoes were being sold in selected supermarkets in June, 1994, and will be marketed nationally by 1996; investors expect them to capture at least 15% of the $5 billion per year tomato market by 1999.

A. B.

FIGURE E.20

Spinachlike plants used as food. (a) A spinach substitute is obtained from *celosia argentea*, a member of the amaranth family and a close relative of these cockscomb plants (*Celosia cristata*). (b) Cariru (*Talinum*), a locally used spinach in Amazonia.

What are Botanists Doing?

Some of the biotechnological improvements in dicots have relied on the transfer of a plasmid of *Agrobacterium tumefaciens* into the plant. However, the *Agrobacterium* system does not generally work well with monocots, because they are not usually susceptible to infection by *Agrobacterium*. How, then, are biotechnologists trying to improve monocots such as corn?

Go to the library and read a recent article about biotechnological improvements in monocots. What other approaches do you think should be tried?

Botanists are also studying a variety of spinachlike plants as new sources of food (fig. E.20). Other plants studied as potential sources of food include saltbush (*Atriplex*) and ironweed (*Bassia scoparia*), both of which contain much protein. These plants grow well in harsh conditions (salty soil, hot weather) and are a promising source of food for livestock.

Botanists have also established gene banks and pollen banks to help conserve rare plants and to increase the world food supply (fig. E.21). For example, to offset a potential disaster due to reliance on only a few types of rice, the International Rice Research Institute (IRRI) was founded in 1961 in the Philippines. It soon became a clearinghouse for the world's rice varieties, storing the seeds of 12,000 natural variants by 1970 and of more than 70,000 by 1983. Representatives of other important crops are being banked as well. For example, the National Germplasm System in the United States is the world's largest distributor of germplasm and seeds: in 1994 it distributed more than 300,000 samples to more than 100 nations. Similarly, potato cells are stored at the International Potato Center in Sturgeon Bay, Wisconsin, and wheat cells are banked at the Kansas Agricultural Experimental Station. Pollen and seeds from 250 endangered species of flowering plants have been frozen at a plant gene bank at the University of California at Irvine.

Writing to Learn Botany

In an effort to conserve and perhaps increase the genetic diversity of crop plants, many nations donate seeds to gene banks. Sometimes, however, the material donated by a developing nation is used in developed countries to breed or engineer new plant varieties. The new plants—based on the genetic material from the poor nation but the technology of the more wealthy nation—are then sold to the developing nation. Can you think of a more equitable way for nations to cooperate in the development of new plant varieties?

Plant banks offer three priceless services to humanity: a source of variants in case a major crop is felled by disease or an environmental disaster; the return of endangered or extinct

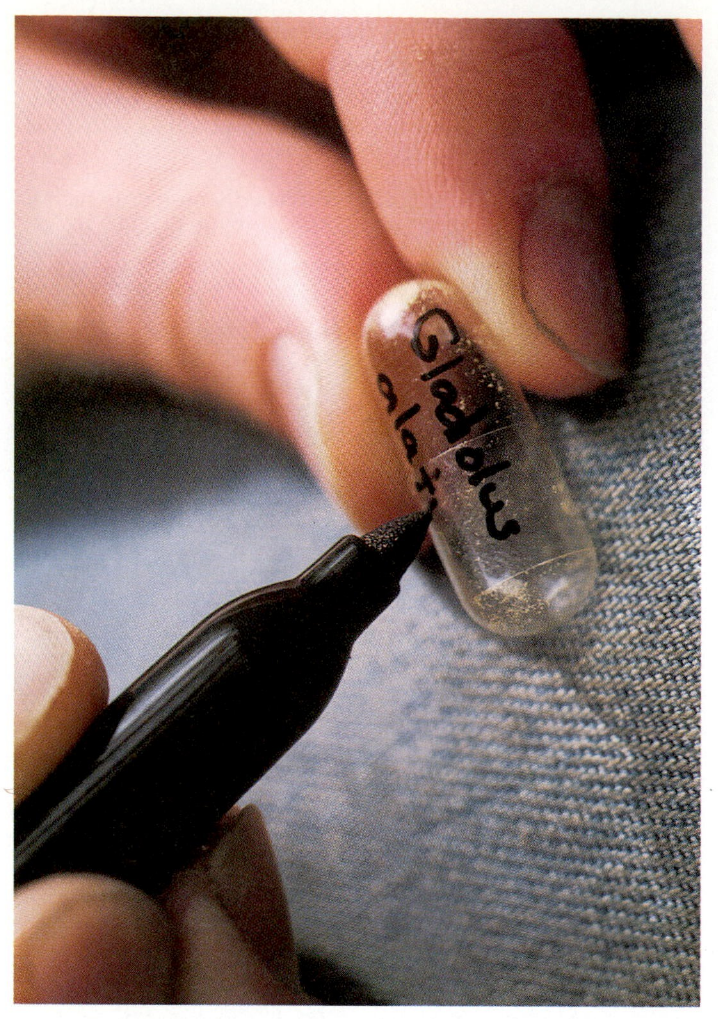

FIGURE E.21

Banking plant genes at the cryogenic gene bank at the University of California at Irvine. Pollen is frozen in small plastic ampules.

varieties to their native lands; and, perhaps most important, a supply of genetic material from which researchers can fashion useful plants in the years to come, even after the species represented in the bank have become extinct.

Establishing a seed or pollen bank is one solution to a growing problem in traditional plant agriculture—the decrease in genetic diversity that makes a crop vulnerable to disease or a natural disaster because it does not have resistant variants. For example, the genetic uniformity of the Irish potato crop made it susceptible to *Phytophthora infestans*, which almost wiped out the crop between 1846 and 1848. (Today in Ireland more than sixty varieties of potatoes are actively farmed to prevent a similar kind of disaster.) However, such lessons are not always heeded. For example, in 1970 more than 70% of the North American corn crop was restricted to just six varieties. A new strain of southern corn leaf blight fungus destroyed 15% of that crop, at a cost of a billion dollars. Gene banks and pollen banks provide the diversity to avoid such disasters in the future.

Our ability to feed ourselves will depend largely on our ability to preserve the genetic diversity of our plants, increase the quality of our crops, produce new hybrids, and develop new crops. Although biotechnology will play a major role in this, we will never be able to eliminate world hunger unless governments reform their food-distribution policies, and we greatly slow the growth of our population. Plant technology alone is not the answer to world hunger.

Chapter Summary

Plants help provide us with food, clothing, shelter, and medicine. About 12,000 to 10,000 years ago, groups of people gradually changed from a hunter-gatherer life-style to an agricultural way of life, intentionally saving and planting seeds from the strongest individual plants of crops that could be used as food. Encouraging the propagation of certain individuals artificially selected particular traits; in this way, humans have influenced plant evolution.

Cereals such as corn, wheat, and rice are members of the grass family, whose edible seeds (grains) can be stored for long periods. Modern wheats contain two, four, or six sets of chromosomes, and they probably evolved from a natural cross between ancient Einkorn wheat (a diploid) and a wild grass (also a diploid), followed by chromosome doubling in some germ cells about 8,000 years ago. This eventually yielded tetraploid Emmer wheat. Emmer wheat also crossed with a wild grass to produce the first hexaploid wheats, which today are used to make bread and other baked goods.

Corn probably evolved from an ancient grass relative, teosinte. Corn and teosinte each have twenty chromosomes, they sometimes grow in the same fields, and they can crossbreed. About 7,500 years ago, an environmental stress may have selected teosinte plants with large kernels, and early farmers may then have cultivated these individual plants. Breeding experiments led to the development of hybrid corn, which results from crosses between separate inbred lines to yield bountiful crops.

Rice is perhaps the most widely consumed modern cereal crop, and fossils of the plants date back some 130 million years. Today, rice grows in a wide range of environments. In 1961, a rice seed bank was started in the Philippines to preserve different varieties and prevent reliance on a few types. Seed and pollen banks have since been founded for other valuable plant species.

For centuries, humans and other animals have used plants for their medicinal properties. Today, natural-products chemists use a combination of laboratory techniques and information from folklore and herbal medicine to make compounds with effects similar to those of plant-derived compounds, in the search for new and more effective drugs.

Questions for Further Thought and Study

1. International agricultural officials have debated for a decade about whether Western countries should pay royalties on genes that germplasm banks have given them for free. The genes can be extremely valuable; for example, a barley gene imported for free from Ethiopia protects the $160 million U.S. barley crop from the yellow dwarf virus. Should countries pay royalties for such genetic resources? Why or why not?
2. Describe the ways in which scientists hypothesize that modern varieties of wheat, corn, and rice arose.
3. Why is reliance on only a few varieties of a crop plant dangerous? Cite an example of when such reliance led to devastation.
4. Artificial selection has produced a modern corn plant that has tasty, nutritious kernels. However, modern corn cannot reproduce successfully without the help of people; the kernels are so tightly protected within the ears that a human or animal must disperse the kernels to start the next year's crop. Do you think that this intervention with a natural process is justified? Why or why not? Are there limits to our intervention? If so, what are they?
5. Devise an experiment to test whether the scenario of corn evolving from teosinte is possible.
6. You are planning your vegetable garden and intend to purchase five packets of carrot seeds. The seed catalog describes a new variety of carrot that is resistant to nearly every known garden pest and produces long, highly nutritious carrots in a variety of climates. It sounds too good to be true, but the company that publishes the catalog has had a lot of experience in plant breeding. If you want to obtain as many carrots as possible, would you be better off buying five packets of the new variety or five different types of carrot seeds? Give a reason for your answer.
7. Botanists and others once believed in the so-called doctrine of signatures, which stated that the shape of a plant part indicated its usefulness in treating a particular ailment. Thus, walnuts, which resemble brains, were used to treat brain disorders, and liverworts, which resemble liver, were used to treat liver ailments. Some people still believe such claims. How would you test such claims?
8. Early settlers fattened their Thanksgiving turkeys and hogs on beechnuts. Perhaps not surprisingly, writer A. A. Milne gave Piglet a home in a beech tree in *Winnie the Pooh*. In what other fictional characters do plants play a role?
9. We use different plants for different purposes. For example, wood of poplar does not splinter, and therefore is used to make toys, tongue depressors, and Popsicle sticks. How are other plant products suited for their function?
10. Many tropical cultures depend heavily on plants such as sugar cane, bananas, and coconuts. Why haven't these plants had the same worldwide impact as cereals?

Suggested Readings

Articles

Doebley, John. 1990. Molecular evidence for gene flow among *Zea mays*. *BioScience* (June). Corn evolved from teosinte: Now genes engineered into corn are being transferred to teosinte.

Feldman, M., and E. R. Sears. 1981. The wild gene resources of wheat. *Scientific American* 244(1):102–113.

Kubo, Isao. 1985. The sometimes dangerous search for plant chemicals. *Industrial Chemical News* (April). Natural-products chemists can sometimes be found in the laboratory and are called modern-day medicine men.

Pimentel, D., et al. 1989. Benefits and risks of genetic engineering in agriculture. *BioScience* 39:606–614. A serious problem arising from genetic engineering would hinder future development of the technology.

Scrimshaw, N. S., and L. Taylow. 1980. Food. *Scientific American* 243 (3):78–99.

Shulman, Seth. 1986. Seeds of controversy. *BioScience* (November). Who owns and should have access to stored plant genetic material?

Tucker, Jonathan B. 1986. Amaranth: The once and future crop. *BioScience* (January). Amaranth is an ancient crop with much future promise.

Vietmeyer, N. D. 1981. New harvests for forgotten crops. *National Geographic* 159(5):702–712.

Woolf, Norma B. 1990. Biotechnologies sow seeds for the future. *BioScience* 40:346–348. How biotechnology is being used to conserve rare plants.

Books

Lewington, A. 1990. *Plants for People*. New York: Oxford University Press.

Simpson, B. B., and M. Conner-Ogorzaly. 1986. *Economic Botany: Plants in Our World*. New York: McGraw-Hill.

Zohary, D., and M. Hopf. 1993. *Domestication of Plants in the Old World*. Oxford: Oxford University Press.

Appendix C

Diversity of Living Organisms

The only thing that is constant about the classification of living organisms is that it is always changing. At the turn of the century, biologists were content to view all living organisms as being either plant or animal. By the 1950s, however, biologists worldwide had begun to accept the idea that living things comprised more than two kingdoms. Today, most classifications divide living organisms into at least five kingdoms: **Monera** (bacteria), **Plantae** (plants), **Fungi** (fungi), **Animalia** (animals), and **Protoctista** (algae, protozoans, slime molds, and other kinds of living things that do not fit into the other kingdoms). Other classifications, some of which are beginning to gain acceptance, divide the bacteria into two kingdoms and the protoctists into several kingdoms. At least one classification divides all of life into thirteen kingdoms.

For this text, we have used the five-kingdom system outlined by Lynn Margulis and Karlene Schwartz. We have chosen it mostly for convenience—Margulis and Schwartz have published an easy-to-read, book-length explanation of their classification system. It is an excellent reference for students and instructors (and authors of textbooks!).

Botanical systems of classification differ in certain respects from the Margulis-Schwartz system in categories below the level of kingdom. A botanically oriented hierarchy of classification, for example, includes the main categories of kingdom, division, class, order, family, genus, and species. In contrast, the Margulis-Schwartz system uses the term phylum in place of division, although phylum is not a formally recognized name in botanical classification. In addition, botanists prefer different names for some divisions of plants other than those used in the Margulis-Schwartz system. In such cases we have adopted the prevailing usages by botanists.

The discussion of classification that follows includes only living plants or organisms that have historically been called plants and that are still presented in most botany texts. This treatment excludes animals and several major groups of protoctists (e.g., protozoans, amoebas). The approximate number of species in each group refers to those that have been named. However, biologists estimate that thousands, or even tens of thousands, of species are yet to be discovered; new species are discovered and named almost every day.

Kingdom Monera

At least 10,000 species of bacteria have been named, although bacteriologists argue that duplicate names for the same organisms might reduce the number of valid species of bacteria to about 2,500. Bacteria are characterized by having a prokaryotic cell type (i.e., no membrane-bound organelles), DNA that is not associated with protein in chromosomes, cell division by fission, and genetic recombination by nonsexual means. Bacteria are the most metabolically diverse of all groups of organisms. Some species can respire aerobically, others anaerobically. Different bacteria can use sulfide, iron, methane, or carbon dioxide in their energy metabolism. Many bacteria can fix atmospheric nitrogen (N_2) into other molecules (e.g., nitrates, ammonia) that are metabolically useful for eukaryotes such as plants. Certain bacteria undergo photosynthesis like plants, while other bacteria undergo photosynthesis that is unique to prokaryotes.

The Monera are divided into two main groups that are usually referred to as the subkingdoms **Eubacteria** and **Archaebacteria.** Differences between these groups are being discovered continually, to the extent that some biologists see these differences as a basis for classifying each group as a separate kingdom.

The Eubacteria include three divisions that are distinguished mainly by properties of their cell walls. Division **Gracilicutes** contains gram-negative bacteria, Division **Firmicutes** contains gram-positive bacteria, and Division **Tenericutes** contains bacteria that have no cell wall.

Division Gracilicutes is divided into three classes, one of which is important in botany because it includes the oxygen-producing photosynthetic bacteria. Bacteriologists call this group the class **Oxyphotobacteria,** but botanists distinguish at least two separate groups in it, either at the class or division level. These are the **Cyanobacteria** and the **Chloroxybacteria.** Both groups use chlorophyll *a* for photosynthesis, as

do algae and plants. In addition, the Chloroxybacteria use chlorophyll *b*, which is otherwise restricted to plants and some algae.

The Archaebacteria conswist of a single division, **Mendosicutes.** Members of this division differ from other bacteria in having eukaryotelike genes with introns, cell walls of glycoproteins and polysaccharides, a eukaryotelike RNA polymerase, ether-linked lipids, and the ability to metabolize methane.

Kingdom Fungi

The fungi are typically filamentous, eukaryotic, spore-producing organisms that lack chlorophyll. They have cell walls made of chitin combined with other complex carbohydrates, including cellulose (chitin is also the main component of the exoskeletons of insects, spiders, and crustaceans). Unlike plants, the main storage carbohydrate of fungi is glycogen, which is also the main storage carbohydrate of animals. All species of fungi are either saprobes or symbionts (i.e., living with other organisms). As symbionts, they may be parasitic, they may provide a benefit to their host, or they may be parasitized by their host.

Nearly 100,000 species of fungi are known, and descriptions of more than 1,000 new species are published each year. Moreover, biologists believe that more than half of all existing fungi have yet to be described. We also suspect that unknown numbers of undiscovered species have already become extinct, due primarily to human activities such as destroying fungal habitats and polluting the air.

Fungi are most often classified into three sexually reproducing groups and one asexually reproducing group, variously treated as divisions, subdivisions, or classes. As divisions, they are the **Zygomycota** (ca. 750 species; e.g., black bread mold, dung fungi, and parasites of amoebas, nematodes, and small animals), the **Ascomycota** (ca. 30,000 species; e.g., yeasts, bread molds, truffles, morels, and ergot fungi), the **Basidiomycota** (ca. 25,000 species; e.g., mushrooms, stinkhorns, puffballs, jelly fungi, and smut and rust diseases of plants), and the **Deuteromycota** (ca. 17,000 species; e.g., penicillin mold, root-rot fungi, vaginal yeast fungi, and athlete's-foot fungus). The distinguishing features of each of these groups involve differences in reproduction:

1. Zygospores characterize the Zygomycota.
2. Spore-containing sacs are formed by the Ascomycota.
3. Spores of the Basidiomycota are produced on basidia.
4. Spores of the Deuteromycota are only produced asexually.

Approximately 25,000 species names have been given to lichens, which are organisms that consist of a fungal body that hosts green algae or cyanobacteria or both. The fungus in this symbiotic relationship seems to be parasitic, since it gets nutrients from the algae and cyanobacteria, apparently without providing any benefit in return. The lichens are often classified in existing fungal divisions, the division depending on the type of fungus in the symbiosis. However, Margulis and Schwartz separate all lichens into their own division, the **Mycophycophyta.**

Kingdom Protoctista

The protoctists that have traditionally been of most interest to botanists are the slime molds, the water molds, and the algae. None of the slime molds or water molds can photosynthesize, and these organisms were long considered to be fungi. Conversely, algae usually contain chlorophyll *a* and can photosynthesize like plants. Botanists have historically classified algae as plants, and many botanists still consider most algae to be plants, although this trend is changing.

Slime Molds and Water Molds

These molds are usually classified into four divisions: the **Myxomycota** (plasmodial slime molds), the **Acrasiomycota** (cellular slime molds), the **Oomycota** (biflagellate water molds), and the **Chytridiomycota** (uniflagellate water molds). The diversity of genera in these groups is such that some taxonomists recognize seven divisions instead of the usual four.

The 70 or so species of cellular slime molds are characterized by a phagotrophic mode of nutrition, by carbohydrate storage as glycogen, by cellulosic cell walls, and by amoeboid cell motility. The plasmodial slime molds (ca. 500 species) are also phagotrophic and store glycogen, but differ from the cellular slime molds by the absence of cell walls and by motile cells that swim by two whiplash flagella. Both kinds of slime molds live in terrestrial habitats.

The water molds absorb nutrients and have flagellated motile cells. The motile cells of the Chytridiomycota (ca. 575 species) have a single whiplash flagellum, whereas those of the Oomycota have two flagella, one of which is whiplash and the other of which is tinsel. The uniflagellate water molds live in freshwater or marine habitats, store glycogen, and form cell walls of chitin or glucan. Conversely, the biflagellate water molds live only in freshwater habitats, store either glycogen or a glucose polymer called mycolaminarin, and form cell walls of cellulose, chitin, glucan, or some combination of these.

Algae

Algae are an informally defined group of photosynthetic eukaryotes that are classified as seven to eleven divisions. These divisions are at least partially distinguished by their pigments, their energy storage polymers, their cell-wall components, and the number and types of their flagella. The major features of the seven main divisions of algae are summarized below.

Division Chlorophyta: Green Algae

The green algae, which include about 7,500 species, are organisms whose pigments include chlorophylls *a* and *b* and various carotenoids and xanthophylls. They store starch like that of plants, contain cellulose in their cell walls, and have either nonmotile cells or motile cells with anywhere from one to about 120 flagella at or near the apex of each swimming cell. Green algae can be unicellular, multicellular, or colonial.

Most green algae live in fresh water, but different species also occur in marine habitats, clouds, snow banks, or soil, or on the shady moist sides of trees, buildings, and fences. Green algae also live symbiotically with lichen fungi and with several different kinds of animals.

Among the many distinctive groups of green algae, three have received the most attention and are most often classified as classes of the Chlorophyta. Class **Chlorophyceae** includes predominantly freshwater algae that undergo mitosis in a persistent nuclear envelope and cell division by forming a phycoplast. Class **Charophyceae,** which includes mostly freshwater species, and class **Ulvophyceae,** which consists of marine organisms, both undergo cell division by forming a phragmoplast like that of plants. The Ulvophyceae also have a persistent nuclear envelope during mitosis. However, in the Charophyceae, as in plants, the nuclear envelope disintegrates as mitosis proceeds.

Division Phaeophyta: Brown Algae

The approximately 1,500 species of brown algae derive their color mostly from the xanthophyll fucoxanthin, but they also contain chlorophylls *a* and *c* and the orange carotenoid β-carotene. Brown algae store mannitol and a glucose polymer called laminarin. The cell walls of brown algae consist of a cellulosic matrix containing alginic acid, and the motile cells have two lateral flagella, one of which is a forward-projecting tinsel type and the other of which is a trailing whiplash type. Brown algae are all multicellular and include microscopic forms as well as the most complex of all the algae, the giant kelps. Most brown algae are marine organisms, but a few species live in fresh water. The brown algae are all included in one class, the **Phaeophyceae,** which contains four orders and five families.

Division Rhodophyta: Red Algae

Like the brown algae, the red algae are mostly marine organisms that are either microscopic filaments or macroscopic forms with complex, leafy branches. The red algae, however, also include some unicellular forms. The 3,900 species of red algae are characterized by proteinaceous pigments called phycobilins. Red algae are red because of their phycoerythrins, which are red phycobilins; they also contain blue-green phycobilins called phycocyanins. Photosynthesis in red algae depends on chlorophyll *a*, but there is no chlorophyll *b*. These algae also contain carotenoids and xanthophylls. The cell walls of red algae usually consist of cellulose and pectins, but the cells walls of some species also form calcium carbonate. Although red algae live almost exclusively in water, mostly marine, they form no cells that swim. All of the red algae are classified into one class, the **Rhodophyceae,** which is divided into five orders and six families.

Division Chrysophyta: Diatoms, Golden-Brown Algae, and Yellow-Green Algae

This is the largest division of algae, with more than 11,000 species. The diatoms, golden-brown algae, and yellow-green algae are regarded by some taxonomists to be sufficiently distinct from one another to be classified as separate divisions. At this time, however, most botanists still treat them as three classes of the Chrysophyta: class **Bacillariophyceae** (diatoms), class **Chrysophyceae** (golden-brown algae), and class **Xanthophyceae** (yellow-green algae). Members of all three classes contain chlorophylls *a* and *c*, carotenoids, and fucoxanthin. The main storage product in these algae is an oil, called chrysolaminarin, that contains a polymer of glucose. Most of the species in this division are unicellular, although some of the golden-brown and yellow-green algae are filamentous and others are colonial.

Algae that are classified in the three orders and four families of diatoms, which include about 10,000 of the species of the Chrysophyta, are best known for their glass cell walls. Each cell wall consists of two parts that fit together like a pillbox. Although diatoms have no flagellated cells, they apparently swim by a mechanism similar to jet propulsion: each cell secretes mucilage-containing fibrils that propel it forward.

Yellow-green algae (ca. 500 species) are grouped into three orders and four families. These algae either have cell walls of cellulose or they have silicon-containing scales instead of a well-defined cell wall. Most yellow-green algae have motile cells that have one or two flagella; biflagellate cells have one leading tinsel flagellum and one trailing whiplash flagellum, as do the motile cells of brown algae. Some species lack motile cells and others have cells that are amoeboid.

Golden-brown algae (ca. 325 species), which are classified into one order with two families, usually have cellulosic cell walls that are rich in pectins, but in certain species the cells are covered with silicon-containing scales. Unlike the other classes of the Chrysophyta, members of the Xanthophyceae lack fucoxanthin. Most golden-brown algae are nonmotile, although some have biflagellated cells like those of the brown algae; still other species have cells that are amoeboid.

Divisions of Unicellular Algae

Three divisions of algae consist exclusively of flagellated unicellular organisms. They are the euglenoids **(Euglenophyta),** the dinoflagellates **(Pyrrhophyta),** and the cryptomonads **(Cryptophyta).** These three divisions are traditionally studied by botanists and zoologists alike because of the plantlike and animal-like features of different species. Each division includes species that are nonphotosynthetic and species that have chlorophyll *a* and carotenoids. In addition, the euglenoids have chlorophyll *b* and the dinoflagellates and cryptomonads have chlorophyll *c*. Cryptomonads also have phycobilins. A major feature of these divisions is that they are characterized by chloroplasts that are surrounded by an extra membrane, apparently derived from the endoplasmic reticulum.

Euglenoids (ca. 800 species) and cryptomonads (ca. 100 species) lack cell walls. Their cells are bounded by a flexible **periplast,** which is a plasma membrane that has extra inner layers of proteins and a grainy outer surface. Dinoflagellates (ca. 1,100 species) also usually lack cell walls, but most species have armorlike cellulosic plates interior to the plasma membrane. The divisions of unicellular algae are characterized by cells that are motile by one to three flagella. Dinoflagellates have the most distinctive flagellar arrangement among the unicellular algae. Both flagella are lateral, but one coils around the cell and undulates so that the cell spins as it moves forward. The other flagellum trails the cell as a rudder.

The cryptomonads have a nucleuslike organelle between the outer chloroplast membrane and the chloroplast ER. This organelle, called a **nucleomorph,** is absent in other algae. Nuclear division among dinoflagellates is unusual because dinoflagellate chromosomes are permanently condensed and have no histones. Finally, both the euglenoids and the dinoflagellates have a persistent nuclear envelope during mitosis.

Kingdom Plantae

Plants are multicellular eukaryotes having cellulose-rich cell walls, chloroplasts containing chlorophylls *a* and *b* and carotenoids, and starch as their primary food-reserve. Plants reproduce sexually by the formation on sporophytes of meiotically derived spores that grow into multicellular gametophytes. In plants, as in some green algae, fertilization produces an embryo that is retained in the parent gametophyte. In all but the bryophytes, the sporophyte is the dominant generation. Plants are informally recognized as three main groups: bryophytes (three divisions), seedless vascular plants (four divisions), and seed plants (five divisions). The main features of each of these twelve divisions of plants are summarized below.

The only swimming cells among plants are the flagellated sperm cells that occur in nine of the twelve divisions. The exceptions are the Pinophyta, Gnetophyta, and Anthophyta (also referred to as Magnoliophyta), none of which forms motile cells.

Division Bryophyta: Mosses

Division Bryophyta is the largest and most familiar division of bryophytes, with more than 10,000 species. Mosses thrive alongside the more conspicuous vascular plants in terrestrial habitats such as on soil, on the trunks of trees, and on shady rock walls. Peat mosses grow more commonly in pools, bogs, and swamps. Some mosses have a central strand of conducting cells that are functionally equivalent to xylem and phloem. This means that, although the bryophytes are referred to as nonvascular plants, mosses do form a simple type of vascular tissue.

Mosses, as well as all other bryophytes, are homosporous. Mosses have leafy gametophytes, multicellular rhizoids, and stomata on the sporophytes. The diversity of these kinds of plants is such that bryologists recognize three classes in the division: the **Sphagnopsida** (peat mosses), which has one order and one family; the **Andreaeopsida** (rock mosses), which also has one order and one family; and the **Bryopsida** ("true" mosses), which includes at least twelve orders and nineteen families.

Division Hepatophyta: Liverworts

Almost 8,500 species of liverworts have been named, ranging in size from tiny, leafy filaments less than 0.5 millimeter in diameter to a thallus more than 20 centimeters wide. Liverworts typically have unicellular rhizoids, no cuticle, no specialized conducting tissue, and no stomata. In addition, their spores are shed from sporangia for a relatively short time. Liverworts are the simplest of all living plants. The liverworts are classified in just one class, which is divided into seven orders and twenty-six families.

Division Anthocerotophyta: Hornworts

The hornworts are the smallest of the three groups of bryophytes. There are only about 100 species in six genera, which are divided between two families in the same class and order. The hornworts get their name from the horn-shaped sporophyte. The sporophyte has an intercalary meristem that seems capable of indefinite

growth; it also has stomata. The gametophyte is thallose, not leafy, and has no specialized conducting tissue. Female sex organs are embedded in the thallus and contact the surrounding vegetative cells. The thallus also has stomatalike structures, which are absent on gametophytes of all other plants.

Division Psilotophyta: Whisk Ferns

The Psilotophyta are the simplest vascular plants, primarily because they have no apparent roots and most of the species have no obvious leaves. Instead of roots with root hairs, whisk ferns have rhizomes with rhizoids. Instead of leaves, they have photosynthetic stems and flattened branches that look and function like leaves. Stems are protostelic and are usually dichotomously branched. Like the bryophytes, all members of the Psilotophyta are homosporous. The bisexual gametophytes are small, inconspicuous and nonphotosynthetic.

The only class, order, and family in this division includes two genera: *Psilotum* (3 species), and *Tmesipteris* (7 species). *Psilotum* is widespread in subtropical regions of the southern United States and Asia; it is also a popular and easily cultivated plant that is grown in greenhouses worldwide. *Tmesipteris* is restricted to islands in the South Pacific, where it often occurs as an epiphyte on the trunks of tree ferns. It is rarely cultivated.

Division Lycopodophyta: Club Mosses and Spike Mosses

The Division Lycopodophyta consists of more than 1,100 species worldwide. They are primarily tropical plants that live in terrestrial habitats, either on soil or as epiphytes, but they also form a conspicuous part of the flora in temperate regions. Most of the species are included in two genera: *Lycopodium* (club mosses; ca. 400 species) and *Selaginella* (spike mosses; ca. 700 species). Each genus is classified in a separate class. Their common names derive from their small, mosslike leaves and the club-shaped or spike-shaped cones at the tips of fertile branches. However, club mosses and spike mosses are differentiated into leaves, stems, and roots, so they are not really mosses. The stems are protostelic, and the leaves each have a single, unbranched vein. The club mosses, like bryophytes and whisk ferns, are homosporous. In contrast, the spike mosses are heterosporous. By producing two kinds of spores, the spike mosses always produce unisexual gametophytes—either male or female.

Division Equisetophyta: Horsetails

Equisetum, with about 15 species, is the only living genus in this division. The Equisetophyta, therefore, have only one class, one order, and one family. Some species have branched stems that, with a good imagination, look like horses' tails. Horsetail-type species, as well as unbranched species, are also referred to as scouring rushes because their epidermal tissue contains abrasive particles of glass. They were used by American Indians to polish bows and arrows and by early colonists and pioneers to scrub pots and pans. *Equisetum* occurs worldwide in most habitats along streams or the edges of forests.

Although horsetails have true leaves, their stems and branches are the dominant photosynthetic parts of the plant body. Horsetails are homosporous, and their gametophytes are photosynthetic.

Division Pteridophyta: Ferns

Ferns include approximately 12,000 living species, making this division by far the largest among the seedless vascular plants. Ferns are primarily tropical plants, but some species inhabit temperate regions and some even live in deserts.

The classification of ferns is in constant turmoil because of the very different views among fern taxonomists. Nevertheless, a reasonable system of classification of the ferns groups them into three classes: **Ophioglossopsida, Marattiopsida,** and **Filicopsida.** The Ophioglossopsida, which includes but a single order and family, is distinguished by having two kinds of leaves. One kind of leaf is exclusively vegetative and the other is reproductive—that is, it bears the sporangia. The Marattiopsida also consists of just one order and one family. This class includes the tree ferns, which have the largest and most complex fronds (leaves) in the division. All members of the Ophioglossopsida and Marattiopsida are homosporous.

Class Filicopsida, known as "true" ferns, is the largest and most diverse group of ferns. The approximately 10,000 species in this class are grouped into three orders and at least ten families. The Filicopsida is distinguished by a type of spore development that is unique to this group of ferns. Such true ferns include homosporous as well as heterosporous species. The heterosporous species are generally aquatic, whereas the homosporous species are terrestrial.

Division Ginkgophyta: Maidenhair Tree

Division Ginkgophyta is one of five divisions of seed plants. Four of these—Ginkgophyta, Cycadophyta, Pinophyta, Gnetophyta—form seeds that are born on cones or are exposed singly at the tips of fertile branches. Together these four divisions are informally referred to as the gymnosperms (*gymno* = naked, *sperm* = seed). The fifth division, the Anthophyta (i.e., Magnoliophyta) is characterized by seeds that form in fruits. Members of the Anthophyta are informally referred to as angiosperms (*angio* = container). All species of seed plants are heterosporous.

The maidenhair tree, *Ginkgo biloba,* is the only species in this division. Botanists are unsure as to whether this species occurs in natural stands, but it is widely cultivated in temperate regions of the world.

Ginkgo is characterized by distinctive, dichotomously veined, fan-shaped leaves, which are produced on two types of shoots. Relatively fast-growing long shoots and seedlings bear leaves with a distinct apical notch (hence *biloba* in the binomial), while slow-growing spur shoots produce leaves without a notch. The trees are deciduous.

Maidenhair trees are exclusively dioecious. Pollen is born in anthers that occur in loose clusters. The ovules of *Ginkgo* are exposed and occur singly at the tips of fertile branches. Ovules develop into seeds with a massive integument that consists of a fleshy outer layer, a hard and stony middle layer, and an inner layer that is dry and papery. The sperm cells of *Ginkgo* are flagellated, although the flagella are not needed for carrying the sperm to the egg.

Division Cycadophyta: Cycads

The approximately 185 species of cycads are divided among three families that are all classified in the same class and order. Cycads live primarily in the tropical and subtropical regions of the world. All species of cycads are dioecious.

Cycads are characterized by their palmlike leaves that bear no resemblance to leaves of other living gymnosperms. Pollen and ovules are formed in sporangia that occur in simple cones on separate plants. Ovules develop into seeds like those of *Ginkgo* in that they have a three-layered integument, but the inner layer of cycad seeds is soft instead of papery. Also like *Ginkgo*, the sperm cells of cycads have flagella that are not needed for carrying the sperm to the egg. The sperm cells of cycads can be up to 400 micrometers in diameter, which are the largest sperm cells among plants.

Division Pinophyta: Conifers

The informal name of this group, conifers, refers to plants that bear cones, even though some species do not bear cones and other divisions of gymnosperms also include cone-bearing species. This division includes pines, which are the most abundant trees in the forests of the northern hemisphere. Most of the approximately 550 species of conifers live in temperate climates, although many also live in alpine habitats or in deserts.

Pines and many other conifers have needle-shaped leaves. Still others have scalelike leaves or leaves with flat blades. Although a few conifers are deciduous, most members of this division are evergreen. Conifers have a wide variety of pollen-bearing and ovule-bearing organs, but none form sperm cells that have flagella.

The Pinophyta consists of two classes of extant species, with a total of two orders and usually six families. The largest class, the **Coniferopsida,** contains pines, firs, cypresses, larches, spruces, hemlocks, and cedars in the northern hemisphere, and araucarias and podocarps in the southern hemisphere. Members of the single order and family in class **Taxopsida,** such as yews and torreyas, are also native to the northern hemisphere.

Division Gnetophyta: Gnetophytes

The gnetophytes include some of the most distinctive, if not bizarre, of all seed plants. The three genera and 71 species of gnetophytes are grouped into one class, but each genus belongs to a separate family in its own order. Gnetophytes are the only gymnosperms that possess vessels, and they are the only gymnosperms that undergo double fertilization. However, unlike double fertilization in angiosperms, double fertilization in gnetophytes is not followed by the formation of endosperm. Instead, the diploid cell from fertilization by the second sperm disintegrates. Like the Pinophyta, the Gnetophyta forms nonmotile sperm cells.

Division Anthophyta (Magnoliophyta): Flowering Plants

Botanists estimate that there are at least 260,000 species of angiosperms, which makes the Anthophyta by far the largest division of plants. As a group, the flowering plants are defined by the formation of flowers, by double fertilization that results in a zygote as well as a nutritive endosperm tissue, by the presence of vessels, by the formation of ovules in organs that develop into fruits, and by few-celled gametophytes.

Flowering plants are presently divided into two classes, the **Magnoliopsida** (ca. 180,000 species) and the **Liliopsida** (ca. 80,000 species). The Magnoliopsida are informally called dicots, which refers to seeds that have two cotyledons (seed leaves). Dicots are also characterized by flowers whose parts are usually in fours or fives, by netlike venation in leaves, by primary vascular bundles occurring in a ring in the stem, and by the presence of a vascular cambium and true secondary growth in many species. Conversely, the Liliopsida are called the monocots because they form seeds that have a single cotyledon. Also in contrast to dicots, monocots have flowers whose parts usually occur in threes, whose leaves have parallel venation, and whose stems have primary vascular bundles that are scattered. Monocots also lack a vascular cambium and true secondary growth.

There are several comprehensive classifications of flowering plants, which differ according to how genera are lumped into families and families into orders. There is no general agreement among botanists as to which is the best, but four classifications have received the greatest acceptance. These are referred to by the names of the botanists who have proposed them. As

summarized in the following table, the different views among these botanists are reflected in the numbers of orders and families that are included in their respective classifications.

Classifier	Number of Orders	Number of Families
Cronquist	83	387
Dahlgren	73	403
Takhtajan	92	410
Thorne	69	440

Suggested Readings

Beck, C. B. 1988. *Origin and Evolution of Gymnosperms.* New York: Columbia University Press.

Birge, E. A. 1992. *Modern Microbiology: Principles and Applications.* Dubuque, IA: Wm. C. Brown Publishers.

Bold, H. C., and M. J. Wynne. 1985. *Introduction to the Algae,* 2nd ed. Englewood Cliffs, NJ: Prentice-Hall, Inc.

Carr, N. G., and B. A. Whitton, eds. 1982. *The Biology of Cyanobacteria.* Berkeley, CA: University of California Press.

Chopra, R. N., and P. K. Kumra. 1988. *Biology of Bryophytes.* New York: John Wiley.

Cronquist, A. 1988. *The Evolution and Classification of Flowering Plants,* 2nd ed. New York: The New York Botanical Garden.

Hale, M. E. 1983. *The Biology of Lichens,* 3rd ed. Baltimore, MD: University Park Press.

Margulis, L., and K. V. Schwartz. 1988. *Five Kingdoms: An Illustrated Guide to the Phyla of Life on Earth.* New York: W. H. Freeman and Company.

Tryon, R. M., and A. F. Tryon. 1982. *Ferns and Allied Plants.* New York: Springer-Verlag.

Glossary

A

A Horizon (Topsoil) The uppermost layer of soil, usually about 10 to 20 cm. thick
ABA(*see* **Abscisic Acid**)
Abaxial Away from the axis
Abby Abbreviation of Abington, which is one of 19 strains of viruses found only in gypsy moth caterpillars
Abiotic Something that is non-living and never has been alive
Abscisic Acid (ABA) A plant hormone (growth regulator) associated with water stress and the inhibition of growth; also induces stomatal closing and seed dormancy in many plants
Abscission The detachment of leaves, flowers, or fruits, usually at a weak area termed the abscission zone
Absorption Spectrum The spectrum of light absorbed by a particular pigment
Accessory Pigment A pigment that captures light energy and transfers it to chlorophyll *a;* beta-carotene is an example of an accessory pigment
Acetyl Coenzyme A (acetyl-CoA) A two-carbon organic acid whose hydroxyl group has been replaced with coenzyme A
Acid-Growth Hypothesis The hypothesis that acidification of the cell wall leads to the breakage of restraining bonds within the wall, thereby leading to cellular elongation that is driven by turgor pressure
Acropetally Toward the apex
Actin a type of globular protein that makes up actin filaments
Actin Filament the smallest (4–7 nm in diameter) of the three types of filaments that comprise the cytoskeleton
Action Spectrum The spectrum of light that elicits a particular response
Active Transport Movement of solutes across a membrane against their concentration gradient; active transport required energy from cellular metabolism
Acylglyceride Linkage The covalent bond between the organic acid group, such as in a fatty acid, and one of the three hydroxyl groups of glycerol
Adaptive Radiation Evolution of divergent forms of a trait in several species that developed from an unspecialized or primitive common ancestor
Adaxial Toward the axis
Adenosine Triphosphate (ATP) A nucleotide consisting of adenine, ribose, and three phosphate groups; the major source of usable chemical energy in metabolism; when hydrolyzed, ATP loses a phosphate to become adenosine diphosphate (ADP) and releases usable energy
Adenovirus 2 A type of virus that causes human respiratory disease; its role in genetic research involved the discovery of introns
Adventitious Root A root that arises from a leaf or stem (i.e., not from another root)
Aerenchyma A tissue containing large amounts of intercellular spaces
Aerobic Respiration (*see* **Respiration**)
Agar A slimy polysaccharide, consisting mostly of a specific mixture of alpha-galactose sulfates that surround the cell walls of certain red algae; in the United States it is harvested for commerce primarily from *Gelidium robustum*
Akinete A thick-walled dormant cell derived from a vegetative cell
Albuminous Cell Certain ray and axial parenchyma cells in the phloem of gymnosperms; these cells are closely associated with sieve cells, both morphologically and physiologically
Aleurone Layer A group of protein-rich cells located at the outer edge of the endosperm of many grains
Alkaloid A nitrogen-containing base in which at least one nitrogen is part of a ring; examples include nicotine, caffeine, cocaine, and strychnine; alkaloids are often bitter and affect the physiology of vertebrates and other animals
Allele One of the alternative forms of a gene; a gene may have two or more alleles
Allopatric Occurring in different places
Allopatric Speciation Speciation induced by geographical or physical separation of the ancestral population
Allopolyploid A polyploid with multiple sets of chromosomes that originated from more than one species
Allosteric Regulation Regulation that results from a change in the shape of a protein that occurs when the protein binds a nonsubstrate molecule; in its new shape, the protein usually has different properties
Allozymes Enzymes that are coded for by different alleles of the same locus; each form is encoded by different alleles
Alpha-Amylase An enzyme that breaks down starch into smaller units by cleaving the 1,4 linkages between molecules of alpha-glucose
Alpha-Glucose The form of glucose whose structure, when drawn in flat plane, has a hydroxyl group at the first carbon that points down
Alpha-Ketoglutarate (Alpha-Ketoglutaric Acid) A five-carbon organic acid that loses a molecule of carbon dioxide and gains an acetyl-CoA group in the fourth step of the Krebs cycle, thereby being converted to succinyl-CoA; also during this conversion, one molecule of NAD^+ is reduced to NADH
Alpha-Tubulin A type of globular protein that is a main component of microtubules
Amino Acid Acceptor Site A sequence of nucleotides that recognizes and binds to a specific amino acid at the 3′ end of a molecule of transfer RNA
Aminoacyl-tRNA Synthetase A type of enzyme that catalyzes the binding of an amino acid to the amino acid acceptor site on a molecule of transfer RNA
Amylase (*see* **Alpha-Amylase**)
Amylopectin A highly branched polymer of up to 50,000 molecules of alpha-glucose
Amyloplast A type of plastid that stores starch
Amylose An unbranched chain of up to several thousand molecules of alpha-glucose
Anabolism Biosynthesis; the constructive part of metabolism
Anaerobic Respiration (*see* **Respiration**)
Anaphase The period of mitosis during which centromeres split and sister chromatids become separate chromosomes that begin to move toward opposite poles of the spindle apparatus
Anaphase I The first anaphase of meiosis; in anaphase I, homologous chromosomes move to opposite poles of the meiotic spindle apparatus, resulting in a halving of the number of chromosomes going to each daughter nucleus
Anaphase II The second anaphase of meiosis; in anaphase II, the centromeres divide, thereby allowing the separation of sister chromatids into independent chromosomes
Androecium (plural, **androecia**) Collectively, all of the stamens of a single flower
Angiosperm Any plant whose seeds are born in a fruit; an informal name for flowering plant (Division Anthophyta or Magnoliphyta)
Anther The pollen-bearing part of a stamen

Antheridiophore In some liverworts, the stalk that bears antheridia

Antheridium (plural: **Antheridia**) A unicellular or multicellular structure in which sperms are produced; may be multicellular or unicellular

Anthocyanin Any red or blue pigment that is a flavonoid; anthocyanins are the primary pigments of blue and red plant parts (e.g., flowers, fruits)

Antibody A protein whose formation is induced by an antigen and that binds to the antigen that induced it

Anticodon A sequence of three nucleotides in a molecule of transfer RNA, which is complementary to a codon sequence

Antigen A large, foreign molecule, such as a protein or polysaccharide, that induces its host to form antibodies against it

Antiparallel Refers to double-stranded DNA, in which the direction of each strand is opposite its complementary strand

Antipodal Cell Cells that form at the chalazal of the embryo sac, opposite the micropylar end

Antisense Strand In DNA, the antisense strand of a gene is the one that does not contain a coding sequence for a molecule of RNA; the antisense strand is not transcribed

Apical Dominance The influence exerted by a terminal bud in suppressing the growth of lateral buds

Apical Meristem The meristem at the tip of a root or shoot in a vascular plant

Apomictic Asexual production of seeds

Apoplastic Movement The movement of water and solutes in the free space of the tissue; the free space includes cell wells and intercellular spaces

Apothecium An open ascocarp; it is usually cup- or saucer-shaped

Archaebacteria Primitive prokaryotes with distinctive chemical and structural features

Archegoniophore In some liverworts, the stalk that bears archegonia

Archegonium A multicellular organ that produces an egg; found in bryophytes and some vascular plants

Aril A fleshy structure that may partially envelop a seed

Artificial Selection Selection by humans of specific traits in organisms being bred to produce desired characteristics

Ascocarp A reproductive structure of ascomycetes, in which asci are formed

Ascogenous Hyphae Hyphae with paired male and female nuclei; ascogenous hyphae eventually produce asci

Ascogonium (plural: **Ascogonia**) The female sexual structure of ascomycetes

Ascomycetes A large group of true fungi with septate hyphae; they produce conidiospores asexually and ascospores sexually within asci

Ascospore A spore produced within an ascus

Ascus (plural: **Asci**) A saclike cell of ascomycetes in which, following meiosis, a specific number (usually 8) of ascospores is produced

ATP Synthase A type of membrane-bound enzyme in mitochondria that phosphorylates ADP to ATP by using energy from the diffusion of protons through the enzyme

ATP (*see* **Adenosine Triphosphate**)

ATP Phosphohydrolase (ATPase) A type of transport protein that uses energy from the hydrolysis of ATP to actively transport ions or other solutes against their concentration gradient

Autopolyploid A polyploid with multiple sets of chromosomes that originated from more than one species

Autotroph An organism that produces its own food, usually by photosynthesis; virtually all plants are autotrophs

Auxin A plant hormone (growth regulator) that influences cellular elongation, among other things; also referred to as indole-3-acetic acid, or IAA

Axil The upper angle between a twig of leaf and the stem from which it grows

Axillary Bud Buds that occur in the axil of a leaf

Axillary Placentation Refers to the attachment of ovules along the central axis of an ovary that has more than one ovule-bearing chamber; lily is an example plant that has axillary placentation

B

B Horizon (Subsoil) The layer of soil immediately beneath the topsoil, usually about 25 to 50 cm. thick

Bacteriochlorophyll A modified chlorophyll that is the primary light-trapping pigment in green and purple photosynthetic bacteria

Bacteriophage A type of virus that parasitizes bacteria

Bar A unit of pressure; one bar is the atmospheric pressure of air at sea level and room temperature

Bark The part of the stem or trunk exterior to the vascular cambium

Basidiocarp A reproductive structure of basidiomycetes, in which basidia are formed

Basidiomycetes A large and diverse group of true fungi with septate hyphae; they produce basidiospores externally on basidia

Basidium (plural: **Basidia**) A club-shaped structure upon which, following meiosis, a specific number (usually 4 or 2) of basidiospores is produced

Basipetally Toward the base

Bedrock Solid rock beneath the layers of soil

Beta-Carotene An orange pigment that is made of eight isoprene units; it occurs in most plants as an accessory pigment to photosynthesis

Beta-Glucose The form of glucose whose structure, when drawn in flat plane, has a hydroxyl group at the first carbon that points up

Beta-Oxidation A sequence of biochemical reactions that oxidize fatty acids into a series of two-carbon compounds that are converted to acetyl-CoA

Beta-Tubulin A type of globular protein that is a main component of microtubules

Bilayer In referring to phospholipids, a bilayer is a spontaneously formed double layer of lipid, with an interior of hydrophobic hydrocarbons and an exterior of hydrophilic phosphate groups

Binomial System of Nomenclature A system of applying two-part scientific names to organisms, each name consisting of the genus (generic) name and a species (specific) epithet

Bioassay A quantitative assay of a substance using a part of or an entire organism

Biochemical Organic and inorganic chemicals that occur in living organisms and are involved in the processes of life

Biochemical Cytology Study of the biochemical properties of cell components in conjunction with techniques of microscopy to unravel the details of cell structure and function

Biochemical Reactions of Photosynthesis The temperature-dependent (i.e., "dark") reactions of photosynthesis that reduce carbon dioxide to carbohydrate; occur in the stroma of chloroplasts

Bioenergetics The energy relationships of living organisms

Biogeography The study of geographic distributions of organisms past and present, and the mechanisms that caused these distributions

Biological Clock An internal biological timing system that influences cyclic phenomena

Biological Species Concept A species consists of groups of actually or potentially interbreeding natural populations that produce viable offspring

Biomass The collective dry weight of all the organisms in a population, area, or sample

Biotic Pertaining to living organisms

Biotic Community All the populations of interactive living organisms sharing a common environment

Biotic Potential The inherent rate of natural increases, as exhibited by an individual's total number of offspring that survive long enough to reproduce

Bivalent A pair of synapsed homologous chromosomes in prophase I

Blade The broad, expanded part of a leaf

Blowout A barren area in arctic tundra caused by wind ripping out part of a vegetation mat whose edges became exposed when pulled up by grazing animals

Body Cell One of two cells produced when the generative cell of a gymnosperm male gametophyte divides; the body cell itself later divides, producing two sperm cells

Bordered Pit A pit in which the secondary wall arches over the pit membrane

Botany The scientific study of plants and plantlike organisms

Bract A structure that is usually leaflike and modified in size, shape, or color

Bracteole Diminutive form of bract

Branch Root A root that arises from another, older root; also called a branch root

Bryophyte Member of a division of nonvascular plants; the mosses, hornworts, and liverworts

Bud Scale Modified leaves that surround and protect a bud

Bulb A short underground stem covered by fleshy leaf-bases that store food

Bulliform Cells Large epidermal cells that occur in groups on the upper surface of leaves of many grasses; loss of turgor pressure in these cells causes leaves to roll up during water stress

Bundle Sheath A layer or layers of cells surrounding the vascular bundle; in C_4 plants, the bundle sheath is photosynthetic and prominent

C

C Horizon (Parent Material) The layer of soil between bedrock and the B horizon. It varies in thickness between about 10 centimeters and several meters, or it may be absent

C3 Plant Plant in which the first fixation of carbon is via the Calvin cycle; the first stable product of C_3 photosynthesis is a three-carbon compound

C4 Plant Plant in which the first fixation of carbon produces a four-carbon acid

Callose A complex carbohydrate in sieve tubes of sieve tube members; callose is especially abundant in injured sieve tubes

Calmodulin A type of protein that is activated when it binds to calcium ions (Ca^{++}); calmodulin activates enzymes in membranes; as much as 2% of the plasma membrane may be calmodulin

Calorie (Cal) 1,000 calories; the amount of heat required to raise the temperature of 1 liter of water 1° C; a slice of apple pie contains about 365 Cal.

calorie (cal) A unit of heat; one calorie is the amount of heat required to raise the temperature of 1 g of water 1° C; 1 cal = 4.12 J

Calvin Cycle Series of enzymatic reactions in which CO_2 is reduced to 3-phosphoglyceraldehyde (a three-carbon compound) and the CO_2 acceptor (ribulose, 1,5-bisphosphate) is regenerated

Calyptra The covering that partially or entirely covers the capsule of some species of mosses

Calyx Collectively, all of the sepals of a single flower

CAM (*see* **Crassulacean Acid Metabolism**)

Capsule 1) the sporangium of a bryophyte; 2) a dehiscient, dry fruit that develops from two or more carpels; 3) a slimy layer around the cells of certain bacteria

Carbohydrate An organic compound consisting of a chain of carbons with hydrogen and oxygen attached, usually in a ration of 2:1; glucose, sucrose, and starch are carbohydrates

Carotenoid Any compound in a class of yellow, orange, or red fat-soluble accessory pigments that are derived from eight isoprene units linked together; the most widespread carotenoid in plants is beta-carotene

Carpel The ovule-bearing organ of a flower; the flower of many species has more than one carpel, collectively called the gynoecium

Carpellate Flower A flower whose reproductive parts consist only of carpels; the kernel-bearing flowers on corn cobs are examples of carpellate flower; synonymous with pistillate flower

Carpellate Plant An individual plant whose flowers bear carpels but not stamens; a fruiting mulberry is an example of a plant that is exclusively carpellate (mulberries can form fertile fruits only when pollen is transferred from a staminate plant to a carpellate plant)

Carrageenan Aa slimy polysaccharide, consisting mostly of a specific mixture of alpha-galactose sulfates that surround the cell walls of certain red algae; the main commercial sources of carrageenan are species of the genus *Chondrus*

Carrion Flower A type of flower that is foul-smelling (carrion odor) and attracts flies or beetles as pollinators

Carrying Capacity The maximum number of individuals in any population of an ecosystem that can survive and reproduce

Casparian Strip The suberized layer covering the radial and transverse walls of endodermal walls

Catabolism The chemical reactions that break down complex materials

Catastrophism The concept that geologic changes result from sudden, violent, large-scale, worldwide catastrophic events

cDNA (*see* **Complementary DNA**)

Cell The structural unit of organisms; plant cells consist of a cell wall and protoplast

Cell Cycle Collectively, the repeating processes of cellular growth and division, including mitosis; the complete cell cycle occurs only in cells that divide, other cells being arrested in development at one of the phases of the cycle

Cell Fractionation The isolation of different organelles or parts of cells by centrifuging a homogenized cell extract in a concentration gradient of, for example, sucrose

Cell Membrane (*see* **Plasma Membrane**)

Cell Plate The disk-shaped structure that forms from the fusion of vesicles at the equator of the spindle apparatus during early telophase in plants and some algae; when mature, the cell plate becomes the middle lamella

Cell Theory A set of postulates describing how cells are the fundamental organizational units of living organisms

Cellulase An enzyme that breaks down cellulose into smaller units by cleaving the 1,4 linkages between molecules of beta-glucose

Central Dogma of Molecular Biology Refers to how genes work to make proteins; each protein-coding gene is transcribed into a molecule of mRNA, which is translated into a sequence of amino acids that comprise a polypeptide (i.e., a protein)

Central Placentation Refers to the attachment of ovules along the central axis of an ovary that has just one ovule-bearing chamber; primrose is an example plant that has central placentation

Centromere A constricted region of a chromosome where sister chromatids are held together

Chalazal Pole The region of the ovule where the stalk of the ovule fuses with the integument; usually the end of the embryo sac that is opposite the micropylar end

Chaparral Dense Mediterranean scrub of lower elevations of the North American Pacific Coast. The unique vegetation, which is either evergreen or deciduous in summer, is well-adapted to fires

Chemiosmosis The coupling of oxidative phosphorylation to electron transport via a proton pump

Chitin A tough, resistant, nitrogen-containing polymer of high molecular weight found in the exoskeletons of arthropods, in the cell walls of many fungi, and in a few other animals and protists

Chlorophyll The pigment responsible for trapping light energy in the primary events of photosynthesis

Chloroplast Organelle specialized for photosynthesis; chloroplasts occur in cells of aboveground parts of plants

Chromatid One of the two threads of a chromosome that has been duplicated in the S phase of the cell cycle; sister chromatids are held together at their centromere

Chromatin A DNA-protein complex that forms chromosomes

Chromatin Fiber A tightly wound coil of chromatin that is believed to consist of six nucleosomes per turn of the coil

Chromosome Theory of Heredity A set of postulates that accounts for the association of genes with chromosomes

Cisterna (plural, **Cisternae**) The flattened tubes and saclike regions of the endoplasmic reticulum and of dictysomes

Citrate (citric acid) A six-carbon organic acid that is converted to isocitric acid in the second step of the Krebs cycle

Citric Acid Cycle (*see* **Krebs Cycle**)

Cladistics A method of classifying and reflecting phylogenetic relationships among organisms, based on an analysis of shared features

Cladogram A line diagram portrayal of a branching pattern of evolution, using the concepts and methods of cladistics

Cladophyll A stem or branch that resembles a leaf

Clamp Connection A looplike lateral connection between adjacent cells, occurring in the mycelium of certain basidiomycete fungi

Class A taxonomic category ranking between division and order

Classical Species Concept (*see* **Morphological Species Concept**)

Cleistothecium A closed, more or less spherical ascocarp

Climacteric Rise Point during the ripening of some fruit in which respiratory rates rise to extremely high levels

Climax Community A self-perpetuating community that becomes established at the completion of succession; its composition is strongly influenced by local climate and soils

Climax Vegetation The vegetation of a climax community

Cline Gradual differences in characteristics within a population across a geographic region

Clone An individual or group of individuals that develop vegetatively from cells or tissues of a single parent individual

***cms-T* cytoplasm (Texas cytoplasm)** Refers to the phenotype of a variety of corn that is male-sterile due to mitochondrial (cytoplasmic) genes

CO2 Compensation Point Concentration of CO_2 at which the uptake of CO_2 equals the release of CO_2; that is, the point at which photosynthesis equals respiration

Coated Pit A bristle-like structure that occurs in clusters in certain regions of the plasma membrane; these regions form vesicles that pinch off into the cell, thereby removing excess plasma membrane; this process recycles excess plasma membrane in animal cells and it is suspected to do the same in plant cells that have coated pits

Codominance A condition that occurs when both alleles of a heterozygous gene are expressed equally

Codon A sequence of three nucleotides in a gene or molecule of mRNA that corresponds to a specific amino acid or to a stop signal at the end of a gene; of the 64 possible codons, 61 are codes for amino acids and three are stop codons

Coenocytic An organism, or part of an organism, that is multinucleate, the nuclei not being separated by membranes or crosswalls

Coenzyme An organic cofactor of enzyme-catalyzed reactions; NAD^+ and coenzyme A are examples of coenzymes

Coenzyme Q (*see* **Ubiquinone**)

Cofactor A nonprotein substance required by enzymes for proper function

Coleoptile The protective sheath around the embryonic shoot in grass seeds

Coleorhiza The protective sheath around the embryonic root in grass seeds

Colony Hybridization A technique that uses probes to find bacterial colonies that contain a gene of interest

Columella Cells Cells in the center of the root cap; characterized by the presence of numerous amyloplasts that sediment in response to gravity

Companion Cell A small cell adjacent to a sieve tube; thought to control the function of sieve tube member

Complementary DNA (cDNA) DNA that is made by reverse-transcribing mRNA into its DNA complement; the collection of vector-cloned cDNA fragments of an organism are its cDNA library

Complete Dominance A condition that occurs when the phenotype of one allele completely masks the phenotype of another allele for a heterozygous gene

Complete Flower A flower that has all four of the main parts: calyx, corolla, androecium, and gynoecium

Compost Partially decayed organic matter used in farming and gardening to enrich the soil and increase its water-holding capacity

Compound Leaf A leaf consisting of two or more independent blades called leaflets

Compression Wood Reaction wood of conifers; compression wood forms along the lower side of leaning stems; compression wood expands and pushes the stem up against gravity

Concentration Gradient The difference in concentration of a substance over a certain distance

Conidiophore A hypha on which one or more condidia are produced

Conidium (plural: **Conidia**) An externally produced, asexual fungal spore

Conjugation Pilus (*see* **Pilus**)

Continuous Synthesis Refers to the uninterrupted synthesis of DNA in the 5′ to 3′ direction; continuous synthesis occurs in the same direction as a growing replication fork

Convergence The independent evolution of similar structures in organisms that are not closely related

Cork The outermost part of the periderm; the secondary tissue produced by the cork cambium

Corm An elongate, upright, underground stem

Corolla Collectively, all of the petals of a single flower

Cortex Ground tissue located between the epidermis and vascular bundles of stems and roots

***Cot* Curve** A graph of the reassociation of DNA that has been denatured in a salt solution; "cot" comes from initial concentration of DNA (C_0) multiplied by the time required for complete reassociation (t)

Cotyledon Seed leaf; the first leaf formed in a seed; monocots have one cotyledon, and dicots have two

Coupled Reactions Reactions in which energy-requiring chemical reactions are linked to energy-releasing reactions

Coupled Cotransport System A set of active and passive transport proteins that work to actively move ions across a membrane against their gradient, then passively allow the same type of ions to diffuse back down their gradient while coupled to another type of solute that is being transported against its concentration gradient; an example of such a system is the active transport of protons against their concentration gradient by ATPase, followed by the co-transport of protons with sucrose through passive transport proteins back across the membrane

Crassulacean Acid Metabolism (CAM) A type of photosynthesis in which CO_2 is fixed at night into four-carbon acids; during the day, the stomata close and the carbon is fixed via the Calvin cycle; CAM helps plants conserve water and is often characteristic of xerophytic plants

Cristae (sing., **Crista**) The tubular or vesicle-shaped folds of the inner membrane of mitochondria; cristae contain cytochromes and other components of the electron transport chain that are involved in the synthesis of ATP

Crossing-Over The exchange of genetic material between the chromatids of homologous chromosomes during prophase I of meiosis

Cultivar A variety of plant that is selected for cultivation through hybridization and not found in nature

Cupule Refers to the seed-bearing structure of an extinct group of plants called the seed ferns

Cuticle The waxy coating on the epidermis of all aboveground parts of a plant

Cuticular Wax Wax that is embedded in a cuticle

Cutin The main waxy substance in a cuticle; it consists of hydroxylated fatty acids that are linked together in a complex array

Cyanophycin A polypeptide functioning as an energy reserve in cyanobacteria

Cyclic Photophosphorylation The light-induced flow of electrons originating from and returning to photosystem I; cyclic photophosphorylation produces ATP but no reduced NADP

Cyclosis Movement of the cytosol and the cellular components that are suspended in it; cyclosis is usually circular around a central vacuole

Cytochemistry (*see* **Biochemical Cytology**)

Cytochrome Heme-containing proteins that carry electrons in respiration and photosynthesis

Cytochrome b-c_1 Complex A cluster of cytochromes that carry electrons in the electron transport chain; the complex probably also pumps protons across the inner mitochondrial membrane

Cytochrome Oxidase Complex A cluster of cytochrome oxidases that function as the terminal electron carrier in the electron transport chain; this complex donates electrons to oxygen, which is then reduced to form water

Cytokinesis The division of cytoplasm into distinct cells; together with mitosis, cytokinesis comprises the phases of the cell cycle involved in cell division

Cytokinin Group of hormones (growth regulators) that promote growth by stimulating cellular division

Cytology The study of cell structure and function

Cytophotometry A method of studying cells by staining selected parts, such as the nucleus, and measuring how much light they absorb; the absorbance of stained chromatin in a nucleus is proportional to the amount of DNA it contains

Cytoplasmic Inheritance Refers to the inheritance of genes that occur in chloroplasts and mitochondria; it is cytoplasmic because it is not nuclear

Cytoplasmic Male Sterility (*cms*) A male-sterile condition in which sterility is controlled by mitochondrial (cytoplasmic) genes

Cytoskeleton A network of microscopy filaments that form a mechanical support system in the cell

D

Day-Neutral Plant Plant whose flowering is not affected by the length of day

Debranching Enzyme A type of enzyme that hydrolyzes the branched linkages of starch

Deciduous Plant Plant that loses all of its leaves during autumn

Decomposer An organism, such as bacterium or a fungus, that facilitates recycling of nutrients in an ecosystem through the breakdown of complex molecules to simpler ones

Denature To break bonds that maintain the three-dimensional structure of proteins or nucleic acids; also to break the hydrogen bonds that hold DNA together in a double helix

Dendrochronology The study of growth rings of trees to determine past conditions

Dentrifying Bacteria Bacteria that convert nitrates or nitrites to gaseous nitrogen

Deoxyribonucleic Acid (DNA) The nucleic acid containing four different nucleotides whose simple sugar is deoxyribose; genes are made of DNA; DNA exists as a double helix that can be unwound to replicate itself or to make RNA

Desert A biome characterized by low annual precipitation and/or porous soil, low humidity, wide daily fluctuations in temperature, high radiation, and living organisms adapted to these conditions

Desertification The conversion of non-desert biomes into deserts

Desynaapsis The unpairing and separation of homologous chromosomes upon the disintregation of the synaptomeal complex

Deuteromycetes Fungi that have no known sexual reproduction; most reproduce by conidia, and most otherwise have characteristics of ascomycetes. Deuteromycetes are also called Fungi Imperfecti

Dicot A type of angiosperm that belongs to a class whose members are characterized by having two cotyledons (seed leaves) per seed; Class Magnoliopsida

Dictyosome A stack of flattened, membranous vesicles that are often branched; dictyosomes are the sites where precursors of cell wall materials and other cellular components are assembled and prepared for secretion from the cell; dictyosomes are also called Golgi bodies

Differentiation Physical and chemical changes associated with the development and/or specialization of an organism or cell

Diffuse-Porous Wood Wood in which the vessels are distributed uniformly throughout the growth layers

Diffusion The net movement of particles, either dissolved or suspended, from a region of higher concentration to a region of lower concentration; the energy of diffusion is derived from the random motion of particles that is caused by molecular motion; diffusion tends to cause the distribution of particles to become homogenous throughout a medium

Dihybrid Cross A hybridization experiment that follows the inheritance of phenotypes that are controlled by two different genes

Dikaryotic Fungi whose hyphal cells each have two nuclei, the nuclei usually being derived from two different parents

Dioecious Having the pollen-producing and the ovule-producing organs on different individuals of the same species; mulberry is an example of a dioecious species

Diploid The condition of having two sets of chromosomes in a nucleus

Directional Selection Selection for a phenotype that is either higher or lower in frequency than the most abundant phenotype

Disaccharide A carbohydrate composed of two monosaccharides that are linked by a covalent bond; sucrose and maltose are examples of disaccharides

Discontinuous Synthesis Refers to the synthesis of DNA that occurs in the opposite direction of a growing replication fork; in discontinuous synthesis, DNA polymerase jumps ahead on one strand in the direction of fork movement (in this case, the 3′ to 5′ direction), then builds a new chain "backward" in the 5′ to 3′ direction

Disulfide Bond A type of covalent bond between the sulfur atoms of separate amino acids in the same protein; disulfide bonds strengthen the tertiary structure of proteins

Diterpene A compound that consists of four isoprene units linked together; gibberellins are examples of diterpenes

Diversifying Selection Selection for the low-frequency (extreme) phenotypes above and below the norm of the population; or selection against the high-frequency phenotype (norm)

Division A taxonomic category between kingdom and class

DNA Polymerase A type of enzyme that catalyzes the covalent bonding between nucleotides into a nucleic acid chain

DNA Ligase A type of enzyme that joins adjacent nucleotides together by catalyzing the formation of sugar-phosphate bonds in a strand of DNA

Dolipore A complex central pore occurring in the hyphal septa of many basidiomycete fungi; it is covered by a cap on both sides of the septum

Domain A structural and functional portion of a polypeptide, which may be encoded separately by a specific exon; a portion of a protein that has a globular tertiary structure

Dominant A trait that masks an alternative (recessive) trait when the gene for these traits is heterozygous

Dormancy A condition in which plant parts such as buds and seeds are temporarily arrested in their development; dormancy is typically seasonal and is broken as environmental conditions change during the year

Double Helix The spiral shape of a double strand of DNA

Double Fertilization In angiosperms, the process by which one sperm cell fertilizes the egg to a zygote and another sperm cell fertilizes the polar nuclei to form a primary endosperm nucleus

Dynein A large contractile protein that forms the connecting sidearms and spokes between microtubules in flagella

E

Ecological Race A race composed of many similar variants of the same species in several local populations distributed over a relatively large geographic area

Ecology The study of the interactions of organisms with one another and with their environment

***Eco*RI** An example restriction enzyme that comes from the bacterium *Eschericia coli;* this restriction recognizes the DNA sequence GAATTC, then cleaves it between the guanine and the adenine

Ecosystem A major system of organisms that are interacting with one another and with their physical environment

Ecotype An individual taxon of plants adapted to a specific community within its overall distribution

Ectomycorrhizae Mycorrhizae that develop externally and do not penetrate to the interior of the cells they surround

Edaphic Factor A soil factor

Egg Apparatus A group of usually three cells in an embryo sac, one of which is the egg and two of which are synergids

Electrochemical Gradient The combination of a concentration gradient and an electrical gradient of ions across a membrane

Electrogenic Pump An active transport protein that transports (pumps) ions against their concentration gradient; the main electrogenic pumps in plants are proton pumps

Electron Microscope An instrument that uses an electron beam to magnify images of specimens

Electron Transport Chain A sequence of electron carriers that use the energy from electron flow to transport protons against a concentration gradient across the inner mitochondrial membrane

Embryo In plants, the part of the seed that will form the growing seedling after germination; includes a radicle, apical meristem, and embryonic leaf or leaves

Embryo Sac The common name for the female gametophyte of flowering plants

Endergonic A reaction that requires an input of energy before it will occur; endergonic reactions never occur spontaneously

Endocarp The innermost layer of simple fleshy fruits; the endocarp can be soft, as in tomatoes, or hard and stony, as in peaches

Endocytosis The process by which the plasma membrane invaginates and forms vesicles whose contents from outside of the cell can be brought into the cell

Endodermis Layer of cells inside the cortex and outside the pericycle of roots; radial walls of the endodermis are suberized by the casparian strip

Endomycorrhizae Mycorrhizae that develop within the interior of cells

Endoplasmic Reticulum (ER) An extensive network of sheetlike membranes distributed throughout the cytosol of eukaryotic cells; portions that are densely coated with ribosomes are the rough ER; other regions, with fewer ribosomes, are the smooth ER

Endosperm The nutritive, storage tissue that grows from the fusion of a sperm cell with polar nuclei in the embryo sac

Endosymbiotic Hypothesis An explanation for the origin of chloroplasts and mitochondria from the descendants of prokaryotes that lived symbiotically in larger prokaryotic hosts

Entrainment The process by which a periodic repetition of some signal (e.g., light, dark) produces a circadian rhythm that remains synchronized with the same cycle as the entraining (i.e., modifying) factor

Entropy The degree of orderliness in a system

Enzyme A biological catalyst, usually a protein, that can speed up a chemical reaction by lowering its energy of activation; amylase is an example of an enzyme

Epicuticular Wax The outermost layer of wax in a cuticle

Epidermis The outermost layer of cells that covers a plant

Epinasty The differential growth of petioles that causes the leaf blade to curve downward

Epiphyte An organism that is attached to another organism without parasitizing it

Epistasis The interaction of two or more genes that act together to make a phenotype; epistasis is best known for serial gene systems that control the multistep biosynthesis of a complex molecule or the sequence of steps in a metabolic pathway

EPSP Synthetase A type of enzyme that catalyzes a step in the synthesis of enolpyruvyl-shikimic acid-3-phosphate, which is a precursor to aromatic amino acids; the herbicide glyphosate works by inhibiting the activity of this enzyme

Ethylene A gaseous plant hormone (growth regulator) that promotes fruit ripening and other physiological responses

Etiolation The abnormal elongation of stems caused by insufficient light; etiolated stems usually lack chlorophyll

Eubacteria The majority of all bacteria; their cell walls contain muramic acid, certain lipids, and other features that distinguish them from archaebacteria

Euchromatin Lightly staining portion of chromatin, not easily visible by light microscopy

Evolutionary Species An ancestral-descendant sequence of populations evolving separately from others and forming a single unit

Exergonic A reaction that releases energy and occurs spontaneously

Exine The outermost layer of a spore or pollen grain; the exine consists of a resistant polymer that protects the male gametophyte from desiccation

Exocarp The outermost layer (usually the skin) of simple fleshy fruits

Exocytosis The process of expelling the contents of a vesicle from a cell by fusing the vesicle membrane with the plasma membrane and opening the inside of the vesicle to the outside of the cell

Exon A sequence of DNA within a gene that codes for an amino acid sequence

Exon-Shuffling Hypothesis An explanation for how complex new genes arise from the joining of independent exons into new combinations

Extensin A family of related glycoproteins that are structural proteins in cell walls

F

Facilitated Diffusion Passive transport through a transport protein

Fascicle A cluster of pine leaves (needles) or other needlelike leaves of gymnosperms

Fascicular Cambium The part of the vascular cambium that forms between the xylem and phloem within a vascular bundle

Fatty Acid A long, mostly hydrocarbon chain that has an organic acid group at one end; the most common fatty acids in plants are oleic acid, linoleic acid, and linolenic acid

Feedback Inhibition Control mechanism in which the increasing concentration of a molecule inhibits the further synthesis of that molecule

Fermentation A process by which energy is obtained from organic compounds without the use of oxygen as an electron acceptor

Fiber An elongated, thick-walled sclerenchyma cell; helps support or protect the plant

Fiddlehead The curled fern frond prior to unrolling and elongation; also known as a crozier

Field Capacity The water-storage capacity of soil; the amount of water in soil after gravitational percolation stops

Filament The stalk of a stamen; or, the vegetative body of filamentous algae and fungi

Filial Refers to a generation of offspring; the first set of offspring from a hybridization experiment is the first filial generation (F_1), the second set is the second filial generation (F_2), etc.

First Gap (*see* **G_1 phase**)

Fission The asexual division and formation of two similar new cells within mitosis, as seen in prokaryote reproduction

Fitness A measure of an individual's evolutionary success; number of its surviving offspring relative to the number of surviving offspring of other individual's within the population

Flagellum (plural, **Flagella**) A hairlike locomotor organelle that protrudes from the cell into the medium surrounding it; flagella enable cells to swim, but the only swimming cells of plants are the sperm cells of some plant groups; flagella also occur in algae, fungi, bacteria, and animals

Flavin Mononucleotide (FMN) The first electron acceptor in the electron transport chain; FMN takes electrons from NADH in the mitochondrial matrix, plus one proton from NADH and one proton from the matrix to become $FMNH_2$; protons from $FMNH_2$ are released into the mitochondrial intermembrane space

Flavonoid Any phenylpropanoid-derived compound that is linked to three acetate units and condensed into a multiple-ringed structure; the most common flavonoid is rutin; flavonoids also include naringin, which is a bitter substance in grapefruits

Flora The plants or organisms (other than animals) of a particular region; also: a publication devoted to the taxonomy of plants of a particular region

Florigen The hypothetical flowering hormone; florigen has never been identified or isolated

Fluid Mosaic Model A model for the structure of membranes as a fluid phospholipid bilayer through which proteins float in a continually shifting mosaic pattern

Fluorescence The release of energy at a longer wavelength than the energy that was absorbed

Food Web An interlocking flow of energy, involving producers and consumers in an ecosystem

Foot Basal part of a moss sporophyte; the foot is embedded in the gametophyte

Free Energy Energy available to do work

Free-Nuclear Embryo An early stage of embryo development in a gymnosperm, in which the zygote nucleus divides repeatedly without walls forming around the nuclei

Frond Photosynthetic leaf blade of a fern

Fumarate (Fumaric Acid) A four-carbon organic that takes on a molecule of water and becomes malic acid in the seventh step of the Krebs cycle

Functional Megaspore The megaspore that, in some types of embryo sac development, is the only one of the four meiotic products to grow into a female gametophyte; the other three spores disintegrate

Fusiform Initials Vertically elongated cells in the vascular cambium that produce cells of the axial system in the secondary xylem and secondary phloem

G

G_1 Phase During interphase, the portion of the cell cycle that occurs between the end of mitosis and the onset of DNA synthesis; G_1 refers to first gap

G_2 Phase During interphase, the portion of the cell cycle that begins at the end of the S phase and lasts until the beginning of mitosis; G_2 refers to the second gap

Gametangium (plural: **Gametangia**) A cell or structure in which gametes are produced

Gamete A haploid reproductive cell that fuses with another gamete to form a zygote; the female gamete is an egg and the male gamete is a sperm; in certain kinds of algae and fungi, however, the gametes are neither male or female

Gametophyte The phase of the plant life cycle that produces gametes

Gametophytic Self-Incompatibility A type of self-incompatibility that is imposed by gametophyic tissues or organs; an example would be incompatibility that is imposed by the pollen tube, which is a gametophytic structure

Gas Vacuole A membrane-bound bubble of gas that enables aquatic bacteria to float

Gel Electrophoresis A technique by which nucleic acids or proteins are separated in a gel that is placed in an electric field

Gemmae (sing. **Gemma**) Asexual plantlets in certain liverworts that can form new gametophytes; often form in gemmae cups

Gemmules An erroneous concept of inheritance; described as packets of heritable information produced throughout a mature organism, transported to the reproductive organs, and packaged into gametes before fertilization

Gene A sequence of DNA that codes for a molecule of mRNA, tRNA, or rRNA, or that regulates the transcription of such codes; a gene is the basic unit of heredity

Gene Conversion The change of one allele to another during crossing over

Gene Flow Introduction of genetic material into the gene pool of one population from another population

Gene Gun An instrument that shoots tiny beads coated with DNA directly into cells; some cells treated this way integrate the foreign DNA that is shot into them, thereby becoming transgenic; early models of the instrument used .22 caliber cartridges, hence the name gene gun.

Gene Pool All of the alleles within a population that are available to future generations

Generation Time In plants, the length of time it takes from seed germination to reach sexual maturity

Generative Cell The cell in the pollen grains of angiosperms that divides to form two sperm cells, or the cell in the pollen grains of gymnosperms that divides to form a sterile cell and another cell that divides to form to sperm cells

Genetic Species Concept Two species are considered distinct if their genetic makeup is sufficiently different from one another

Genetic Drift Random changes in gene frequencies within the gene pool of a population

Genetic Distance Measure of the degree of genetic difference between different populations or species

Genetic Code A system of codons (nucleotide triplets) in DNA or RNA that together code for a sequence of amino acids in a polypeptide; 61 of the possible codons are codes for amino acids, the remaining three being stop codons that are not translated

Genetic Engineering The artificial manipulation of genes, or the transfer of genes from one organism to another; synonymous with recombinant DNA technology

Genomic Library The set of fragments of an organism's genome that are cloned in a virus or bacterial plasmid

Genophore A bacterial chromosome; its DNA is not associated with the histone proteins of eukaryotic chromosomes

Genotype An organisms's genes, either individually or collectively

Genus (plural: **Genera**) A taxonomic category between family and species; it forms the first part of the binomial of the scientific name of organism

Gibberellin A type of plant hormone that affects, for example, stem elongation and seed germination

Gill One of the fleshy plates that radiate out from the stipe beneath the cap of a mushroom basidiocarp

Gliadin A storage protein in the grains of wheat

Glucose (*also see* **Alpha-Glucose** and **Beta-Glucose**) A common monosaccharide whose empirical formula is $C_6H_{12}O_6$

Glutelins A complex mixture of storage proteins in the grains of wheat

Glycogen A carbohydrate food reserve similar to starch in many organisms other than plants

Glycolysis The anaerobic metabolic pathway by which glucose is broken down into two molecules of pyruvic acid in the cytosol; the substrate-level phosphorylation of two molecules of ADP to ATP and the reduction of two molecules of NAD^+ to NADH occur for the breakdown of each molecule of glucose

Glycoprotein A type of protein that has sugars attached to it; extensin in cell walls is an example of a family of glycoproteins

Glyoxylic Acid Cycle A sequence of biochemical reactions that converts acetyl-CoA into carbohydrate

Glyoxysome A type of microbody that is common is germinating oilseeds and seedlings that arise from them; glyoxysomes contain enzymes that catalyze the breakdown of fatty acids into acetyl-CoA

Glyphosate The generic name of one of the most commonly used herbicides in agriculture

Golgi Body (*see* **Dictyosome**)

Gram Stain A crystal violet stain that is retained by gram-positive bacteria and not retained by gram-negative bacteria, after alcohol or a similar solvent is applied

Gram-Negative (*see* **Gram Stain**)

Gram-Positive (*see* **Gram Stain**)

Grana (sing., **Granum**) Stacks of thylakoids where the photochemical (i.e., "light") reactions of photosynthesis occur

Grassland A biome characterized by the predominance of grasses

Gravitropism The curvature of roots or stems in response to gravity

Ground Meristem The fundamental tissue of the apical meristem; produces the cortex

Growth Ring A growth layer in secondary xylem or secondary phloem, as seen in cross section

GTP Cap A molecule of 7-methylguanosine triphosphate (GTP) that is attached to the 5′ end of a molecule of RNA as transcription begins; the GTP cap protects the RNA from degradation as it is being synthesized

Guard Cells Two specialized epidermal cells that form a stomatal apparatus

Gum A hemicellulose that is secreted by plants, which consists of several kinds of monosaccharides; an example is gum arabic, which is a mixture of the monosaccharides arabinose, galactose, glucose, and rhamnose

Gum Arabic A gum produced by the plant species *Acacia senegal;* this gum is a hemicellulose, which in this case is a complex branched chain consisting of arabinose, galactose, glucose, and rhamnose

Guttation The exudation of liquid water from leaves; caused by root pressure

Gynoceium (plural, **Gynoecia**) Collectively, all of the carpels of a single flower

H

Habitat The location, with its own specific set of environmental conditions, where an organism naturally occurs

Habitat, Operational The soil components and moisture, shade, associated organisms, and other habitat features that directly affect an organism

Half-Life Time required in a chemical reaction for half the original reactant material to decay or be consumed

Halon A bromine-based compound that is especially destructive of the ozone layer

Haploid The condition of having only one set chromosomes in a nucleus

Hardpan A hard soil with disrupted structure that may develop through the gradual accumulation of salt residues when inorganic fertilizers are applied annually without the addition of organic matter; it generally restricts the downward movement of water and roots

Hardwood A woody dicot

Heartwood Wood in the center of a tree trunk; usually darker due to the presence of resins, oils, and gums; does not transport water and solutes

Helicase A type of enzyme that breaks hydrogen bonds between complementary base pairs of DNA, thereby causing the double strand to split into separate single strands

Helix Anything of a spiral shape; in biology it refers to the shape of DNA molecules, which occur as double helices

Heme A complex organic ring structure, called a protoporphyrin, to which an iron atom is bound; heme occur in the cytochromes of all organisms and in the hemoglobin of animals

Hemicellulose Primarily a cell wall polysaccharide of variable composition and structure; hemicellulose that is secreted by plants is also called a gum (*see* **Gum** and **Gum Arabic**)

Herbarium A systematically arranged collection of dried, pressed, and mounted plant specimens

Heterochromatin A condensed, darkly staining portion of chromatin, easily visible by light microscopy

Heterocyst A relatively large, unpigmented, thick-walled, nitrogen-fixing cell that is produced within the filaments of certain cyanobacteria

Heterogenous Nuclear RNA (hnRNA) The pool of primary RNA transcripts in the nucleus, which are of various, usually large sizes

Heterosis A condition in which crossbred organisms are more fit than inbred organisms because they have more heterozygotic loci

Heterotroph An organism that obtains its food from other organisms

Heterozygote Superiority A condition in which individuals heterozygous at one or more loci have higher fitness than an individual with fewer heterozygous loci

Heterozygous A condition in which a gene has two different alleles in a diploid individual

Hill Reaction The photolysis of water and the photoreduction of an artificial electron-acceptor by chloroplasts in the absence of CO_2

Histone A type of protein that comprises the protein component of chromatin

hn RNA (*see* **Heterogenous Nuclear RNA**)

Homologous Refers to a pair of chromosomes that have alleles for the same genes

Homology A condition in which a common trait possessed by different species was derived from a common ancestor

Homozygous A condition in which both alleles of a gene are the same in a diploid individual

Hormone An organic molecule made in one plant that exerts an effect in another part of the plant; effective in small concentrations

Humus The organic portion of soil; derived from partially decayed plant and animal material

Hybrid Vigor (*see* **Heterosis**)

Hybridization Production of offspring from crossing different species or between genetically different populations

Hydrolysis Any chemical reaction that proceeds by the addition of water to break down a molecule; the breakdown of starch by amylase and the breakdown of sucrose by invertase are examples of enzyme-catalyzed hydrolyses

Hydrophilic Refers to chemicals that are freely soluble in water; sugars are examples of hydrophilic compounds

Hydrophobic Refers to chemicals that are not soluble in water but are soluble in nonpolar solvents; lipids and hydrocarbons are generally hydrophobic

Hydrophyte A plant that is adapted to submersion in water or an aquatic environment for at least part of its growing season

Hydrotropism Growth of a root toward water

Hypertonic Refers to a solution of high solute concentration in comparison with one of low solute concentration; a hypertonic solution tends to gain water across a membrane from a solution of lower solute concentration

Hypha (plural: **Hyphae**) A single tubular thread of the mycelium of a fungus or similar organism

Hypocotyl The region of an embryo that is between the radicle and the attachment point of the cotyledons

Hypodermis One or more layers of cells just beneath the epidermis that are distinct from the underlying cortical or mesophyll cells

Hypothesis A proposed solution to a scientific problem that must be tested by experimentation; a working explanation based on evidence and suggesting some principle; if disproved, a hypothesis is discarded

Hypotonic Refers to a solution of low solute concentration in comparison with one of high solute concentration; a hypotonic solution tends to lose water across a membrane to a solution of higher solute concentration

I

Ice-Minus Bacteria Genetically engineered bacteria that contain a foreign gene whose polypeptide inhibits the formation of ice crystals

Imbition The adsorption of water onto the internal surfaces of materials

Imperfect flower A flower that lacks either an androecium or a gynoecium

Imperfect Fungi (*see* **Deuteromycetes**)

In-Group Analysis The assumption in cladistics that the most prevalent character state is primitive

Inbreeding Mating within the same plant or between the offspring of an inbred parent

Incomplete Flower A flower that has one or more of the parts absent (calyx, corolla, androecium, or gynoecium)

Incomplete Dominance A condition that occurs when the phenotype of one allele only partly masks the phenotype of another allele for a heterozygous gene

Indeterminate Growth Growth that is not inherently limited, as with a vegetative apical meristem that produces an unrestricted number of organs indefinitely

Indole-3-Acetic Acid (IAA) A naturally occurring auxin (*see* **Auxin**)

Inferior Ovary An ovary located below the other flower parts on a floral axis

Inflorescence A cluster of flowers that are arranged on their axis in a specific pattern

Inorganic Compound A type of molecule that either lacks carbon or contains carbon but not hydrogen; carbon dioxide and water are examples of inorganic compounds

Integument The layer or layers of tissue that surround the megasporangium (nucellus) in an ovule; the integument becomes the seed coat

Intercalary Meristem Meristem at the base of a blade and/or sheath of many monocots

Interfasicular Cambium The part of the vascular cambium that forms between vascular bundles and connects with the fascicular cambium

Intermediate Filament The middle-sized (8–12 nm in diameter) of the three types of filaments that comprise the cytoskeleton

Internode Part of the stem between two successive nodes

Interphase Collectively, all of the phases of cell growth apart from cell division

Intine The inner layer of a spore or pollen grain; the intine consists of cellulosic and pectic material that is exported from the microspore

Introgression Back-crossing; mating of fertile hybrids with parent populations

Intron A sequence of DNA within a gene that does not code for an amino acid sequence

Invertase A type of enzyme that catalyzes the breakdown of sucrose by hydrolysis into glucose and fructose

Island Biogeography A theory explaining the relationship between defined habitat area (such as an island) available for organisms and the number and diversity of species in that area

Isocitrate (Isocitric Acid) A six-carbon organic acid that loses a molecule of carbon dioxide in the third step of the Krebs cycle, thereby being converted to alpha-ketoglutaric acid; also during this conversion, one molecule of NAD^+ is reduced to NADH

Isoprene The basic five-carbon subunit of terpenoid polymers

Isozymes Enzymes that have the same function but are encoded from different genes

J

Joule (J) The amount of energy needed to move one kilogram through one meter with an acceleration of one meter per second per second; 10^7 ergs; one watt-second; a slice of apple pie contains about 1.5×10^6 J

K

Kinetic Energy The energy of motion; a solute that moves down its concentration gradient has kinetic energy

Kinetin A purine that acts as a cytokininin

Kinetochore A disc-shaped complex of proteins that is bound on one side to a centromere and on the other side to a spindle fiber

Kingdom The highest taxonomic category

Kranz Anatomy Specialized leaf anatomy characteristic of C_4 plants; characterized by having vascular bundles surrounded by a photosynthetic bundle sheath

Krebs Cycle The metabolic pathway by which acetyl-CoA is oxidized in mitochondria to carbon dioxide; each turn of the Krebs cycle also forms one ATP by substrate-level phosphorylation, reduces one NAD^+ to HADH, and reduces one ubiquinone to ubiquinol; the Krebs cycle is also called the citric acid cycle or the tricarboxylic cycle

L

Lateral Meristem Meristem that produces secondary tissue; the vascular cambium and cork cambium are examples of lateral meristems

Leader Sequence A short non-coding sequence of DNA, immediately upstream from the beginning of a gene, that is transcribed into RNA

Leaf Buttress A lateral protrusion below the apical meristem; the initial stage in the development of a leaf primordium

Leaf Gap Region of parenchyma tissue in the primary vascular cylinder above a leaf trace

Leaf Primordium A lateral outgrowth from the apical meristem that will eventually form a leaf

Leaf Scar A scar left on a twig when a leaf falls from a stem

Leaf Trace The part of a vascular bundle that extends from the base of a leaf to its connection with a vascular bundle of a stem

Leaflet An individual blade of a compound leaf

Lectin A type of protein that binds to carbohydrates on cell surfaces; many lectins are glycoproteins; lectins occur in all parts of the cell but are mostly associated with the endoplasmic reticulum and other membranes, including the plasma membrane

Lenticel Spongy areas in the cork surfaces of stems and roots of vascular plants; allows gas to exchange to occur across the periderm

Liana A woody vine that is supported by other plants

Light Microscope An optical instrument that uses light to magnify images of specimens

Light-Compensation Point Light level at which photosynthesis equals respiration

Lignin A complex phenylpropanoid polymer that makes cell walls stronger, more waterproof, and more resistant to pests, herbivores, and disease organisms

Lilium-Type Embryo Sac Development A type of embryo sac development that entails all four spores of an ovule; in this type of development, the antipodal cells and one of the polar nuclei are triploid; the other polar nucleus and the egg apparatus are haploid (*also see* **Polygonum-Type Embryo Sac Development**)

Linkage The condition of having genes on the same chromosome (linked); alleles of genes that are linked tend to be inherited together

Locus (plural, **Loci**) The position of a gene on a chromosome

Long-Day Plant Plant that flowers when the length of dark is shorter than some critical value; long-day plants flower in spring and summer

Looped Domain A fold or loop in a region of packed chromatin fibers, which extends out from the main axis of the chromosome; looped domains may consist of 20,000 to 100,000 nucleotide pairs

M

Macroevolution Evolutionary changes that refer to the development of new species

Macronutrients Inorganic elements required in large amounts for plant growth (e.g., nitrogen, calcium, sulfur)

Magnification Enlargement of an object

Malate (Malic Acid) A four-carbon acid that is oxidized by the reduction of NAD^+ to NADH in the eighth step of the Krebs cycle; malic acid is also formed by the reduction of oxaloacetic acid that is derived from fixing carbon dioxide to phosphoenolpyruvic acid in C_4 and CAM photosynthesis

Marginal Placentation Refers to the attachment of ovules (placentation) along the edge (margin) of a suture; garden pea pods have marginal placentation

Matric Potential The component of water potential caused by the attraction of water molecules to a hydrophilic matrix

Mediterranean Scrub The often dense, shrubby vegetation that occurs in areas with wet winters and dry summers; it is dominated by evergreen bushes, or those that are deciduous in the summer

Megapascal (MPa) A unit of pressure; one million (10^6) pascals; 1 MPa = 10 atmospheres of pressure; a car tire is typically inflated to about 0.2 MPa, whereas the water pressure in home plumbing is 0.2–0.3 MPa

Megaspore A spore that will grow into a female gametophyte

Megaspore Mother Cell A cell that will undergo meiosis and cytokinesis to produce megaspores

Megasporophyll Refers to a leaf-like organ that bears megasporangia

Meio-Blastospore A spore that arises by budding from a haploid, meiotically produced spore

Meiosis Nuclear division in which chromosomes are doubled, then divided twice; the daughter nuclei from meiosis have half the number of chromosomes of the parent nucleus; in plants, meiosis forms spores

Meiosis I The first of two nuclear divisions that, in plants, form spores; in meiosis I, homologous chromosomes synapse, cross over, and move to opposite poles of the meiotic spindle apparatus; the separation of homologous chromosomes in meiosis results in a reduction in chromosome number by one-half in daughter nuclei

Meiosis II The second of two nuclear divisions that, in plants, form spores; in meiosis II, centromeres divide and sister chromatids become independent chromosomes that move to opposite poles of the spindle apparatus

Membrane Potential The potential electrical energy of ions across a membrane; membrane potential is measured in volts

Membrane Selectivity The control that a membrane exerts over how much and what kinds of materials pass through it

Membrane System The interconnected membranes of a cell, including the plasma membrane and the various organellar membranes

Mendelian Inheritance Refers to patterns of inheritance that were discovered by Gregor Mendel

Meristem Regions of specialized tissue whose cells undergo cell division

Mesocarp The middle layer (often fleshy) of simple fleshy fruits; the mesocarp occurs between the exocarp and the endocarp

Mesophyll Parenchyma tissue between the epidermal layers of a leaf; is usually photosynthetic

Mesophyte A plant that requires a relatively humid atmosphere and abundant soil water

Messenger RNA (mRNA) A class of RNA that carries the genetic message of genes to ribosomes, where the message is translated into the amino acid sequence of a polypeptide

Metabolism The sum of all chemical reactions occurring in a cell or organism

Metaphase The period of mitosis during which chromosomes become attached to spindle fibers, which align the chromosomes in a circular plane that is perpendicular to the microtubules of the spindle apparatus

Metaphase I The first metaphase of meiosis; in metaphase I, pairs of homologous chromosomes align along an equatorial plane that is perpendicular to the axis of the spindle apparatus

Metaphase II The second metaphase of meiosis; in metaphase II, chromosomes align along an equatorial plane that is perpendicular to the axis of the spindle apparatus

Metaphase Plate The plane of alignment of chromosomes during metaphase; the metaphase plate is perpendicular to the axis of the spindle apparatus

Metaxylem Primary xylem that differentiates after the protoxylem; reaches maturity after the part of the plant in which it is located has stopped elongating

Microbody A vesicle-like organelle that is bounded by a single membrane and is generally associated with the endoplasmic reticulum; glyoxysomes and peroxisomes are types of microbodies

Microevolution Evolutionary changes that occur within a population; may eventually lead to the formation of a new species, but not as a one-time event

Microfibril A complex of cellulose molecules that are twisted together into a strong, threadlike component of cell walls

Microhabitat The particular part of a habitat occupied by an individual

Micronutrients Inorganic elements required in small amounts for plant growth (e.g., boron, copper, zinc)

Micropyle The opening in a ovule through which the pollen tube will enter in angiosperms, or through which the pollen grains will enter in gymnosperms

Microsporangium (plural, **Microsporangia**) A microspore-containing sporangium

Microspore A spore that will grow into a male gametophyte

Microspore Mother Cell A cell that will undergo meiosis and cytokinesis to produce microspores

Microsporophyll Refers to a leaf-like organ that bears microsporangia

Microtome An instrument that is used for slicing specimens into microscopically thin sections

Microtubule The largest (18–25 nm in diameter) of three types of filaments that comprise the cytoskeleton; microtubules also move chromosomes during nuclear division and make up the internal structure of flagella

Middle Lamella The pectin-containing layer between cells that probably acts as the glue to hold cells together

Midrib The large central vein of a leaf

Mitosis The process of nuclear division in which chromosomes are first duplicated, followed by the separation of daughter chromosomes into two genetically identical nuclei; the division of nuclei; together with cytokinesis, mitosis comprises the phases of the cell cycle involved in cell division

Molecular Phylogeny A phylogeny based on molecular data

Monocot A type of angiosperm that belongs to a class whose members are characterized by having one cotyledon (seed leaf) per seed; Class Liliopsida

Monoecious Having the pollen-producing and the ovule-producing organs on the same individuals

Monokaryotic Fungi whose cells each contain a single nucleus

Monophyletic A taxon and all its descendants

Monomer The smallest subunit that is a building block of a polymer

Monosaccharide A simple sugar that cannot be broken down by hydrolysis; glucose is an example monosaccharide

Monoterpene A compound that consists of two isoprene units linked together; menthol is an example monoterpene

Morphological Species Concept Traditional concept of taxonomic species surmising that two species are considered distinct if they are sufficiently different morphologically

Morphological Plasticity Condition in which environmental factors induce different phenotypes from the same genotype

mRNA (*see* **Messenger RNA**)

Mucigel Slimy material secreted by root tips to facilitate growth of the root through soil

Multigene Family A set of duplicated genes; many genes occur in multigene families; an example is the family of genes in which each gene codes for the small subunit of ribulose-1, 5-bisphosphate carboxylase/oxygenase

Mutation A genetic change; mutations include changes in DNA sequences of genes, rearrangements of chromosomes, and the movements of transposable elements

Mycelium (plural: **Mycelia**) Collective term for the hyphae of a fungus

Mycolaminarin A carbohydrate food reserve of water molds (oomycetes)

Mycorrhizae (plural: **Mycorrhizae**) A mutualistic association between a fungus and the roots of a plant

Myxobacteria A group of complex, gram-negative soil bacteria that often form upright, multicellular, reproductive bodies

N

NADH dehydrogenase complex A complex of enzymes whose function is to transport protons from NADH across the inner mitochondrial membrane

Nastic Movement A movement that occurs in response to a stimulus, but whose direction is independent of the direction of the stimulus

Natural Selection Differential reproduction of phenotypes; genotypes and phenotypes vary among organisms and some of these phenotypes promote reproduction more than other phenotypes

Nectar A sweet exudate secreted by plants to attract insects (e.g., for pollination)

Nectary A structure in angiosperms that secretes nectar; usually (but not always) associated with flowers

Net Movement The amount of movement that goes in one direction more than another; particles diffuse in all directions, but net movement occurs away from where particles are most concentrated to where they are least concentrated

Net Productivity The energy produced in an ecosystem by photosynthesis minus the energy lost through respiration

Niche The ecological role of a species within a community

Nitrification The oxidation of ammonium ions or ammonia to nitrate, done by certain free-living bacteria in the soil

Nitrogen Fixation Incorporation of atmospheric nitrogen into nitrogenous compounds; done by certain free-living and symbiotic bacteria

Nitrogen-Fixing Bacteria Bacteria that convert gaseous nitrogen to nitrates or nitrites

Nitrogenase A complex of enzymes that convert atmospheric nitrogen gas into ammonia

Nivea Gene (*niv*) A gene in snapdragons that, when homozygous recessive, blocks the synthesis of flower pigments; plants that are homozygous recessive for this gene have white flowers

Node Point where one or more leaves attach to a stem

Nodule Tumorlike swelling on roots of certain higher plants (e.g., legumes) that houses nitrogen-fixing bacteria

Noncyclic Photophosphorylation The light-driven flow of electrons from water to $NADP^+$ in oxygen-evolving photosynthesis; requires both photosystems I and II

Nonvascular Plants Plants that lack vascular tissue (e.g., liverworts)

Northern Blotting A procedure by which molecules of RNA are separated by gel electrophoresis, transferred to a filter, and probed with DNA that is complementary to the RNA sequence of interest; the location of the target sequence is found because it becomes radioactive when the probe anneals to it (*also see* **Southern Blotting**)

Nuclear Envelope The double membrane that surrounds the nucleus

Nucleic Acid An organic acid that is a polymer of mostly four different nucleotides; deoxyribonucleic acid (DNA) and ribonucleic acid (RNA) are the two kinds of nucleic acids

Nucleosome The basic beadlike unit of chromatin in eukaryotes, consisting of DNA that is wound around a core of histone proteins

Nucleotide The subunit of a nucleic acid, consisting of a phosphate group, a simple sugar (either ribose or deoxyribose), and a nitrogen-containing base that is either a purine or a pyrimidine

Nyctinasty The "sleep movements" of leaves in response to changes in turgor of cells at the base of their petioles

O

Occam's Razor A principle of logic that holds that the best explanation of an event is the simplest, using the fewest assumptions of hypotheses

Operculum The lid of the sporangium in mosses

Opposite Phyllotaxis Leaves occurring in pairs at a node

Organelle A specialized part of the cell, usually bounded by a membrane; nuclei, chloroplasts, and mitochondria are membrane-bound organelles; ribosomes are membrane-free organelles

Organic Evolution Changes in the genetic composition of a population of organisms across generations

Organismal Theory A set of postulates describing how whole organisms, not cells, are the fundamental organizational units of living organisms; according to this theory, organisms develop by compartmenting the whole organism into cells, not by building the organism from cells

Osmosis The diffusion of water or other solvent through a differentially permeable membrane

Osmotic Potential The potential of solutes to cause osmotic pressure; also called solute potential

Osmotic Pressure The water potential of pure water across a membrane; osmotic pressure is an indicator of how concentrated a solution is on the other side of a membrane from pure water

Osmotically Active Solutes that can cause a change in a cell's osmotic potential; potassium (K^+) and other ions are osmotically active

Out-Breeding Mating with unrelated individuals

Out-Group Analysis The assumption in cladistics that the most prevalent character state of plants outside of a given group is primitive

Outcrossing Mating between different individual plants

Ovary The enlarged, ovule-bearing portion of a carpel or of a cluster of fused carpels; after fertilization, an ovary matures into a fruit

Ovule The structure that contains the female gametophyte in seed plants; the female gametophyte is surrounded by a nucellus (megasporangium tissue), which is covered by one or two integuments; when mature, an ovule is called a seed

Oxaloacetate (Oxaloacetic Acid) A four-carbon organic acid that is converted to citric acid by the addition of an acetyl group in the first step of the Krebs cycle; oxaloacetic acid is also the product of the carbon dioxide fixation of phosphoenolpyruvic acid in C_4 and CAM photosynthesis

Oxidation The loss of electrons from an atom or molecule that is involved in an oxidation-reduction (redox) reaction; oxidation removes energy from one substance, which is coupled with the simultaneous addition of energy to another substance by reduction (*also see* **Beta-Oxidation**)

Oxidative Phosphorylation Phosphorylation of ADP to ATP that uses energy from a proton pump fueled by the electron transport system

Ozone A form of oxygen (O^3) in the stratosphere that, when compared with ordinary oxygen (O^2), more effectively shields living organisms from intense ultraviolet radiation

P

Paleospecies A species defined only by fossil morphology

Palisade Mesophyll The vertical photosynthetic cells below the upper epidermis of a leaf

Parallelism In cladistics, a pattern of character evolution where the same character state arises from the primitive state more than once

Parapatric Occurring in adjoining places

Parapatric Speciation Speciation that occurs between contiguous populations, often induced by low dispersal range of the individuals

Paraphyletic Term applied to a group of organisms that does not contain all the descendants of a single ancestor

Paraphyses (sing. **Paraphysis**) Sterile filaments that grow among the reproductive cells of certain fungi and brown algae

Parenchyma The tissue type characterized by relatively simple, living cells having only primary walls

Parietal Placentation Refers to the attachment of ovules (placentation) along the wall of an ovary (i.e., parietal); violets are example plants that have parietal placentation

Parsimony In cladistics, the shortest hypothetical pathway that provides the most likely explanation of an evolutionary event

Parthenocarpy Development of fruit without fertilization

Pascal (Pa) The pressure unit (i.e., energy per unit volume) used to measure water potential; one pascal equals the force of one newton per square meter; one atmosphere of pressure equals 1.0×10^5 Pa

Passage Cell Endodermal cells of root that have a thin wall and casparian strip when other endodermal cells develop thick secondary walls

Passive Transport The unrestricted movement of a substance through a biological membrane; the energy for passive transport is the kinetic energy of movement down a concentration gradient; it is passive because it does not require energy from cellular metabolism

Pectin A gluey polysaccharide that holds cellulose fibrils together; pectins are mostly polymers of galacturonic acid monomers with alpha-1,4 linkages

Pedicel The stalk of a flower in an inflorescence

Peduncle The stalk of a flower or of an inflorescence

Pentose Phosphate Pathway A series of chemical reactions that start with glucose-6-phosphate from glycolysis and involve several five-carbon sugars (pentoses); during this pathway, NADP is reduced to NADPH, but no ATP is produced

Peptide Bond A carbon-nitrogen bond that links amino acids together in a chain

Peptidoglycan A large carbohydrate polymer found in the walls of true bacteria; it is composed of long chain molecules interconnected by short chains of peptides

Peptidyl Transferase A type of enzyme in the large ribosomal subunit that catalyzes the formation of a peptide bond between the amino acid at the end of a growing polypeptide and the next amino acid to be added to the chain

Perfect Flower A flower that has an androecium and a gynoecium

Pericarp Refers collectively to the layers of ovary tissue in a fruit; pericarp is the preferred term for fruits whose layers cannot be easily distinguished from one another

Pericycle The layer of cells surrounding the xylem and phloem of roots; produces branch roots

Periderm The protective tissue that replaces epidermis; includes cork (phellem), cork cambium (phellogen), and phelloderm

Peripheral Cells Outermost cells of the root cap that secrete mucigel; are sloughed from the root cat as the root grows through the soil

Peristome The "teeth" around the opening of the sporangium of mosses

Perithecium A flask-shaped or spherical ascocarp with a terminal opening

Permanent Wilting Point The moisture content of soil at the point when a particular plant's root system cannot absorb water, even when given water and placed in a humid chamber

Peroxisome A type of microbody that occurs primarily in leaves and contains enzymes that metabolize hydrogen peroxide and glycolic acid

Petal One of the parts of the flower that are attached immediately inside the calyx; collectively, the petals of a single flower are called the corolla; the corolla is usually the part of the flower that is conspicuously colored

Petiole The stalklike part of a leaf that connects the blade to the stem

Phagotropic Ingesting solid food particles
Phellem Cork; produced by the phellogen
Phelloderm The inner part of the periderm; forms inside of the phellogen
Phellogen Cork cambium
Phenolic Any compound that contains a fully unsaturated, six-carbon ring that is linked to an oxygen-containing side group
Phenotype An organism's observable features, either individually or collectively; the phenotype results from the interaction of the genotype of an organism with its environment
Phenylpropanoid A complex phenolic that has a three-carbon side chain; phenylpropanoids are generally derived from the amino acids phenylalanine and tyrosine; myristicin, the main flavor ingredient of nutmeg, is a phenylpropanoid
Phloem Vascular tissue that transports water and organic solutes
Phospholipid Aa lipid that has two fatty acids and a phosphate group bound to a molecule of glycerol; phospholipids are important components of membranes
Photochemical Reactions The "light" reactions of photosynthesis; these reactions occur on the grana of chloroplasts and produce ATP and reduced NADP
Photon The elementary particle of light
Photoperiodism Response to the duration and timing of day and night; the system within plants that measures seasons and coordinates seasonal events such as flowering
Photorespiration The light-dependent formation of glycolic acid in chloroplasts and its subsequent oxidation in peroxisomes
Photosynthesis The production of carbohydrates by combining CO_2 and water in the presence of light energy; occurs in chloroplasts and releases oxygen
Photosystem A complex of chlorophyll and other pigments embedded in the thylakoids of chloroplasts and involved in the photochemical (i.e., "light") reactions of photosynthesis
Phototropism Growth of a stem or root toward or away from light
Phragmoplast A set of microtubules oriented parallel to the axis of the spindle apparatus (perpendicular to the plane of cell division), which will form a cell plate; phragmoplasts occur in plants and in most green algae
Phycobilins Water-soluble accessory pigments occurring in the red algae and cyanobacteria
Phycocyanin A blue photosynthetic pigment in cyanobacteria and red algae
Phycoerythrin A red phycobilin
Phycoplast A set of microtubules oriented perpendicular to the axis of the spindle apparatus (parallel to the plane of cell division), which will form a cell plate; phycoplasts occur only in a few green algae
Phyllotaxis The arrangement of leaves on a stem
Phylogenetic Reflecting evolutionary relationships
Phylum A taxonomic category between kingdom and class in animals; it is equivalent to *division* in plants
Phytochrome A group of proteinaceous pigments involved in phenomena such as photoperiodism, the germination of seeds, and leaf formation; absorbs red and far-red light
Pigment Molecule that reflects and absorbs light at particular wavelengths
Pilus (plural: **Pili**) A minute tube between two bacterial cells, through which transfer of genetic material may occur
Pioneers The first plants to become established on new soil
Pistillate Flower (*see* **Carpellate Flower**)
Pistillate Plant (*see* **Carpellate Plant**)
Pith Parenchyma tissue in the center of a stem; located interior to the vascular bundles
Placenta (plural, **Placentae**) The area inside a carpel where the ovules are attached
Plasma Membrane The semipermeable membrane that surrounds the cytoplasm and is next to the cell wall; also called the cell membrane or the plasmalemma
Plasmid A small, circular fragment of DNA in bacteria; a plasmid can be integrated into and replicated with the rest of the bacterial genome; because of their ability to take up foreign DNA, bacterial plasmids are used as vectors for genetic engineering and research
Plasmodesma (plural, **Plasmodesmata**) A tiny, membrane-lined channel between adjacent cells
Plasmolysis Shrinkage of cytoplasm away from the cell wall due to the loss of water by osmosis
Plastid A type of organelle that is bounded by a double membrane and is associated with different pigments and storage products; chloroplasts are green, photosynthetic plastids; amyloplasts are storage plastids that contain starch
Pleiotropic Gene A gene that affects more than one phenotypic character; an example of a pleiotropic gene occurs in tobacco, in which a single gene controls the size and shapes of leaves, flowers, anthers, and fruits
Pneumatophor Upward-growing roots of some plants that grow in swamps; contain much aerenchyma and function in gas exchange
Polar Fiber A spindle fiber that does not bind to a kinetochore
Polar Nuclei Nuclei that come from opposite poles of the embryo sac and fuse with a sperm cell to form the primary endosperm nucleus
Polarity Establishment of poles of specialization at opposite ends of a cell, tissue, organ, or organism; for example, polarity leads to the differentiation of roots and shoots
Pollen Grain A male gametophyte that is surrounded by a microspore wall in seed plants
Pollen Tube The germination tube of a pollen grain, which grows from the stigma, through the style, and into the micropyle of the ovule; the pollen tube carries the sperm cells to the embryo sac
Pollination The transfer of pollen from microsporangia to the stigma in angiosperms or to directly to the ovule in gymnosperms
Pollination Droplet A sticky exudate at the mouth of the micropyle of a gymnosperm ovule; pollen grains catching in it are slowly withdrawn to the interior (pollen chamber) as the droplet recedes
Poly-A Tail A chain of adenylic acid molecules that is added to a molecule of RNA immediately after it has been transcribed and cleaved from its DNA template
Polyembryony, Cleavage The development of multiple embryos in a gymnosperm seed as a result of the differentiation of certain cells of a single embryo
Polyembryony, Simple The development of multiple embryos in a gymnosperm seed as a result of the development of two or more zygotes
Polygene A set of genes that act together, without dominance, to control a continuously variable phenotype; length, width, and oil content are examples of continuously variable phenotypes that are most like to be under polygenic control
Polygonum-Type Embryo Sac Development A type of embryo sac development from a functional megaspore that forms eight free nuclei, three of which become an egg apparatus, two of which are polar nuclei, and two of which become antipodal cells
Polymer A molecule consisting of many identical or similar monomers linked together by covalent bonds
Polymerase Chain Reaction (PCR) A procedure by which free nucleotides are assembled into a nucleic acid chain in a test tube by enabling the activity of a bacterial DNA polymerase to bind them together; the PCR is cycled 30 or more times to produce a million-fold amplification of the target DNA sequence
Polypeptide A chain of amino acids linked together by peptide bonds
Polyploid A condition in which a nucleus has more than two complete sets of chromosomes
Polysaccharide A carbohydrate polymer composed of many monosaccharides that are linked covalently into a chain; polysaccharides include starch, glycogen, and cellulose
Polysome A cluster of ribosomes on a single molecule of mRNA
Population A group of interbreeding individuals of the same species usually occupying the same territory at the same time
Population Density The number of individuals of a population within a given area
Population Genetics The application of genetic laws and principles to entire populations; assumes that evolution is the result of progressive change in the genetic composition of a population rather than individuals
Population, Local A population within a relatively small geographic area
Postulate A basic or necessary assumption; a set of postulates that address the same phenomenon can be taken together as a theory
Potential Energy The energy stored by matter because of its location or configuration; regarding a solute, the higher its concentration, the steeper is its concentration gradient and the greater is its potential energy; energy available to do work
Preprophase Band A band of microtubules that rings the cell just beneath the plasma membrane in a plane that is perpendicular to the axis of the future mitotic spindle apparatus; the preprophase band also corresponds to the orientation of the future metaphase plate and cell plate

Pressure Potential The component of water potential caused by the force created by turgor pressure against a membrane

Primary Cell Wall The usually thin cell wall that forms during cell division; it is part of all but some sperm cells in plants

Primary Consumer A consumer that feeds directly on producers

Primary Growth Growth resulting from the activity of apical meristems

Primary Pit-Field A thin area in a cell wall where clusters of plasmodesmata occur

Primary RNA Transcript A molecule of RNA that includes the GTP cap, the leader sequence, the gene sequence, the trailer sequence, and the poly-A tail

Primary Structure The sequence of amino acids in a protein

Primary Succession Succession that is initiated on bare rock or in water after a disturbance has occurred

Primary Thickening Meristem In some monocots, the meristem that increases the thickness of the shoot axis

Primer A short sequence of single-stranded DNA (e.g., 10-30 nucleotides long) that is complementary to one end of a target gene of interest; primers are annealed to their complementary sequences so that DNA polymerase can begin replicating the target gene in the polymerase chain reaction

Probe In genetic research, a sequence of radioactive DNA or RNA that is used to find the complementary sequence of a gene of interest in a culture of clones or cells

Procambrium A meristem that produces the primary vascular tissues

Producer An autotrophic organism; producers form the base of food chains in an ecosystem

Progymnosperms A group of extinct plants believed to be the ancestors of gymnosperms

Prop Roots Adventitious roots that form on a stem above the ground; help support the plant (as in corn)

Prophase The period of mitosis during which chromosomes condense, first appearing as a mass of elongated threads and later as individual chromosomes

Prophase I The first prophase of meiosis; in prophase I, homologous chromosomes condense, synapse, cross over, and desynapse; chiasmata move to the ends of chromosomes by the end of prophase I

Prophase II The second prophase of meiosis; in prophase II, chromosomes condense, the nuclear envelope disintegrates, and a spindle apparatus is assembled; in many organisms, prophase II is bypassed if telophase I is also bypassed, in which case the meiotic nuclei go directly from anaphase I to metaphase II

Protease Inhibitor Any chemical that inhibits the activity of enzymes that digest proteins (i.e., proteases); protease inhibitors can also be proteins

Protenema (pl. **Protenemata**) The early, filamentous growth of the gametophyte of bryophytes and ferns

Prothallial Cells Two of the four cells produced during the development of a gymnosperm microspore into a pollen grain. The prothallial cells are functionless.

Protoderm The outermost tissue of an apical meristem; produces the epidermis

Protoxylem The first xylem cells formed in the primary xylem

Pseudoplasmodium A phase of cellular slime molds in which the myxamoebae do not fuse but aggregate into a sluglike body that moves as a unit

Pulvinus Jointlike thickening at the base of a petiole; involved in movements of a leaf (or leaflet)

Punctuated Equilibrium A model stating that long geological time periods with little or no evolutionary change are punctuated by periods of rapid evolution

Punnett Square A gametic grid that is used to show the expected genotypic and phenotypic ratios resulting from a hybridization experiment

Purine A two-ringed nitrogen-containing base that is part of a nucleotide; the most common purines are adenine and guanine

Pyrimidine A one-ringed nitrogen-containing base that is part of a nucleotide; the most common pyrimidines are thymine, cytosine, and uracil

Q

Quaternary Structure The way that different subunits attach to one another in a multi-subunit protein

Quiescent Center The relatively inactive region in the apical meristem of a root

R

R Group A general term for the side group of a molecule, such as a methyl group, a hydroxyl group, or a monosaccharide

R-loop A sequence of DNA within a gene that is displaced into a loop-like projection when the gene is annealed to its complementary mRNA; the R-loop does not anneal with the mRNA because it is an intron whose complementary sequence has been spliced out of the mRNA molecule

Radicle The root of an embryo

Ray Initials Cells in the vascular cambium that produce the ray cells of secondary xylem and secondary phloem

Reaction Wood Wood produced in response to a stem that has lost its vertical position; reaction wood straightens the stem

Receptacle The region of the floral shoot where the parts of the flower are attached

Recessive A trait that is masked by an alternative (dominant) trait when the gene for these traits is heterozygous

Recombinant DNA Technology (*see* **Genetic Engineering**)

Recombination Nodule A cluster of enzymes in a synaptonemal complex, which are believed to act in concert to bring matching segments of homologous chromosomes together

Reduction The gain of electrons by an atom or molecule that is involved in an oxidation-reduction (redox) reaction; reduction involves the addition of energy to one substance, which is coupled with the simultaneous removal of energy from another substance by oxidation

Reduction Division A synonym for meiosis, specifically for meiosis I

Reductionism The approach of studying simpler components in order to understand the functions of complex systems

Release Factors A group of cytoplasmic proteins that bind to a stop codon on a molecule of mRNA and interrupt translation by hydrolyzing the bond between the final amino acid in a polypeptide and its transfer RNA

Repetitive DNA Sequences of DNA that occur in many copies in a genome; some sequences of repetitive DNA can occur in a million copies per nucleus

Replication Bubble A region of DNA that has been separated into single strands between opposing replication forks

Replication Fork The region where a DNA double strand is split into separate strands, creating a fork-like appearance in electron micrographs; once replication begins at a replication origin, two replication forks proceed along the double helix in opposite directions from one another

Replication Origin The point of initiation of DNA synthesis along the double helix; two replication forks form at the replication origin and move in opposite directions from one another during DNA synthesis

Replicon A block of DNA between two adjacent replication origins

Reproductive Barriers Various mechanisms that prevent reproduction between individuals, usually from different species

Resin A thick, translucent, combustible, organic fluid usually secreted into resin ducts in pines and many other seed plants

Resin Duct An elongate intercellular space lined with resin-secreting cells and containing resin

Resolving Power The minimum distance necessary to distinguish two points from each another

Respiration The process by which organic compounds are oxidized with the release of energy; respiration is aerobic if oxygen is required as the terminal electron acceptor; it is anaerobic if oxygen is not used as an electron acceptor

Restriction Enzyme A type of enzyme that recognizes a specific sequence of DNA and catalyzes the cleavage of the double helix at that site; most restriction enzymes recognize DNA sequences whose complementary sequence reads the same in the reverse direction; synonym with restriction endonuclease

Reverse Transcriptase A type of enzyme from viruses that catalyzes the synthesis of DNA from an RNA template; in genetics, reverse transcriptase is used for making cDNA of eukaryotic genes

Rhizome A fleshy, horizontal, underground stem

Ribonucleic Acid (RNA) The nucleic acid containing four different nucleotides whose simple sugar is ribose; molecules of RNA, which are made as complements of DNA segments called genes, function in protein synthesis

Ribosome An organelle that is responsible for protein synthesis; ribosomes consist of ribosomal RNA (rRNA) and proteins that are arranged into two subunits, one large and one small

Ribosomal RNA (rRNA) The type of RNA that is a component of ribosomes

Ribozyme A sequence of RNA that has enzymatic properties; first named from a self-splicing intron

Ring-Porous Wood Wood having larger vessels in early wood than in late wood, thereby producing a ring when viewed in a cross-section of wood

RNA Polymerase A type of enzyme that catalyzes the synthesis of RNA as the complement to a specific sequence of DNA

RNA Processing The trimming of larger primary RNA transcripts in the nucleus into smaller, coding sequences that are exported into the cytosol; synonymous with RNA splicing

RNA splicing (*see* **RNA Processing**)

Root Cap An organ that covers the root meristem; helps the growing root penetrate the soil

Root Hairs Epidermal cells just behind the zone of elongation in roots; increase the absorptive surface area of the root

Rosin The hard, brittle component of resin remaining after volatile parts have been removed

rRNA (*see* **Ribosomal RNA**)

Rubber A large polymer of up to 6,000 isoprene units

Runner (*see* **Stolon**)

S

S phase During interphase, the portion of the cell cycle in which DNA synthesis occurs; S refers to the synthesis of DNA

Saprobic Obtaining food directly from nonliving organic matter

Sapwood Wood found between the vascular cambium and the heartwood, transports water and solutes

Saturated Refers to fatty acids or other hydrocarbon-containing chemicals whose carbon-carbon bonds are all single bonds; palmitic acid is an example of a saturated fatty acid

Savanna A grassland with scattered trees. Many savannas are located in tropical or subtropical areas.

Scanning Electron Microscopy (SEM) Microscopy that focuses an electron beam that is reflected from a specimen, thereby showing fine details of its surface structure

Scarification The cutting, abrading, or otherwise softening of the seed coat to induce the seed to germinate

Scientific Method A way of analyzing the physical universe; observations are used to construct a hypothesis that predicts the outcome of future observations or experiments; something that cannot be verified cannot be accepted as part of a scientific hypothesis

Sclereids Sclerenchyma cells found in tissues varying from pear fruits to the hard shells of some nuts

Scutellum The cotyledon of a grass seed; the scutellum is specialized for absorbing nutrients from the endosperm as the seed germinates

Second Gap (*see* **G_2 phase**)

Secondary Cell Wall The cell wall that forms interior to the primary cell wall only after cell division is completed; it is restricted to certain cells and often contains lignin

Secondary Consumer A consumer that feeds on primary consumers

Secondary Growth Growth derived from lateral meristems (e.g., the vascular cambium and cork cambium)

Secondary Metabolism The metabolism of chemicals that occur irregularly or rarely among different plants and that usually have no known metabolic role in cells

Secondary Structure The portion of a protein's shape that is maintained by hydrogen bonds between amino acids

Secondary Succession Succession in habitats where the climax community has been disturbed or removed

Secondary Xylem Xylem formed by the vascular cambium; wood

Seed A mature ovule, consisting of a seed coat that surrounds the embryo and associated tissues

Seed Bank The ungerminated but still viable seeds that occur in natural storage in soil

Seed Coat The outer layer of a seed; the seed coat develops from the integument of the ovule

Seed Ferns An extinct group of plants that were characterized by frond-like leaves and seed-bearing structures; classified together in the Division Pteridospermophyta

Selection Pressures Those environmental factors that promote or retard reproductive success of a phenotype

Selectively Permeable Refers to a membrane that restricts the passage of some solutes through it

Self-Compatible Refers to the potential for successful reproduction between flowers of the same plant or between stamens and carpels of the same flower

Self-Incompatible Incapable of successful reproduction between flowers of the same plant or between stamens and carpels of the same flower

Self-Replication Refers to the ability of DNA to make exact copies of itself

Selfish DNA Refers to DNA that can perpetuate itself by semi-autonomous replication; transposons are considered to be selfish DNA because they can move copies of themselves to several sites in a genome

Semiconservative Replication Refers to the replication of a DNA molecule wherein half of each new double strand consists of one newly synthesized strand and one strand from the parent double helix

Sense Strand In DNA, the sense strand of a gene is the one that contains the coding sequence for a molecule of RNA and, in the case of mRNA, indirectly for a polypeptide

Sepal One of the outermost parts of a flower; collectively, the sepals of a single flower are called the calyx

Septum (plural: **Septa**) A crosswall in a fungal hypha

Serotype A protein that is a unique antigen; it induces and binds to antibodies that are specific to it alone. Serotypes are used in a classification system applied to viruses

Sessile Leaf Leaf Lacking a petiole; blades of sessile leaves attach directly to the stem

Seta The stalk that supports the capsule of a moss sporophyte

Short-Day-Plant Plant that flowers when the length of dark is longer than some critical value; short-day plants usually flower in autumn

Sieve Area Part of the wall of a sieve element containing many pores through which the protoplasts of adjacent sieve elements are connected

Sieve Cell A long sieve-element having unspecialized sieve areas and tapering end-walls that lack sieve plates; sieve cells occur in the phloem of gymnosperms and lower vascular plants

Sieve Elements Cells in the phloem that transport organic solutes; sieve cells and sieve-tube members are examples of sieve elements

Sieve Plate The part of a wall of a sieve-tube member that has one or more sieve areas

Sieve Tube A series of sieve tube members arranged end-to-end and connected by sieve plates

Simple Leaf A leaf having one blade which may be lobed or dissected

Single-Strand Binding Proteins Proteins that prevent the fusion and rewinding of DNA once the double strands are split apart for replication

Sink Where organic solutes are being transported by the phloem; where metabolites such as sugar are used or stored

Sister Chromatids A pair of chromatids in a duplicated chromosome

Sliding-Microtubule Hypothesis An explanation for how chromosomes are moved during anaphase; this hypothesis holds that opposing polar spindle fibers slide past one another, creating a force that pushes the poles of a spindle apparatus apart

Slug (*see* **Pseudoplasmodium**)

Small Nuclear Ribonucleoprotein (snRNP) A complex of small RNA molecules condensed with specific proteins in the nucleus; a snRNP is the basic unit of a spliceosome

snRNP (*see* **Small Nuclear Ribonucleoprotein**)

Softwood Coniferous gymnosperm

Solute Potential The component of water potential caused by the presence of solutes in water

Solute A substance dissolved in a solution

Solvent A liquid that dissolves solutes

Source Where organic compounds such as sugar are being made and loaded into the phloem

Southern Blotting A procedure by which fragments of DNA are separated by gel electrophoresis, transferred to a filter, and probed with DNA that is complementary to the gene of interest; the location of the target gene is found because it becomes radioactive when the probe anneals to it (*also see* **Northern Blotting**)

Speciation Evolutionary formation of a new species

Species (plural: **Species**) A species is a group of similar organisms capable of, or potentially capable of, freely interbreeding. The scientific names of species are binomials consisting of a genus (generic) name and a specific epithet.

Species Diversity The number of species and the number of individuals per species in an ecosystem

Spindle Apparatus Refers to the elliptically shaped collection of spindle fibers in a cell

Spindle Fibers Nearly parallel microtubules that form between the poles of dividing cells; some spindle fibers attach to chromosomes but fibers from opposite poles mostly interact with each other; spindle fibers are believed to move chromosomes both by pulling homologous chromosomes in opposite directions and by pushing poles apart

Spliceosome A cluster of snRNPs; a spliceosome binds to a large primary RNA transcript, cuts out certain parts of the RNA, then splices the rest of the RNA back into a continuous strand

Spongy Mesophyll Leaf tissue consisting of loosely arranged photosynthetic cells

Sporangium A structure within which the protoplasm becomes converted into an indefinite number of spores

Spore A small reproductive structure, usually consisting of a single cell, which is capable of developing independently into a much larger, mature and often multicellular body

Sporophyte The phase of the plant life cycle that produces spores

Sporophytic Self-Incompatibility A type of self-incompatibility that is imposed by sporophytic tissues or organs; an example would be incompatibility that is imposed by the stigma, which is a sporophytic structure

Spring (Early) Wood Wood produced in the spring; usually characterized by relatively large cells

Stabilizing Selection Selection for a phenotype within the norm of a population, or selection against extreme phenotypes

Stable Isotope Tracing A technique based on the typical ratio of carbon^{12} to carbon^{13} in tissue samples, which enables ecologists to determine food sources and consumption in food webs

Stalk Cell One of two cells produced when the generative cell of a gymnosperm male gametophyte divides. Immediately before fertilization the body cell divides, becoming two sperms.

Stamen The pollen-producing part of a flower, usually consisting of an anther and a filament; collectively, the stamens of a single flower are called the androecium

Staminate Flower A flower whose reproductive parts consist only of stamens; the tassels at the tops of corn plants are examples of staminate flowers

Staminate Plant An individual plant whose flowers bear stamens but not carpels; a "fruitless" mulberry is an example of a plant that is exclusively staminate (mulberries can reproduce only when pollen is transferred to a carpellate plant)

Starch-Branching Enzyme (SBEI) A type of enzyme that converts straight chains of amylose to the branched polymers of amylopectin; "I" refers to an isoform of the enzyme

Starch Phosphorylase A type of enzyme that cleaves a molecule of glucose from one end of a glucose polymer by phosphorylating the glucose that is removed from the chain

Start Codon The codon at the beginning of a polypeptide-coding gene; the start codon codes for the first amino acid in the polypeptide, which is usually methionine

Stele The central vascular cylinder of roots and stems

Sterol A compound derived from six isoprene units linked together in a multiple-ringed structure; beta-sitosterol is an example of a plant sterol; cholesterol is a widely known example of an animal sterol

Sticky End The single-stranded portion of a DNA sequence after it undergoes a zigzag by a restriction enzyme

Stigma The surface of a carpel that is receptive to pollen grains and upon which the pollen grains germinate; a photosensitive eyespot found in certain kinds of algae

Stipule A leaflike appendage that occurs on either side of the base of a leaf (or encircles the stem) in several kinds of flowering plants

Stolon A stem that grows horizontally along the ground

Stoma (pl. **Stoma**) The epidermal structure consisting of two guard cells and the pore between them

Stop Codon A codon that occurs at the end of a gene and signals where translation stops

Stratification The exposure of seeds to extended cold periods before they will germinate at warm temperatures

Strobilus, Compound An axis with lateral branches bearing sporophylls

Strobilus, Simple An unbranched axis bearing sporophylls

Stroma The matrix between the grana in chloroplasts; the site of the biochemical (i.e., "dark") reactions of photosynthesis

Structural Polysaccharide A polysaccharide that holds cells and organisms together; cellulose is the most abundant structural polysaccharide in plants

Style A column of carpel tissue arising from the top of an ovary and upon which is the stigma; the style raises the stigma to a receptive position for pollen grains whose pollen tubes must grow through it

Suberin Aa waxy substance that occurs in cork cells and in the cells of underground plant parts; it consists of hydroxylated fatty acids that are linked together in a complex array

Subsidiary Cella Epidermal cells that are structurally distinct from other epidermal cells and associated with guard cells

Substrate-Level Phosphorylation The transfer of a phosphate group from a substrate, such as phosphoenol pyruvic acid, to ADP, thereby making ATP

Subunit A polypeptide that combines with other polypeptides to comprise a multi-subunit protein

Succession An orderly progression of population replacements in a specific geographic area; it is initiated with pioneer species and completed with the establishment of a stable climax community

Succinate (Succinic Acid) Aa four-carbon organic acid that is oxidized by the reduction of ubiquinone to ubiquinol in the sixth step of the Krebs cycle; the product of this oxidation is fumaric acid

Succinyl-CoA An acetylated four-carbon acid that is converted to succinic acid by losing its acetyl-CoA group, thereby driving the substrate-level phosphorylation of one molecule of ADP to ATP in the fifth step of the Krebs cycle

Succulent A plant having think, fleshy leaves or stems; succulence is usually an adaptation to water or salt stress

Sucker A sprout on the roots of some plants that forms a new plant

Sucrose Synthase A type of enzyme that catalyzes the reversible breakdown of sucrose from starch by hydrolysis into free fructose and bound glucose; the glucose is bound to a carrier molecule called uridine diphosphate (UDP)

Summer (Late) Wood Wood produced in the summer; characterized by relatively small cells

Superior Ovary An ovary located above the other flower parts on a floral axis

Suspensor A group of cells at the base of the embryo of many seed plants that expands and moves the embryo into the endosperm

Suture The line along which a fruit splits when it is mature

Symbiont One of tow (or more) dissimilar organisms that live in close association with each other. The association may be beneficial to both organisms (*mutualism*) or harmful to one organism (*parasitism*)

Sympatric Having the same or overlapping geographic distribution but separated by reproductive or biotic barriers

Sympatric Speciation Formation of a new species entirely within the geographical range of its parental form

Symplast The interconnected living mass of an organism; the symplast is a continuous unit that is comprised of cells that are connected by plasmodesmata throughout the organism

Symplastic Movement of water and solutes through tissues by passing through interconnected protoplasts and their plasmodesmata

Synapsis The pairing of homologous chromosomes by their attachment along a synaptonemal complex; crossing over occurs during synapsis

Synaptonemal Complex A complex of proteins that forms a chromosome-length axis linking homologous chromosomes between the same gene loci

Synergid A type of cell that occurs next to the egg in an embryo sac; sperm cells entering the embryo sac first pass through one of the synergids

Synonymous Codon Refers to codons that code for the same amino acid

Syringomycin A toxic polypeptide that is secreted by *Pseudomonas syringae*, a species of bacteria that infects corn, beans, and many other kinds of plants

Systematics The classification of organisms into a hierarchy of categories (taxa) based on evolutionary interrelationships

T

Taiga A coniferous forest biome adjacent to arctic tundra in large areas of North America and Erasia.

Tandem Repeat The occurrence of two or more copies of a gene in a row; ribosomal RNA genes typically occur as tandem repeats

Tangential Section A longitudinal section that does not pass through the center of the structure

Tapetum A tissue of sterile cells that surrounds the microspores in a microsporangium; the tapetum acts as a nutritive tissue for the spores and pollen grains while they remain in a sporangium

Taproot A relatively large primary root that produces secondary roots

Tassel The downward-hanging inflorescence of some plants; in corn, tassel refers to an inflorescence of pollen-bearing flowers at the top of the plant

Taxol A drug obtained from the Pacific yew, and also from a fungus that grows on the yew, with potential for treating certain forms of cancer

Taxon (plural: **Taxa**) Any taxonomic category, such as species or family

Taxonomy The classification, description, and naming of organisms

Telomere Sequences of DNA at the tip of a chromosome that counteract its condensation before the onset of nuclear division

Telophase The period of mitosis during which chromosomes seem to mimic prophase in reverse; chromosomes steadily elongate and decondense back into diffuse chromatin, and each new daughter nucleus becomes surrounded by a nuclear envelope

Telophase I The first telophase of meiosis; in telophase I, chromosomes decondense, the spindle apparatus disintegrates, and a new nuclear envelope forms around each daughter nucleus; in many organisms, telophase I is bypassed and the meiotic nuclei go directly from anaphase I to metaphase II

Telophase II The second telophase of meiosis; in telophase II, chromosomes decondense, the spindle apparatus disintegrates, and a new nuclear envelope forms around each of the four new daughter nuclei

Temperate Deciduous Forest A biome dominated by deciduous hardwood trees, and located in areas with temperate climates

Tendril A modified leaf or stem in which only a slender strand of tissue constitutes the entire structure

Tensile Strength The maximum amount of lengthwise pull that a substance can bear without tearing apart

Tension Wood Reaction wood that forms along the upper side of leaning stems; straightens the stem by contracting and "pulling" the stem up

Terpenoid Any compound that is derived from five-carbon precursors called isoprene units; examples include menthol (two isoprene units), beta-carotene (eight isoprene units), and rubber (up to 6,000 isoprene units)

Tertiary Structure The portion of a protein's shape that is maintained by disulfide bonds, ionic interactions, or hydrophobic attraction between amino acids

Tetrad Scar A scar on a primitive spore at the point of attachment to three other spores, all four having developed after meiosis; germination takes place in the vicinity of the scar

Thallophytes A term once used to designate fungi and algae collectively

Thallus (plural: **Thalli**) A body not differentiated into roots, stems, or leaves, as seen in lichens and liverworts

Thigmotropism A response to contact with a solid object

Thorn A modified woody stem that terminates in a sharp point

Thylakoid A saclike membranous structure of the grana of chloroplasts; thylakoids house chlorophylls

Tissue Culture A technique for growing and manipulating pieces of tissue in a medium after their removal from the organism

Tonoplast The membrane that surrounds a vacuole; also called a vacuolar membrane

Topoisomerase A type of enzyme that relieves the kinks in DNA that would otherwise block the movement of replication forks; topoisomerases work by breaking one or both strands, thereby allowing the strands to uncoil by swiveling around one another; after uncoiling, the strands are also linked back together by topoisomerases

Totipotent Refers to the notion that every cell has the same genes and therefore the same genetic potential to make all cells other cell types

Trace Elements (*see* **Micronutrients**)

Tracheid Elongated, spindle-shaped cell that transports water in the xylem; also helps support the plant; occur in virtually all vascular plants

Trailer Sequence An extra amount of non-coding DNA that is transcribed into RNA beyond the end of the gene

Transcription The synthesis of a molecule of RNA as a complement to a specific sequence of DNA; transcription occurs in the nucleus

Transduction The transfer of genetic material between bacteria by bacteriophages

Transect, Line A straight line extending between two points, usually established arbitrarily for the purpose of studying vegetation immediately adjacent to it

Transfer RNA (tRNA) A type of small RNA molecule that binds to a specific amino acid and to a codon on messenger RNA; it is called transfer RNA because it is associated with the transfer of amino acids to mRNA in ribosomes; more than 40 different tRNA molecules have been found, at least one for each protein amino acid

Transformation A form of genetic transfer in bacteria by which a fragment of DNA is taken up and incorporated into the DNA of the recipient cell

Transgenic Refers to cells or organisms that contain genes that were inserted into them from other organisms by genetic engineering

Translation The synthesis of an amino acid sequence from a specific sequence of codons along a molecule of mRNA; translation occurs in ribosomes

Transmission Electron Microscopy (TEM) Microscopy that focuses an electron beam through the thin section of a specimen to study its internal structure

Transpiration Evaporation of water from leaves and stems; occurs mostly through stomata

Transport Protein Aa type of membrane protein that enables the transport of specific solutes across the membrane

Transposable Element A fragment of DNA that is able to multiply and move spontaneously among an organism's chromosomes

Tricarboxylic Cycle (*see* **Krebs Cycle**)

Trichogyne A receptive, slender outgrowth for spermatia or similar reproductive cells in red algae and ascomycete fungi

Trichome An epidermal outgrowth (e.g., a hair or scale)

Triplet Refers to a sequence of three nucleotides that together comprise a codon

Triterpene A compound that consists of six isoprene units linked together; sterols, such as beta-sitosterol, are triterpenes

tRNA (*see* **Transfer RNA**)

Tropical Rain Forest An endangered tropical biome with exceptional diversity of species

Tropism A response to an external stimulus in which the direction of the response is determined by the direction from which the stimulus comes; tropisms such as phototropism and gravitropism are produced by differential growth

True-Breeding Refers to purebred strains for a given trait, which means that the gene for that trait is homozygous

Tube Cell The cell in the pollen grains of seed plants that develops into the pollen tube

Tuber A fleshy, underground stem having an enlarged tip (e.g., potato tuber)

Tubulin (*see* **Alpha-Tubulin** and **Beta-Tubulin**)

Tundra A vast biome primarily above the arctic circle, and above the timberlines of mountain ranges further south, whose vegetation includes no typical trees

Tunica-Corpus The organization of the shoot apex of most angiosperms and some gymnosperms; consists of one or more peripheral layers (i.e., tunica layers) and an interior corpus

Turgid Full of water taken in by osmosis

Turgor Pressure The pressure on a cell wall that is created from within the cell by the movement of water into it

Turpentine The volatile, combustible component of resin

Type Specimen A specimen upon which the original description of species is based

U

Ubiquinol The reduced form of ubiquinone; ubiquinol donates electrons to cytochrome *b* in the electron transport chain

Ubiquinone A lipid-soluble quinone whose function is to accept electrons from electron donors like NADH and from the oxidation of fatty acids; also called coenzyme Q

Ultracentrifuge A high-speed centrifuge that is capable of spinning at more than 100,000 revolutions per minute

Unequal Crossing-Over Refers to the exchange of unequal amounts of DNA between homologous chromosomes that are not perfectly aligned

Unsaturated Refers to fatty acids or other hydrocarbon-containing chemicals whose carbon-carbon bonds include double bonds as well as single bonds; oleic acid (one double bond), linoleic acid (two double bonds), and linolenic acid (three double bonds) are examples of unsaturated fatty acids

Uridine Diphosphate (UDP) A uracil-containing nucleotide that acts as a carrier molecule for glucose and similar monosaccharides; the UDP-sugar complex is also an intermediate compound for the interconversion of one monosaccharide to another (e.g., glucose to galactose)

Vacuole A membrane-bound organelle that is filled mostly with water but also may contain water-soluble pigments and other substances; the vacuolar membrane is called the tonoplast

Vascular Bundle Strand of tissue containing primary phloem and primary xylem (and possibly procambrium); often enclosed by a bundle sheath

Vascular Tissue Tissue specialized for long-distance transport of water and minerals; xylem and phloem

Vector In genetics, a virus or bacterial plasmid that can take up a foreign gene and integrate it into the genome of a target organism; in plant reproduction, an animal that carries pollen (a pollinator) from a pollen sac to a stigma or to an ovule

Vegetative Cell A cell that is neither sexually reproductive nor divides to form cells that are sexually reproductive; this term particularly refers to the tube cell of angiosperm pollen grains, which is the only vegetative cell of the male gametophyte

Vein Vascular bundle that forms part of the connecting and supporting tissue of a leaf or other expanded organ

Velamen Multiple epidermis covering the aerial roots of some orchids and aroids

Venter The swollen base of an archegonium containing the egg

Vernalization The induction of flowering by cold

Vessel A tubelike structure in the xylem that consists of vessel elements placed end-to-end and connected by perforations; vessel elements conduct water and minerals; found in nearly all angiosperms and a few other vascular plants

Vessel Element One of the cells forming a vessel

Vesticular-Arbuscular (V-A) Mycorrhizae Treelike or bulblike mycorrhizae

Viroid A plant-infecting, viruslike particle with a single circular strand of RNA that is not associated with any protein

Virusoid A particle similar to a viroid but located inside the protein coat of a true virus

Water Potential The potential energy of water to move down its concentration gradient; water potential is expressed in units of pressure instead of units of energy, because pressure is simpler to measure

Whorl A circular group of at least three leaves or flower parts all attached to an axis at the same level

Wood Secondary xylem

Xanthophyll A yellow carotenoid; one xanthophyll, zeaxanthin, in the blue-light photoreceptor in shoot phototropism

Xerophyte A plant adapted for growth in arid conditions

Xylem Vascular system specialized for transporting water and dissolved minerals upward in the plant; characterized by the presence of tracheary elements

Yeast Artificial Chromosome (YAC) A yeast chromosome into which large fragments of foreign DNA (millions of base pairs) have been inserted; YACs can be replicated like native chromosomes in yeast cells, thereby cloning large amounts foreign DNA as well

Z

Zeatin A natural cytokinin isolated from corn (*Zea mays*)

Zein A storage protein in the kernels of corn

Zygomycetes A large group of fungi with primarily coenocytic mycelia; they reproduce asexually by spores produced within sporangia; sexual reproduction includes the formation of zygosporangia

Zygosporangium (plural: **Zygosporangia**) A sporangium containing a thick-walled, multinucleate zygospore that develops in zygomycetes after the fusion of isogametes

Zygote The diploid cell that is formed by the fusion of two gametes

Credits

Photographs

Part Openers

Part 7: © William E. Ferguson, photo by Stephen Ferguson; **Part 8:** © Jeff Gnass; **Part 9:** © Wolfgang Kaehler

Chapter Twenty-Two

Opener: © Runk/Schoenberger/Grand Heilman Photography, Inc.; **22.1:** © Bill Kamin/Visuals Unlimited; **22.2:** © William E. Ferguson; **22.8:** Courtesy of Hunt Institute for Botanical Documentation, Carnegie Mellon University, Pittsburgh, PA; **22.9:** © Science VU-USM/Visuals Unlimited; **22.10:** © Wm. Ormerod/Visuals Unlimited; **22.11A:** © Heather Angel/Biofotos; **22.11B:** © John M. Thager/Visuals Unlimited; **22.12A:** © Bruce Berg/Visuals Unlimited; **22.12B:** © Heather Angel/Biofotos; **22.12C:** © William E. Ferguson; **22.14:** © R. Van Nostrand/Photo Researchers, Inc.; **22.16:** © Robert & Linda Mitchell; **22.20:** © Runk/Schoenberger/Grant Heilman Photography; **22.21:** © Heather Angel/Biofotos; **22.22A:** Courtesy of DEKALB Genetics Corp., DeKalb, IL; **22.22B:** Courtesy of DEKALB Genetics Corp., DeKalb, IL.

Chapter Twenty-Three

Opener: © Dwight Kuhn; **23.1A:** © Robert E. Lyons/Photo/Nats; **23.1B:** © Jon Bertsch/Visuals Unlimited; **23.1C:** © Hans Reinhard/Bruce Coleman, Inc.; **23.2 A,B:** Courtesy of Donald J. Pinkava; **23.4:** © William E. Ferguson; **23.5A:** © James L. Amos/Photo Researchers, Inc.; **23.5B:** © Kjell B. Sandved/Visuals Unlimited; **Box 23.1A:** Courtesy of Renee Van Buren; **23.6:** © Dede Gilman/Visions From Nature; **23.7:** © J. Alcock/Visuals Unlimited; **23.8:** © William McPherson/Bruce Coleman, Inc.; **23.11:** © Richard Thom/Visuals Unlimited; **23.15:** Courtesy of Darrell Vodopich; **23.18:** Courtesy of Loren Rieseberg; **23.21:** © Jim Strawser/Grant Heilman Photography; **23.22:** © Runk/Schoenberger/Grant Heilman Photography; **Page 565:** © Kjell B. Sandved/Visuals Unlimited

Chapter Twenty-Four

Opener: Courtesy of Hunt Institute for Botanical Documentation, Carnegie Mellon University, Pittsburgh, PA; **24.1:** © Runk/Schoenberger/Grant Heilman Photography; **Box 24.1:** © Kingsley Stern; **Page 574:** © Heather Angel/Biofotos; **24.2:** © Tom McHugh/Photo Researchers, Inc.; **24.3, 24.4:** © Robert & Linda Mitchell; **24.9A:** © Thomas Hovland/Grant Heilman Photography; **24.9B:** © Jeff Foott/Bruce Coleman, Inc.; **24.11:** © K. G. Vock/Okapia/Photo Researchers, Inc.

Chapter Twenty-Five

Opener: © Masana Izawa/Nature Production; **25.1A:** © David M. Phillips/Visuals Unlimited; **25.1B:** © M. I. Walker/Photo Researchers, Inc.; **25.1C:** Courtesy of Dennis Clark; **25.1D:** © P. L. Grilione & J. Pangborn, San Jose State University; **25.2 A-1:** © David M. Phillips/Visuals Unlimited; **25.2 A-2:** © R. Kessel-G.Shih/Visuals Unlimited; **25.2 B-1, B-2:** © Runk/Schoenberger/Grant Heilman Photography; **25.3:** © George B. Chapman/Visuals Unlimited; **25.4:** © G. Musil/Visuals Unlimited; **25.5A:** © Dr. Dennis Kunkel/Phototake; **25.6:** © Cabisco/Visuals Unlimited; **25.9:** © Imagery; **25.10A:** © Philip Sze/Visuals Unlimited; **25.10B:** © Bruce Iverson; **25.10C:** © George J. Wilder/Visuals Unlimited; **25.11:** © Cabisco/Visuals Unlimited; **25.12A:** © Ray Coleman/Photo Researchers, Inc.; **25.12B:** © E. R. Degginger/Bruce Coleman, Inc.; **25.12C:** © Richard Walters/Visuals Unlimited; **25.12D:** © Heather Angel/Biofotos; **25.12E:** © Bill Keogh/Visuals Unlimited; **25.12F:** © Hans Reinhard/Bruce Coleman, Inc.; **25.13 A-1:** © Dr. Jeremy Burgess/SPL/Photo Researchers, Inc.; **25.13 B-1:** © M. F. Brown/Visuals Unlimited; **Box 25.2:** © E. R. Degginger/Bruce Coleman, Inc.; **25.14A:** © Dwight Kuhn; **25.14B:** © R. Hussey/Visuals Unlimited; **25.16A:** © John D. Cunningham/Visuals Unlimited; **25.16B:** © Doug Sherman/Geofile; **25.17:** © Gary T. Cole, Ph.D., University of Texas, Austin/ Biological Photo Service; **Box 25.3:** © Manfred Kage/Peter Arnold, Inc.; **25.19:** © M. Eichelberger/Visuals Unlimited; **25.20A:** Courtesy of Mylan Laboratories, Inc. Photo by Rienzi & Rienzi Communications, Inc.; **25.20B:** Courtesy of the Roquefort Societe. Photo by P. Lauthier; **25.21A:** Courtey of G. S. Ellmore; **25.21B:** © Stephen Sharneff/Visuals Unlimited; **25.22A:** © Robert & Linda Mitchell; **25.22B:** © William E. Ferguson; **25.22C:** © Heather Angel/Biofotos; **25.24:** © John D. Cunningham/Visuals Unlimited; **25.26B:** © Stanley L. Flegler/Visuals Unlimited; **Box 25.4A:** © Stephen P. Parker/Photo Researchers, Inc.; **Box 25.4B:** © Barbara A. Roy, reprinted by permission from *Nature* **362:57; 25.28A:** © M. I. Walker/Photo Researchers, Inc.; **25.28B:** © Bruce Iverson/Visuals Unlimited; **25.29:** © Holt Studios, Ltd./Animals Animals/Earth Scenes; **25.31A:**

© Bruce Iverson; **25.31B:** © John Shaw/Bruce Coleman, Inc.; **25.31C:** © Cabisco/Visuals Unlimited; **25.31D:** © A. M. Siegelman/Visuals Unlimited; **25.32A:** © Russell Steere/Visuals Unlimited; **25.32B:** © Cabisco/Visuals Unlimited; **25.33:** Courtesy of Ikuo Kimura, Faculty of Agriculture, Hokkaido University, Japan.

Chapter Twenty-Six

Opener: © Randy Morse/Tom Stack & Associates; **26.1A:** © Heather Angel/Biofotos; **26.1B:** © Gary Retherford/Photo Researchers, Inc.; **26.2:** © M. I. Walker/Photo Researchers, Inc.; **26.3A:** © John D. Cunningham/Visuals Unlimited; **26.3B:** © Philip Sze; **26.4:** © Phil Gates, University of Durham/Biological Photo Service; **26.7:** © W. L. Dentler, University of Kansas/Biological Photo Service; **26.10A:** © Richard H. Gross; **26.10B, 26.11:** © Bruce Iverson; **26.12:** © K. G. Murti/Visuals Unlimited; **26.13:** © Philip Sze; **26.14:** © M. Eichelberger/Visuals Unlimited; **26.16A:** © Manfred Kage/Peter Arnold, Inc.; **26.16B:** © M. I. Walker/Photo Researchers, Inc.; **26.17:** © Heather Angel/Biofotos; **26.18:** © Carolina Biological Supply/Phototake; **26.19:** © Brian Parker/Tom Stack & Associates; **26.21:** © Cabisco/Visuals Unlimited; **26.24:** © Tammy Peluso/Tom Stack & Associates; **26.26:** © Sea Studios, Inc./Peter Arnold, Inc.; **26.27:** Reproduced by permission of the National Research Council of Canada from the *Canadian Journal of Botany*, Vol. 60, pages 85–97, 1982.; **26.30A:** © Cabisco/Visuals Unlimited; **26.30B:** © Veronika Burmeister/Visuals Unlimited; **26.31A:** Courtesy of Milton R. Sommerfield; **26.31B:** © T. E. Adams/Visuals Unlimited; **26.33:** © R. Kessel-G. Shin/Visuals Unlimited; **26.34:** © Biophoto Associates/Photo Researchers, Inc.; **26.35:** © John D. Cunningham/Visuals Unlimited; **26.38:** © Roland Birke/Peter Arnold, Inc.; **Box 26.2:** © Philip Sze; **26.39A:** © Heather Angel/Biofotos; **26.39B:** © E.C.S. Chan/Visuals Unlimited.

Chapter Twenty-Seven

Opener: © Richard J. Green/Photo Researchers, Inc.; **27.1A:** © Robert & Linda Mitchell; **27.1B:** © E. Webber/Visuals Unlimited; **27.1C:** © Robert & Linda Mitchell; **27.1D:** © David Cavagnaro/Peter Arnold, Inc.; **27.2 A-1:** © Robert & Linda Mitchell; **27.2 A-2:** © Ed Reschke/Peter Arnold, Inc.; **27.2 B-1:** © George J. Wilder/Visuals Unlimited; **27.2 B-2:** © Bruce Iverson; **27.4A:** © Jeff Gnass; **27.4B:** © Heather Angel/Biofotos; **27.4C:** © George Holton/Photo Researchers, Inc.; **27.4D:** © William E. Ferguson; **27.4E:** © Kingsley Stern; **27.5:** © Bruce Iverson; **27.6:** © R. Calentine/Visuals Unlimited; **27.7:** © G. Shih-R.Kessel/ Visuals Unlimited; **27.10:** © Heather Angel/Biofotos; **27.11B:** © J. H. Troughton. Provided by Industrial Research, Ltd.; **27.11C:** © Bruce Matheson/Photo/Nats, Inc.; **Box 27.1:** © Dwight Kuhn; **27.12A:** © Kjell Sandved/ Visuals Unlimited; **27.12 B,C,D,F:** Courtesy of Howard Crum, University of Michigan Herbarium; **27.12E:** © David M. Dennis/Tom Stack & Associates; **27.13A:** © John D. Cunningham/Visuals Unlimited; **27.13B:** © Robert & Linda Mitchell; **27.14:** © Triarch/Visuals Unlimited; **27.16A:** © Heather Angel/Biofotos; **27.16B:** © E. R. Degginger/ Bruce Coleman, Inc.; **27.16C:** © Ed Reschke/Peter Arnold, Inc.; **27.16D:** © Bruce Iverson; **27.17A:** © Ed Reschke/ Peter Arnold, Inc.; **27.17B:** © John D. Cunningham/Visuals Unlimited; **27.17C:** © Bruce Iverson; **27.17D:** © Robert & Linda Mitchell; **27.17E:** © J. H. Troughton. Provided by Industrial Research, Ltd.; **27.18:** © Robert & Linda Mitchell; **27.19A:** © William E. Ferguson; **27.19C:** Courtesy of Ray Evert; **27.19D:** Courtesy of Damian S. Neuberger; **27.20 A,B:** Courtesy of Howard Crum, University of Michigan Herbarium; **27.21:** © M. C. F. Proctor; **27.22:** Courtesy of Dr. Jonathan Shaw; **Box 27.2:** © Ira Block; **27.23A:** © M. P. Gadomski/Bruce Coleman, Inc.; **27.23B:** © W. Ormerod/Visuals Unlimited; **27.24A:** © BIOS/Peter Arnold, Inc.; **27.24B:** © Fred Bavendam; **27.24C:** © Dan Budnik/Woodfin Camp & Associates; **27.25 A-E:** From *Bryologist* 70:102–104, 1967 by R. N. Chopra and Urmilla Gupta.

Chapter Twenty-Eight

Opener: © Kjell B. Sandved; **28.1:** © Grant Heilman/Grant Heilman Photography; **28.4A:** © John D. Cunningham/ Visuals Unlimited; **28.4B:** © Triarch/Visuals Unlimited; **28.5:** © Edwin Reschke; **28.6 A,B:** © Bruce Iverson; **28.8:** © E. R. Degginger/Animals Animals/Earth Scenes; **28.9A:** © William Ormerod/Visuals Unlimited; **28.9B:** © Runk/ Schoenberger/Grant Heilman Photography; **28.9C:** © James W. Richardson/Visuals Unlimited; **28.12 A,B, 28.14:** © Robert & Linda Mitchell; **28.18:** © Cabisco/ Visuals Unlimited; **28.19:** © Barry Runk/Grant Heilman, Inc.; **28.20B:** © Patti Murray/Animals Animals/Earth Scenes; **28.24:** © James W. Richardson/Visuals Unlimited; **28.26A:** © W. H. Hodge/Peter Arnold, Inc.; **28.26B:** © William E. Ferguson; **28.27:** © Kjell B. Sandved; **28.28:** © R. F. Ashley/Visuals Unlimited; **28.29A:** © Runk/ Schoenberger/Grant Heilman Photography; **28.29 B,:** © Kjell B. Sandved; **28.29C:** © Kjell B. Sandved/Visuals Unlimited; **28.30:** © John D. Cunningham/Visuals Unlimited; **28.31A:** © Stan Elems/Visuals Unlimited; **28.31B:** © Hugh A. Johnson/Visuals Unlimited; **28.33:** © Heather Angel/Biofotos; **28.34A:** © W. H. Hodge/Peter Arnold, Inc.; **28.35 A,** Courtesy of Warren H. Wagner, Jr., University of Michigan; **28.36 A,B:** Courtesy of Spencer Barrett, University of Toronto; **28.37:** © George J. Wilder/ Visuals Unlimited; **Box 28.2:** © Russ Peppers, Ph.D., Geologist, Illinois State Geological Survey; **Box 28.3 A,B:** © William E. Ferguson; **Box 28.3C:** © Stephen J. Krasemann/Peter Arnold, Inc.; **28.39A:** The Field Museum, #GEO85637C, Chicago; **28.39B:** The Field Museum, #CK-49TC, Chicago, painting by Charles Knight.

Chapter Twenty-Nine

Opener: © R. J. Erwin/Photo Researchers, Inc.; **29.3A:** Courtesy of Kathleen B. Pigg; **29.3B:** © George J. Wilder/Visuals Unlimited; **29.4A:** © William E. Ferguson; **29.4B:** © Robert & Linda Mitchell; **29.4 C,D:** © William E. Ferguson; **29.5A:** © Runk/Schoenberger/Grant Heilman Photography; **29.5B:** © Grant Heilman/Grant Heilman Photography; **29.5C:** © Doug Sokell/Visuals Unlimited; **29.5D:** © Alan G. Nelson/Animals Animals/Earth Scenes; **Box 29.1:** Courtesy of Gary A. Strobel. Photo © William Hess, Brigham Young University; **29.6A:** © Robert & Linda Mitchell; **29.6B:** © Stan W. Elems/Visuals Unlimited; **29.7:** © Robert & Linda Mitchell; **29.10:** © Knut Norstog; **29.11:**

© Jeff Gnass; **29.12 A,B, 29.13 A,B, 29.14A:** © Robert & Linda Mitchell; **29.14B:** Science VU/Visuals Unlimited; **29.15A:** © Michael S. Thompson/Comstock; **29.15B:** © Rosamond Purcell; **29.17:** The Field Museum, photo by Diane Alexander White, #FEO85723C, Chicago.

Chapter Thirty

Opener: © Bob Hasenick; **30.1A:** © Fritz Prenzel/Animals, Animals/Earth Scenes; **30.1B:** © Ed Reschke/Peter Arnold, Inc.; **30.1C:** © Hal Horwitz/Photo/Nats; **30.2A,B:** Courtesy of the Peabody Museum of Natural History, Yale University; **30.3A:** © Robert & Linda Mitchell; **30.3B:** © William E. Ferguson; **30.7A:** © John Shaw/Tom Stack & Associates; **30.8:** © Wolfgang Kaehler; **30.9:** © Kjell B. Sandved/ Visuals Unlimited; **30.11B:** Courtesy of Michael G. Simpson; **30.13:** © E. R. Degginger/Photo Researchers, Inc.

Chapter Thirty-One

Opener: © Gary Meszaron/Bruce Coleman, Inc.; **31.1:** Courtesy of Hunt Institute for Botanical Documentation, Carnegie Mellon University, Pittsburgh, PA; **31.2,:** © Heather Angel/ Biofotos; **31.3:** © Bruce Iverson; **31.4:** © Heather Angel/ Biofotos; **31.5:** © Richard Kolar/Animals, Animals/Earth Scenes; **31.6:** © William E. Ferguson; **31.10:** © Kingsley Stern; **31.11:** © E. R. Degginger/Photo Researchers, Inc.; **31.12:** © John Cancalosi/Peter Arnold, Inc.; **31.13:** © William E. Ferguson; **31.14:** © D. Rintoue/Visuals Unlimited; **31.16:** © Jack Wilburn/Animals, Animals/Earth Scenes; **31.19:** © Don Johnston/Photo/Nats; **31.20:** © William E. Ferguson; **31.21:** © Runk/Schoenberger/Grant Heilman Photography; **31.22:** © J. Lichter/Photo Researchers, Inc.; **31.23:** © Kim Todd/Photo/Nats; **31.24:** © John D. Cunningham/Visuals Unlimited; **31.25:** © B. Heidel/Visuals Unlimited; **31.26:** © Raymond A. Mendez/ Animals, Animals/Earth Scenes; **31.27:** © Breck P. Kent/ Animals, Animals/Earth Scenes; **31.28:** © S. Maslowski/ Visuals Unlimited.

Chapter Thirty-Two

Opener: © Jeff Gnass; **32.4A:** © John Shaw/Bruce Coleman, Inc.; **32.4B:** © John D. Cunningham/Visuals Unlimited; **32.5:** © B. J. O'Donnell/Biological Photo Service; **32.7:** © Brian Parker/Tom Stack & Associates; **32.9:** © Milton H. Tierney, Jr./Visuals Unlimited; **32.10:** © Jeff Gnass; **32.11:** © C. P. Hickman/Visuals Unlimited; **32.12:** © Patti Murray/ Animals, Animals/Earth Scenes; **32.13:** © Nada Pecnik/ Visuals Unlimited; **32.15A:** © John D. Cunningham/Visuals Unlimited; **32.15B:** © Breck P. Kent/Animals, Animals/ Earth Scenes; **32.15C:** © Ed Reschke/Peter Arnold, Inc.; **32.16:** © Luiz Claudio Marigo/Peter Arnold, Inc.; **32.17:** © Don W. Fawcett/Visuals Unlimited; **32.18:** © Ron Spomer/Visuals Unlimited; **32.19:** © Jeff Gnass; **32.20:** © David C. Fritts/Animals, Animals/Earth Scenes; **32.21:** © Jane Thomas/Visuals Unlimited; **32.23:** © David R. Frazier; **32.24:** © Jeff Gnass; **32.26:** © Jeff Gnass; **32.27:** © Dede Gilman/Visions From Nature; **32.29, 32.30:** © Jeff Gnass; **32.31:** © John D. Cunningham/Visuals Unlimited; **32.32:** © D. Newman/Visuals Unlimited; **32.34:** © Heather Angel/Biofotos; **32.36:** © Wolfgang Kaehler; **32.37:** © Kjell B. Sandved/Visuals Unlimited; **32.38:** © Kevin Schafer/Tom Stack & Associates.

Epilogue

Opener: © Mark Edwards/Still Pictures; **E.1A:** © Norfolk Lavender; **E.1B:** © Benjamin H. Kaestner, McCormick & Co., Inc.; **E.2A:** © Walter H. Hodge/Peter Arnold, Inc.; **E.2 B-1 (left):** © Walter H. Hodge/Peter Arnold, Inc.; **E.2 B-2 (right):** © Robert & Linda Mitchell; **E.3:** © L. L. T. Rhodes/Animals Animals/Earth Scenes; **E.4A:** © Grant Heilman/Grant Heilman Photography; **E.4B:** © William H. Allen, Jr.; **E.4C:** © R. Konis/JACANA/Photo Researchers, Inc.; **E.4D:** © Jim Steinberg/Photo Researchers, Inc.; **E.4E:** © D. Cavagnaro/Visuals Unlimited; **Box E.1:** © Scott Camazine/Photo Researchers, Inc.; **Box E.2:** © Alford W. Cooper/Photo Researchers, Inc.; **E.5A:** © David M. Stone/Photo/Nats; **E.5B:** © John D. Cunningham/Visuals Unlimited; **E.5C:** © Heather Angel/ Biofotos; **E.5D:** © Brian Moser/The Hutchison Library, London; **E.6:** © Mark J. Plotkin; **Box E.3:** Courtesy of Hunt Institute of Botanical Documentation, Carnegie Mellon University, Pittsburg, PA; **E.7A:** © Michael J. Balick/Peter Arnold, Inc.; **E.7B:** © Walter H. Hodge/Peter Arnold, Inc.; **E.8:** From Dr. Daniel Klayman, "Quinghausa (Artemysinin): An Anti-malarial Drug From China," *Science* 228:1051, May 3, 1985. Copyright 1985 by the AAAS; **E.9:** © James P. Blair; **E.11:** © Edward Parker/Oxford Scientific Films/ Animals Animals/ Earth Scenes; **E.12A:** © Scott Camazine/ Photo Researchers, Inc.; **E.12 B-1** (coconut trees): © W. H. Hodge/Peter Arnold, Inc.; **E.12 B-2** (coconuts): © Patricia Goycolca/The Hutchison Library, London; **E.13 A-1** (background): © Grant Heilman Photography; **E.13 A-2** (inset): © Cecile Brunswick/Peter Arnold, Inc.; **E.13B:** © Michael Holford; **E.13C:** © E. S. Ross; **E.15:** © Walter H. Hodge/Peter Arnold, Inc.; **E.17B:** © Smithsonian/Carl Hansen; **E.17C:** © Smithsonian/ Antonio Montaner; **Box E.4A:** © Runk/Schoenberger/Grant Heilman Photography; **Box E.4B:** © A. Kerstitch/Visuals Unlimited; **Box E.4C:** © Walter H. Hodge/Peter Arnold, Inc.; **E.18A:** NOAA; **E.18B:** © George Holton/Photo Researchers, Inc.; **E.18C:** © 1989 Silvester/Black Star; **E.18D:** © Luiz C. Marigo/Peter Arnold, Inc.; **E.19:** © Ghillean Prance; **E.20A:** © Kjell B. Sandved; **E.20B:** © Ghillean Prance; **E.21:** © Douglas Kirkland/Sygma.

Line Art

Brian Evans/Marjorie Leggitt:

Figures 28.22B, 28.32; 29.1, 29.2, 29.18A-B, 29.19A-E, 29.21, 29.22A-D.

Brian Evans/Gail Kohler Opsahl:

Figures 27.11A, 27.19B&E.

Illustrious, Inc./Elizabeth Morales:

Figures 22.3, 22.4, 22.6, 22.7, 22.13A-D, 22.15, 22.17, 22.18, 22.19; BOX 23.1B, 23.3A-C, 23.9, 23.10, 23.12, 23.14, 23.17, 23.19, 23.20A, 23.23; 24.5, 24.7B, 24.8, 24.10A-C, 24.12, 24.13, 24.14; 25.5B, 25.7, 25.8, 25.13AB2, 25.26A, 25.30; BOX 26.1A-B, 26.5, 26.6, 26.8, 26.9A-B, 26.20, 26.22, 26.29, 26.32, 26.36, 26.37A-B; 27.8, 27.9A-B, 27.26, 27.27; BOX 28.1, 28.2, 28.3A-B, 28.7, 28.10, 28.13, 28.17A-B, 28.21, 28.23, 28.34B, 28.38, 28.43; 29.9, 29.23,

29.24A-B, 29.25; 30.6, 30.10A-B, 30.11A, 30.12, 30.14, 30.15; 31.7; 32.1, 32.2; BOX 32.2A-B, 32.3, 32.6, 32.8, 32.14, 32.22, 32.25, 32.28, 32.33; E.2C, E.10, E.16, E.17A.

Marjorie Leggitt:

Figures 22.5; 24.7A; BOX 28.3D, 28.20A; 29.16, 29.20; 30.4A-E; 31.8; 32.35; E.14.

Marjorie Leggitt/Gail Kohler Opsahl:

Figures 25.15, 25.18, 25.23, 25.25, 25.27; 26.15, 26.23, 26.25, 26.28; 27.3, 27.15, 28.11; 29.8; 30.7B.

Gale Mueller:

Figures 28.42; 30.5.

Gale Mueller/Marjorie Leggitt:

Figures 28.25, 28.40, 28.41A-F.

Gale Mueller/Gail Kohler Opsahl:

Figures 28.15A-B, 28.16A-B.

Laurie O'Keefe/Marjorie Leggitt:

Figures 31.9, 31.15, 31.17, 31.18.

Laurie O'Keefe/Gail Kohler Opsahl:

Figures BOX 22.1; 23.13, 23.16, 23.20B; 24.6.

Text/Line Art

Chapter Twenty-Two

Figure 22.4: From Ricki Lewis, *Life*. Copyright © 1992 Wm. C. Brown Communications, Inc., Dubuque, Iowa. All Rights Reserved. Reprinted by permission; **Fig. 22.5:** From Stephen A. Miller and John P. Harley, *Zoology*, 2d ed. Copyright © 1994 Wm. C. Brown Communications, Inc., Dubuque, Iowa. All Rights Reserved. Reprinted by permission.

Chapter Twenty-Three

Figure 23.14: From Douglas Futuyma, *Evolutionary Biology*, 2d ed. Copyright © 1986 Sinauer Associates, Inc., Sunderland, MA. Reprinted by permission.

Chapter Twenty-Seven

Figure 27.9a (bottom): From Raven, Evert, and Curtis, *Biology of Plants*, 3d ed. Copyright © 1981 Worth Publishers, New York. Reprinted by permission; **Fig. 27.9b (bottom):** From Raven, Evert, and Curtis, *Biology of Plants*, 3d ed. Copyright © 1981 Worth Publishers, New York. Reprinted by permission; **Fig. 27.25b:** Source: Data from R. N. Chopra and P. K. Kumra, *Biology of Bryophytes*. Copyright © 1988 Wiley Eastern Ltd., Daryaganj, New Delhi, India.

Chapter Thirty-Two

Figure 32.1: From Eldon D. Enger and Bradley F. Smith, *Environmental Science*, 4th ed. Copyright © 1992 Wm. C. Brown Communications, Inc., Dubuque, Iowa. All Rights Reserved. Reprinted by permission. Map based on Robinson Project.

Epilogue

Boxed Reading E.1: From *The House at Pooh Corner* by A. A. Milne. Illustrations by E. H. Shepard. Copyright 1928 by E. P. Dutton, renewed © 1956 by A. A. Milne. Used by permission of Dutton Children's Books, a division of Penguin USA Inc; **Fig. E.10:** Source: Data from Ricki Lewis, *Life*, W. C. Brown Communications, Inc., Dubuque, Iowa, 1992; **Fig. E.16:** From Ricki Lewis, *Life*. Copyright © 1992 Wm. C. Brown Communications, Inc., Dubuque, Iowa. All Rights Reserved. Reprinted by permission; **Fig. E.17a:** From Ricki Lewis, *Life*. Copyright © 1992 Wm. C. Brown Communications, Inc., Dubuque, Iowa. All Rights Reserved. Reprinted by permission.

Index

A

B

C

D

E

F

G

H

I

N

O

P

Q

R

S

T

U

V